Light Scattering Reviews 9

Light Scattering and Radiative Transfer

Alexander A. Kokhanovsky (Editor)

Light Scattering Reviews 9

Light Scattering and Radiative Transfer

Springer

Published in association with
Praxis Publishing
Chichester, UK

PRAXIS

Editor
Dr. Alexander A. Kokhanovsky
Institute of Environmental Physics
University of Bremen
Bremen
Germany

ISBN 978-3-642-37984-0 ISBN 978-3-642-37985-7 (eBook)
DOI 10.1007/978-3-642-37985-7
Springer Heidelberg New York Dordrecht London

Library of Congress Control Number: 2014950797

Cover design: Jim Wilkie
Project copy editor: Christine Cressy
Author-generated LaTex, processed by EDV-Beratung Herweg, Germany

Printed on acid-free paper

Springer is a part of Springer Science+Business Media (www.springer.com)

Contents

Part III Polarimetry

Part IV Radiative Forcing

List of Contributors

Ackerman, Andrew A.
Laboratory for Atmospheres
NASA Goddard Space Flight Center
Greenbelt, MD
USA
andrew.ackerman@nasa.gov

Adam, John A.
Department of Mathematics & Statistics
Old Dominion University
Norfolk, VA 23529
USA
jadam@odu.edu

Cairns, Brian
NASA Goddard Institute for Space Studies
New York, NY
USA
brian.cairns@nasa.gov

Cho, Hyoun-Myoung
Joint Center for Earth Systems Technology
Baltimore, MD
USA
hmcho@umbc.edu

Gritsevich, Maria
Finnish Geodetic Institute
Box 14
02431 Masala
Finland
maria.gritsevich@fgi.fi

Jethva, Hiren
NASA Goddard Space Flight Center
Greenbelt, MD
USA
hiren.jethva@nasa.gov

Kacenelenbogen, Meloe
BAER Institute
NASA Ames Research Center
Moffett Field, CA
USA
Meloe.Kacenelenbogen@nasa.gov

Kandler, Konrad
Institut für Angewandte Geowissen-schaften-Fachgebiet Umweltmineralogie
Technische Universität Darmstadt
Schnittspahnstr. 9
D-64287 Darmstadt
Germany
kzk@gmx.de

Knobelspiesse, Kirk
NASA Ames Research Center
Moffett Field, CA
USA
kirk.d.knobelspiesse@nasa.gov

Lanconelli, Christian
Institute of Atmospheric Sciences and Climate (ISAC)
National Council of Research (CNR)
Area della Ricerca
Via Piero Gobetti 101
I-40129 Bologna
Italy
c.lanconelli@isac.cnr.it

Lupi, Angelo
Institute of Atmospheric Sciences and Climate (ISAC)
National Council of Research (CNR)
Area della Ricerca
Via Piero Gobetti 101
I-40129 Bologna
Italy
a.lupi@isac.cnr.it

Mazzola, Mauro
Institute of Atmospheric Sciences and Climate (ISAC)
National Council of Research (CNR)
Area della Ricerca
Via Piero Gobetti 101
I-40129 Bologna
Italy
m.mazzola@isac.cnr.it

Nousiainen, Timo
Finnish Meteorological Institute
Earth Observation
P.O. Box 503
FI-00101 Helsinki
Finland
timo.nousiainen@fmi.fi

Peltoniemi, Jouni I.
Finnish Geodetic Institute
Box 14
02431 Masala
Finland
jouni.peltoniemi@fgi.fi

Platnick, Steven
Laboratory for Atmospheres
NASA Goddard Space Flight Center
Greenbelt, MD
USA
steven.platnick@nasa.gov

Puttonen, Eetu
Finnish Geodetic Institute
Box 14
02431 Masala
Finland
eetu.puttonen@fgi.fi

Savenkov, Sergey N.
Department of Radiophysics
Taras Shevchenko National University
Vladimirskaya Str. 60
01033 Kiev
Ukraine
sns@univ.kiev.ua

Segal-Rosenheimer, Michal
BAER Institute
NASA Ames Research Center
Moffett Field, CA
USA
michal.segalrozenhaimer@nasa.gov

Sharma, Subodh K.
S. N. Bose National Centre for Basic Sciences
Block JD, Sector III
Salt Lake
Kolkata 700098
India
sharma@boson.bose.res.in

Tomasi, Claudio
Institute of Atmospheric Sciences and Climate (ISAC)
National Council of Research (CNR)
Area della Ricerca
Via Piero Gobetti 101
I-40129 Bologna
Italy
c.tomasi@isac.cnr.it

Torres, Omar
NASA Goddard Space Flight Center
Greenbelt, MD
USA
omar.torres@nasa.gov

Zhang, Zhibo
Physics Department
University of Maryland Baltimore County
Baltimore, MD
USA
zhibo.zhang@umbc.edu

Preface

This volume of *Light Scattering Reviews* is aimed at the presentation of recent advances in light scattering, polarimetry, remote sensing, and radiative forcing. It consists of eight chapters. The first chapter of the volume, prepared by *Timo Nousiainen* and *Konrad Kandler*, is devoted to the presentation of recent results related to light scattering by atmospheric mineral dust particles. These particles originate mostly from the arid and semi-arid regions, particularly from the deserts and their margins. The authors consider the physical properties of dust particles including chemical composition, their shape and structure, and also particle size distributions. Light-scattering measurements and modeling of light-scattering properties of atmospheric dust are reviewed at great depth. In particular, the impact of morphological details and anisotropy on scattering is discussed. *Subodh K. Sharma* gives a review of approximate analytical results for the scattering phase functions of various small particles. The closed-form solutions are of importance for the studies of radiative transfer processes in particulate matter and also for the aims of remote sensing, where the speed of calculations is of importance due to the large volume of data to be processed. *John A. Adam* gives a survey of literature related to the analytical solutions of the radial *TE* and *TM* mode electromagnetic equations for radially inhomogeneous media. The author gives also a brief discussion of the ray-theoretic approach to propagation in radially inhomogeneous media. The problems related to the satellite remote sensing of cloud droplet effective radii are discussed by *Zhibo Zhang et al. Kirk Knobelspiesse et al.* discuss the application of light scattering and radiative transfer to remote sensing of aerosol layers located above clouds. The next two chapters are devoted to the polarimetric studies of various objects. *Sergey Savenkov* discusses the principles of the Mueller matrix measurements, while *Jouni I. Peltoniemi et al.* present results of measurements of the intensity and polarization of light reflected from various vegetated surfaces. The results are of importance for the remote sensing of atmosphere and underlying surface using airborne and space-borne instrumentation. The concluding section, prepared by *Claudio Tomasi et al.*, is aimed at the discussion of the direct aerosol-induced radiative forcing from clear-sky field measurements performed during seven regional experiments.

Bremen, Germany
December, 2013

Alexander A.Kokhanovsky

Part I

Light Scattering

1 Light scattering by atmospheric mineral dust particles

Timo Nousiainen and Konrad Kandler

1.1 Introduction

When discussing atmospheric aerosol particles, mineral dust refers to suspended soil-constituting mineral particles that originate mainly from arid and semi-arid regions, particularly from deserts and their margins. These particles constitute one of the most prominent aerosol classes in Earth's atmosphere and exert a considerable impact on radiation in the atmosphere. In addition, mineral dust particles act as ice nuclei and under some conditions as condensation nuclei, thus also indirectly impacting radiation and contributing to the global water cycle. Furthermore, dust particles are the main source of iron for ocean surface waters outside continental margins. Mineral dust is therefore a highly important atmospheric constituent.

To quantify the radiative effect of mineral dust, to monitor their presence and abundance in the atmosphere with remote sensing methods, or to correct for their impact in other types of atmospheric remote sensing applications, dust particles' single-scattering properties are needed. These properties depend on the sizes, shapes, and compositions of the dust particles, as well as the wavelength of incident radiation. Computing the single-scattering properties accurately is a great challenge, in part due to the great complexity of the particles, and in part for the lack of suitable, exact light-scattering methods that could be applied to such targets.

This chapter aims at reviewing the current understanding of the dust particle properties, and critically assessing different modeling approaches adapted to model their single-scattering properties. To keep the chapter from getting overly extensive, we do not elaborate on the merits of different light-scattering codes or the underlying theories. On the other hand, we will shortly introduce light-scattering measurements, because of their central role in assessing the performance of the modeling approaches. Regarding measurements, we mainly consider those carried out in a laboratory, where the physical properties of the target particles can also be analyzed. Therefore, a vast amount of literature related to remote sensing of mineral dust has been left out. Also, we only focus on the single-scattering properties and do not consider how uncertainties in them translate into uncertainties in remote sensing or radiative forcing estimates. Finally, we only consider single-scattering properties related to elastic scattering.

The chapter is organized as follows. The physical dust particle properties are reviewed in section 1.2, with separate subsections for shape, composition, structure (1.2.1), and size distributions (1.2.2). Section 1.3 shortly outlines controlled light-scattering measurements and their role in validation of the approaches used for dust particles' light-scattering modeling. Those appoaches are assessed in section 1.4, with separate subsections for the theoretical concepts and definitions (1.4.1), approaches based on simple (1.4.2) and complex (1.4.3) geometries, as well as pure modeling studies assessing the impact of different morphological details, namely surface roughness, internal inhomogeneity, and material anisotropy on scattering (1.4.4). The discussion, conclusions, and an outlook for the future are presented in section 1.5.

1.2 Physical properties of dust particles

To properly model the single-scattering properties of dust particles, it is important to understand the physical properties, namely the structure, shape, and composition of these particles. This section provides, in the first part, an introduction to the radiation-relevant compositional properties of mineral particles, including characteristic structure types and their influence on particle shape. The second part deals with particle size distributions and their evolution.

1.2.1 Composition, structure, and shape

The major constituents of mineral dust particles are different mineral species. A mineral is a naturally occurring solid substance with specific chemical composition and ordered atomic structure. For our purposes, mineral dust can be defined as atmospheric aerosol derived from minerals constituting the soil. As such, it may consist of any mineral species present in the soil, but excludes the organic compounds. It should be noted, however, that there exist other definitions – that is, i.e., some authors exclude soluble species like sulfates or nitrates, while others define mineral dust by analysis technique and location, such as the refractory, crystalline, or the insoluble fraction of the aerosol, collected downwind from a known mineral dust source. Here, we follow the definition of the soil-derived matter, but discuss also soluble material, undistinguished as to whether it comes from the original soil or from atmospheric processing. As the strongest dust sources are the warm deserts (e.g. Zender et al., 2003), mineral dust is referred to as desert dust by many authors.

The dust composition can be assessed by a number of methods, none of which provides the complete information on its own. Bulk samples – for example, filter samples – are commonly analyzed using X-ray fluorescence (XRF), proton-induced X-ray fluorescence (PIXE), X-ray diffraction (XRD), Fourier-transform infrared (FTIR), or Raman spectroscopy. From XRF and PIXE, elemental concentrations and their ratios are obtained, from which the composition in terms of mineral species has to be derived, making assumptions based on a known soil composition or common frequency of occurrence. XRD, in contrast, is able to detect crystal lattice characteristics and, thus, provides information on mineral species directly. However, it is not equally sensitive to all minerals (in particular, less to clay minerals), and some mineral groups are difficult to distinguish. Also, it has a rather high detection limit (absolute as well as relative), and it can only detect (well-)crystallized species (e.g. no glassy or micro-crystalline materials). FTIR can also provide information on particular mineral species by identifying stretching vibrations in bonds but, like XRD, it suffers from ambiguity and sometimes lacks reference spectra. Raman spectroscopy can also be applied to assess the mineralogical composition (e.g. Stefaniak et al., 2006; Sobanska et al., 2012), but apparently has never been used extensively for atmospheric dust.

Samples, from which individual particles can be analyzed, are usually subject to electron-microscopic analysis, but are sometimes also analyzed by micro-XRF/PIXE or Raman microscopy. These methods are rather labor-intensive, so the data basis is small. From electron microscopy, usually also an XRF signal is obtained. It has larger uncertainties than bulk XRF but, on the other hand, provides high-resolution images from which characteristic morphology can be taken into account. If transmission electron microscopy is used, the electron diffraction pattern can be analyzed and information on the crystal structure extracted, so a true mineral species determination can be performed. Raman or FTIR microscopy yields a considerably lower image resolution than electron microscopy, but provides information on the bonding state, allowing conclusions on the mineral species to be drawn.

Based on the variability of the source soils, the composition of atmospheric dust may also vary considerably. The only components that have been reported at every location studied are quartz and phyllosilicates in general. Of the large phyllosilicates group, illite and/or kaolinite are most common, but also chlorite, muscovite, montmorillonite, biotite, palygorskite, smectites, and inter-stratified clay minerals are often reported (e.g., Formenti et al., 2011; Scheuvens et al., 2013; and references therein). Note that most of these mineral denominations still refer only to mineral groups, as the actual mineral species were not determined. In many cases, additional silicate minerals are reported: feldspars like albite, anorthite, and potassium feldspars, less frequently orthoclase, or other phyllosilicates like chrysotile. In varying abundance and depending on the source region, calcite, dolomite, and, more rarely, apatite are found. Also in its abundance depending on the source, the iron compounds hematite and goethite are reported, less frequently also ilmenite.

The most common soluble species accompanying the insoluble ones are sulfates, nitrates, and chlorides, which are not reported with a specific mineral denomination as they usually recrystallize in the atmosphere quickly and fractionally, depending on the environmental conditions. In addition to the above-mentioned major dust components, a multitude of rarer mineral species are reported in the literature, namely biological debris like diatomite; metal oxides like rutile, periclase, baddeleyite, or spinel; other iron-rich minerals like lepidocrocite or limonite; different carbonates such as aragonite or magnesite; more or less soluble sulfates like anhydrite, gypsum, thenardite, mirabilite, mascagnite, and glauberite; and silicates like chloritoid, leucite, forsterite, zircon, or enstatite (Glaccum and Prospero, 1980; Leinen et al., 1994; Merrill et al., 1994; Molinaroli, 1996; Caquineau et al., 2002; Shao et al., 2007; Jeong, 2008; Journet et al., 2008; Kandler et al., 2009; Shen et al., 2009; Kendler et al., 2011b; Wang et al., 2012).

With respect to its natural variability, it is not possible to calculate a representative average dust composition. Instead, there have been regionally resolved compilations of dust composition as a function of provenance (e.g. Formenti et al., 2011; Scheuvens et al., 2013) or modeled compositions derived from the more or less well-known soil compositions (e.g. Claquin et al., 1999; Nickovic et al., 2012). These compilations present some general trends for the dust composition, but the variability can be very high even on a small scale (see, e.g. data of Bristow et al. (2010) for the Bodélé depression, in which the calcium-to-iron ratio varies over the same range as for the whole Saharan Desert; Scheuvens et al., 2013). As an example for atmospheric measurement data, Fig. 1.1 illustrates the temporal variation of dust composition in Morocco and Cape Verde. While quartz, K-feldspars, and illite usually dominate the aerosol in Morocco, kaolinite and K-feldspars are the major components at Cape Verde, with additional marine contributions in halite and gypsum that are expectedly absent in Morocco. Besides the major difference in clay minerals, the feldspars of the plagioclase group are more common in Cape Verde. The temporal variability becomes obvious between dusty and cleaner periods as well as within single intense dust periods: in Morocco, the dominant compound switches between quartz, K-feldspars, and illite, but calcite is also a major component on certain days. A similar behavior is visible for plagioclases, K-feldspars, and kaolinite at Cape Verde.

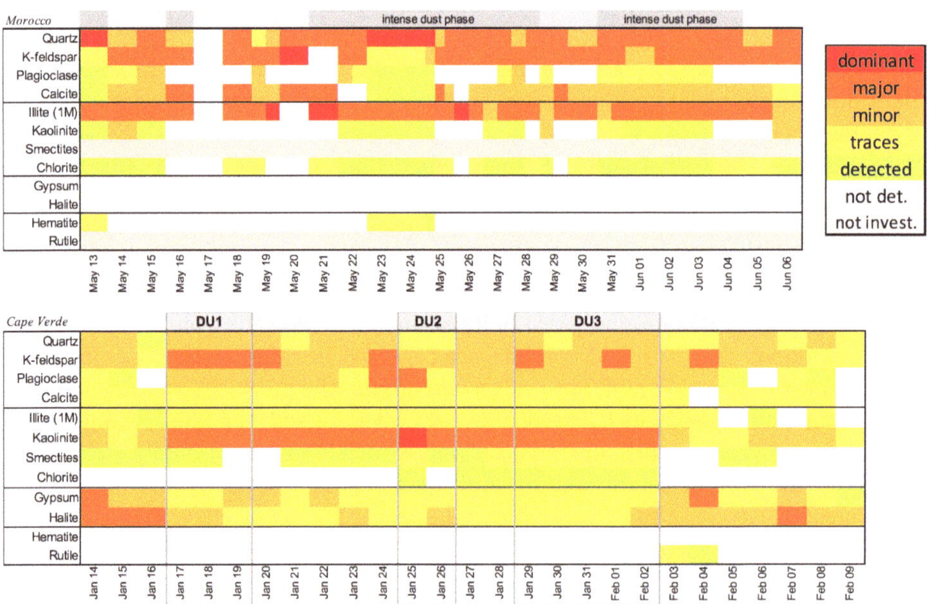

Fig. 1.1. Crystalline aerosol components observed in Morocco in 2006 and at Cape Verde in 2008, determined by X-ray diffraction of filter samples. Dust concentration of the several periods is shown as the graylevel on top with the most intense periods marked. For details about the locations, sampling and analysis, see Kandler et al. (2009, 2011b).

Beyond the variability in bulk composition, there is variation between single dust particles. Desert dust in particular is a mixture of single- and multi-mineral grains, where a single grain can consist of a nearly arbitrary combination of the minerals mentioned above. Figure 1.2 shows some discriminating elemental ratios for a few hundred individual dust particles from a single sample. We can observe some characteristic differences between Morocco and Cape Verde, such as lower calcium and higher iron contents as well as lower sodium and higher magnesium contents at Cape Verde than in Morocco. However, there is a very high inter-particle variation, which becomes especially obvious when we compare these data sets to the more uniform volcanic mineral particles, where most of the particles are supposed to have the same source (the melt) and the same age. Even there, a considerable variation in the calcium/iron ratio exists, but the variability in the other elemental ratios is much lower than for desert dust.

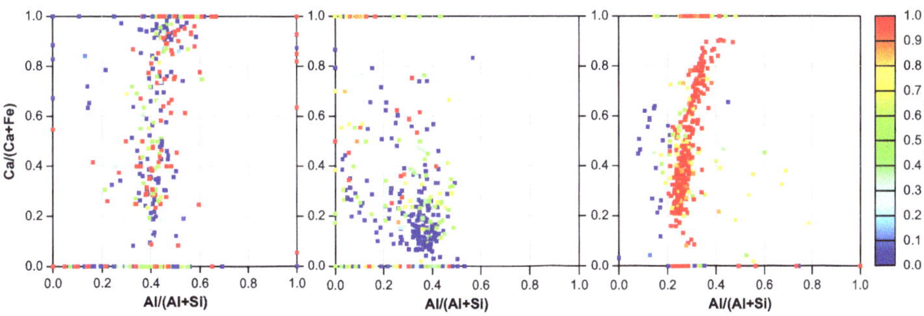

Fig. 1.2. Discriminating elemental atomic ratios for the mineral dust component of samples collected in Morocco (May 27, 2006; left panel), Cape Verde (Jan 29, 2008; center panel), and during the Eyjafjell volcano eruption (May 17, 2010; right panel); graphs drawn from data by Kandler et al. (2009, 2011a) and Schumann et al. (2011).

The composition of the mineral dust depends also on the particle size. Usually, it is assumed that the mechanically more stable minerals are less subject to abrasion and, thus, have larger particle sizes, while the less stable species subsequently exist in smaller particles. For example, Kandler et al. (2009) demonstrated that the largest particles with diameters $D > 50\ \mu$m are dominated by quartz. Smaller ones consist of feldspars and clay minerals; the clay minerals usually dominate, and their fraction increases with decreasing particle size (Kandler et al., 2011a; Schütz and Rahn, 1982; Shi et al., 2005). If present, carbonates such as calcite or dolomite are usually found between $D = 1\ \mu$m and 20 μm (Kandler et al., 2009).

Looking even closer, at the single-particle level, we observe that, for particles consisting of more than one mineral species, the compounds are anything but evenly distributed. In particular, clay minerals tend to form aggregates of several micrometers in diameter, in which grains of other substances are frequently embedded. Figure 1.3 (left) shows such a compact clay mineral aggregate. From the localized characteristic XRF, we find the presence of iron oxi(hydroxi)des and titanium oxides in small grains. Also, we can see that a quartz or feldspar grain must exist inside the particle, as there are no visible features corresponding with the elevated silicon signal. Finally, phosphates and probably sulfates also exist within this aggregate.

Besides the occurrence of mixed particles, which are already formed prior to emission, mineral dust may be processed in the atmosphere by clouds (Sullivan et al., 2007a; Matsuki et al., 2010) or by non-cloud processes like condensation (Deboudt et al., 2010), heterogeneous reactions (Ullerstam et al., 2002; Usher et al., 2002), or sea-salt mixing (Zhang and Iwasaka, 2004). Depending on the composition of the individual particle, atmospheric processing might result in a coating or adhering of usually soluble substances (Kandler et al., 2011a; Deboudt et al., 2010; Li and Shao, 2009), or in a thoroughly processed particle (Krueger et al., 2003, 2004; Matsuki et al., 2005). While, in the latter case, nearly nothing is preserved from the original particle structure, an addition of a soluble substance might just cover some surface features of the original particle; also, it can be present in a single location or between insoluble mineral grains (e.g. Fig. 1.3 (right panel) and Fig. 1.4).

Subsequently, in an atmospheric mineral dust sample, we expect to find a mixture of different particle structures, depending on the parent soil and the atmospheric history. The structure types can be described as 'mono-grain', 'main grain with minor adhesions', 'agglomerate' and 'aggregate'. Mono-grain particles might show an explicit crystal structure, but might also be more or less featureless. Mono-

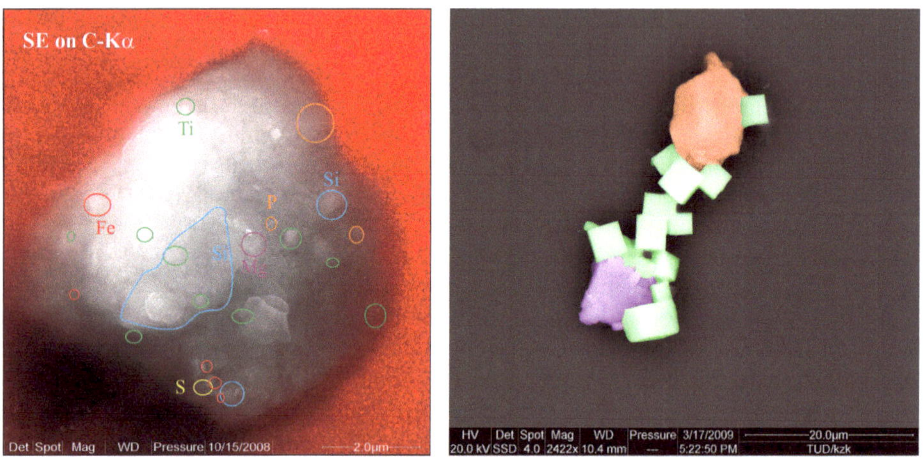

Fig. 1.3. *Left:* clay mineral aggregate (secondary electron image) collected over Morocco in 2006 (for details, see Scheuvens et al. (2011)). The colored spots show regions with enhanced elemental concentrations, indicating the presence of titanium oxides and iron oxi(hydroxi)des, as well as quartz, phosphates, and sulfates. The red background is the carbon fluorescence signal from the substrate and illustrates the thickness of the particle in the lower left corner by 'shadowing' through X-ray shielding from the detector. *Right:* Internal mixture of an alumosilicate (marked in red), sodium chloride (green), and calcium sulfate (violet) collected at Cape Verde (for information on location and sampling, see Kandler et al. (2011a)). In the atmospheric state prior to sampling, the sodium chloride most probably was in solution and recrystallized after sampling; in contrast, the calcium sulfate probably attains its original structure.

grains are not very frequent. The larger, supermicron ones consist usually either of mechanically stable, well-crystallizing minerals like quartz, feldspars, carbonates, and calcium sulfates, or of substances which may have (re-)crystallized in the atmosphere (or even after sampling), like sodium chloride or sulfate. Though the latter compounds might be acquired by atmospheric processing, there is some evidence that they are present in abundance also in desert soils (Osada, 2013). The smaller mono-grains commonly observed are single clay mineral flakes, but may also be metal oxides like rutile or silicates like zircon. More frequently, the type 'main grain with minor adhesions' is found. While the main grains usually consist of the larger, insoluble mono-grains, the adhesions are usually clay flakes which cover the surface of the main grain. Also, a mono-grain particle that acquires a coating through atmospheric processing can be assigned to this category, in which case its abundance is largely dependent on the atmospheric history of the dust sample. The agglomerate and aggregate types are both made of many small grains without a dominating one. Discrimination between them is usually difficult, as the differences are rather gradual in strength of cohesion and compactness: agglomerates are stable in airborne state and might disintegrate on impact or submersion, while aggregates would not. Also, agglomerates can exhibit a higher variability on composition of the single grains, while in aggregates usually one mineral species dominates. Presently, no systematic assessment of the structure-type abundance is available. However, from literature data of single-particle measurements, it seems safe to assume that agglomerates/aggregates dominate over the other types. Figure 1.4 shows examples of the structure types. From quartz (panel a) to K-feldspar with single flakes of clay minerals (b) and large clay mineral grain (c), we can observe a transition from the mono-grain type to the main grain with adhesions; (d) shows a calcite main grain with small clay minerals on top, while (e) and (f) present clay mineral aggregates with rather homogeneous matrix compositions. The particle in (e) is atmospherically aged and contains sodium salts (chloride/nitrate) at the lower and right end. Panels (g) to (i) show the transition from aggregates to agglomerates with more heterogeneous compositions, which are demonstrated by the elemental compositions marked in panel (i).

Mineral dust particles transported in the atmosphere can be processed and become mixed with non-dust material. We can observe this in Fig. 1.4e and i, where sodium compounds – from heterogeneous chemistry and sea-salt mixing – are contained within the particles. The reported abundances of mixed particles range from a few percent (Kandler et al., 2007, 2011a; Matsuki et al., 2010; Zhang et al., 2003) to more than half of the particles (Sullivan et al., 2007a; Zhang et al., 2006; Sullivan and Prather, 2007b). For these partly soluble particles, the structure under higher humidities is unknown; the soluble fraction will accumulate water and form a solution, which then might cover the particle or adhere to it. In the case of agglomerates, the single grains might redistribute and the particles get compacted by surface tension, when the liquid water evaporates under lower humidities.

According to their variety in structures, dust particles can have very different shapes, of which most are irregular or angular. Practically only the mono-grain type can have a symmetric and regular crystal structure, but, as soil material usually suffers physical stress during dust emission (e.g. Shao et al., 2011), most particles show at least damages or irregularities, as can be seen in Fig. 1.4. Clear crystal

Fig. 1.4. Secondary (panels a-c, e, h) and backscatter (d, f, g, i) electron microscope images of particle structure types found in mineral dust. Panels (a)–(h) were taken from two samples of the same airmass over Morocco (Scheuvens et al., 2011), while (i) was sampled at Praia, Cape Verde from transported Saharan dust (Kandler et al., 2011a). The presence of certain elements marked in image (i) was determined by X-ray fluorescence.

structures can be observed for all mono-grain compounds, but usually not as mono-crystal, instead mostly of a combination of crystals. Particles with a pronounced crystal structure are usually rather smooth, while the main grain particles often possess pronounced surface roughness, in particular in the form of adhering smaller grains, such as clay mineral flakes. A crystal structure in the form of a general particle outline can sometimes still be observed on the main grain type (e.g. Fig. 1.4d). While the mono-grain particles can have smooth, angular surfaces, the main grain particles exhibit roughness due to adhering smaller grains, to the degree that only a preferred orientation of the adhesions according to the underlying surface might remain. The aggregates also have rough surfaces – that is, they possess on their surface irregular oriented regions in the size range of a few to hundreds of nanometers, and no smooth or angular surfaces. In contrast, the agglomerates can possess larger angular surfaces in irregular orientations (e.g. Fig. 1.4h and i), deriving from their primary grains, but also can have a mixture of these properties. Not much is currently known of the surface roughness of atmospheric dust particles (Formenti et al., 2011). From the few actual surface roughness measurements (Chou et al., 2008) and electron microscopy observations, we might deduce that the size scale of the surface roughness is in the range of hundred(s) of nanometers. However, there is no statistically reliable information available, particularly not on whether the surface roughness depends on the base mineral and whether there are typical grain sizes for certain minerals, which might produce a uniform surface roughness for a particle type.

More information has been collected on the simplified overall particle shape, which is mostly derived by electron-microscopic methods followed by image analysis (e.g. Kandler et al., 2009; Okada et al., 2001; Reid et al., 2003; Coz et al., 2009). A major drawback of this approach is that the image information is usually 2D, while, for optical modeling, 3D shape information is needed. Extrapolation from a 2D to a 3D simplified shape then either requires assumptions on the particles' orientations relative to the image plane or, alternatively, the extrapolation can be used as additional degree of freedom in data inversion (Otto et al., 2009). In particular, the assumptions can introduce a major error in the shape description, as, for example, platelets like clay minerals when oriented flat on a substrate would be described as near spherical. Many simplifying shape descriptors are available (Hentschel and Page, 2003; Rosin, 2003), of which mostly the 2D aspect ratio is chosen, being least dependent on image resolution (Podczeck et al., 1999; Almeida-Prieto et al., 2007). Nevertheless, literature data from different sources are not truly comparable, as there are varying methods in use for aspect ratio calculation, probably biasing the results (Almeida-Prieto et al., 2007). The 2D aspect ratio distribution can be well represented by a modified log-normal distribution (Kandler et al., 2007) with median values in general between 1.5 and 1.7. No significant variation is found between different dust sources; instead, there is a slight dependency on dust composition (Kandler et al., 2007; Coz et al., 2009). Aspect ratios also depend on the particle size (Kandler et al., 2009; Chou et al., 2008; Okada et al., 2001), such that usually the smallest $D < 1$ μm particles have decreasing aspect ratios with decreasing particles size. Also, the large particles $D > 10$ μm have been observed to have lower aspect ratios than those in the 1 μm $< D <$ 10 μm range, where the aspect ratio peaks. This can be explained by particles a few micrometers across consisting of platelets like clay minerals, while towards larger sizes more roundish aggregates or abraded mono-grain particles prevail, with the submicron aerosol particles being predominantly non-dust particles with low aspect ratios. However, we have to keep in mind that the observed 2D values have not shown to be representative for the 3D shape in general. Apart from that, it is safe to say that the highest aspect ratios (needle-like shapes) are formed by mono-grain particles, probably due to their mechanical stability or, in the case of soluble substances, due to their later atmospheric crystallization after the mechanical stress at the emission stage. Three-dimensional simplifying determinations of particle shape are only available in a few cases, mostly as examples (Osada, 2013; Chou et al., 2008). Also, Okada et al. (2001) analyzed several thousand particles of Asian dust with a shadowing technique in the electron microscope to obtain the particle height; they found a height-to-length ratios between 0.1 and 0.4, implying that the third dimension of the particles is usually smaller than those two readily seen in 2D images, suggesting that indeed the imaged particles tend to be preferentially oriented on the substrate.

A newer approach to measurement of dust particle shape is the application of electron-microscopical stereogrammetry (Lindqvist et al., 2011). With this technique, 3D information of the upper particle half can be obtained in detail, as well as its distance from the grid on the background. These can then be incorporated into a detailed particle model as a basis for calculations of the single-scattering (optical) properties. As of today, only a few particles have been investigated by this approach.

1.2.2 Mineral dust size distribution

Particle size is one of the major parameters determining its optical properties. To assess the radiative properties of an aerosol, the knowledge of its particle size distribution is thus of is of primary importance. For mineral dust, however, the full particle size distribution is challenging to measure, as the particle diameters range from below $D = 100$ nm to larger than 100 μm. Instruments for sizing particles with $D < 10$ μm are readily available (e.g. electrical mobility particle sizers, optical spectrometers, aerodynamic particle sizers); the question at these sizes is rather whether one should – or can – differentiate dust from other particles. In contrast, only a few methods exist to measure size and number concentrations for large airborne particles. This is mainly caused by the 'inlet problem', namely the difficulty of producing an aerosol inlet able to sample representatively particles with diameters considerably larger than $D = 10$ μm. For that reason, size distributions for larger particles are available rather from inlet-free instruments. For example, optical instruments for measuring cloud droplets can be used on board aircraft to measure dust particle size distributions (e.g. Weinzierl et al., 2009), provided that the particle optical properties are well known; otherwise, considerable errors might occur in particular when using forward-scattering instruments (Schumann et al., 2011; Weinzierl et al., 2009). Those instruments usually cannot be used for ground-based measurements, as they need to be moved relatively to the aerosol in a free stream. For ground-based measurements of $D > 30$ μm particles, specialized optical (see an instrument comparison by Mikami et al. (2005)) and acoustical instruments (e.g. Van Pelt et al., 2009)) are available. Furthermore, inlet-free particle collection followed by light-microscopic size analysis can be used for particles with $D > 5$ μm (Kandler et al., 2009). As all of the latter techniques are work-intensive, only small data sets are available for the large particles.

The size distribution of an aerosol is not a conservative property; instead, it is continuously modified by new-particle formation (gas-to-particle and bulk-to-particle conversion as well as heterogeneous chemistry) and removal processes. Subsequently, all available size distribution measurements describe only a certain point in space and time and do not represent a 'general' mineral dust size distribution. Nevertheless, we can observe a systematic behavior in their development. Variability of particle concentrations is highest for the smallest and largest particles (Tanré et al., 2001; Williams et al., 2002); in the case of mineral dust, practically only the large-particle variability is of interest, because the small-particle variability is mostly due to other aerosol species. For freshly emitted mineral dust, removal is dominated by sedimentation and therefore impacts the largest particles most. Figure 1.5 shows the development of the aerosol size distributions as observed during westward transport out of Africa. First, we observe that the total volume (and mass) concentration is decreasing by about four to five orders of magnitude, governed by the removal of particles from about 200 μm down to 10 μm in diameter. As a result, the volume/mass median diameter shifts from the 100 μm range to the 1 μm range. Second, the variability in the source region can be more than one order of magnitude in concentration, depending on the meteorological and soil conditions. Third, owing to this source variation, concentrations at 1000 km distance can be as high as close to the sources for particles with $D < 20$ μm (see also Kandler et al. (2011b)). When $D > 10$ μm particles have been almost completely removed, the variation decreases. For example, Maring et al. (2003) report that the concentrations of particles smaller than $D = 7$ μm do not change significantly during trans-atlantic transport. Similarly, Reid et al. (2008) observed only minor variability for particles smaller than $D = 10$ μm despite the different emission and transport conditions in the Arabian Gulf region. We may sub-summarize that the form of the size distribution varies considerably near the source area and should always be determined case by case, but further away a priori assumptions might be sufficiently accurate.

Another connected question is how much of the aerosol as a function of particle size actually consists of soil-derived material. This can be addressed by several techniques; distinction between dust and non-dust can be made by hygroscopicity, volatility, optical or chemical properties, assuming that dust is non-hygroscopic and non-volatile, shows a hematite absorption pattern or consists of typical soil minerals, respectively. By all of these techniques, it has been shown for a representative African dust situation that particles with $D > 500$ nm on average consist predominantly of soil material, whereas smaller ones are dominated by non-dust components (Kandler et al., 2009; Weinzierl et al., 2009; Kaaden et al., 2009; Müller et al., 2009). Similar transitional behavior has been reported for Asian dust (Sullivan et al., 2007a). However, this relationship should not be treated as constant: the relative dust abundance between $D = 200$ nm and 3 μm can be quite variable (Kandler et al., 2009, 2011a; Lieke et al., 2011; Schladitz et al., 2011). Not much data are available on the size-resolved dust/non-dust relationship, most likely because the determination is laborious.

Recently, a compilation of measured size distributions of desert-dominated aerosol has become available (Formenti et al., 2011). Assuming that a size distribution is a composite of several lognormally distributed modes – with dust predominant amongst the largest and a minor constituent amongst the smallest particles – allows extracting the 'dust modes' from the available measurements. This assumption seems reasonable in general. However, the observed internal mixing between the components is neglected by this approach. For desert dust close to the source, but neglecting particles with $D > 10$ μm (i.e. an estimate of the long-range transport fraction), these dust modes now have characteristic count median diameters between $D = 1$ and 2 μm and around $D = 5$ and 9 μm (Formenti et al., 2011). Further downwind from the sources, the coarser mode can only be detected

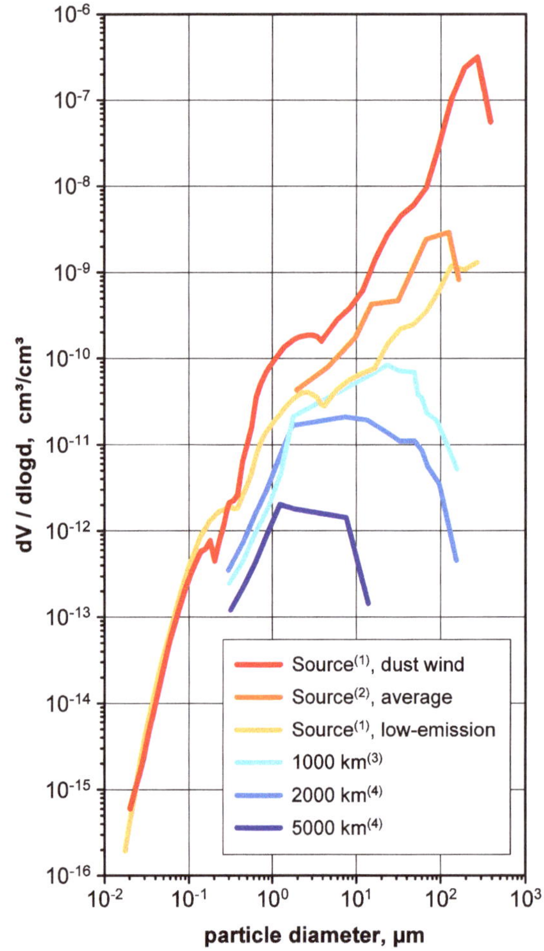

Fig. 1.5. Development of aerosol volume size distributions for Saharan dust with transport distance. [(1)] Morocco (Kandler et al., 2009); [(2)] Libya (Schütz and Jaenicke, 1974; Schütz et al., 1981); [(3)] Cape Verde (Schütz et al., 1981; Jaenicke and Schütz, 1978); [(4)] Meteor cruises (Schütz et al., 1981).

in a few cases; instead, more frequently, only one dust mode is found with count median diameters of 300 nm to 2 μm.

The mean and the variability of the measured size distributions over northwest Africa (near the source) and Cape Verde (long-range transport) are shown in Fig. 1.6. Note that size distributions for total aerosol are shown, so the small-particle modes are not necessarily representative of dust. It is obvious from the figure that, overall, the variability of the size distributions is fairly low. Further, it can be seen that the variability is higher for transported dust than over the source region. This shows that, on one hand, the dust concentration (strength of the dust modes) is rather a matter of mixing with dust-free airmasses or maybe regional wet removal, and thus exhibits a higher variability in a downwind region not continu-

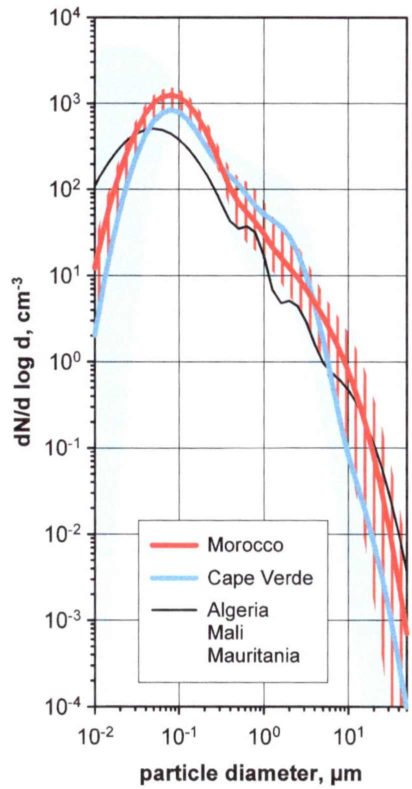

Fig. 1.6. The mean (solid lines) and 3% to 97% variation ranges (shading) of the size distributions determined over Morocco and over Cape Verde (Weinzierl et al., 2011) as well as over Algeria, Mali, and Mauritania (Ryder et al., 2013). For the Cape Verde region, only measurements without considerable biomass-burning influence were taken into account.

ously influenced by dust outbreaks. On the other hand, the faster removal of larger particles by sedimentation predominantly controls the median size of a mode, decreasing this size with increasing transport distance. In contrast, over Morocco, the size distributions have a lower variation, as turbulence tends to mix the single dust outbreaks into a dust reservoir over the continent (Schütz, 1980; Engelstaedter et al., 2006). Nevertheless, also here we still observe a variation in the concentration of more than a factor of five up to more than an order of magnitude at $D = 10$ μm.

1.3 Light-scattering measurements

Dust particles are irregularly shaped and inhomogeneous, and composed of minerals which are typically anisotropic. No analytical, exact solution exists to solve the single-scattering properties of such particles. The solution that comes closest is probably that by Petrov et al. (2011), which applies to any shape that can be described with Laplace series expansion in spherical coordinates, but is limited to homogeneous targets composed of isotropic material. For other cases for which exact, analytical solutions have been derived, see, for example, Mishchenko et al. (2000). Further, even with exact analytical solutions, there may be practical obstacles for the usage, such as numerical instability of the implementation.

To model dust particles' single-scattering properties, we must therefore choose one of the following three approaches:

- Simplify the problem. Replace dust particles by targets conforming to one of those special geometries for which exact solutions can be obtained.
- Simplify the solver. Use an approximate method which does not solve the problem rigorously, but allows treating the target particle properties accurately.
- Apply brute force. Solve the underlying equations directly using numerical techniques. These can be applied to nearly arbitrary single-scattering problems, but limitations in computer memory and computing power set practical limitations for cases where accurate solutions can be obtained. In general, particles much larger than the wavelength are beyond these methods.

Especially for the first two options, but also for the third option if the accuracy criteria are relaxed or the target particle characteristics are not sufficiently well known, the obtained solutions should be validated. Single-scattering properties measured in a laboratory can provide the reference data needed for this.

The laboratory facilities and the measurement data available for this have been recently reviewed by Muñoz and Hovenier (2011), and we will not go into details here. Rather, we will provide only a short introduction to some such resources. We will not consider field measurements from remote sensing instruments such as lidars or radiometers, because the target particles of these measurements are usually not well characterized (if at all). In addition, often these data are obtained through mathematical inversion of the single-scattering process, where forward single-scattering modeling is required as part of the inversion. Such data would be ill suited for validating (forward) single-scattering methods. The laboratory data we consider present measured single-scattering properties that have been obtained without any modeling or analytical single-scattering computations. Unfortunately, this does not mean that these data could always be applied for validation purposes without some modeling.

Of particular relevance for validation are measurements of the full scattering matrix (see section 1.4.1, Eq. (1.1)), because it fully describes the single-scattering event, except for absorption. The scattering matrix contains up to seven independent elements, which are all functions of the physical properties of the scatterers and the scattering geometry. With the measured scattering matrix, both the intensity and polarization characteristics (including depolarization) of the scattered radiation, predicted by the chosen modeling approach, can be validated. For various types of dust particles, full scattering matrices measured in the laboratory are available at the Amsterdam-Granada light-scattering database (Muñoz et al., 2012). The database also contains the measured size distribution for all the samples, as well as an estimate for the refractive index, which are needed when trying to reproduce the measured scattering matrix by modeling.

When comparing simulated and measured scattering matrices, one must realize that the measured matrices are often in arbitrary units, so they differ from the simulated matrices by an unknown normalization coefficient. To renormalize the measured matrices, the normalization integral (section 1.4.1, Eq. (1.3)) can be applied. This requires, however, that the measurements cover the whole angular range to be integrated, which usually is not the case. If the size distribution of the sample is known and the refractive index can be estimated reasonably well, it may be possible to fill the gaps in the measurements by modeled values. This is thus one of those instances where validation based on measurements requires some light-scattering simulations before the measured data can be applied for the purpose. The impact of possibly erroneous renormalization on the validation must then be kept in mind. In particular for dust, probably all size distribution measurements are subject to uncertainties. For different renormalization procedures suggested, see, for example, Liu et al. (2003) and Kahnert and Nousiainen (2006, 2007).

The measurement of the whole scattering matrix is quite demanding (Muñoz et al., 2010), requiring multiple measurements with varying polarization states for the incident radiation. It is much simpler to measure only part of the scattering matrix, and the resulting data can be supplemented with other independent data to still provide a reasonable base for validation. For example, simultaneous infrared extinction spectrum measurements have been used (Meland et al., 2012; Alexander et al., 2013). The requirement of simultaneously reproducing scattering and extinction data provides a much more stringent test for the model than either set of data could, alone.

1.4 Light-scattering modeling

A wide variety of modeling approaches have been used for the purpose of estimating, via numerical computations, how natural dust particles interact with electromagnetic radiation. The purpose of this section is to review these approaches. The first subsection (1.4.1) introduces the basic concepts that are used to characterize how dust particles, either individually or as an ensemble, scatter and absorb light. The next two subsections (1.4.2 for simple and 1.4.3 for complex model particles) address different modeling approaches and how they perform in mimicking scattering by real dust particles. Section 1.4.4 focuses on how certain physical characteristics of the dust particles, in particular the surface roughness, internal inhomogeneity, or material anisotropy of the component minerals of the dust particles, influence their single-scattering properties.

1.4.1 Definitions

The interaction of electromagnetic radiation with particles is described by the single-scattering properties. Often, in the literature, these are also called optical properties, although this is a less preferable term due to its non-consistent use in different fields.

Single-scattering events are commonly described by scattering matrices that relate the properties of the incident and scattered radiation. Here we adapt the Mueller matrix formalism which relates the incident and scattered radiation expressed as Stokes vectors $[I, Q, U, V]^T$, resulting in a 4×4 scattering matrix. The components of the Stokes vector, the Stokes parameters, describe the intensity (I), linear polarization (Q, U), and circular polarization (V) of radiation and are all measurable quantities and, for incoherent light, additive.

There are many different Mueller matrices. The one often used is the phase matrix $\mathbf{P}$, with which the scattering event can be written as

$$\begin{bmatrix} I_s \\ Q_s \\ U_s \\ V_s \end{bmatrix} = \frac{C_{sca}}{4\pi d^2} \begin{bmatrix} P_{11} & P_{12} & P_{13} & P_{14} \\ P_{21} & P_{22} & P_{23} & P_{24} \\ P_{31} & P_{32} & P_{33} & P_{34} \\ P_{41} & P_{42} & P_{43} & P_{44} \end{bmatrix} \begin{bmatrix} I_i \\ Q_i \\ U_i \\ V_i \end{bmatrix}, \tag{1.1}$$

where the subscripts 'i' and 's' refer to the incident and scattered radiation, respectively; C_{sca} is the scattering cross-section for unpolarized incident light and d is the distance from the scatterer.

The scattering cross-section describes the amount of scattered radiation. However, one should note that it is expressed in units of area. To obtain the total rate of energy scattered, one needs to multiply C_{sca} by the flux density of the incident radiation. Thus, C_{sca} actually describes a surface area perpendicular to the incident radiation, upon which power equal to that scattered is incident. This definition makes C_{sca} a useful descriptor of the scattered power without being dependent on the incident flux density. One can similarly define the absorption cross-section C_{abs} and the extinction cross-section $C_{ext} = C_{sca} + C_{abs}$, the latter of which desribes the rate of energy removed from the incident radiation (again in the units of area) through scattering and absorption in a single-scattering event. The single-scattering albedo,

$$\varpi = \frac{C_{sca}}{C_{ext}}, \tag{1.2}$$

describes the relative portions of scattering and absorption in extinction.

The first element of the phase matrix, P_{11}, is called the phase function. It describes the angular distribution of scattered intensity for incident unpolarized light, and is normalized such that

$$\int_0^{2\pi} \int_0^{\pi} P_{11} \sin\theta \, d\theta \, d\phi, = 4\pi, \tag{1.3}$$

where θ and ϕ are the scattering and azimuthal angles, respectively. The θ angle describes the angle between the propagation directions of the incident and scattered radiation at the scattering plane, which is defined by the unit vectors specifying these propagation directions, whereas ϕ refers to the orientation of the scattering plane. From the phase function, one can compute the asymmetry parameter g, which describes how the scattered power is distributed between the forward and backward scattering hemispheres, as

$$g = \frac{1}{4\pi} \int_0^{2\pi} \int_0^{\pi} P_{11} \cos\theta \sin\theta \, d\theta \, \mathrm{d}\phi. \tag{1.4}$$

For a cloud of randomly oriented particles, scattering is independent of the choice of the scattering plane, so the ϕ dependence vanishes. The phase function and asymmetry parameter, together with the single-scattering albedo, are among the most commonly used single-scattering properties due to their central role in scalar radiative transfer computations.

For particle ensembles, one needs the single-scattering properties in an additive form. The phase matrix can be converted into an additive scattering matrix, **S**, by

$$\mathbf{S} = \frac{C_{\mathrm{sca}} k^2}{4\pi} \mathbf{P}, \tag{1.5}$$

where k is the wave number. To average asymmetry parameters for different particles, each must be multiplied by the corresponding scattering cross-section. The cross-sections are additive as they are and, to compute the single-scattering albedo for an ensemble of particles, one needs only to add up the C_{sca} and C_{ext} of each particle separately before taking their ratio.

The scattering matrices are often much simpler than the one presented in Eq. (1.1). For example, if the particle ensemble is composed of particles in random orientations and each particle is either plane symmetric or the ensemble contains an equal number of particles and their mirror particles, the phase matrix (and other types of Mueller matrices) has the form

$$\mathbf{P} = \begin{bmatrix} P_{11} & P_{12} & 0 & 0 \\ P_{12} & P_{22} & 0 & 0 \\ 0 & 0 & P_{33} & P_{34} \\ 0 & 0 & -P_{34} & P_{44} \end{bmatrix} \tag{1.6}$$

with only six independent, non-zero scattering matrix elements. Even though the second assumption is often artificial for dust particles in the atmosphere, their ensemble-averaged scattering matrices still closely resemble this form: the vast number of different kinds of particles under different orientations effectively mimic these assumptions: for complex particles, there is not necessarily much difference in the single-scattering properties between a particle and its mirror particle; or, even if exact mirror particles are absent, most particles may have counterparts which scatter similarly to how their real mirror particles would scatter. The simple form of the scattering matrix is thus often approximately obtained. This is illustrated in Fig. 1.7, showing how the elements in the off-diagonal blocks are much smaller (albeit not zeros) than those in the diagonal block for randomly oriented complex particles. Thus, most of the literature with dust particle scattering matrices, either measured or simulated, consider only these six scattering matrix elements.

Of the scattering matrix elements, some have clear physical interpretation and are more often analyzed. In particular, $-P_{12}/P_{11}$ describes the degree of linear polarization of the scattered radiation for incident unpolarized light, and $1 - P_{22}/P_{11}$ the depolarization of incident, linearly polarized light. Depolarization arises from anisotropy of the scatterer, due to nonspherical shape or anisotropic composition, and is often used as an indicator for the presence of nonspherical scatterers.

In addition to these basic quantities, there are also many derived quantities relevant for specific applications. For example, lidars are often employed to measure atmospheric dust. Lidar quantities are related mostly to backscattering characteristics. For example, instead of the scattering cross-section, lidar applications employ

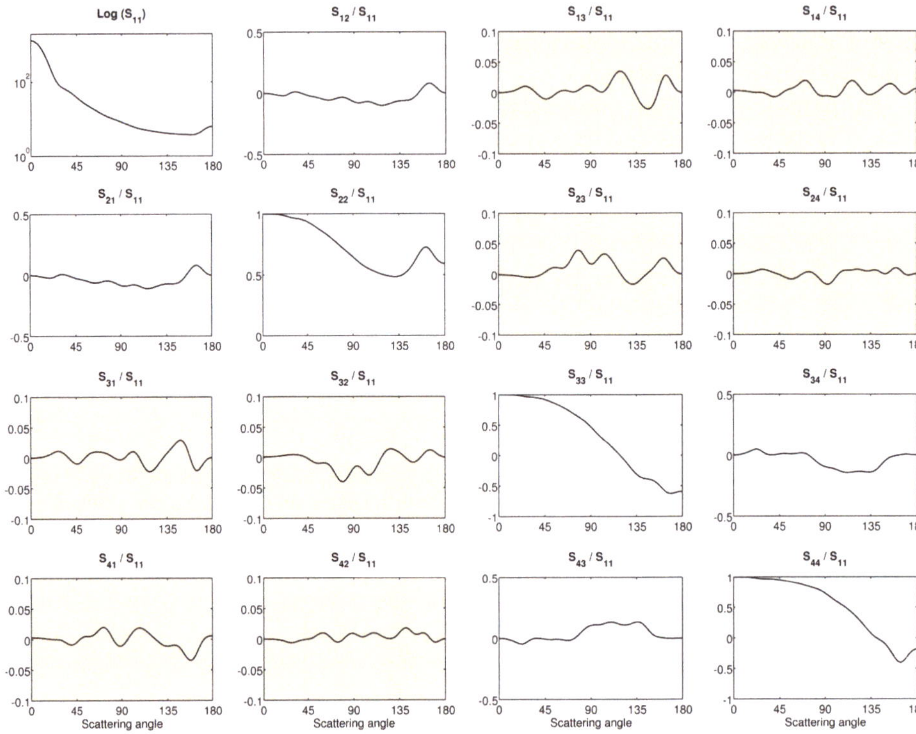

Fig. 1.7. The full 4×4 scattering matrix **S** for a small ensemble of 10 irregular model particles with size parameters $x = 6$, generated using an aggregate of 10 randomly positioned silicate spheres which has then been coated with ice using the concave-hull transformation (Lindqvist et al., 2009). Note the differences in the y-axis scaling for different elements.

the backscattering cross-section, defined by

$$C_{\mathrm{b}} = \frac{C_{\mathrm{sca}} P_{11}(\pi)}{4\pi} \,. \tag{1.7}$$

Another lidar quantity of high relevance is the lidar ratio, which describes the ratio of the extinction (beam attenuation) and backscattering cross-sections (proportional to the measured backscattered power):

$$R = 4\pi \frac{C_{\mathrm{ext}}}{C_{\mathrm{sca}} P_{11}(\pi)} \,. \tag{1.8}$$

The third commonly encountered lidar quantity is the linear depolarization ratio,

$$\delta_L = \frac{P_{11}(\pi) - P_{22}(\pi)}{P_{11}(\pi) + P_{22}(\pi)} \,, \tag{1.9}$$

which descibes how the degree of linear polarization of the transmitted light decreases through scattering for backscattered light.

The single-scattering properties are functions of particle size, shape, and composition. The particle size is often specified with the size parameter

$$x = \frac{2\pi r}{\lambda}, \tag{1.10}$$

where λ is the wavelength of the incident radiation and r the radius of the particle. For nonspherical particles, some kind of equivalent size must be established: the most common choices are the maximum diameter and volume or surface-equivalent radii. The size parameter is also an important parameter when considering the suitability and applicability of different light-scattering methods. The composition specifies the complex refractive index $m(\lambda)$ of a particle, or, in the case of anisotropic materials, the dielectric tensor. These describe how the material responds to the time-harmonic electric and magnetic fields of the incident radiation. The imaginary part of m is related to the absorptivity of the material.

1.4.2 Models with simple homogeneous particles

For light-scattering purposes, dust particles are often modeled using simple geometries, assuming internal homogeneity, even though real dust particles are very complex targets. In this section, the performance of such models, illustrated in Fig. 1.8, is assessed.

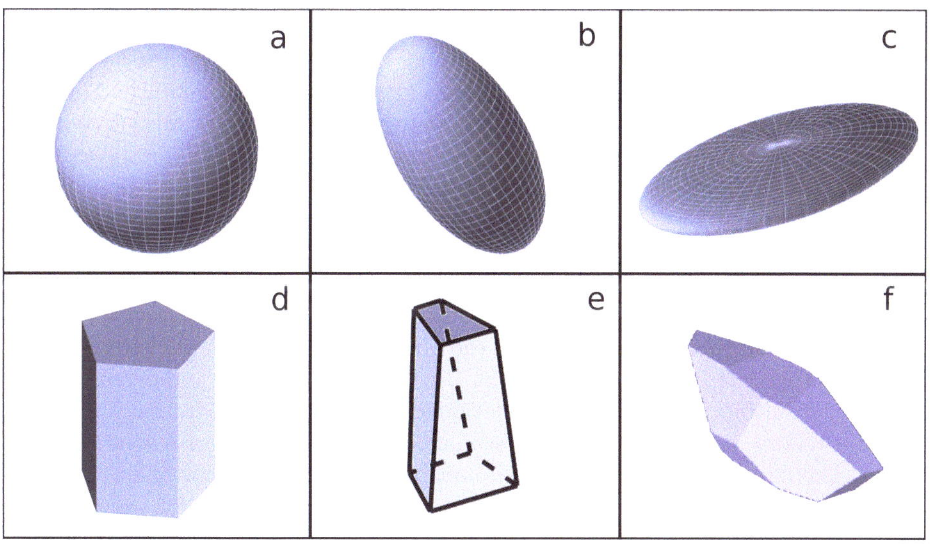

Fig. 1.8. Simple, regular model geometries adapted for mineral dust particles in light-scattering simulations: spheres (a), spheroids (b), ellipsoids (c), polyhedral prisms (d), nonsymmetric hexahedra (e), and convex polyhedra (f).

1.4.2.1 Spheres

Probably the most common approach to model single-scattering properties of dust particles is to assume the dust particles to be isotropic, homogeneous spheres (Fig. 1.8a), so that the Mie theory (Mie, 1908) can be applied. This is in general also the least accurate modeling approach of those considered here.

The most obvious weaknesses related to using Mie spheres for size-shape distributions of mineral dust particles are the following (e.g. Nousiainen et al., 2006; Bi et al., 2010; Merikallio et al., 2011):

- Whereas P_{11} of dust particles are typically flat at the side and backscattering angles, with only a modest enhancement at the backscattering direction, Mie spheres produce low values at side scattering and very pronounced backscattering. Therefore, Mie spheres underestimate scattering at side-scattering angles and overestimate it at backscattering angles. The overestimation at backscattering can exceed a factor of 10.
- While $-P_{12}/P_{11}$ of dust particles tend to be positive, with nearly symmetric angular dependence peaking close to $\theta = 90°$, Mie spheres produce mainly negative values, typically with a pronounced minimum close to $\theta = 150°$. The predicted degree of linear polarization is often tens of percent in error, and has the wrong sign.
- Dust particles tend to be efficient in depolarizing incident radiation, with P_{22}/P_{11} values reaching down to 0.5 or even below. Mie spheres do not depolarize, producing P_{22}/P_{11} values of unity. This makes Mie spheres useless for depolarization lidar applications, for example.

Regardless of these deficiencies, the Mie theory also has its uses. First of all, its numerical implementations are generally very fast and stable. Second, the Mie solution depends only on the particle size parameter and the complex refractive index, so most Mie codes are very easy to set up and use. In addition, the Mie theory often provides quite reasonable estimates for certain single-scattering properties, such as ϖ, C_{sca}, C_{abs}, or C_{ext}. The asymmetry parameter values may also be reasonably accurate, especially when considering size distributions and spectral averages, because the Mie solution typically underestimates g at small and overestimates it at large parameters for dust particles (Kahnert et al., 2007; Haapanala et al., 2012). For example, for atmospheric dust at the shortwave domain (wavelengths below 5 μm), the particle size distribution typically covers both under and overestimated g values, resulting in an error cancellation that improves the accuracy. Further, the Mie theory provides a good approximation for the forward-scattering peak of P_{11}, making it useful for sizing also nonspherical particles. For example, when comparing laboratory-measured and modeled scattering matrices for samples with size distributions measured using both the Mie theory and the standard Fraunhofer diffraction theory, the former performs clearly better (Merikallio et al., 2013; Dabrowska et al., 2013).

1.4.2.2 Spheroids

Of the simple models for mineral dust particles, spheroids are probably the most interesting and, among the nonspherical models, by far the most commonly adapted.

Spheroids (Fig. 1.8b) are ellipsoids of revolution, with two of their three axes identical. The particularly curious aspect of spheroids is that, even though they are among the simplest nonspherical shapes and much unlike the real dust particle shapes, they have been unsurpassed in their ability to reproduce the laboratory-measured Mueller matrices of real dust particles, until very recently.

Spheroids were adapted to study the single-scattering properties of atmospheric dust particles in the mid 1990s (e.g., Mishchenko et al., 1995, 1997; Mishchenko and Travis, 1997). When the laboratory-measured Mueller matrices became available (Volten et al., 2001), their performance to reproduce the full 4×4 scattering matrices could be properly assessed. The first such investigation (Nousiainen and Vermeulen, 2003) already showed the great potential of spheroids to mimic scattering by dust particle ensembles realistically, with subsequent studies, using both laboratory data and radiation measurements from atmospheric dust, establishing that scattering by real mineral dust particles is best mimicked with spheroid distributions where strongly elongated spheroids are predominant at the expense of nearly spherical spheroids (Nousiainen et al., 2006; Merikallio et al., 2011; Dubovik et al., 2006). Facilitated by the efficient computational methods, in particular the T-matrix method implementation by (Mishchenko and Travis, 1998), and the very good performance, spheroids have become widely applied. For example, they are operationally used in the Aerosol Robotic Network (AERONET) retrievals (Dubovik et al., 2006), they have been applied to estimate the broadband dust radiative effects (Otto et al., 2009; Haapanala et al., 2012), and used to replace spheres as models for dust particles in climate simulations (Räisänen et al., 2013).

But why do spheroids, being such simple and symmetric model particles, perform so well in reproducing scattering by dust particles that certainly are not homogeneous spheroids? There are essentially two key reasons for this. First, it appears that ensembles of quite different model particles can have very similar single-scattering properties. This fact will become quite obvious by the end of this chapter. Second, spheroids with different aspect ratios scatter light very differently, and this gives shape distributions of spheroids much flexibility for fitting different types of scattering matrices. Even scattering by cubes can be mimicked by a suitable set of spheroids (Nousiainen et al., 2011).

It follows that the extent of particle shape information that can be derived from measurements of light scattered by particle ensembles may be quite limited, and is likely to be particularly challenging when using spheroids. For example, Kahnert (2004) reports poor correlation between the aspect ratios of the reference particles and spheroids optimally fitted to reproduce their scattering. Some contrary evidence is, however, also reported, where simultaneous scattering and extinction measurements of dust samples were best explained by a set of spheroids that is consistent with aspect ratios reported in the literature for dust particles of similar composition (Meland et al., 2012; Alexander et al., 2013). Unfortunately, the real aspect ratios of the particles in the dust samples were not measured. Nevertheless, this good correlation was associated with small particles with strongly elongated flake-like (oblate) shapes; Nousiainen et al. (2009) suggest that scattering properties of such shapes could be exceptional, not reproducable by any set of spheroids limited to moderate aspect ratios. It is quite possible that such extreme shapes have very specific single-scattering properties that very elongated spheroids also share. Then again, scattering by very elongated prolate-type dust particles studied by Lindqvist et al. (2013) could not be reproduced with corresponding spheroids. Therefore, even if in some cases spheroids with matching aspect ratios perform well in reproducing scattering by non-spheroidal target particles, it seems rather an exception than the rule.

It also turns out that different sets of spheroids are needed to optimally reproduce different single-scattering properties (e.g. different scattering matrix elements or the asymmetry parameter), and even the same properties for the same target particles at different wavelengths (Nousiainen et al., 2006; Merikallio et al., 2011). This lack of consistency suggests that the dust single-scattering properties predicted using spheroids may not be entirely reliable, even though the model could be tuned to closely match scattering by the target particles. In particular, it is difficult to choose what kind of shape distribution of spheroids to use in the absence of reference data. Nousiainen et al. (2006) suggest that the aspect ratios of spheroids, compared to the target particles, should be exaggerated, because the elongation is the only factor in which spheroids deviate from the spherical symmetry. This may allow spheroids to better mimic scattering by real, irregular dust particles. Likewise, the wavelength dependence of the scattering properties predicted by spheroids is clearly not correct. This is well demonstrated by Zubko et al. (2013). Still, spheroids are far superior to Mie spheres, which they often replace when single-scattering treatments in practical applications are improved.

1.4.2.3 Ellipsoids

As model particles, ellipsoids are very similar to spheroids. An example ellipsoid is depicted in Fig. 1.8c. The principal differences to spheroids are that (i) there is one additional free parameter to describe their shapes and (ii) the resulting shapes are generally not rotationally symmetric, while still being mirror symmetric. The reduced symmetry and the additional free parameter suggest that ellipsoids can better reproduce scattering by dust particles, as demonstrated by Bi et al. (2009), Meng et al. (2010) and Merikallio et al. (2013), when ellipsoids were applied to reproduce the laboratory-measured scattering matrices for terrestrial and Mars analog dust. Indeed, nearly perfect matches for the laboratory-measured matrices were achieved. Ellipsoids have also been applied to investigate the radiative effects of atmospheric dust by Yi et al. (2011). The application of ellipsoids to mineral dust is facilitated by the pre-computed lookup tables in the form of an optical database made available by Meng et al. (2010).

The close similarity of spheroids and ellipsoids suggests that ellipsoids probably suffer from the same flaws. Even though no studies apparently exist to test the applicability of ellipsoids for predicting the spectral dependence of dust particle single-scattering properties, Merikallio et al. (2013) derived best-fit ellipsoidal distributions separately for each scattering matrix element measured in the laboratory for the Mars analog sample and observed a clear inconsistency in the obtained distributions, exactly as was observed for spheroids by Nousiainen et al. (2006). Consequently, care should be taken when using ellipsoids for predicting single-scattering properties of dust particles.

1.4.2.4 Polyhedra

Several studies have employed varying types of polyhedra as model particles to simulate the single-scattering properties of dust particles. In Nousiainen et al. (2006), dust particles were modeled as symmetric polyhedra (Fig. 1.8d), defined by two shape parameters: the aspect ratio and the number of corners in the polygonal cross-section. The number of corners was varied from four to seven, while the aspect ratios were varied roughly within a factor of two, with different sets for polyhedra with different cross-sections. A T-matrix code specifically tailored for different types of polyhedra was used for light-scattering computations; recently, this code has been made publicly available (Kahnert, 2013).

To test the performance of polyhedra as model particles for mineral dust, the simulations were compared against a scattering matrix for a feldspar sample measured by Volten et al. (2001). The feldspar particles in this sample have been artificially grounded and are rather angular (see, e.g. Fig. 2 of Volten et al. (2001)). Nevertheless, the symmetric polyhedra were found to perform unsatisfactorily: scattering matrices for individual shapes did not closely match those of the feldspar sample, and different polyhedra turned out to have very similar scattering matrices. The latter is most likely because the change in the aspect ratio does not change the angles between the facets, and increasing the number of corners in the polygonal cross-section makes the scattering resemble more and more that of a circular cylinder. In the parameter range considered, neither shape parameter was capable of altering scattering strongly. Consequently, the agreement with the measurements could not be significantly improved by considering different sets of symmetric polyhedra: symmetric polyhedra do not present a flexible base for fitting. Of the scattering matrix elements, the phase function was fairly well reproduced, except for the backscattering direction where the scattered intensity by the polyhedra was too high. Overall, the performance was clearly better than that by Mie spheres, but also clearly worse than that by an ensemble of spheroids, albeit spheroids are morphologically simpler and resemble the shapes of the feldspar particles less. Clearly a better morphological resemblance does not automatically mean better performance in terms of scattering.

Nonsymmetric polyhedra have been considered by Bi et al. (2010). For simplicity, they confined their model shapes to convex hexahedra with quadrilateral faces, eight vertices, and 12 edges. An example shape is shown in Fig. 1.8e. The hexahedra were generated by starting from a regular polyhedron with specific size lengths and then randomly tilting the facets. Tilts that would disrupt the topology were not allowed. The hexahedra they considered thus differ from the polyhedra considered by Nousiainen et al. (2006) in two ways: only polyhedra with a polygonal cross-section of four corners are considered but, in their polyhedra, the cross-sections vary within each particle due to the tilted facets. The angles between the different facets also vary from particle to particle.

To test the performance of the nonsymmetric hexahedra for mimicking scattering by dust particles, Bi et al. (2010) compared the simulations with a laboratory-measured scattering matrix of quartz particles (Volten et al., 2001). By using only three different hexahedra, Bi et al. (2010) were able to closely match the measured scattering matrix. One would expect that the better performance of the nonsymmetric hexahedra compared to the symmetric polyhedra would arise, at least in part, from the larger variability in scattering by different nonsymmetric hexahedra, due to each particle having different angles between their facets but, at least for the three random polyhedra considered, the scattering matrices were actually very similar. Thus, even a single hexahedron could mimic scattering by an ensemble of irregular quartz particles very well. Clearly, the nonsymmetric hexahedra are superior to symmetric polyhedra as model particles for mineral dust. Unfortunately, Bi et al. (2010) did not test whether their hexahedra also realistically reproduce the wavelength dependence of the scattering matrix. They did, however, compare the hexahedra also against a laboratory-measured scattering matrix for Pinatubo volcanic ash, and found a good match.

Finally, Gasteiger et al. (2011) considered convex polyhedra generated by selecting the planes that define the particle faces manually and rather arbitrarily. An example shape of such a particle can be seen in Fig. 1.8f. The obtained scattering properties were not compared against laboratory-measured scattering matrices but, instead, they were compared against lidar measurements of atmospheric dust. This restricts the comparisons to fewer parameters, in this case the lidar ratio R and the linear depolarization ratio δ_L. In addition, the convex polyhedra were considered only as part of a particle mixture where complex shapes (see section 1.4.3) were also present, so their individual performance for lidar applications cannot be assessed. Plots for the individual model types, however, show that the convex polyhedra scatter light considerably differently to those with more complex shapes or spheroids. Overall, the mixture of simple and convex model particles, including both absorbing and non-absorbing particles of both types, generally reproduced the measured wavelength-dependent R and δ_L values within the measurement error.

1.4.3 Models with complex anisotropic, and inhomogeneous particles

In addition to simple, regular model geometries, dust particles have been modeled in light-scattering considerations using complex particle geometries as well as having a shape derived directly from stereo images of real dust particles. The complex geometries considered in this section are presented in Fig. 1.9.

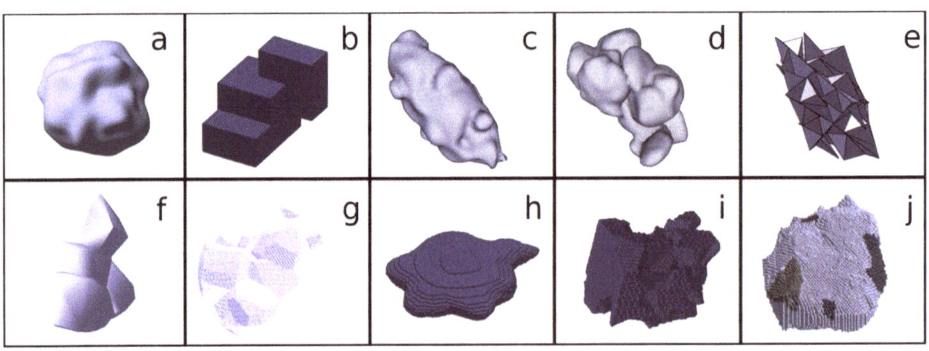

Fig. 1.9. Complex particle shapes adapted in light-scattering considerations: Gaussian random spheres (a), random blocks (b), deformed spheroids (c), deformed aggregates (d), concave fractal polyhedra (e), spatial Poisson–Voronoi tessellation (f), agglomerated debris particles (g), irregular flakes (h), irregular rhombohedra (i), and inhomogeneous stereogrammetric shapes (j).

1.4.3.1 Homogeneous, isotropic models

The majority of complex model geometries for dust particles have been applied with the assumption of internal homogeneity and material isotropy. Thus, these models differ from the simple model geometries considered in section 1.4.2 only in that the particle shapes are more complex.

Generation of complex model shapes may itself be complicated, but one notable exception is the Gaussian random sphere (GRS) geometry by Muinonen et al. (1996), which is a stochastic geometry based on deforming a sphere by a series expansion of spherical harmonics in a statistically controlled way. The statistical shape is defined by the covariance function of radius or logradius, from which individual particles are obtained by randomizing the weights of the spherical harmonics expansion, following the statistics set by the covariance function. A wide variety of different types of shapes can be generated by altering the covariance function, and a practically endless number of different individual shapes, obeying the given statistics, can be generated with different sets of random numbers. In particular, shapes greatly resembling those of roundish mineral dust particles can be generated. An example GRS particle is shown in Fig. 1.9a.

The GRS geometry has been applied to mineral dust particles, such as by Nousiainen et al. (2003, 2011a), Veihelmann et al. (2006), and Muñoz et al. (2007). Of these, Veihelmann et al. (2006) is of particular interest, because they compared semi-exact discrete-dipole approximation (DDA) simulations against a laboratory-measured scattering matrix for wavelength-scale particles, whereas the other works consider particles much larger than the wavelength, with modifications to account for the surface roughness and/or internal inhomogeneity, the incorporation of which then influences scattering more than the model shape assumed, so that the performance of the shape model itself cannot be well established.

According to Veihelmann et al. (2006), the GRS model does not provide as good an agreement between simulated and measured scattering matrices as, for example, spheroids did in Nousiainen et al. (2006) for the same measured sample. This is surprising, because the GRS model supposedly resembles more closely the real, irregular dust particle shapes. Indeed, the covariance function of the radius employed had been derived from the observed shapes of the sample particles. There are several possible explanations for this apparent controversy:

- The good performance of spheroids is probably artificial, as discussed in section 1.4.2.
- The size distribution of the measured sample was truncated at a fairly small particle size in the simulations for the GRS particles due to the computational burden. The simulations therefore are not entirely representative of the measured sample.
- The GRS geometry may be unusually ill suited for this particular sample studied, because the sample particles were generated artificially by grinding. The resulting shapes are much more angular and edgy compared to the roundish shapes that the GRS geometry produces. Of course, spheroids should be even more ill suited for this sample.
- The GRS geometry is based on deforming spheres. Even with considerable deformations, some light-scattering characteristics resembling those of spheres may

remain; for example, ray optics computations for GRSs show remnants of rainbow features that become more diffuse with increasing deformation but do not completely disappear (Muinonen et al., 1996).
- Model shapes that better match the overall shape of the target particles do not necessarily scatter light more similarly.

Even though the covariance function of the radius can be derived from observed shapes of the target particles, it is not obvious how this should be done. The covariance function defines the statistical shape of the whole population rather than an individual model particle, so it seems reasonable to compute the covariance function from an ensemble of target particles. However, if the ensemble contains particles of very different characteristics, such as both roundish, fairly equi-dimensional shapes and thin, flake-like shapes, the resulting covariance function will be an average of those, and the resulting model shapes may not resemble anything in the target ensemble: one cannot average shapes and expect to get averaged scattering. Also, specific features such as sharp edges will be averaged out in the covariance function, and will not be present in the generated shapes. Thus, there are some practical complications in applying the GRS geometry to dust particles. In addition, its performance record, as of now, is not very good. On the other hand, by a somewhat refined approach, such as deriving the shape statistics separately for sub-ensembles of clearly different shape types, much improved performance might be achievable. Considering the potential benefits of this geometry, such investigations might be worthwhile.

A considerably different approach to modeling shapes of dust particles is that adapted by Kalashnikova et al. (2005), where a small number of individual dust particle shapes have been mimicked by shapes built from cubic blocks. An example shape is shown in Fig. 1.9b. These model shapes are actually not much more complex than some of the simple shapes considered in section 1.4.2, but have been assigned here in the complex category because the resulting shapes are concave, unlike all the section 1.4.2 geometries, and because they coarsely mimic real dust particle shapes. The performance of this modeling approach has not been tested with comparisons against laboratory-measured scattering matrices, but they have been incorporated into the operational retrieval algorithm of the Multiangle Imaging SpectroRadiometer (MISR) satellite instrument (Kalashnikova and Kahn, 2006), where they improved both the coverage of successful retrievals and the agreement of the retrieved spectral aerosol optical depths with nearby AERONET measurements, compared to the previous version of the retrieval algorithm which had been based on spheroids.

Deformed spheroids and aggregates have been considered as model particles for mineral dust by Gasteiger et al. (2011). The monomers in the aggregates are overlapping, deformed spheroids. Example shapes of these particles are shown in Fig. 1.9, panels (c) and (d). The study compared simulated scattering properties against lidar measurements, but only in such a way that model ensembles always included multiple model particle types. It is therefore not possible to judge how well each shape model performs individually. It is noteworthy, however, that Gasteiger et al. (2011) found good spectral agreement for the lidar ratio to require mixtures of non-absorbing and absorbing particles, and that model particles with small aspect ratios (below about 1.4) resulted in too strong backscattering compared to the measured values. Good agreement with the lidar measurements required that only a modest portion of the model dust particles had small aspect ratios. Linear depolarization ratios similar to the measured values could be obtained with many different model particle ensembles considered, such as varying types of spheroids, but the irregularly shaped models were more consistent with the measurements than the spheroidal models.

Concave fractal polyhedra (regular and distorted Koch fractals) have been proposed as a model for dust particles by Liu et al. (2012). An example shape is shown in Fig. 1.9e. The results for the distorted Koch fractals are particularly interesting, because they are shown to reproduce the laboratory-measured scattering matrix of feldspar particles, measured by Volten et al. (2001), very well. Interestingly, different random realizations of the distorted Koch fractals with the same irregularity parameter scatter light almost identically. Indeed, the good agreement with the feldspar scattering matrix is obtained by using only one model shape, so the good performance is not artificial, based on a suitable set of differently scattering model particles. Unfortunately, spectral consistency of the model was not tested by fitting the scattering matrix of feldspar particles at the other wavelength where the measurements would have been also available.

Shape models based on 3D Voronoi cells have been adapted for simulating light-scattering properties of dust particles by Ishimoto et al. (2010) and Zubko et al. (2013). The shape model is called spatial Poisson–Voronoi tessellation by Ishimoto et al., while Zubko et al. call it an agglomerated debris particle. The principal idea of the Voronoi cells is that seed points are located within a 3D volume, and then the space is divided into subregions, cells, depending on which seed point is closest. The shape models proposed in Ishimoto et al. (2010) and Zubko et al. (2013) are not, however, identical. In Ishimoto et al. (2010), a 3D space is divided into Voronoi cells, and then those cells whose seed points are located within a specific spherical or spheroidal volume are chosen to form the particle. In contrast, Zubko et al. (2013) begins with a spherical volume, which is then divided into Voronoi cells; they then assign some cells to be empty and only some occupied with material; further, they treat separately and differently the surface layer and the interior of the spherical volume, using much smaller cells in the surface layer. The cells in the surface layer do not extend beyond the boundary of the original sphere, so this boundary acts as one facet in the Voronoi cells. Both methods lead to fairly equi-dimensional shapes, but there are notable differences. Most strikingly, shapes by Ishimoto et al. are compact, while Zubko et al. produce quite porous, sparse structures. The surfaces are also different: the shapes in Zubko et al. (2013) show the remnants of the original spherical geometry, while the shapes in Zubko et al. (2013) do not. Example shapes of these geometries are shown in Fig. 1.9, panels (f) and (g).

In Ishimoto et al. (2010), light-scattering simulations for a single spatial Poisson–Voronoi tessellation particle, generated using a spheroidal volume, for two distinct size parameters, were compared with laboratory-measured scattering matrices for small quartz and olivine particles. The monodisperse nature of the simulations makes it challenging to quantify the performance of the model, but clearly the model particles can reproduce the generic characteristics in the laboratory-measured scattering matrices and, in general, a good agreement between the simulations and measurements is observed. In contrast, Zubko et al. (2013) use a power-law fit for the measured size distribution of a feldspar sample for their model particles. They only tune the distribution by adjusting the lower cut-off of the power-law distribution that was not measured by the experimentalists to optimize the agreement of the simulated $-P_{12}/P_{11}$ with the measurements. All other parameters were fixed. They then applied this distribution also to a second wavelength. The agglomerated debris particles reproduce the scattering matrix simultaneously at both wavelengths (442 nm and 633 nm) very well. This model geometry therefore appears to correctly model also the wavelength dependence of scattering.

1.4.3.2 Inhomogeneous or anisotropic models

Real dust particles often have complex shapes and inhomogeneous composition. Most of the commonly occurring mineral species present in atmospheric dust particles are also anisotropic. Yet, as of now, there appear to be no studies where all the common dust particle characteristics are taken into account. There are, however, a few studies where either the inhomogeneity or the material anisotropy has been accounted for with a complex shape model.

Birefringent flakes and irregular rhombohedra were adapted by Dabrowska et al. (2013) to reproduce the laboratory-measured scattering matrix of a dust sample consisting of calcite particles. Both of these shape models, illustrated in Fig. 1.9 (panels (h) and (i)), are based on the Poisson–Voronoi tessellation. In the case of flakes, the generation is started with an oblate spheroid volume, while irregular rhombohedra are generated from a regular rhombohedral volume. First, these volumes are divided into surface and core layers, and both layers are further divided into volume elements. The tessellation is applied to the surface layers, converting the volume elements to either material or void type, depending on which type of seed cell is the closest. In the case of flakes, some additional smoothing is applied, by turning volume elements of the material type into voids if too few of their neighboring volume elements are also of the material type. The irregular rhombohedra are not smoothed to retain their sharp, angular features. In the light-scattering computations, carried out with the DDSCAT DDA code by Draine and Flatau (1994), the strong, natural birefringence of calcite was explicitly accounted for.

As the size-parameter range of the simulated particles does not cover more than about half of the total scattering cross-section of the sample particles, quantitatively good fits cannot be expected, but comparisons still give qualitative information about the performance of the shape models. For example, irregular rhombohedra produce stronger depolarization and thus smaller P_{22}/P_{11} values than flakes, matching the measured values much better, whereas flakes produce linear polarization ($-P_{12}/P_{11}$) with the same kind of angular dependency as the measurements, but with too high an amplitude, while for rhombohedra the values are too small. It is shown that the best agreement with measurements is achieved when both flake-like and rhomboidal model particles are included in the model particle ensemble, which is consistent with the presence of both types of particles in the measured sample. Indeed, the best matches obtained are, overall, very good, considering how much the size distribution was truncated for the simulations.

The GRS geometry with internal inhomogeneity has been considered a model for large Saharan dust by Nousiainen et al. (2003, 2011a) and Muñoz et al. (2007). Of these, Nousiainen et al. (2003) and Muñoz et al. (2007) incorporated the inhomogeneity by a simple, phenomenological ad hoc scheme where internal Lambertian screens were put inside the model particles, while Nousiainen et al. (2011a) applied a more physical approach where the particle interiors were filled with a cloud of independent scatterers with pre-defined single-scattering properties. The simulations were compared against laboratory-measured scattering matrices, in each study for a different sample, and, in each case, the agreement between the simulations and the measurements improved when the internal scatterers were employed. Very similar improvements were, however, also achieved by adding small-scale surface roughness to the particles, so it is unclear whether the improved agreement with measurements tells us anything about the importance of accounting for the internal inhomogeneity, or the merits of the applied approaches. In addition, the light-scattering methods adapted in the studies are far from being exact, and the link between the optical treatment for the inhomogeneity and actual physical inhomogeneities could not be established. Indeed, very little is known about internal inhomogeneity of real dust particles.

By far the most interesting study is, however, that by Lindqvist et al. (2013), where stereogrammetry and energy-dispersive XRF from scanning electron microscope images were used to obtain model particles with 3D shapes and spatial mineral compositions resembling those of the real dust particles analyzed. An example model shape is shown in Fig. 1.9j. The single-scattering properties were computed using the DDA method. Only four target particles were analyzed and simulated, and there are no corresponding light-scattering measurements to validate the results, but the accuracy of the DDA compares favorably with exact numerical methods (Yurkin and Kahnert, 2013), so the accuracy of the results is primarily limited by the accuracy of the shape model. Even though the derived shapes and composition distributions of the model particles are unlikely to exactly match those of the target particles, the model shapes are nevertheless reasonable and realistic models for similar dust particles. For the particles considered, most of the constituent minerals had fairly similar refractive indices, with only trace amounts of hematite. Consequently, when the computed single-scattering properties were compared against those obtained with particles that had identical shapes but were homogeneous, the differences were minor. This may suggest that the inhomogeneity becomes optically important only if constituents with sufficiently different refractive indices are present in sufficiently large quantities. It is, however, noted that inhomogeneity may also arise from structural features such as fractures or porous cavities inside the particles, which were not considered. It is envisaged that the modeling approach presented by Lindqvist et al. (2013) could be very useful as a reference for validating faster models suitable also for larger-sized parameters.

In the absence of laboratory-measured single-scattering properties for individual particles, the model results provided by Lindqvist et al. (2013) are probably the best currently available source for estimating how much scattering by different types of dust particles can differ from each other. To demonstrate this, Fig. 1.10 shows the mean scattering matrix as well as the range of variation among the four modeled, stereogrammetrically derived model particles, integrated over the considered size distribution. As can be seen, the model particles based on stereogrammetrically derived shapes and realistic spatial distributions for their mineralogical composition possess quite different scattering matrices. In monodisperse cases, these differences tend to be even larger.

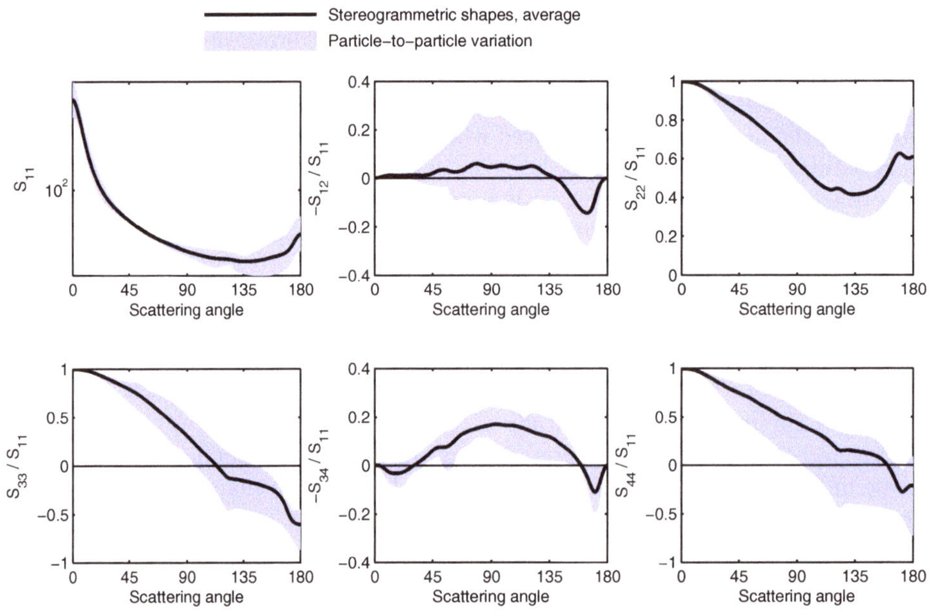

Fig. 1.10. The mean (solid line) and the range on variation (shaded area) for scattering matrix **S** elements S_{11}, $-S_{12}/S_{11}$, S_{22}/S_{11}, S_{33}/S_{11}, $-S_{34}/S_{11}$, and S_{44}/P_{11} by the four stereogrammetrically derived model particles, integrated over the size distribution considered (Lindqvist et al., 2013).

1.4.4 Impact of morphological details and anisotropy on scattering

This section considers how the surface roughness, inhomogeneity, and material anisotropy impact the single-scattering properties of dust particles. The studies considered here are pure modeling studies, not validated by comparisons with measurements. Indeed, in many cases, it would be quite a challenge to investigate these through measurements. The microwave analog approach (see, e.g. Gustafson, 2000; Vaillon et al., 2011) would lend itself for this purpose, as it allows measuring scattering by individual targets specifically constructed to have the desired properties at size parameters relevant for mineral dust, but the representativeness of the results would obviously be subject to the representativeness of the target shapes used.

1.4.4.1 Surface roughness

As discussed in section 1.2.1, the dust particle surfaces are seldom smooth. Rather, surface structures in varying scales are often seen. In addition, there are often small dust particles covering the surface of larger dust particles. Strictly speaking, we then have an aggregate particle but, for light-scattering purposes, it may also be possible to treat the aggregated small particles as surface roughness elements.

The impact of dust particles' surface roughness on scattering have been recently reviewed by Nousiainen (2009). It was identified as potentially quite a significant factor, but the findings were somewhat controversial, different studies reporting differing, even conflicting, findings. The most likely explanation for this is that the impact of roughness on scattering depends on the properties of the roughness. Indeed, there were suggestions that it might also depend on the properties of the host particle, such that the host particle and the roughness characteristics are interconnected. These results find support in the most recent literature. Different roughness models employed in these studies are presented in Fig. 1.11.

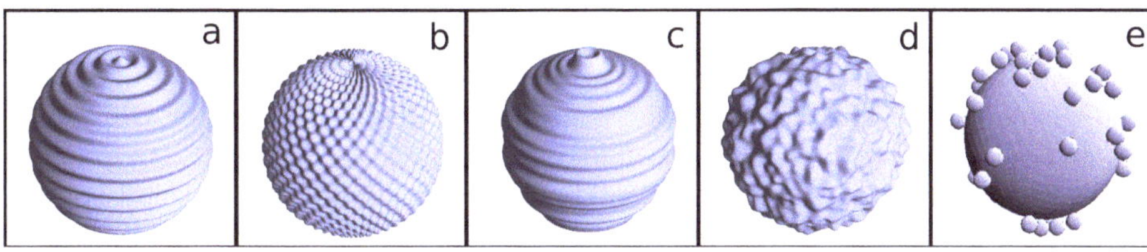

Fig. 1.11. Example model shapes with rough surfaces: 2D Chebyshev particles (a), 3D Chebyshev particles (b), 2D GRS particles (c), 3D GRS particles (d), and dusted spheres (e). The two other roughness models considered are shown in Fig. 1.9, panels (c) and (d).

The impact of surface roughness on absorbing hematite particles was investigated by Kahnert et al. (2011). The surface roughness was modeled using an axisymmetric high-order Chebyshev series of varying amplitudes on a spherical body (Fig. 1.11a). The impact of roughness became evident in the single-scattering properties at about $x = 6$, regardless of the roughness amplitude considered, and increased in effect with increasing particle size parameter. The impacts were generally systematic: to decrease ϖ, g, and C_b from the values for smooth spheres. At the largest size of parameter considered, 14, the roughness (with the largest relative amplitude of 0.05 considered) reduced ϖ by about 0.1, g by about 0.2, and C_b by about a factor of two. The roughness did not contribute much to depolarization, but decreased P_{11} at side- and backscattering angles, and made $-P_{12}/P_{11}$ more positive.

In Kahnert et al. (2012), four different types of surface roughness were considered, all having fairly small amplitudes (a few percent relative to the particle radius). These roughness models are presented in Fig. 1.11, panels (a) to (d). Simulations were carried out for two different size parameters, 5 and 50, and for silicate and hematite compositions. In the case of the silicate composition, the single-scattering properties for the axisymmetric GRS particle (Fig. 1.11c) showed the largest deviations from those for perfect spheres but, even then, only the backscattering quantities were significantly affected by roughness at either size parameter considered. For the hematite composition, the impacts were larger, but also more complicated. Overall, it was found that roughness has a potentially larger relative impact on scattering by absorbing particles (hematite) and with larger x (even though the relative roughness amplitude was much smaller for larger x), but also that the impact can be quite complicated, different for different single-scattering quantities and roughness types, defying simplistic rules. For example, with the hematite composition, the axisymmetric GRS roughness (Fig. 1.11c) typically produced the largest impact on scattering at $x = 5$, but the weakest effect at $x = 50$. The backscattering quantities were found to be generally most affected by roughness.

Deformed spheroids and aggregates studied by Gasteiger et al. (2011) can also be considered as models for surface roughness, where the deformation is expressed in Gardner series. These model shapes are illustrated in Fig. 1.9, panels (c) and (d). The study focused on lidar parameters and noted that, at size parameters below 10, deformation had only a small impact on either the lidar ratio R or the linear depolarization ratio δ_L. For larger particles, the differences were larger, such that deformation typically decreased R, most likely due to increased backscattering, and increased δ_L.

The light-scattering impact of small aggregated spheres on the surface of a larger host sphere, the so-called 'dusted sphere' geometry, were considered by Dlugach et al. (2012) at refractive indices representative of silicates and hematite. An example of this geometry can be seen in Fig. 1.11e. The refractive index of small spheres was also varied to mimic 'dusting' by dust or soot particles. In general, the impact of dusting is to flatten P_{11} by increasing side- and decreasing backscattering, and decreasing P_{22}/P_{11}, the latter making the particles more depolarizing. The impact on $-P_{12}/P_{11}$ depended on the refractive index: for silicate, it was increased by dusting while, for hematite, it was decreased. This is most likely due to the very different $-P_{12}/P_{11}$ elements for the plain host spheres in these cases: negative for the silicate and strongly positive for hematite. In both cases, the dusting thus acted to drive $-P_{12}/P_{11}$ toward zero or slightly positive values. Not surprisingly, the effects of dusting became more pronounced with the increased number or size of the dusting particles. They were also somewhat more pronounced for the hematite than for the silicate composition, but this increase appeared to be connected to the increased real part of the refractive index, not to that of increased absorption. The strongest impact was seen when a silicate sphere was covered with small soot spheres.

The impact of surface roughness on scattering by large dust particles was investigated by Nousiainen et al. (2011a). It was seen that increasing the amount of surface roughness resulted in (i) an increase in scattered intensity at side- and backscattering directions; (ii) decreased $-P_{12}/P_{11}$ at side-scattering angles; and (iii) smaller P_{22}/P_{11}, namely strengthened linear depolarization. In other words, surface roughness promoted those features in scattering which mineral dust particles are often associated with. Indeed, including surface roughness in the simulations allowed better fits with the corresponding light-scattering measurements for the dust samples being simulated. However, it was also noticed that the impact of surface roughness on scattering was qualitatively similar to the impact of the imaginary part of the refractive index on scattering, such that increasing surface roughness and decreasing $\mathrm{Im}(m)$ affected scattering fairly similarly. This would make $\mathrm{Im}(m)$ values inverted from light-scattering measurements dependent on the surface roughness model used.

1.4.4.2 Inhomogeneity

Compared to the surface roughness, not much attention has been given to the optical impacts of particle inhomogeneity. Certainly, the dust particle inhomogeneity is often acknowledged and even accounted for by applying an effective medium approximation and computing the corresponding effective refractive index for the particles, but this treatment results in homogeneous model particles. Without corresponding simulations where the inhomogeneity is explicitly accounted for, the impact of inhomogeneity cannot be addressed. There also exist some pure modeling studies where particle inhomogeneity has been addressed, in particular structural inhomogeneity in the form of porosity.

Most of the relevant literature has been reviewed by Nousiainen (2009). Similarly to the impact of roughness on scattering, internal structures tend to promote positive $-P_{12}/P_{11}$ with Rayleigh-like symmetry, and strenghthen depolarization (reduce P_{22}/P_{11}) for wavelength-scale particles (Vilaplana et al., 2006; Lindqvist et al., 2009). Similar behavior is also observed when an effective medium approximation is applied, so these effects may be partially linked with the refractive index (Lindqvist et al., 2009). Inhomogeneity appears also to increase g. For large dust particles, the opposite is observed, unless inhomogeneity increases absorption and causes g to increase instead (Macke et al., 1996; Muinonen et al., 2009). Overall, the impacts of inhomogeneity on scattering are often similar to those of the surface roughness. Both these features act to make the model particles more complex.

The impact of spatial inhomogeneity on scattering has been considered by Lindqvist et al. (2013). A small number of dust particles were carefully analyzed with an electron microscope and energy-dispersive XRF, and corresponding model particles, including spatial distribution of different minerals present in the particles we produced. These were then accounted for, as is, in semi-exact DDA simulations. Simulations were also carried out for the corresponding cases where the dielectric properties of different constituent minerals are averaged (effective medium), weighted by their relative volumes. This study gives a good estimate on how much even minor inhomogeneity contributes to scattering by these particles. First of all, the effect is size-dependent: the largest differences are found at size parameters from $x = 3$ to 8. Secondly, S_{22}/S_{11} is systematically lower (meaning stronger depolarization) for inhomogeneous particles, and the effective medium generally overestimates forward-scattered intensity. The $-S_{34}/S_{11}$ is also clearly affected by the inhomoneity, and might explain, in part, why this element has proven difficult to mimic with any of the proposed approaches (see, e.g. Fig. 1.12). Thirdly, the asymmetry parameter, single-scattering albedo, lidar ratio, and linear depolarization ratio are not systematically affected but depend on the individual particles (shape and spatial distribution of composition). Finally, all of the findings were strongest for particles including absorbing minerals, even if only up to 2% of the volume. This finding might be connected to the vast, several orders of magnitude difference in the imaginary parts of the refractive index between the typical weakly absorbing minerals and, for example, hematite. This allows for a large impact in the effective refractive index even with modest volume fractions. It is emphasized that the use of effective medium approximations for such particles can lead to considerable errors, but the errors are to some extent averaged out for ensembles of differently shaped particles.

1.4.4.3 Material anisotropy

In the soil, from which atmospheric dust originates, the dominant mineral groups are illite, kaolinite, chlorite, smectites as well as quartz, feldspars, hematite, calcite, and gypsum (Claquin et al., 1999). All these minerals are anisotropic, so in principle one should use the dielectric tensor rather than the scalar refractive index when simulating light scattering by particles composed of these minerals. Of these minerals, the anisotropy of calcite and hematite is quite substantial. They belong to the trigonal system of crystal symmetries, so they both have a single optic axis (McKie and McKie, 1974). Along the optic axis, they have an *extraordinary* refractive index, m_e, and, in the perpendicular direction, an *ordinary* refractive index, m_o. Such materials are called (linearly) *birefringent*. The birefringence arises from the response of the material being dependent on the direction of the electric field. In case the birefringent material is also absorbing, the amount of absorption also depends on the (linear) polarization state and the material is considered to be (linearly) *dicroic* (Bohren and Huffman, 1983).

Birefringence is quantified by the difference of the real part of the ordinary and extraordinary refractive indices:

$$\Delta n = n_o - n_e \,. \tag{1.11}$$

For calcite, Δn varies from 0.17 to 0.2 at visible wavelengths (Ghosh, 1999). For hematite, Δn is even slightly larger (0.28 at 589 nm wavelength; http://www.mindat.org). Hematite is, however, a strongly absorbing substance, meaning that the internal electromagnetic waves, where the impact of birefringence would be manifested, are dampened. We are not aware of studies where the impact of birefringence on hematite would have been investigated but, because of this dampening, the impact is unlikely to be larger than that for calcite.

The impact of birefringence on the single-scattering properties of calcite have been studied by Dabrowska et al. (2012, 2013) and Nousiainen et al. (2009, 2011a). In Nousiainen et al. (2009, 2011a), only thin calcite flakes were considered (the latter simply showing size-averaged results from Nousiainen et al. (2009)), while Dabrowska et al. (2012, 2013) also consider irregular rhomboids, which are more equi-dimensional in shape. The sensitivity of the single-scattering properties on the orientation of the optic axis was investigated by Dabrowska et al. (2012). This is of considerable practical importance, because the orientation of the optic axis within a calcite particle is not arbitrary, but rather has a specific orientation with respect to the crystal lattice, which in turn correlates with the shape, because calcite is most likely to break along the principal crystal faces. Still, it is often anything but trivial to identify the principal crystal faces from the shapes of natural calcite particles. This may lead to some ambiguity in the correct orientation of the optic axis in model particles.

The findings of these studies can be summarized as follows:

- Of the scattering matrix elements, the phase function is least affected by the birefringence. The higher sensitivity of polarization on birefringence is not surprising, considering that birefringence in effect separates the incident waves into two perpendicularly (linearly) polarized waves propagating along different paths within the particle.

- For size parameters close to unity or below, the impact of birefringence on scattering is weak. This can be understood by considering interference between internal waves, which is associated with the phases of these waves. Birefringence contributes to this interference in two ways: on one hand, ordinary and extraordinary waves will have different wavelengths; on the other hand, ordinary and extraordinary waves (usually) propagate along different paths. However, neither mechanism can modify the interference substantially unless the internal path lengths are sufficiently long, which they are not in small particles.
- Scattering depends on the orientation of the optic axis considerably when the particles themselves have preferential orientations; for an ensemble of randomly oriented particles, the impact is much smaller. This suggests that the difference in scattering properties between otherwise identical isotropic and birefringent particles is not systematic, so that the orientation averaging tends to decrease the overall effect.
- Scattering can depend considerably on the orientation of the optic axis within the particle for very elongated targets; for nearly equi-dimensional particles, the impact is small. This is because, in the case of elongated targets, the orientation of the optic axis will control how large the refractive index is along the long and short axes of the particle, whereas, in the case of an equi-dimensional scatterer, it only influences how the refractive indices are relative to smaller-scale shape features. The former effect can obviously induce a larger effect on scattering. This finding is fortuious, because the orientation of the principal crystal faces is easier to identify from elongated particles (one of the principal planes is likely to be parallel to the axis of elongation, because crystals are likely to break along these faces). For those equi-dimensional cases where the identification of the principal crystal faces is difficult, correctly orienting the optic axis seems not to be necessary.

The findings of these studies imply that, in radiative forcing considerations, for example, birefringence can be safely ignored. In remote sensing applications where scattering at specific angles is considered, birefringence can be a factor, especially if polarized quantities are measured (see, e.g. Fig. 2 of Nousiainen et al. (2011a)). One should notice, however, that, for most mineral species present in atmospheric dust particles, birefringence is weaker than for the calcite considered in these studies, often quite considerably. So, in most practical considerations, birefringence is likely to be a minor issue.

1.5 Discussion and conclusions

As can be seen, there is a wide variety of modeling approaches applied to simulating the single-scattering properties of atmospheric dust particles. The principal difference between the approaches considered is the morphological characteristics assumed for the model particles but, due to the inherent limitations of different numerical light-scattering solvers, the choice of particle morphology is often closely linked with the choice of the computational method. For more complex particle models considered, only the brute-force numerical methods are applicable, restricting these approaches to relatively small parameters. For example, methods such as the DDA are in practice restricted to size parameters of about $x < 20$, which corresponds to particles smaller than 4 μm in diameter at the wavelength of 628 nm. This is not sufficient for covering the radiatively important particle sizes at solar wavelengths even far away from the source regions, thus making such methods ill suited for assessing the dust radiative impact on a global scale. It is, of course, possible to apply other methods for larger particles, but this may cause other complications, as the two methods do not necessarily converge (e.g. Nousiainen et al., 2011a). It thus appears that, for wavelength-scale dust particles, computational issues do not impose to us inhibiting simplifications in the model particle morphology; whereas, for larger particles, the available computational methods appear to be severely restricting. The particularly crucial aspect is the treatment wavelength-scale structure in dust particles much larger than the wavelength. Such structures are obviously highly relevant for scattering, but presently cannot be accounted for in a physically rigorous way together with realistically shaped model particles (e.g. Nousiainen et al., 2011a; Nousiainen, 2009; Kahnert et al., 2012; Muinonen et al., 2009). In general, the larger the dust particles are, the less exact, accurate, and realistic our modeling approaches currently can be.

In this review, we have focused on modeling approaches that have been compared with measurements, so that their performance could be assessed. This assessment is, however, anything but easy, because different studies use different measurement data for comparison, and the performance does not depend only on the approach adapted, but also on the target particle properties, the accuracy of the auxiliary information (such as the size distribution measurements), as well as the accuracy of the assumptions about the refractive index. In addition, the performance of an approach greatly depends on what it is being used for: for particle sizing based on forward scattering, even Mie spheres are very suitable, while for backscattering or (de)polarization applications, their use would be a poor choice indeed. Consequently, we do not even attempt to rate the different approaches quantitatively. Rather, we try to draw some general conclusions about their performance, or the lack thereof. In addition, we will assess all the methods using the same criteria.

Let us first consider what would be good criteria to use. Ideally, they should be generic rather than application-specific. Since we are considering single-scattering modeling of dust particles, and the scattering event is described by the scattering matrix, it appears plausible to consider criteria related to the scattering matrix. We know from the laboratory measurements that different dust particle ensembles tend to have fairly similar scattering matrices (at visible wavelength), with characteristic features such as flat, featureless P_{11} at side- and backscattering angles; weak, positive $-P_{12}/P_{11}$ with a maximum close to 90° scattering angle; and considerable depolarization with P_{22}/P_{11} often dropping below 0.5, with a minimum in the backscattering hemisphere (e.g. Volten et al., 2001). Those same measurements also show that the scattering matrices depend, sometimes rather considerably, on the wavelength. We are not aware of measured scattering matrices for individual dust particles, but model simulations based on realistic, stereogrammetrically derived shapes suggest that particle-to-particle differences due to different shapes and compositions are considerable even after integrating over a size distribution (Lindqvist et al., 2013). If these differences extend to real dust particle ensembles, perhaps originating from different sources, the modeled ensemble-averaged single-scattering properties should ideally exhibit similar variability. Incidentally, it might allow identifying different types of dust particles through remote sensing.

Based on these considerations, we arrive at the following three criteria:

1. How accurately they predict the dependence of the full 4×4 scattering matrix of a dust ensemble on the scattering angle.
2. How accurately they predict the dependence of the full 4×4 scattering matrix of a dust ensemble on the wavelength.
3. How well they produce realistic particle-to-particle variability due to shape and composition in the full 4×4 scattering matrices for individual dust particles.

If all these criteria are well met, the approach should be well suited for most types of applications related to light scattering, with reasonable confidence for its reliability and consistency.

The true merit of an approach is, however, in accurately predicting the single-scattering properties of dust particles given their physical properties as input. Good performance with our three criteria does not guarantee this, but testing this properly is not possible with the currently available reference data due to the poorly constrained physical properties of the laboratory-measured samples. We also note that these are not the only suitable criteria. In particular, criteria related to absorption might be reasonable additions in case suitable reference data were available. Also, a criterion for correctly predicting the size dependence of the scattering matrix would have been quite relevant, but very few of the approaches have been tested for this, and there are no suitable validation data available (there are no dust samples with the same shapes and compositions but different sizes). The wavelength dependence, however, works as a reasonable proxy for the size dependence, since they are both related to changing the size parameter. The wavelength dependence includes an additional effect from changing the refractive index as a function of wavelength, however.

Our assessment of the approaches proposed so far, based on the criteria proposed, is summarized in Table 1.1. The references given indicate studies that were used in the assessment, rather than giving a comprehensive list of studies where these approaches have been considered. It is immediately obvious from the table that the currently available data about the different approaches are severely lacking for this kind of assessment. In particular, very few methods have been compared against the measurements at different wavelengths. In many cases, this is clearly connected to computational limitations: comparisons have been limited to the longest wavelength where measurement data are available, where the size parameters are smallest and the computationally heavy methods can reach the largest particle sizes and thus require least truncation in the size distribution. Still, the multi-wavelength comparisons have often been overlooked also when there would have been both computations and measurement data available to do so. We strongly urge that, in future, studies on this aspect of model performance would be given more attention. In addition, some of the methods have not been tested against measured scattering matrices, but, for example, lidar or satellite data, making the assessment based on our criteria impossible. Finally, we point out that different methods have been validated against different data, with different refractive indices, shapes, and size parameters, so the validation data present varying levels of

Table 1.1. Assessment of different approaches to model single-scattering properties of mineral dust based on the suggested criteria. In the assessment, a rough scale of bad–poor–good–very good–excellent is used.

Simple geometries				
Model	Criterion 1	Criterion 2	Criterion 3	References
Spheres	Poor	Poor	Bad	Alexander et al., 2013; Nousiainen et al., 2006; Min et al., 2005
Spheroids	Very good	Poor	Very good	Nousiainen et al., 2006; Merikallio et al., 2011; Zubko et al., 2013
Ellipsoids	Excellent	Unknown	Very good	Merikallio et al, 2013; Bi et al., 2009; Meng et al., 2010
Symmetric polyhedra	Poor	Unknown	Poor	Nousiainen et al., 2006
Nonsymmetric hexahedra	Excellent	Unknown	good	Bi et al., 2010
Convex polyhedra	Unknown	Unknown	Good	Gasteiger et al., 2011
Complex geometries				
Model	Criterion 1	Criterion 2	Criterion 3	References
Gaussian random spheres	good	Unknown	Poor	Veihelmann et al., 2006
Cubic blocks	Good	Unknown	Poor	Kalashnikova et al., 2005
Deformed shapes	Unknown	Unknown	Unknown	Gasteiger et al., 2011
Concave fractal polyhedra	Excellent	Unknown	Poor	Liu et al., 2012
Spatial Poisson–Voronoi	Very good	Unknown	Unknown	Ishimoto et al., 2010
Agglomerated debris	Excellent	Excellent	Unknown	Muinonen et al., 1996
Irregular flakes	Very good	Unknown	Unknown	Dabrowska et al., 2013; Nousiainen et al., 2009
Irregular rhombohedra	Very good	Unknown	Unknown	Dabrowska et al., 2013

challenge in modeling. Since it would be quite difficult to compensate for this, all methods have been simply assessed as they perform against the validation data that had been used.

Regarding the simple model geometries, considered in section 1.4.2, the scattering matrices of laboratory-measured dust samples have been closely reproduced with spheroids, ellipsoids, and nonsymmetric hexahedra. As discussed in section 1.4.2, this good performance is probably artificial for spheroids and ellipsoids: it is not achieved because individual spheroids scatter like individual dust particles, but because spheroids have large particle-to-particle variability in their single-scattering properties, allowing different sets of spheroids to mimic very different types of scattering matrices, but so that the optimally performing sets are different for different single-scattering properties and at different wavelengths. In other words, the physical and single-scattering properties appear to be linked inconsistently. The particle-to-particle variability for the nonsymmetric hexahedra, on the other hand, is weaker than for spheroids or ellipsoids, so its good performance cannot be as artificial. However, the variability might also be too weak to represent that for real dust particles realistically, which might make it too inflexible for mimicking scattering by varying dust particle ensembles. We emphasize that a good model needs sufficiently large particle-to-particle variability in the single-scattering properties to be representative of real dust, but having this property also makes it susceptible to artificially good performance. Testing and validating such approaches require extra care.

We are not aware of any tests of the spectral performance for the simple approaches, except for spheres and spheroids, so the second criterion cannot be used to assess their relative performance. Based on the first and third criteria, ellipsoids and nonsymmetric hexahedra are found to be the most promising. We consider the latter to be more interesting, because its good performance cannot be as artificial, and is therefore likely to be more reliable. It would be interesting to see how ellipsoids and nonsymmetic hexahedra reproduced the spectral dependence of the measured scattering matrices.

Among the approaches based on complex geometries, we have more cases with good performance for the first criterion. Most of the methods considered reproduce the reference scattering matrices at least reasonably well. For the Gaussian random sphere, spatial Poisson–Voronoi tessellation, irregular flake, and irregular rhombohedron geometries, the assessment accuracy suffers from the quite limited size-parameter coverage of the simulations compared to the reference data. The grades should therefore be taken with a grain of salt. Two of the methods, the concave fractal polyhedra (Koch fractals, the irregular type to be precise) and the agglomerated debris, however, show remarkable performance in reproducing the reference scattering matrices. Their performance in reproducing the laboratory-measured scattering matrix for small feldspar particles (Volten et al., 2001) is illustrated in Fig. 1.12. As can be seen, they closely match all the scattering matrix elements considered (note that $-P_{12}/P_{11}$ and $-P_{34}/P_{11}$ have truncated y-axes), and generally provide closer fits than those obtained using spheroids or ellipsoids, also shown for comparison. The agglomerated debris also provides an equally good performance for the spectral dependence. Different concave fractal polyhedra produce very similar scattering properties, however, so it does not seem suitable for reproducing realistic particle-to-particle variability in the scattering properties. For agglomerated debris, this information is not available because, in the approach, orientation and ensemble-averaging are carried out simultaneously: orientation-averaged results for single realizations are therefore not available. We believe, however, that agglomerated debris particles would perform very similarly to concave fractal polyhedra in this respect: both of these approaches use model geometries that are even more complex than real dust particles. In a way, much of the complexity of real dust particle ensembles is included already in single model particles, reducing the particle-to-particle variability of the model particle scattering properties. This may make these methods ill suited for discriminating different types of dust particle ensembles.

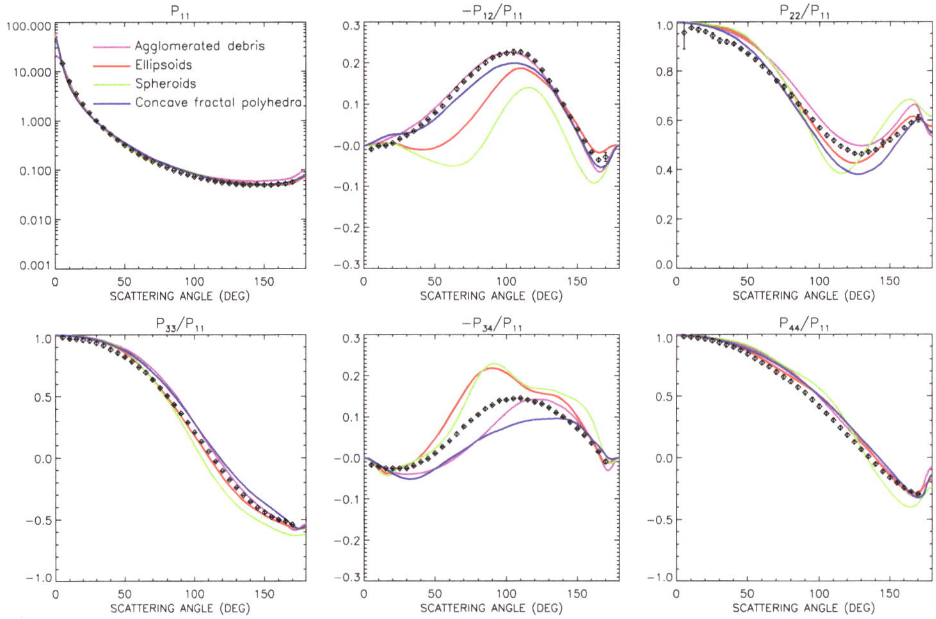

Fig. 1.12. Comparison of the measured scattering matrix of a feldspar sample (diamonds with error bars) at 633 nm wavelength (Volten et al., 2001) and model simulations based on an ensemble of agglomerated debris particles (magenta line; Zubko et al., 2013), a single concave fractal polyhedron (blue line; Liu et al., 2012), and shape distributions of spheroids and ellipsoids (green and red lines, respectively; Bi et al., 2009). All P_{11} elements have been set to unity at 30° scattering angle to assure common normalization.

Overall, we find that none of the modeling approaches proposed so far performs ideally in terms of the criteria suggested, or at least has not been proven to do so. We also find that many of the methods currently used in practical applications are severely lacking in some respects. It is important to be aware of these limitations, and not to use these approaches to generate data for which they are ill suited. In particular, the issue of shape retrieval deserves some discussion. We have seen that dust particles show a remarkable variability in their shapes. Likewise, we have seen that the ensemble-averaged scattering matrices for real dust particles can be mimicked by a wide variety of different model particles from spheroids to irregular fractal polyhedra, and that laboratory-measured scattering matrices for different dust samples, even those for volcanic ash, resemble each other quite considerably. When Nousiainen et al. (2012) investigated how size-shape distributions of different, wavelength-scale model particles scatter light, not only was it found that surprisingly different shape models could result in surprisingly similar scattering properties, but also that seemingly similar shape models could show larger differences in their scattering properties than some other shape models with obviously more different shapes. These findings suggest that, when ensemble-averaged dust particle single-scattering properties are considered, the shapes and the scattering properties do not necessarily correlate as well as one would think, and that information about dust particle shapes is not easily accessible. We often pay attention to the overall shapes of particles, and think that model particles with similar overall shapes would scatter similarly, but clearly this is not the case. The small-scale surface roughness, or internal inhomogeneity, for example, can contribute significantly to scattering, while being almost undetectable in shape analysis. There are indications that some morphological features, such as particles having very thin, platy shapes, would have sufficiently unique scattering properties to be distinguished even from ensemble-averaged data. In contrast, it seems ill advised to suggest that similarity in simple shape parameters such as the aspect ratio or circularity between model particles and real dust particles would guarantee similarity in scattering.

In addition, it has been shown that some of the suggested modeling approaches are very flexible. When an ensemble of model particles with widely varying single-scattering properties is used, it is generally possible to optimize the shape distribution of such model particles so that it can produce a wide variety of different types of scattering matrices. For example, Nousiainen et al. (2011) show that scattering by a cube can be mimicked by a suitable shape distribution of spheroids. From remote sensing point of view, this means that, if we had perfect cubes in the atmosphere and used a retrieval algorithm based on spheroids to interpret scattering measurements from them, we would get very good fits that would suggest a specific distribution of spheroids present in the atmosphere. The good agreement, or the small residual in the fitting algorithm, might be taken to indicate great confidence in the inversion. Yet the result would be wrong. A realistic model for dust particles should have considerable particle-to-particle variability in the scattering properties, but such a model will be difficult to validate because it is very flexible and can easily mimic ensemble-averaged scattering properties of target particles that do not resemble the model particles. We therefore suggest that we should use more stringent criteria when validating our models. Some kind of additional test for the consistency of the model seems to be called for. One way is to consider multiple wavelengths simultaneously. Another option is to use some other fairly independent data, such as extinction spectra adapted by Meland et al. (2012) and Alexander et al. (2013). Ultimately, the choice of validation method depends on the application of interest. If one is looking for a good method for some very specific task, then the validation should be carried out in the context of that task. Here we consider validation broadly, focusing on the scattering matrix, because of its high information content and generality, making it relevant for a wide variety of purposes and applications.

Ultimately, we want to move from ensemble-averaged scattering data to single-particle scattering data in validation. Once the method has been validated against single-particle data, it will automatically work also for ensembles, and single-particle data will be much more difficult to reproduce 'accidentally' correctly. When working with single particles, the characterization of the physical properties of the target particles also becomes easier. Indeed, it may be possible to use single-particle data to study aspects of dust particles that are currently beyond our tools to investigate even in a laboratory, such as their internal structure. The research based on single dust particles and their scattering properties is a frontier where work has barely started.

Acknowledgments

We wish to thank L. Bi, D. Dabrowska, J. Gasteiger, H. Ishimoto, M. Kahnert, O. Kalashnikova, H. Lindqvist, C. Liu, M. Mishchenko, P. Yang, and E. Zubko for prodiving images and data, and Hannakaisa Lindqvist for her help with figures. The constructive comments and suggestions by Gorden Videen were also found helpful. The Academy of Finland and the Deutsche Forschungsgemeinschaft (DFG) are acknowledged for funding.

References

Alexander, J.M., Laskina, O., Meland, B., Young, M.A., Grassian, V.H. and Kleiber, P.D. (2013) 'A combined laboratory and modeling study of the infrared extinction and visible light scattering properties of mineral dust aerosol', *J. Geophys. Res.*, **118**, 435–52, doi:10.1029/2012JD018751.

Almeida-Prieto, S., Blanco-Méndez, J. and Otero-Espinar, F.J. (2007) 'Microscopic image analysis techniques for the morphological characterization of pharmaceutical particles: Influence of the software, and the factor algorithms used in the shape factor estimation', *Eur. J. Pharm. Biopharm.*, **67**, 766–76.

Bi, L., Yang, P., Kattawar, G. and Kahn, R. (2010) 'Modeling optical properties of mineral aerosol particles by using nonsymmetric hexahedra', *Appl. Opt.*, **49**, 334–42.

Bi, L., Yang, P., Kattawar, G.W. and Kahn, R. (2009) 'Single-scattering properties of triaxial ellipsoidal particles for a size parameter range from the Rayleigh to geometric-optics regimes', *Appl. Opt.*, **48**, 114–26.

Bohren, C.F. and Huffman, D.R. (1983) *Absorption and Scattering of Light by Small Particles*, New York, John Wiley & Sons.

Bristow, C.S., Hudson-Edwards, K.A. and Chappell, A. (2010) 'Fertilizing the Amazon and equatorial Atlantic with West African dust', *Geophys. Res. Lett.*, **37**, L14807, doi:10.1029/2010GL043486.

Caquineau, S., Gaudichet, A., Gomes, L. and Legrand, M. (2002) 'Mineralogy of Saharan dust transported over northwestern tropical Atlantic Ocean in relation to source regions', *J. Geophys. Res.*, **107**, 4251, doi:10.1029/2000JD000247.

Chou, C., Formenti, P., Maille, M., Ausset, P., Helas, G., Harrison, M. and Osborne, S. (2008) 'Size distribution, shape, and composition of mineral dust aerosols collected during the African Monsoon Multidisciplinary Analysis Special Observation Period 0: Dust and Biomass-Burning Experiment field campaign in Niger, January 2006', *J. Geophys. Res.*, **113**, D00C10, doi:10.1029/2008JD009897.

Claquin, T., Schulz, M., and Balkanski, Y.J. (1999) 'Modeling the mineralogy of atmospheric dust sources', *J. Geophys. Res.*, **104**, 22243–56.

Coz, E., Gómez-Moreno, F.J., Pujadas, M., Casuccio, G.S., Lersch, T.L., Artíñano, B. (2009) 'Individual particle characteristics of North African dust under different long-range transport scenarios', *Atmos. Environ.*, **43**, 1850–63.

Dabrowska, D.D., Muñoz, O., Moreno, F., Nousiainen, T. and Zubko, E. (2012) 'Effect of the orientation of the optic axis on simulated scattering matrix elements of small birefringent particles', *Opt. Letters*, **37**, 3252–4.

Dabrowska, D.D., Muñoz, O., Moreno, F., Nousiainen, T., Zubko, O. and Marra, A.C. (2013) 'Experimental and simulated scattering matrices of small calcite particles at 647 nm', *J. Quant. Spectrosc. Radiat. Transfer*, **124**, 62–78, doi:10.1016/j.jqsrt.2013.02.010.

Deboudt, K., Flament, P., Choël, M., Gloter, A., Sobanska, S. and Colliex, C. (2010) 'Mixing state of aerosols and direct observation of carbonaceous and marine coatings on African dust by individual particle analysis', *J. Geophys. Res.*, **115**, D24207, doi:10.1029/2010JD013921.

Dlugach, J.M., Mishchenko, M.I. and Mackowski, D.W. (2012) 'Scattering and absorption properties of polydisperse wavelength-sized particles covered with much smaller grains', *J. Quant. Spectrosc. Radiat. Transfer*, **113**, 2351–5.

Draine, B.F. and Flatau, P.J. (1994) 'Discrete-dipole approximation for scattering calculations', *J. Opt. Soc. America*, **11**, 1491–9.

Dubovik, O., Sinyak, A., Lapyonok, T., Holben, B.N., Mishchenko, M., Yang, P., Eck, T.F., Volten, H., Muñoz, O., Veihelmann, B., van der Zande, W.J., Leon, J.-F., Sorokin, M. and Slutsker, I. (2006) 'Application of spheroid models to account for aerosol particle nonsphericity in remote sensing of desert dust', *J. Geophys. Res.*, **111**, D11208.

Engelstaedter, S., Tegen, I. and Washington, R. (2006) 'North African dust emissions and transport', *Earth-Sci. Rev.*, **79**, 73–100, doi:10.1016/j.earscirev.2006.06.004.

Formenti, P., Schütz, L., Balkanski, Y., Desboeufs, K., Ebert, M., Kandler, K., Petzold, A., Scheuvens, D., Weinbruch, S. and Zhang, D. (2011) 'Recent progress in understanding physical and chemical properties of mineral dust', *Atmos. Chem. Phys.*, **11**, 8231–56, doi:10.5194/acp-11-8231-2011.

Gasteiger, J., Wiegner, M., Groß, S., Freudenthaler, V., Toledano, C., Tesche, M. and Kandler, K. (2011) 'Modelling lidar-relevant optical properties of complex mineral dust aerosols', *Tellus*, **63B**, 725–41.

Ghosh, G. (1999) 'Dispersion-equation coefficients for the refractive index and birefringence of calcite and quartz crystals', *Opt. Commun.*, **163**, 95–102.

Glaccum, R.A. and Prospero, J.M. (1980) 'Saharan aerosols over the tropical north Atlantic: Mineralogy', *Mar. Geol.*, **37**, 295–321, doi:10.1016/0025-3227(80)90107-3.

Gustafson, B.Å.S. (2000) 'Microwave analog to light-scattering measurements', in Mishchenko, M.I., Hovenier, J.W. and Travis, L.D. (eds), *Light Scattering by Nonspherical Particles*, London, Academic Press, 367–90.

Haapanala, P., Räisänen, P., Kahnert, M. and Nousiainen, T. (2012) 'Sensitivity of the shortwave radiative effect of dust on particle shape: Comparison of spheres and spheroids', *J. Geophys. Res.*, **117**, D08201, doi:10.1029/2011JD017216.

Hentschel, M.L. and Page, N.W. (2003) 'Selection of descriptors for particle shape characterization', *Part. Syst. Charact.*, **20**, 25–38, doi:10.1002/ppsc.200390002.

Ishimoto, H., Zaizen, Y., Uchiyama, A., Masuda, K. and Mano, Y. (2010) 'Shape modeling of mineral dust particles for light-scattering calculations using the spatial Poisson–Voronoi tessellation', *J. Quant. Spectrosc. Radiat. Transfer*, **111**, 2434–43.

Jaenicke, R. and Schütz, L. (1978) 'Comprehensive study of physical and chemical properties of the surface aerosols in the Cape Verde Islands region', *J. Geophys. Res.*, **83**, 3585–99, doi:10.1029/JC083iC07p03585.

Jeong, G.Y. (2008) 'Bulk and single-particle mineralogy of Asian dust and a comparison with its source soils', *J. Geophys. Res.*, **113**, D02208, doi:10.1029/2007JD008606.

Journet, E., Desboeufs, K.V., Caquineau, S. and Colin, J.-L. (2008) 'Mineralogy as a critical factor of dust iron solubility', *Geophys. Res. Lett.*, **35**, L07805, doi:10.1029/2007GL031589.

Kaaden, N., Maßling, A., Schladitz, A., Müller, T., Kandler, K., Schütz, L., Weinzierl, B., Petzold, A., Tesche, M., Leinert, S. and Wiedensohler, A. (2009) 'State of mixing, shape factor, number size distribution, and hygroscopic growth of the Saharan anthropogenic and mineral dust aerosol at Tinfou, Morocco', *Tellus*, **61B**, 51–63.

Kahnert, F.M. (2004) 'Reproducing the optical properties of fine desert dust aerosols using ensembles of simple model particles', *J. Quant. Spectrosc. Radiat. Transfer*, **85**, 231–49.

Kahnert, M. (2013) 'The T-matrix code Tsym for homogeneous dielectric particles with finite symmetries', *J. Quant. Spectrosc. Radiat. Transfer*, **123**, 62–78, doi:10.1016/j.jqsrt.2012.12.019.

Kahnert, M. and Nousiainen, T. (2006) 'Uncertainties in measured and modelled asymmetry parameters of mineral dust aerosols', *J. Quant. Spectrosc. Radiat. Transfer*, **100**, 173–8, doi:10.1016/j.jqsrt.2005.11.035.

Kahnert, M. and Nousiainen, T. (2007) 'Variational data analysis method for combining laboratory-measured light scattering phase functions and forward-scattering computations', *J. Quant. Spectrosc. Radiat. Transfer*, **103**, 27–42, doi:10.1016/j.jqsrt.2006.07.011.

Kahnert, M., Nousiainen, T. and Mauno, P. (2011) 'On the impact of non-sphericity and small-scale surface roughness on the optical properties of hematite aerosols', *J. Quant. Spectrosc. Radiat. Transfer*, **112**, 1815–24.

Kahnert, M., Nousiainen, T. and Räisänen, P. (2007) 'Mie simulations as an error source in mineral aerosol radiative forcing calculations', *Q. J. R. Meteorol. Soc.*, **133**, 299–307, doi:10.1002/qj.40.

Kahnert, M., Nousiainen, T., Thomas, M.A. and Tyynelä, J. (2012) 'Light scattering by particles with small-scale surface roughness: Comparison of four classes of model geometries', *J. Quant. Spectrosc. Radiat. Transfer*, **113**, 2356–67.

Kalashnikova, O., Kahn, R., Sokolik, I. and Li, W.H. (2005) 'Ability of multiangle remote sensing observations to identify and distinguish mineral dust types: Optical models and retrievals of optically thick plumes', *J. Geophys. Res.*, **110**, D18S14, doi:10.1029/2004JD004550.

Kalashnikova, O.V. and Kahn, R. (2006) 'Ability of multiangle remote sensing observations to identify and distinguish mineral dust types: 2. Sensitivity over dark water', *J. Geophys. Res.*, **111**, D11207, doi:10.1029/2005JD006756.

Kandler, K., Benker, N., Bundke, U., Cuevas, E., Ebert, M., Knippertz, P., Rodríguez, S., Schütz, L. and Weinbruch, S. (2007) 'Chemical composition and complex refractive index of Saharan Mineral Dust at Izaña, Tenerife (Spain) derived by electron microscopy', *Atmos. Environ.*, **41**, 8058–74, doi:10.1016/j.atmosenv.2007.06.047.

Kandler, K., Lieke, K., Benker, N., Emmel, C., Küpper, M., Müller-Ebert, D., Ebert, M., Scheuvens, D., Schladitz, A., Schütz, L. and Weinbruch, S. (2011a) 'Electron microscopy of particles collected at Praia, Cape Verde, during the Saharan Mineral dust experiment: Particle chemistry, shape, mixing state and complex refractive index', *Tellus*, **63B**, 475–96, doi:10.1111/j.1600-0889.2011.00550.x.

Kandler, K., Schütz, L., Deutscher, C., Hofmann, H., Jäckel, S., Knippertz, P., Lieke, K., Massling, A., Schladitz, A., Weinzierl, B., Zorn, S., Ebert, M., Jaenicke, R., Petzold, A. and Weinbruch, S. (2009) 'Size distribution, mass concentration, chemical and mineralogical composition, and derived optical parameters of the boundary layer aerosol at Tinfou, Morocco, during SAMUM 2006', *Tellus*, **61B**, 32–50, doi:10.1111/j.1600-0889.2008.00385.x.

Kandler, K., Schütz, L., Jäckel, S., Lieke, K., Emmel, C., Müller-Ebert, D., Ebert, M., Scheuvens, D., Schladitz, A., Šegvić, B., Wiedensohler, A. and Weinbruch, S. (2011b) 'Ground-based off-line aerosol measurements at Praia, Cape Verde, during the Saharan Mineral Dust Experiment: Microphysical properties and mineralogy', *Tellus*, **63B**, 459–74, doi:10.1111/j.1600-0889.2011.00546.x.

Krueger, B.J., Grassian, V.H., Cowin, J.P. and Laskin, A. (2004) 'Heterogeneous chemistry of individual mineral dust particles from different dust source regions: The importance of particle mineralogy', *Atmos. Environ.*, **38**, 6253–61.

Krueger, B.J., Grassian, V.H., Laskin, A. and Cowin, J.P. (2003) 'The transformation of solid atmospheric particles into liquid droplets through heterogeneous chemistry: Laboratory insights into the processing of calcium containing mineral dust aerosol in the troposphere', *Geophys. Res. Lett.*, **30**, 1148, doi:10.1029/2002GL016563.

Leinen, M., Prospero, J.M., Arnold, E. and Blank, M. (1994) 'Mineralogy of Aeolian dust reaching the North Pacific Ocean. 1. Sampling and analysis', *J. Geophys. Res.*, **99**, 21017–23.

Li, W.J. and Shao, L.Y. (2009) 'Observation of nitrate coatings on atmospheric mineral dust particles', *Atmos. Chem. Phys.*, **9**, 1863–71, doi:10.5194/acp-9-1863-2009.

Lieke, K., Kandler, K., Scheuvens, D., Emmel, C., Von Glahn, C., Petzold, A., Weinzierl, B., Veira, A., Ebert, M., Weinbruch, S. and Schütz, L. (2011) 'Particle chemical properties in the vertical column based on aircraft observations in the vicinity of Cape Verde Islands', *Tellus*, **63B**, 497–511, doi:10.1111/j.1600-0889.2011.00553.x.

Lindqvist, H., Jokinen, O., Kandler, K. and Nousiainen, T. (2011) 'Modelling mineral dust using stereophotogrammetry', *Atti Accad. Pelorit. Pericol.*, **89**, C1V89S81P058, doi:10.1478/C1V89S1P058.

Lindqvist, H., Jokinen, O., Kandler, K., Scheuvens, D., and Nousiainen, T. (2014). Single scattering by realistic, inhomogeneous mineral dust particles with stereogrammetric shapes. *Atmos. Chem. Phys.*, **14**, 143–157, doi:10.5194/acp-14-143-2014.

Lindqvist, H., Muinonen, K. and Nousiainen, T. (2009) 'Light scattering by coated Gaussian and aggregate particles', *J. Quant. Spectrosc. Radiat. Transfer*, **110**, 1398–410, doi:10.1016/j.jqsrt.2009.01.015.

Liu, C., Panetta, R.L., Yang, P., Macke, A. and Baran, A.J. (2012) 'Modeling the scattering properties of mineral aerosols using concave fractal polyhedra', *Appl. Opt.*, **52**, 640–52.

Liu, L., Mishchenko, M.I., Hovenier, J.W., Volten, H. and Muñoz, O. (2003) 'Scattering matrix of quartz aerosols: Comparison and synthesis of laboratory and Lorenz-Mie results', *J. Quant. Spectrosc. Radiat. Transfer*, **79–80**, 911–20.

Macke, A., Mishchenko, M.I. and Cairns, B. (1996) 'The influence of inclusions on light scattering by large ice particles', *J. Geophys. Res.*, **101**, 23311–16.

Maring, H., Savoie, D.L., Izaguirre, M.A., Custals, L. and Reid, J.S. (2003) 'Mineral dust aerosol size distribution change during atmospheric transport', *J. Geophys. Res.*, **102**, D8592, doi:10.1029/2002JD002536.

Matsuki, A., Iwasaka, Y., Shi, G., Zhang, D., Trochkine, D., Yamada, M., Kim, Y.-S., Chen, B., Nagatani, T., Miyazawa, T., Nagatani, M. and Nakata, H. (2005) 'Morphological and chemical modification of mineral dust: Observational insight into the heterogeneous uptake of acidic gases', *Geophys. Res. Lett.*, **32**, L22806, doi:10.1029/2005GL024176.

Matsuki, A., Schwarzenboeck, A., Venzac, H., Laj, P., Crumeyrolle, S. and Gomes, L. (2010) 'Cloud processing of mineral dust: Direct comparison of cloud residual and clear sky particles during AMMA aircraft campaign in summer 2006', *Atmos. Chem. Phys.*, **10**, 1057–69, doi:10.5194/acp-10-1057-2010.

McKie, D. and McKie, C. (1974) *Crystalline Solids*, Exeter, UK, Thomas Nelson and Sons, Ltd.

Meland, B., Alexander, J.M., Wong, C.-S., Grassian, V.H., Young, M.A. and Kleiber, P.D. (2012) 'Evidence for particle size-shape correlations in the optical properties of silicate clay aerosol', *J. Quant. Spectrosc. Radiat. Transfer*, **113**, 549–58.

Meng, Z., Yang, P., Kattawar, G.W., Bi, L., Liou, K.N. and Laszlo, I. (2010) 'Single-scattering properties of tri-axial ellipsoidal mineral dust aerosols: A database for application to radiative transfer calculations', *J. Aerosol Sci.*, **41**, 501–12, doi:10.1016/j.jaerosci.2010.02.008.

Merikallio, S., Lindqvist, H., Nousiainen, T. and Kahnert, M. (2011) 'Modelling light scattering by mineral dust using spheroids: Assessment of applicability', *Atmos. Chem. Phys.*, **11**, 5347–63.

Merikallio, S., Nousiainen, T., Kahnert, M. and Harri, A.-M. (2013) 'Light scattering by the Martian dust analog, palagonite, modeled with ellipsoids', *Opt. Express*, **21**, 17972–85, doi:10.1364/OE.21.017972.

Merrill, J., Arnold, E., Leinen, M. and Weaver, C. (1994) 'Mineralogy of Aeolian dust reaching the North Pacific Ocean. 2. Relationship of mineral assemblages to atmospheric transport patterns', *J. Geophys. Res.*, **99**, 21025–32.

Mie, G. (1908) 'Beiträge zur Optik trüber Medien, speziell kolloidaler Metallösungen', *Ann. Phys.*, **25**, 377–455.

Mikami, M., Aoki, T., Ishizuka, M., Yabuki, S., Yamada, Y., Gao, W. and Zeng, F. (2005) 'Observation of number concentration of desert aerosols in the south of the Taklimakan Desert, China', *J. Meteor. Soc. Japan*, **83A**, 31–43, doi:10.2151/jmsj.83A.31.

Min, M., Hovenier, J.W. and de Koter, A. (2005) 'Modeling optical properties of cosmic dust grains using a distribution of hollow spheres', *Astron. Astrophys.*, **432**, 909–20, doi:10.1051/0004-6361:20041920.

Mishchenko, M.I. and Travis, L.D. (1997) 'Satellite retrieval of aerosol properties over the ocean using polarization as well as intensity of reflected sunlight', *J. Geophys. Res.*, **102**, 16989–7013.

Mishchenko, M.I. and Travis, L.D. (1998) 'Capabilities and limitations of a current Fortran implementation of the T-matrix method for randomly oriented rotationally symmetric scatterers', *J. Quant. Spectrosc. Radiat. Transfer*, **60**, 309–24.

Mishchenko, M.I., Lacis, A.A., Carlson, B.E. and Travis, L.D. (1995) 'Nonsphericity of dust-like tropospheric aerosols: Implications for aerosol remote sensing and climate modeling', *Geophys. Res. Letters*, **22**, 1077–80.

Mishchenko, M.I., Travis, L.D., Kahn, R.A. and West, R.A. (1997) 'Modeling phase functions for dustlike tropospheric aerosols using a shape mixture of randomly oriented polydisperse spheroids', *J. Geophys. Res.*, **102**, 16831–47.

Mishchenko, M.I., Wiscombe, W.J., Hovenier, J.W. and Travis, L.D. (2000) 'Overview of scattering by nonspherical particles', in Mishchenko, M.I., Hovenier, J.W. and Travis, L.D. (eds), *Light Scattering by Nonspherical Particles*, San Diego, Academic Press, Chapter 2, 29–60.

Molinaroli, E. (1996) 'Mineralogical characterisation of Saharan dust', in Guerzoni, S. and Chester, R. (eds), *The Impact of Desert Dust Across the Mediterranean*, Dordrecht, Kluwer Academic Publishers, 153–62.

Muinonen, K., Nousiainen, T., Fast, P., Lumme, K. and Peltoniemi, J.I. (1996) 'Light scattering by Gaussian random particles: Ray optics approximation', J. *Quant. Spectrosc. Radiat. Transfer*, **55**, 577–601.

Muinonen, K., Nousiainen, T., Lindqvist, H., Muñoz, O. and Videen, G. (2009) 'Light scattering by Gaussian particles with internal inclusions and roughened surfaces using ray optics', *J. Quant. Spectrosc. Radiat. Transfer*, **110**, 1628–39, doi:10.1016/j.jqsrt.2009.03.012.

Müller, T., Schladitz, A., Maßling, A., Kaaden, N., Wiedensohler, A. and Kandler, K. (2009) 'Spectral absorption coefficients and refractive index of Saharan dust during SAMUM 2006', *Tellus*, **61B**, 79–95, doi:10.1111/j.1600-0889.2008.00399.x.

Muñoz, O. and Hovenier, J.W. (2011) 'Laboratory measurements of single light scattering by ensembles of randomly oriented small irregular particles in air: A review', *J. Quant. Spectrosc. Radiat. Transfer*, **112**, 1646–57, doi:10.1016/j.jqsrt.2011.02.005.

Muñoz, O., Moreno, F., Guirado, D., Dabrowska, D.D., Volten, H. and Hovenier, J.W. (2012) 'The Amsterdam–Granada light scattering database', *J. Quant. Spectrosc. Radiat. Transfer*, **113**, 565–74.

Muñoz, O., Moreno, F., Guirado, D., Ramos, J.L., López, A., Girela, F., Jerónimo, J.M., Costillo, L.P. and Bustamante, I. (2010) 'Experimental determination of scattering matrices of dust particles at visible wavelengths: The IAA light scattering apparatus', *J. Quant. Spectrosc. Radiat. Transfer*, **111**, 187–96, doi:10.1016/j.jqsrt.2009.06.011.

Muñoz, O., Volten, H., Hovenier, J., Nousiainen, T., Muinonen, K., Guirado, D., Moreno, F. and Waters, L. (2007) 'Scattering matrix of large Saharan dust particles: Experiments and computations', *J. Geophys. Res.*, **112**, D13215, doi:10.1029/2006JD008074.

Nickovic, S., Vukovic, A., Vujadinovic, M., Djurdjevic, V. and Pejanovic, G. (2012) 'Technical Note: High-resolution mineralogical database of dust-productive soils for atmospheric dust modeling', *Atmos. Chem. Phys.*, **12**, 845–55, doi:10.5194/acp-12-845-2012.

Nousiainen, T. (2009) 'Optical modeling of mineral dust particles: A review', *J. Quant. Spectrosc. Radiat. Transfer*, **110**, 1261–79, doi:10.1016/j.jqsrt.2009.03.002.

Nousiainen, T. and Vermeulen, K. (2003) 'Comparison of measured single-scattering matrix of feldspar particles with T-matrix simulations using spheroids', *J. Quant. Spectrosc. Radiat. Transfer*, **79–80**, 1031–42.

Nousiainen, T., Kahnert, M. and Lindqvist, H. (2011a) 'Can particle shape information be retrieved from light-scattering observations using spheroidal model particles?', *J. Quant. Spectrosc. Radiat. Transfer*, **112**, 2213–25.

Nousiainen, T., Kahnert, M. and Veihelmann, B. (2006) 'Light scattering modeling of small feldspar aerosol particles using polyhedral prisms and spheroids', *J. Quant. Spectrosc. Radiat. Transfer*, **101**, 471–87.

Nousiainen, T., Muinonen, K. and Räisänen, P. (2003) 'Scattering of light by large Saharan dust particles in a modified ray optics approximation', *J. Geophys. Res.*, **108**, 4025, doi:10.1029/2001JD001277.

Nousiainen, T., Muñoz, O., Lindqvist, H., Mauno, P. and Videen, G. (2011b) Light scattering by large Saharan dust particles: Comparison of modeling and experimental data for two samples', *J. Quant. Spectrosc. Radiat. Transfer*, **112**, 420–33.

Nousiainen, T., Zubko, E., Lindqvist, H., Kahnert, M. and Tyynelä, J. (2012) 'Comparison of scattering by different nonspherical, wavelength-scale particles', *J. Quant. Spectrosc. Radiat. Transfer*, **113**, 2391–405, doi:10.1016/j.jqsrt.2012.03.032.

Nousiainen, T., Zubko, E., Niemi, J.V., Kupiainen, K., Lehtinen, M., Muinonen, K. and Videen, G. (2009) 'Single-scattering modeling of thin, birefringent mineral dust flakes using the discrete-dipole approximation', *J. Geophys. Res.*, **114**, D07207, doi:10.1029/2008JD011564.

Okada, K., Heintzenberg, J., Kai, K. and Qin, Y. (2001) 'Shape of atmospheric mineral particles collected in three Chinese arid-regions', *Geophys. Res. Lett.*, **28**, 3123–6.

Osada, K. (2013) 'Water soluble fraction of Asian dust particles', *Atmos. Res.*, **124**, 101–8, doi:10.1016/j.atmosres.2013.01.001.

Otto, S., Bierwirth, E., Weinzierl, B., Kandler, K., Esselborn, M., Tesche, M., Schladitz, A., Wendisch, M. and Trautmann, T. (2009) 'Solar radiative effects of a Saharan dust plume observed during SAMUM assuming spheroidal model particles', *Tellus*, **61B**, 270–96.

Petrov, D., Shkuratov, Y. and Videen, G. (2011) 'Electromagnetic wave scattering from particles of arbitrary shapes', *J. Quant. Spectrosc. Radiat. Transfer*, **112**, 1636–45, doi:10.1016/j.jqsrt.2011.01.036.

Podczeck, F., Rahman, S.R. and Newton, J.M. (1999) 'Evaluation of a standardised procedure to assess the shape of pellets using image analysis', *Int. J. Pharm.*, **192**, 123–38, doi:10.1016/S0378-5173(99)00302-6.

Räisänen, P., Haapanala, P., Chung, C.E., Kahnert, M., Makkonen, R., Tonttila, J. and Nousiainen, T. (2013) 'Impact of dust particle non-sphericity on climate simulations', *Q. J. R. Meteorol. Soc.*, doi:10.1002/qj.2084 (in press).

Reid, E.A., Reid, J.S., Meier, M.M., Dunlap, M.R., Cliff, S.S., Broumas, A., Perry, K. and Maring, H. (2003) 'Characterization of African dust transported to Puerto Rico by individual particle and size segregated bulk analysis', J. Geophys. Res., 108, 8591, doi:10.1029/2002JD002935.

Reid, J.S., Reid, E.A., Walker, A., Piketh, S., Cliff, S., Al Mandoos, A., Tsay, S.-C. and Eck, T.F. (2008) 'Dynamics of southwest Asian dust particle size characteristics with implications for global dust research', *J. Geophys. Res.*, **113**, D14212, doi:10.1029/2007JD009752.

Rosin, P.L. (2003) 'Measuring shape: Ellipticity, rectangularity, and triangularity', *Machine Vision and Applications*, **14**, 172–84, doi:10.1007/s00138-002-0118-6.

Ryder, C.L., Highwood, E.J., Rosenberg, P.D., Trembath, J., Brooke, J.K., Bart, M., Dean, A., Crosier, J., Dorsey, J., Brindley, H., Banks, J., Marsham, J.H., McQuaid, J.B., Sodeman, H. and Washington, R. (2013) 'Optical properties of Saharan dust aerosol and contribution from the coarse mode as measured during the Fennec 2011 aircraft campaign', *Atmos. Chem. Phys.*, **13**, 303–25, doi:10.5194/acp-13-303-2013.

Scheuvens, D., Kandler, K., Küpper, M., Lieke, K., Zorn, S., Ebert, M., Schütz, L. and Weinbruch, S. (2011) 'Individual-particle analysis of airborne dust samples collected over Morocco in 2006 during SAMUM 1', *Tellus*, **63B**, 512–30, doi:10.1111/j.1600-0889.2011.00554.x.

Scheuvens, D., Schütz, L., Kandler, K., Ebert, M. and Weinbruch, S. (2013) 'Bulk composition of northern African dust and its source sediments: A compilation', *Earth-Sci. Rev.*, **116**, 170–94, doi:10.1016/j.earscirev.2012.08.005.

Schladitz, A., Müller, T., Nowak, A., Kandler, K., Lieke, K., Massling, A. and Wiedensohler, A. (2011) 'In situ aerosol characterization at Cape Verde Part 1: Particle number size distributions, hygroscopic growth and state of mixing of the marine and Saharan dust aerosol', *Tellus*, **63B**, 531–48, doi:10.1111/j.1600-0889.2011.00569.x.

Schumann, U., Weinzierl, B., Reitebuch, O. and 35 co-authors (2011) 'Airborne observations of the Eyjafjalla volcano ash cloud over Europe during air space closure in April and May 2010', *Atmos. Chem. Phys.*, **11**, 2245–79, doi:10.5194/acp-11-2245-2011.

Schütz, L. (1980) 'Long range transport of desert dust with special emphasis on the Sahara', *Ann. NY Acad. Sci.*, **338**, 515–32, doi:10.1111/j.1749-6632.1980.tb17144.x.

Schütz, L., and Jaenicke, R. (1974) 'Particle number and mass distributions above 10–4 cm radius in sand and aerosol of the Sahara Desert', *J. Appl. Meteorol.*, **13**, 863–70, doi:10.1175/1520-0450(1974)013_0863:PNAMDA_2.0.CO;2.

Schütz, L. and Rahn, K.A. (1982) 'Trace-element concentrations in erodible soils', *Atmos. Environ.*, **16**, 171–6, doi:10.1016/0004-6981(82)90324-9.

Schütz, L., Jaenicke, R. and Pietrek, H. (1981) 'Saharan dust transport over the North Atlantic Ocean', *Geol. Soc. Am. Spec. Paper*, **186**, 87–100, doi:10.1130/SPE186-p87.

Shao, L., Li, W., Yang, S., Shi, Z. and Lü, S. (2007) 'Mineralogical characteristics of airborne particles collected in Beijing during a severe Asian dust storm period in spring 2002', *Sci. China Ser. D*, **50**, 953–9, doi:10.1007/s11430-007-0035-7.

Shao, Y., Ishizuka, M., Mikami, M. and Leys, J.F. (2011) 'Parameterization of size-resolved dust emission and validation with measurements', *J. Geophys. Res.*, **116**, D08203, doi:10.1029/2010JD014527.

Shen, Z., Caquineau, S., Cao, J., Zhang, X., Han, Y., Gaudichet, A. and Gomes, L. (2009) 'Mineralogical characteristics of soil dust from source regions in northern China', *Particuology*, **7**, 507–12, doi:10.1016/j.partic.2009.10.001.

Shi, Z., Shao, L., Jones, T.P. and Lu, S. (2005) 'Microscopy and mineralogy of airborne particles collected during severe dust storm episodes in Beijing, China', *J. Geophys. Res.*, **110**, D01303, doi:10.1029/2004JD005073.

Sobanska, S., Hwang, H., Choël, M., Jung, Eom, H.-J., Kim, H., Barbillat, J. and Ro, C.-U. (2012) 'Investigation of the chemical mixing state of individual Asian dust particles by the combined use of electron probe X-ray microanalysis and Raman microspectrometry', *Anal. Chem.*, **84**, 3145–54, doi:10.1021/ac2029584.

Stefaniak, E.A., Worobiec, A., Potgieter-Vermaak, S., Alsecz, A., Török, S. and Van Grieken, R. (2006) 'Molecular and elemental characterisation of mineral particles by means of parallel micro-Raman spectrometry and scanning electron microscopy/energy dispersive X-ray analysis', *Spectrochim. Acta B*, **61**, 824–30, doi:j.sab.2006.04.009.

Sullivan, R.C. and Prather, K.A. (2007) 'Investigations of the diurnal cycle and mixing state of oxalic acid in individual particles in Asian aerosol outflow', *Environ. Sci. Technol.*, **41**, 8062–9, doi:10.1021/es071134g.

Sullivan, R.C., Guazzotti, S.A., Sodeman, D.A. and Prather, K.A. (2007) 'Direct observations of the atmospheric processing of Asian mineral dust', *Atmos. Chem. Phys.*, **7**, 1213–36, doi:10.5194/acp-7-1213-2007.

Tanré, D., Kaufman, Y.J., Holben, B.N., Chatenet, B., Karnieli, A., Lavenu, F., Blarel, L., Dubovik, O., Remer, L.A. and Smirnov, A. (2001) 'Climatology of dust aerosol size distribution and optical properties derived from remotely sensed data in the solar spectrum', *J. Geophys. Res.*, **106D**, 18205–17, doi:10.1029/2000JD900663.

Ullerstam, M., Vogt, R., Langer, S. and Ljungström, E. (2002) 'The kinetics and mechanism of SO2 oxidation by O3 on mineral dust', *Phys. Chem. Chem. Phys.*, **4**, 4694–9, doi:10.1039/B203529B.

Usher, C.R., Al-Hosney, H., Carlos-Cuellar, S. and Grassian, V.H. (2002) 'A laboratory study of the heterogeneous uptake and oxidation of sulfur dioxide on mineral dust particles', *J. Geophys. Res.*, **107**, 4713, doi:10.1029/2002JD002051.

Vaillon, R., Geddrin, J.-M., Eyraud, C., Merchiers, O., Sabouroux, P. and Lacroix, B. (2011) 'A new implementation of a microwave analog to light scattering measurement device', *J. Quant. Spectrosc. Radiat. Transfer*, **112**, 1753–60.

Van Pelt, R.S., Peters, P. and Visser, S. (2009) 'Laboratory wind tunnel testing of three commonly used saltation impact sensors', *Aeolian Research*, **1**, 55–62, doi:10.1016/j.aeolia.2009.05.001.

Veihelmann, B., Nousiainen, T., Kahnert, M. and van der Zande, W.J. (2006) 'Light scattering by small feldspar particles simulated using the Gaussian random sphere geometry', *J. Quant. Spectrosc. Radiat. Transfer*, **100**, 393–405, doi:10.1016/j.jqsrt.2005.11.053.

Vilaplana, R., Moreno, F. and Molina, A. (2006) 'Study of the sensitivity of size-averaged scattering matrix elements of nonspherical particles to changes in shape, porosity and refractive index', *J. Quant. Spectrosc. Radiat. Transfer*, **100**, 415–28, doi:10.1016/j.jqsrt.2005.11.068.

Volten, H., Muñoz, O., de Haan, J.F., Vassen, W., Hovenier, J.W., Muinonen, K. and Nousiainen, T. (2001) 'Scattering matrices of mineral aerosol particles at 441.6 nm and 632.8 nm', *J. Geophys. Res.*, **106**, 17375–401.

Wang, X., Hua, T., Zhang, C., Lang, L. and Wang, H. (2012) 'Aeolian salts in Gobi deserts of the western region of Inner Mongolia: Gone with the dust aerosols', *Atmos. Res.*, **118**, 1–9, doi:10.1016/j.atmosres.2012.06.003.

Weinzierl, B., Petzold, A., Esselborn, M., Wirth, M., Rasp, K., Kandler, K., Schütz, L., Köpke, P. and Fiebig, M. (2009) 'Airborne measurements of dust layer properties, particle size distribution and mixing state of Saharan dust during SAMUM 2006', *Tellus*, **61B**, 96–117, doi:10.1111/j.1600-0889.2008.00392.x.

Weinzierl, B., Sauer, D., Esselborn, M., Petzold, A., Veira, A., Rose, M., Mund, S., Wirth, M., Ansmann, A., Tesche, M., Gross, S. and Freudenthaler, V. (2011) 'Microphysical and optical properties of dust and tropical biomass burning aerosol layers in the Cape Verde region: An overview of the airborne in situ and lidar measurements during SAMUM-2', *Tellus B*, **63**, 589–618, doi:10.1111/j.1600-0889.2011.00566.x.

Williams, J., de Reus, M., Krejci, R., Fischer, H. and Ström, J. (2002) 'Application of the variability-size relationship to atmospheric aerosol studies: Estimating aerosol lifetimes and ages', *Atmos. Chem. Phys.*, **2**, 133–45, doi:10.5194/acp-2-133-2002.

Yi, B., Hsu, C.N., Yang, P. and Tsay, S.-C. (2011) 'Radiative transfer simulation of dust-like aerosols: Uncertainties from particle shape and refractive index', *J. Aerosol Sci.*, **42**, 631–44.

Yurkin, M.A. and Kahnert, M. (2013) 'Light scattering by a cube: Accuracy limits of the discrete dipole approximation and the T-matrix method', *J. Quant. Spectrosc. Radiat. Transfer*, **123**, 176–83, doi:10.1016/j.jqsrt.2012.10.001.

Zender, C.S., Bian, H. and Newman, D. (2003) 'The mineral Dust Entrainment And Deposition (DEAD) model: Description and 1990s dust climatology', *J. Geophys. Res.*, **108**, D4416, doi:10.1029/2002JD002775.

Zhang, D. and Iwasaka, Y. (2004) 'Size change of Asian dust particles caused by sea salt interaction: Measurements in southwestern Japan', *Geophys. Res. Lett.*, **31**, L15102, doi:10.1029/2004GL020087.

Zhang, D., Iwasaka, Y., Matsuki, A., Ueno, K. and Matsuzaki, Y. (2006) 'Coarse and accumulation mode particles associated with Asian dust in southwestern Japan', *Atmos. Environ.*, **40**, 1205–15, doi:10.1016/j.atmosenv.2005.10.037.

Zhang, D., Zang, J., Shi, G., Iwasaka, Y., Matsuki, A. and Trochkine, D. (2003) 'Mixture state of individual Asian dust particles at a coastal site of Qingdao, China', *Atmos. Environ.*, **37**, 3895–901, doi:10.1016/S1352-2310(03)00506-5.

Zubko, E., Muinonen, K., Muñoz, O., Nousiainen, T., Shkuratov, Y., Sun, W. and Videen, G. (2013) 'Light scattering by feldspar particles: Comparison of model agglomerate debris particles with laboratory samples', *J. Quant. Spectrosc. Radiat. Transfer*, **131**, 175–187, doi:10.1016/j.jqsrt.2013.01.017.

2 A review of approximate analytic light-scattering phase functions

Subodh Kumar Sharma

2.1 Introduction

Consider a plane light wave of wavelength λ incident on an obstacle of characteristic size a and complex relative refractive index m. Let $\mathbf{k_i}$ and $\mathbf{k_s}$ denote the incident and the scattered wave vectors such that $|\mathbf{k_i}| = |\mathbf{k_s}| = k$. The scattering phase function is then defined as

$$\phi(m, k, a, \theta, \varphi) = \frac{4\pi}{k^2} \frac{i(m, x, \theta, \varphi)}{\sigma(m, k, a)}, \tag{2.1}$$

where θ is the scattering angle and φ is the azimuthal angle. The two angles define the direction of the scattered wave in relation to the incident direction. The variation of angular scattered intensity is denoted by $i(m, x, \theta, \varphi)$ and $\sigma(m, k, a)$ is the scattering cross-section of the particle defined as (see, e.g. van de Hulst, 1957)

$$\sigma(m, k, a) = \frac{1}{k^2} \int_o^{2\pi} \int_0^{\pi} i(m, x, \theta, \varphi) \sin\theta \, d\theta \, d\varphi , \tag{2.2a}$$

with

$$x = ka = 2\pi a/\lambda . \tag{2.2b}$$

The size parameter x may be looked upon as a measure of the size of the scatterer in the units of wavelength of the incident radiation. The characteristic size of a sphere is its radius. Commonly used definitions of particle size for other shapes can be found in Jonasz and Fournier (2007). The phase function is normalized as,

$$\frac{1}{4\pi} \int_0^{2\pi} \int_0^{\pi} \phi(m, k, a, \theta, \varphi) \sin\theta \, d\theta \, d\varphi = 1 , \tag{2.3}$$

which allows it to be interpreted as giving the probability that the radiation propagating in a given direction is scattered into an elementary solid angle making an angle θ with the incident direction.

For a random collection of particles, the scattered intensity is simply a sum of scattered intensities from individual scatterers. This assumes that the multiple scattering in electromagnetic wave propagation in the medium is negligible. It is also assumed that each particle scatters independently. That is, the dependent scattering (effect of the presence of other particles) is absent. If the particles in the collection are polydisperse, with a size distribution denoted by $f(a)$, the scattering phase function can be expressed as

$$p(m,k,a_0,a_m,\theta,\varphi) = \frac{\int_{a_0}^{a_m} \phi(p,m,a,\theta,\varphi)\sigma(m,k,a)f(a)\,da}{\int_{a_0}^{a_m} \sigma(m,k,a)f(a)\,da}, \tag{2.4}$$

where $\phi(m,k,a,\theta,\varphi)$, as defined in Eq. (2.1), is the scattering phase function for the scattering of a plane light wave by an isolated scatterer of size a. The lower and upper limits of the size distribution function $f(a)$ are denoted by a_0 and a_m, respectively.

The phase function for a collection of particles is normalized in a manner similar to Eq. (2.3):

$$\frac{1}{4\pi}\int_0^{2\pi}\int_0^{\pi} p(m,k,a,\cos\theta,\varphi)\sin\theta\,d\theta\,d\varphi = 1\,. \tag{2.5}$$

The azimuthal dependence φ of the scattering phase function is often removed, which is possible under the assumption of a spherical symmetry. The normalization condition then becomes

$$\frac{1}{2}\int_0^{\pi} \phi(m,k,a,\theta)\sin\theta\,d\theta = 1 \tag{2.6a}$$

for a single scatterer and

$$\frac{1}{2}\int_0^{\pi} p(m,k,a,\theta)\sin\theta\,d\theta = 1 \tag{2.6b}$$

for a collection of scatterers. In the present article, this would be the case almost all the time unless and until stated otherwise.

If the collection of scatterers is such that the multiple scattering cannot be ignored, the wave propagation is described by an equation known as the radiative transfer equation (RTE) (see, e.g. Liou, 2002). In a collection of scatterers like this, it is always possible to identify a small enough volume in which the multiple scattering is negligible. The scattering phase function can be defined for particles in this volume as described above. This phase function then appears as an input in the RTE.

In principle, the scattering phase function for a single-particle scattering can be obtained by solving the Maxwell equations for interaction between the incident wave and the scatterer. Employing Eq. (2.4), the scattering phase function for a collection of scatterers can also be obtained. Many analytic and numerical techniques are available which provide solutions to the problem of scattering of electromagnetic waves by a scatterer of arbitrary shape and size. These methods include, amongst others, the separation of variable method, T-matrix method, integral equation method, T-matrix method, point matching method, finite element method, finite difference time domain method, discrete-dipole approximation method, etc. These approaches have been reviewed by Wriedt (1998), Mishchenko et al. (2002), Kanhert (2003), etc. The difficulty with the use of the recipe of exact solutions, however, is that the phase functions so obtained are either numerical or are in cumbersome analytic form which is not always conducive for use in the RTE. This has prompted development of many approximate phase functions aimed at reducing the complexities of the RTE solutions.

The electromagnetic wave-scattering technique is an interdisciplinary tool. The need for its use occurs in a variety of situations—generally to characterize a scatterer or a population of scatterers. As a consequence, simple approximate analytic phase functions have been developed in very diverse fields such as atmospheric and aerosol or environmental optics, astrophysics and astronomy, biomedical optics, ocean optics, computer graphics, and in many industrial applications dealing with electromagnetic wave propagation in a collection of particles. In fact, the approximate analytic phase functions are of interest even in problems related to acoustic scattering and thermal transport (Ying and Truell, 1956; Prasher, 2004; Kim and Majumdar, 2006; Zukerman and Lukes, 2008). A review of approximate scattering phase functions developed in different contexts, therefore, is needed and should serve as a useful reference and resource for workers in a wide spectrum of disciplines. Thus, the main purpose of this review article is to take stock of various approximate analytic scattering phase functions that have been developed over the years in various contexts. Although this review includes a wide variety of phase functions, it restricts itself to scattering phase functions developed over the whole angular scattering domain. Those designed for a particular angular range are out of the scope of this review. Indeed, many phase functions have been developed which describe the small angle scattering very well (see, e.g. Ramsauer, 1921; Box, 1983; Turcu, 2004, 2006; Kokhanovsky, 2004; Sharma and Somerford, 2006; Louedec and Urban, 2012; Sharma and Banerjee, 2012) but are not a part of this article.

For the purpose of organizing this review, I classify the scattering phase functions into two main categories. (i) This category consists of those obtained by parametrizing an observed or a measured scattering phase function. This approach aims at obtaining simple mathematical expressions for an observed phase function. As a result, the origin of the curve or the context of its origin is not of real consequence. In other words, we need not differentiate between ϕ and p for this class. (ii) This category consists of scattering phase functions which reflect geometrical as well as optical properties of the individual scatterers. A deeper understanding and knowledge of scattering phase functions and radiative properties of individual particle and nature of size distribution are required for constructing the phase function of a collection of particles. In sections 2 and 3, we examine scattering phase functions of type (i). Scattering phase functions of type (ii) are treated in section 4. Modifications in the scattering phase function due to dependent scattering have been expressed in section 5. Section 6 briefly deliberates on the role of phase function in ray tracing Monte Carlo solution of the RTE. Some distribution-specific phase functions are considered in section 7. The review ends with brief concluding remarks in section 8.

2.2 Scattering phase function as a series expansion

2.2.1 Expansion in terms of Legendre polynomials

Any scattering phase function, $p(\cos\theta)$, can be expressed as an infinite series in terms of orthogonal basis functions. The Legendre polynomials $P_l(\cos\theta)$ are widely used for this purpose. A phase function may thus be expressed as

$$p(\cos\theta) = \sum_{l=0}^{N-1} a_l P_l(\cos\theta) . \tag{2.7}$$

The larger the value of N, the greater the accuracy of the phase function. Most phase functions can be represented as a series expansion given in Eq. (2.7) (Siegel and Howell, 1981). The expansion coefficients a_l in the above equation are nothing but the moments of the phase function and may be written as

$$a_l = \frac{2l+1}{2} \int_{-1}^{1} P_l(\mu) p(\mu)\, d\mu . \tag{2.8}$$

The first few Legendre polynomials are given by the following:

$$P_0(\mu) = 1 , \tag{2.9a}$$
$$P_1(\mu) = \mu , \tag{2.9b}$$
$$P_2(\mu) = \frac{1}{2}(3\mu^2 - 1) , \tag{2.9c}$$
$$P_3(\mu) = \frac{1}{2}(5\mu^3 - 3\mu) , \tag{2.9d}$$
$$P_4(\mu) = \frac{1}{8}(35\mu^4 - 30\mu^2 + 3) , \tag{2.9e}$$
$$P_5(\mu) = \frac{1}{8}(65\mu^5 - 70\mu^3 + 15\mu) , \tag{2.9f}$$

where $\mu = \cos\theta$.

An alternative form which is used to cast Eq. (2.7) is:

$$p(\mu) = \sum_{l=0}^{N-1} (2l+1)\chi_l P_l(\mu) , \tag{2.10}$$

where

$$\chi_l = \frac{1}{2} \int_{-1}^{1} P_l(\mu) p(\mu)\, d\mu . \tag{2.11}$$

A comparison of Eqs (2.7) and (2.10) shows that

$$a_l = (2l+1)\chi_l . \tag{2.12}$$

For $l = 1$, $a_1 = 3\chi_1$ and hence the asymmetry parameter $\langle\cos\theta\rangle = \chi_1$.

Yet another expansion employed to represent the exact phase function is a series in powers of μ:

$$p(\mu) = \sum_{n=0}^{N-1} c(n)\mu^n \,. \tag{2.13}$$

The coefficients $c(n)$ are related to the coefficients a_n via the relation (Pegoraro et al., 2010)

$$c(n) = \frac{1}{4\pi} \sum_{k=n}^{\leq N-1} a_k \frac{(-1)^{\frac{k-n}{2}}}{2^k} \frac{(k+n)!}{\left(\frac{k+n}{2}\right)!\left(\frac{k-n}{2}\right)!n!} \,. \tag{2.14}$$

A power series for the Mie scattering in the powers of $\sin^2(\theta/2)$ has also been developed (Box, 1983) which should prove to be useful whenever the diffraction peak is of primary interest.

A plot of the phase function against scattering angle shows many oscillations for a particle large compared to the wavelength of the incident radiation. To accurately reproduce such a phase function, a sizable number of terms are needed in the polynomial expansion. The value of N could reach hundreds to approximate a phase function (Dave, 1970). The more terms used the more accurate the reconstruction of phase function. A computer code for obtaining the phase function for a spherical particle to the desired order of moments has been developed by Edwards and Slingo (1996). For polydispersions of randomly oriented, homogeneous rotationally symmetric nonspherical particles, an efficient technique for computing Legendre expansion coefficients has been described in detail by Mishchenko and Travis (1998). The question that poses itself is: how many terms are necessary for different applications? The issue has been addressed by many authors. For the phase function of an isolated homogeneous sphere, a simple empirical criterion for choosing N could be (Mishchenko et al., 2002)

$$N = x + 4.05x^{1/3} + 8 \,.$$

Clearly, N increases as the size of the scatterer increases. Slightly differing criteria have been given by Deirmendjian (1969), Wiscombe (1980), and Bohren and Huffman (1983). For the phase function for a collection of particles, no such relation seems to exist. In general, however, larger asymmetry requires a larger number of terms.

The phase function does not exhibit considerable oscillations for a collection of particles with a size distribution even when the size distribution is narrow. The inquiry regarding how many terms are necessary for different applications has been addressed for some applications. The bidirectional reflectances associated with dust aerosols, water clouds, and ice clouds have been simulated and it is found that, to have numerical errors less than 5% in the visible spectrum, 16 Legendre polynomials should be used for dust aerosols and 32 Legendre polynomials should be used for both water and ice clouds (Ding et al., 2009). In the infrared spectrum, however, four terms were sufficient in the radiative transfer computation for practical applications. Rinzema et al. (1993) have described a procedure to find the phase function which has a maximum value of $\langle\mu\rangle$ for a given N, where

$$\langle\mu\rangle \equiv \langle\cos\theta\rangle = \frac{1}{2}\int_{-1}^{1} \mu p(\mu)\, d\mu \tag{2.15}$$

describes the asymmetry of the phase function around the scattering angle $\theta = 90°$. It was found that, to attain $\langle\mu\rangle = 0.9$, at least a polynomial of degree 9 is needed. For $\langle\mu\rangle = 0.95$, which is a typical value for soft biomedical tissues, 12 terms are needed in the expansion.

For particles small compared to the wavelength of the radiation ($x \le 1$), the variation of the phase function with the scattering angle does not show oscillations even for scattering by an isolated particle (see Fig. 2.1). In addition, the anisotropy of the phase function is not large. In such a situation, the phase function may be approximated by the first few terms of the expansion Eq. (2.7). It is easy to see that the zeroth moment of the phase function, a_0, is always 1 because of the normalization condition (Eq. (2.6b)). The first term of expansion Eq. (2.7) then gives an isotropic phase function

$$p(\mu) = 1\,. \tag{2.16}$$

Next, the first moment of the phase function is

$$a_1 = \frac{3}{2}\int_{-1}^{1} \mu p(\mu)\, d\mu = 3\langle\mu\rangle\,. \tag{2.17}$$

Retaining the first two terms in the expansion Eq. (2.7), one obtains

$$p(\cos\theta) = 1 + 3\langle\mu\rangle\cos\theta\,. \tag{2.18}$$

This phase function is sometimes also referred to as the Eddington phase function. Many authors have obtained simple phase functions by direct expansion of exact scattering matrices in generalized spherical functions directly (see, e.g. Sekera, 1952; Chu and Churchill, 1955, 1960; Clarke et al., 1957; Domke, 1975; Bugaenko, 1976; Fowler, 1983; de Rooij and van der Stap, 1984).

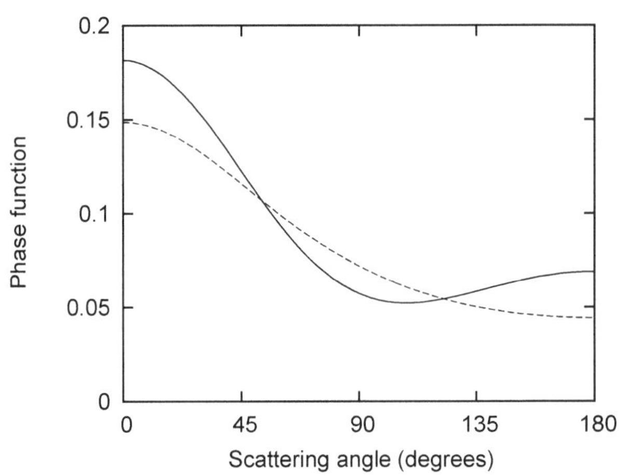

Fig. 2.1. Solid line: ϕ_{ex}; dashed line: ϕ_{hgpf} for $x = 1.0$ and $m = 1.5$.

2.2.2 The Rayleigh phase function (RPF)

The term Rayleigh scattering refers to the scattering of light by particles that are small compared to the wavelength of the incident radiation ($x \ll 1$). The resulting phase function for such particles is (see, e.g. Chandrasekhar, 1950; BenZvi et al., 2007)

$$p_{rpf}(\mu, \gamma) = \frac{3}{4(1+2\gamma)}\left((1+3\gamma)+(1-\gamma)\mu^2\right), \tag{2.19a}$$

where γ accounts for the effect of molecular anisotropy on the Rayleigh scattering and is given by the relation

$$\gamma = \frac{\rho_0}{2-\rho_0}, \tag{2.19b}$$

where ρ_0 is the depolarization factor. Because ρ_0 varies with the wavelength, the effect of anisotropy also changes with the wavelength. Studies have shown that the effect of molecular anisotropy is small and weakly wavelength-dependent. At $\lambda = 360$ nm, $\gamma = 0.015$ (Bucholtz, 1995). An alternative expression for the RPF, after taking into account the effect of γ, has been given by Penndorf (1957) as

$$p_{rpf}(\mu) = 0.7629(0.9324 + \mu^2). \tag{2.19c}$$

For isotropic scattering, γ is zero and Eq. (2.19a) reduces to

$$p_{rpf}(\mu) = \frac{3}{4}(1+\mu^2), \tag{2.20}$$

which is nothing but Eq. (2.7) with $N = 3$ ($a_0 = 1,\ a_1 = 0,\ a_2 = 1/2$). For small particles, Eq. (2.20) has been found to agree reasonably well with exact results.

Although the RPF is not valid for large particles, closed-form analytic phase functions, similar in form to the RPF, have been derived in the context of hazy and murky atmospheres whose constituents are large particles (Nishita et al., 1987). For hazy atmosphere, the phase function reads

$$p_{hzpf} = \frac{3}{4}\left[1 + 9\cos^{16}(\theta/2)\right], \tag{2.21a}$$

and for murky atmosphere it reads

$$p_{mrpf} = \frac{3}{4}\left[1 + 50\cos^{64}(\theta/2)\right]. \tag{2.21b}$$

These are, however, empirical results and not derived with any basic physics underneath. The two atmosphere models differ in that the murky atmosphere model is much more directional than the haze model.

2.2.3 The $\delta - M$ phase function approximation

The phase function for a large particle is strongly forward-peaked and thus its representation requires a large number of terms in the Legendre expansion. As the particle size increases, the forward peak becomes sharper. A typical phase function can be seen in Fig. 2.2 ($x = 10$ in this figure). The $\delta - M$ method takes advantage

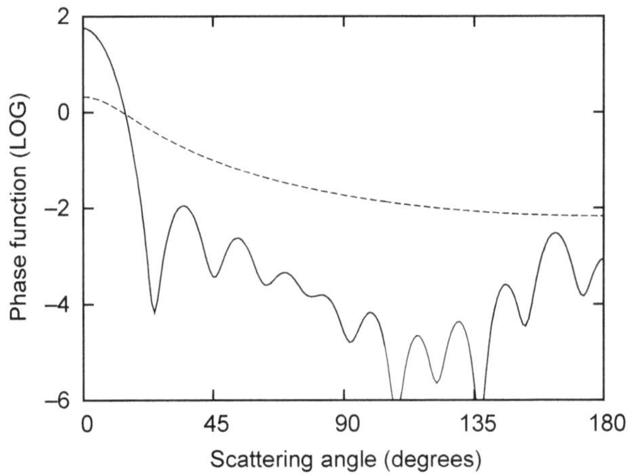

Fig. 2.2. Solid line: ϕ_{ex}; dashed line: ϕ_{hgpf} for $x = 10.0$ and $m = 1.5$.

of the fact that higher-order Legendre polynomial terms contribute primarily to the δ function like forward peak. Thus, a phase function with strong forward peak may be formulated as (Wiscombe, 1977)

$$p_{dmpf}(\mu) = 2\alpha\delta(1-\mu) + (1-\alpha)p'(\mu)\,. \tag{2.22}$$

The term $p'(\mu)$ is a normalized phase function with the forward peak removed and may be expressed as

$$p'(\mu) = \sum_{l=0}^{(2M-1)} (2l+1)a_l' P_l(\mu)\,, \tag{2.23}$$

where M is essentially the order of approximation. Substitution of p_{dmpf} given by Eqs (2.22) and (2.23) in Eq. (2.8) for $p(\mu)$ results in the following general relationships:

$$(1-\alpha)a_l' = a_l - \alpha, \quad l \leq 2M-1 \tag{2.24a}$$

$$a_l = \alpha, \quad l \geq 2M, \tag{2.24b}$$

which may be solved uniquely to give

$$a_l' = \frac{a_l - \alpha}{1-\alpha}, \quad l = 0, \ldots, 2M-1, \tag{2.24c}$$

but there is freedom in how to determine α. Wiscombe (1977) fixes α as equal to a_{2M} to keep consistency with the delta-Eddington approximation (section 3.2.3).

Numerical computations of the delta-M phase function for large dielectric spheres of $x = 100$ and $x = 300$ reveal that this phase function is very effective in reducing the number of terms needed to parametrize the phase function (Crosbie and Davidson, 1985). The 11-term approximation for $x = 100$ and $m = 1.33$ follows the trends of the actual phase function which otherwise requires 240 terms. The four-term approximation was noted to be a fair approximation of the scattering by a bubble in ice ($x = 300$, $m = 0.75$). The approximation may be even more suitable for absorbing spheres where the forward peak is greatly increased.

Large scattering angles are the most important viewing angles for most satellite observations. For ice clouds, hundreds of Legendre polynomial expansion terms were needed to simulate the backscattering portion of cloud phase function accurately (Hu et al., 2000). An alternative method for computing the moments a_l' for getting accurate phase function at backscattering angles with a smaller number of terms is by using least-squares fitting to minimize the relative difference between the original and the approximated phase function (Hu et al., 2000). Thus, one defines a merit function

$$\epsilon = \sum_i w_i \left(\frac{p'(\cos\theta_i)}{p(\cos\theta_i)} - 1 \right)^2 , \tag{2.25}$$

where θ_i is the scattering angle and w_i is the weight for each scattering angle. a_l' is determined by solving the least-squares problem. If the fitting is performed with $w_i = 0$ for small angles (e.g. $\theta < \theta_c = 3°$), then the forward peak is automatically truncated. Using this method, the phase function as well as the reflected radiances were accurately computed for both water and ice clouds with fewer than 30 terms. The method has been referred as the δ-fit method in the literature.

Mengüc and Subramaniam (1990) used a step function in place of the delta function. The accuracy of the resulting phase function was examined by comparing it against the exact phase function for spherical particles and by comparing the RTE solutions based on exact, delta- and step phase functions. Comparisons of the angular distribution of approximate phase functions yielded acceptable agreement with true phase functions. Studies of the effect of these approximate phase functions on radiative transfer solutions showed that the recovered coefficients can be used to calculate the radiative fluxes accurately.

A detailed comparison of the accuracy of various truncation approximations for radiative transfer calculations can be found in Iwabuchi and Suzuki (2009) and in Rozanov and Lyapustin (2010). It has been concluded in the latter paper that the delta-M method, combined with the single-scattering correction, provides the best overall accuracy for the intensity computations.

2.2.4 Peak truncated phase functions

Since the photons scattered into the forward peak deviate very little from the original direction, it may be a good approximation to treat them as being not scattered at all (Potter, 1970). This allows removal of the peak from the phase function. Generally, the error introduced by this simplification has been noted to be small in most applications. The reflection computations from conservative atmosphere obtained using approximate phase function were nearly identical to those obtained using the complete cloud phase function (Hansen, 1969). The error in reflectivity was less than 1% for most angles of incidence and emergence. The exceptions were for direct backscattering, for near grazing emergent angle, and for near forward-scattering. Potter (1970) showed that the accuracy of the approximation for transmitted and reflected fluxes is within 1% for widely differing values of the incident angle and optical depths. Nakajima and Asano (1977) estimated the errors introduced by the peak truncation approximation in the emerging fluxes from the cloud and the hydrosol layers. The errors were identified as small for a wide range of values of the incident angle, the optical thickness, and the absorption.

The effect of this approximation in RTE computations is that the single-scattering albedo defined as

$$\omega_0 = \frac{\text{scattering cross-section}}{\text{total cross-section}}$$

and the optical thickness defined as

$$\tau = \text{scattering coefficient} \times \text{geometrical thickness}$$

are reduced to ω_{0t} and τ_t for the truncated phase function (Hansen, 1969; Potter, 1970; Nakajima and Asano, 1977)

$$\omega_{0t} = \frac{(1 - S_t)\omega_0}{1 - S_t\omega_0}, \quad \tau_t = (1 - S_t\omega_0)\tau, \tag{2.26}$$

where S_t is a truncation ratio defined as

$$S_t = \frac{1}{2}\int_0^\pi (p_{epf} - p_{tpf}) \sin\theta \, d\theta . \tag{2.27}$$

Subscripts *epf* and *tpf* refer to exact and truncated phase functions respectively.

2.3 Parametrized phase functions

An alternative approach to the Legendre expansion is to use simple parametrized phase functions which model the scattering characteristics of a wide variety of scatterers. In this section, we take stock of the phase functions that have been developed with this approach in mind.

2.3.1 One-parameter phase functions

2.3.1.1 The Henyey–Greenstein phase function (HGPF)

The HGPF originated in 1941 in connection with astrophysics problems (Henyey and Greenstein, 1941). It is a single-parameter phase function expressed as

$$p(\theta)_{hgpf} = \frac{1 - g^2}{(1 + g^2 - 2g\cos\theta)^{3/2}}, \tag{2.28}$$

where the single parameter g turns out to be the asymmetry parameter of the HGPF. It can be easily checked that

$$g = \int_{-1}^{1} \mu p_{hgpf}(\mu) \, d\mu . \tag{2.29}$$

The asymmetry parameter, as mentioned earlier, is a measure of the asymmetry of the phase function around the scattering angle $\theta = 90°$. It is also the first moment of the phase function. The value of g lies between -1 and 1. While $g = 1$ indicates complete forward-scattering, $g = -1$ implies total backscattering. For $g = 0$, this is an isotropic phase function.

A more general relationship between the power of g and the order of the moment of the phase function is

$$g^n = \int_{-1}^{1} P_n(\mu) p_{hgpf}(\mu, g)\, d\mu\,. \tag{2.30}$$

This can be verified by recalling that the generating function of the Legendre polynomial is

$$(1 - 2g\mu + g^2)^{-1/2} = \sum_{n=0}^{\infty} g^n P_n(\mu)\,. \tag{2.31}$$

Differentiating both sides with respect to μ, it is easy to see that

$$p_{hgpf}(\theta) = \sum_{n=0}^{\infty} g^n (2n+1) P_n(\mu)\,. \tag{2.32}$$

Recalling that

$$\int_{-1}^{1} P_n(\mu) P_m(\mu)\, d\mu = \frac{2}{2n+1} \delta_{nm}\,, \tag{2.33}$$

Eq. (2.32) can be easily seen to reduce to Eq. (2.30). For $n = 1$, Eq. (2.30) gives $g = \langle \cos\theta \rangle$.

The analysis of the HGPF accuracy for an isolated scatterer as well as for a polydisperse collection of particles has been done in many contexts and a variety of results have been obtained. Figures 2.1 and 2.2 show a comparison of the HGPF with exact phase function for $x = 1.0$ ($g = 0.2$) and $x = 10.0$ ($g = 0.72$), respectively, for a homogeneous spherical particle (Sharma et al., 1998). In both figures, the relative refractive index of the scatterer is $m = 1.5$. It may be seen that the HGPF generally underestimates the important region of forward-scattering. In contrast, for biological cells ($g \sim 0.95$), the values given by the HGPF were noted to be elevated too much from values given by Mie theory but reasonably good agreement with the Mie phase function was noted for an asymmetry parameter less than 0.8 (Liu, 1994). The reason for this divergence in comparison to the observations of Sharma et al. (1998) perhaps lies in the difference in the refractive indices in the two applications. The relative refractive index for bioparticles ($m = 1.107$) in Liu (1994) is much closer to unity in comparison to $m = 1.5$ for spheres considered in Sharma et al. (1998).

The oscillations observed in the scattering phase function of a large scatterer average out to give a smoother phase function for a polydispersion of particle sizes. Hansen (1969) computed angular distributions for haze and clouds using the HGPF and demonstrated that they are not very accurate for thin layers. The outcome was much better for thick layers. Zhao and Sun (2010) computed scattering characteristics of clear sky cumulus using the Monte Carlo method. The study showed that the errors in using the HGPF can be very large. At wavelengths and grain-size distributions relevant to interstellar medium, the HGPF is not very accurate (see, e.g. Draine, 2003; Bianchi et al., 1996). In the context of biomedical applications, Reynolds and McCormick (1980) found that the HGPF showed significant discrepancy from solutions of Maxwell equations. However, use of the HGPF in Monte

Carlo simulations for the tissue diffuse reflectance and the fluence shows excellent agreement with those obtained using the Mie scattering phase function (Sharma and Banerjee. 2003). The tissue was modeled as turbid medium with particles of fractal size distribution including small as well as large particles. Mobley et al. (2002) have contrasted commonly used Petzold phase function (Petzold, 1972) with the HGPF. The HGPF underestimates the phase function in the near forward direction and the backscattering rise seen in most phase functions is also not observed. It was also noted that shape of the total phase function at intermediate and backward angles can have a significant effect on computed underwater radiance and reflectance. However, the exact shape of the phase function in backscatter direction does not greatly affect the light field, as long as the overall shape of the phase function does not deviate significantly from the correct shape. A comparison of the Mie phase function and the HGPF and their use in the Monte Carlo simulations for polydisperse aerosols has also been performed by Bai et al. (2011). Numerical observations show that using the HGPF leads to significant underestimating of the transmittance. It was shown that the root mean square error

$$\sqrt{\frac{\left[\sum_{\theta=0}^{180}\left(p(\theta)-p_{hgpf}(\theta)\right)^2\right]}{180}} \tag{2.34}$$

can be used as a criterion for the validity of use of the HGPF in multiple scattering computations. It was concluded that, when the nature of the phase function is close to isotropic shape, the root mean square error is small and the use of the HGPF in simulations provides adequate accuracy.

Single-scattering properties of dust aerosols and their effect in radiative flux calculations for spherical and spheroidal particles have been studied recently for log-normal size distribution by Fu et al. (2009). These studies were done at a wavelength of $\lambda = 0.55$ μm. The size distribution was defined as

$$\frac{dN(r_\nu)}{dr_\nu} = \frac{N_0}{r_\nu ln(10)\sigma\sqrt{2\pi}} \exp\left(-\frac{[log(r_\nu/r_{\nu m})]^2}{2\sigma^2}\right), \tag{2.35}$$

where $r_{\nu m}$ is the particle radius corresponding to the peak of the size distribution and σ is the geometric standard deviation of r_ν; N_0 is the number density of scatterers. The size range was $0.05\ \mu\text{m} \le r_\nu \le 0.15\ \mu\text{m}$ and $\sigma = 0.4$. The equivalent sphere was defined as one giving the same (volume)/(projected area) ratio. Two refractive indices and four effective radii were considered. The results showed that the errors in extinction efficiency and albedo in approximating spheroids with spheres is less than 1%. The errors in the asymmetry parameter were less than 2%. The phase function obtained by approximating spheroids by spheres was found to be better than that predicted by HGPF in the angular range 0° to 90°. In the range 90° to 180°, the HGPF was systematically smaller than the spheroidal scattering phase function but the spherical scattering phase function was smaller in the range 90° to 145° and larger in the range 145° to 180°.

The widely used procedure for determining the parameter g for use in the HGPF is to set g equal to the asymmetry factor of the given phase function. However, this method does not seem to yield the best fit (Kamiuto, 1987). A more accurate way to determine g is to minimize the following integral

$$\Gamma(g) = \int_{-1}^{1} \left[p(\mu) - p_{hgpf}(\mu) \right]^2 d\mu \,. \tag{2.36}$$

Test computations were performed for dielectric spheres of refractive index 1.33 and size parameter between 1 and 25. It was determined that the g obtained by least mean square fit yields better results than that obtained by using the asymmetry parameter of the exact phase function. Pomraning (1988) has argued that, instead of the least-squares procedure, it is more appropriate to minimize the error in the solution of the equation of transfer to determine the value of g. However, despite availability of these sophisticated procedures for determining g, in most cases, its value is set as the asymmetry parameter of the phase function to be parametrized.

The HGPF is perhaps the most widely used phase function in radiative transfer problems despite its shortcomings in accuracy and absence of a definite criterion regarding its validity even after it has been in practice for more than 70 years. It has been employed even in the studies relating to ultrasound wave propagation. The overriding reason for its widespread operation is perhaps its simplicity and ease of its treatment in problem solving. In particular, in Monte Carlo computations of ray tracing in a particulate medium, it results in a simple analytic expression for determining the scattered direction (section 6).

2.3.1.2 Combined Henyey–Greenstein (HGPF) and Rayleigh phase function

Even though the HGPF can approximate a given phase function in a variety of situations, it may be noted that it does not reduce to the Rayleigh phase function in the small scatterer limit (equivalently $g \to 0$ limit). Hence it is not expected to be a good approximation for the phase functions of particles that are very small compared to the wavelength of the radiation. One way to overcome this deficiency of the HGPF is to modify it in such a way that, in the limit $g \to 0$, the modified phase function reduces to the Rayleigh phase function. Cornette and Shanks (1992) demonstrated that this can be achieved by defining a phase function

$$p_{cspf}(\mu, g) = \frac{3}{2}\frac{1+\mu^2}{2+g^2}\phi_{hgpf}(\theta, g) = \frac{3}{2}\frac{1-g^2}{2+g^2}\frac{1+\mu^2}{(1+g^2-2g\mu)^{3/2}} \,, \tag{2.37}$$

where

$$\langle \mu \rangle = g\frac{3(4+g^2)}{5(2+g^2)} \,, \tag{2.38a}$$

with

$$g = \frac{5}{9}\langle \mu \rangle - \left(\frac{4}{3} - \frac{25}{81}\langle \mu \rangle^2 \right) y^{-1/3} + y^{1/3} \,, \tag{2.38b}$$

and

$$y = \frac{5}{9}\langle \mu \rangle + \frac{125}{729}\langle \mu \rangle^3 + \left(\frac{64}{27} - \frac{325}{243}\langle \mu \rangle^2 + \frac{1250}{2187}\langle \mu \rangle^4 \right)^{1/2} \,. \tag{2.38c}$$

This phase function is still a one-parameter (g) phase function. It converges to the Rayleigh phase function as $\langle\mu\rangle \to 0$ and approaches the HGPF for $\langle\mu\rangle \to 1$. It may be pointed out that essentially the same scattering phase function was re-derived independently 14 years later by Liu and Weng (2006).

Numerical comparisons of the HGPF and the CSPF have been performed for single-particle scattering by Cornette and Shanks (1992), Toublanc (1996), and Sharma et al. (1998), and for polydispersion of particles by Cornette and Shanks (1992). In both the situations, the CSPF provides a more realistic approximation to the phase function. A numerical comparison of angular variation of the HGPF and the CSPF with the exact phase function for a sphere has been performed by Sharma et al. (1998). A typical plot for $m = 1.5$ and $x = 1.0$ ($g = 0.2$) is shown in Fig. 2.3. It is clear from the figure that, while the CSPF shows good improvement at larger angles, its agreement with exact phase function at small angles is still poor and does not constitute improvement over the HGPF. However, as expected, for large g, the CSPF as well as the HGPF yields similar results (Cornette and Shanks, 1992; Toublanc, 1996).

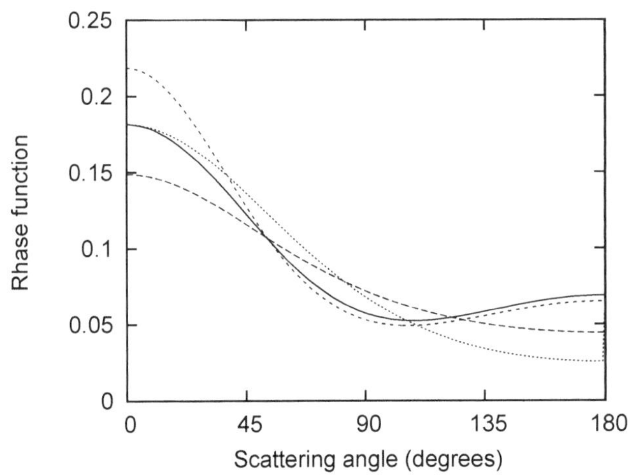

Fig. 2.3. Solid line: ϕ_{ex}; dashed line (large dashes): ϕ_{hgpf}; dashed line (small dashes): ϕ_{cspf}; dotted line: ϕ_{lsf} for $x = 1.0$ and $m = 1.5$.

The HGPF and the CSPF have been compared with the Mie phase function for three particle size distributions corresponding to Haze C, Haze M, and Cloud C.1 by Cornette and Shanks (1992). For Haze C, the CSPF provides an improved fit to the polydisperse Mie phase function. The fit is extremely good for wavelengths above 2 μm. For the Haze M distribution, the CSPF provides some improvement over the HGPF. The fit improves at longer wavelengths. Neither phase function gives good fit for the Cloud C.1 model where the scattering particles are of large size. As wavelength increases, the effective particle size decreases and the phase functions provide an improved fit. The Legendre expansion for this phase function is (Cornette and Shanks, 1992)

$$p(\theta, g) = \frac{3}{4+2g^2} \sum_{n=0} \left[\frac{n(n-1)}{2n-1} g^{n-2} + \frac{(n+2)(n+1)}{2n+3} g^{n+2} + \left(\frac{(n+1)^2}{2n+3} + \frac{5n^2-1}{2n-1} \right) g^n \right] P_n(\cos\theta) . \quad (2.39)$$

2.3.1.3 The Neer–Sandri phase function (NSPF)

An often-used phase function in atmospheric optics problems is the Deirmendjian phase function (Deirmendjian, 1969). One limitation of this phase function is that it exists only in the form of tabulated values restricted to specific wavelengths and haze types. A modified HGPF has been suggested by Fishburne et al. (1976), which looks more like the Deirmendjian phase function in its appearance. This modified phase function has been sometimes referred to as the Neer–Sandri phase function and is written as

$$p_{nspf}(\theta) = \left[\frac{1-g^2}{(1+g^2-2g\mu)^{3/2}} + \lambda'(3\mu^2-1) \right] . \quad (2.40)$$

The second term in Eq. (2.40) is proportional to the second Legendre polynomial. This added term is symmetric about $\theta = 90°$ and hence does not affect the asymmetry parameter and the normalization of the HGPF. The parameter λ' is chosen such that the phase function is positive at all angles. Setting the derivative of Eq. (2.40) to zero, λ' can be expressed in terms of the scattering angle θ_0 which corresponds to the minimum value of the phase function

$$\lambda' = \left[\frac{-g(1-g^2)}{2\cos\theta_0(1+g^2+2g\cos\theta_0)^{5/2}} \right] . \quad (2.41)$$

The Neer–Sandri phase function given by Eq. (2.40) can then be expressed as

$$p_{nspf}(\theta) = \left[\frac{1-g^2}{(1+g^2-2g\cos\theta)^{3/2}} + \frac{g(1-g^2)(3\mu^2-1)}{2|\cos\theta_0|(1+g^2+2g\cos\theta_0)^{5/2}} \right] . \quad (2.42)$$

Riewe and Green (1978) have shown that, by a judicious choice of λ', the Mie phase function can be reproduced quite accurately. It is found that a value of $\mu = 1/7$ not only guarantees that the phase function is non-negative everywhere, but it also ensures that it looks similar to the Deirmendjian's phase function.

2.3.1.4 Kagiwada–Kalaba phase function (KKPF)

This phase function, introduced by Kagiwada and Kalaba (1967), reads as

$$p_{kkpf}(\mu) = \frac{K}{b-\mu} , \quad (2.43a)$$

where

$$K = 2\left[\ln\left(\frac{b+1}{b-1}\right)\right]^{-1} . \tag{2.43b}$$

The shape parameter b, in contrast to the shape parameter g in the HGPF, is determined by the relation

$$b = \frac{r+1}{r-1} , \tag{2.44}$$

where r is the ratio of forward to backward scattering phase function. The simulations using the KKPF and the HGPF have been compared for cloudy and hazy planetary atmospheres (Hansen, 1969). For thick atmospheres, the HGPF has been found to be better than KKPF for albedo and planetary magnitudes. None of the two phase function reproduces the angular distribution accurately for thin clouds. The outcome is better for thick layers—particularly for hazes.

2.3.1.5 The Schlick phase function

Blasi and coworkers (1993) have developed a phase function in the context of computer graphics. This phase function avoids the fractional exponent that occurs in the HGPF. It is expressed as

$$p_{shpf}(\mu) = \frac{1-k^2}{(1+k\mu)^2} , \tag{2.45a}$$

where $-1 \leq k \leq 1$ is a parameter analogous to g in the HGPF. This phase function may be used efficiently in problems where the accurate shape of the phase function is not very important.

For intermediate values, k can be related to g as (Pharr and Humphreys, 2004)

$$k = 1.55g - 0.55g^3 . \tag{2.45b}$$

2.3.1.6 The binomial phase function

The binomial scattering phase function was introduced by Kaper et al. (1970):

$$p_{bph} = \frac{N+1}{2^N}(1+\cos\theta)^N . \tag{2.46a}$$

This scattering law has the advantage that N can be related to a_l in Eq. (2.7) by the recursion relation (McCormick and Sanchez, 1981)

$$a_l = \left(\frac{2l+1}{2l-1}\right)\left(\frac{N+1-l}{N+1+l}\right)a_{l-1} , \tag{2.46b}$$

for $l = 1, 2 \ldots$ with $a_0 = 1$.

2.3.1.7 The delta-hyperbolic phase function

Another theoretical one-parameter phase function in the context of radiance distribution at a depth in seawater was proposed by Haltrin (1988). This phase function has the following form:

$$p_{dhpf} = 2g\delta(1-\mu) + \frac{1-g}{\sqrt{2(1-\mu)}} = \sum_{n=0}^{\infty}(1+2ng)P_n(\mu), \quad 0 \le g \le 1\,, \tag{2.47}$$

where $\delta(1-\mu)$ is a Dirac-delta function and g is a shape parameter. The distinguishing feature of this phase function is that it leads to an exact analytic solution for the radiance distribution in the depth of scattering medium in terms of inherent optical properties of the medium.

2.3.1.8 The transport phase function (TPF)

A simple phase function which is often used in neutron transport theory is known as the transport phase function (Davison, 1957). It is a sum of an isotropic part and a forward delta function part:

$$p_{tpf} = 2B + 2(1-2B)\delta(1-\mu)\,, \tag{2.48a}$$

where

$$B = \frac{1}{2}\int_{-1}^{0} p_{tpf}(\mu)\,d\mu \tag{2.48b}$$

is the backscattering probability. The TPF has the advantage of reducing the RTE to the easily solvable case of isotropic scattering (Haltrin, 1997). This phase function has been generalized to include purely backward and low-order anisotropic scattering, too (Siewert and Williams, 1977; Devaux et al., 1979).

2.3.2 Two-parameter phase functions

2.3.2.1 The modified Henyey–Greenstein phase function

Jacques et al. (1987) found that

$$p_{jpf}(\theta) = \alpha p_{hgpf}(\theta, g) + (1-\alpha) \quad 0 \le \alpha \le 1, \tag{2.49}$$

which describes the phase function of the human dermis well. This phase function is a combination of a HGPF and an isotropic phase function. Note that, for $\alpha = 1$, $p_{jpf} = p_{hgpf}$. The factor α is necessary to account for the backscattering in excess of what is provided by the HGPF. The same phase function agrees reasonably well with the scattering phase function of aorta (Yoon, 1988) and dental enamel (Fried et al., 1995). The values of the parameters α and g are, of course, different in the three applications.

Bevilacqua and Depeursinge (1999) proposed two hybrid phase functions in a form similar to that in Eq. (2.49). The first of these hybrid phase functions is expressed as

$$p_{bdpf1} = \alpha p_{hgpf} + 3(1-\alpha)\cos^2\theta \,. \quad 0 < \alpha < 1 \tag{2.50}$$

The added term is now proportional to $\cos^2\theta$. The first three moments of the phase function Eq. (2.50) are

$$\langle\mu\rangle \equiv \langle\cos\theta\rangle = \alpha g \,, \tag{2.51a}$$

$$\langle\mu^2\rangle \equiv \langle\cos^2\theta\rangle = \alpha g^2 + \frac{2}{5}(1-\alpha) \,, \tag{2.51b}$$

and

$$\langle\mu^3\rangle \equiv \langle\cos^3\theta\rangle = \alpha g^3 . \tag{2.51c}$$

These equations clearly allow independent adjustment of the parameters g and α with restriction that only those values of g and α are admissible for which the phase function is positive.

The other phase function was also constructed in a similar way. It replaces p_{hgpf} by p_{bpf}:

$$p_{bdpf2} = \alpha p_{bpf} + 3(1-\alpha)\cos^2\theta \,, \quad 0 < \alpha < 1 \,, \tag{2.52}$$

where p_{bpf} is the binomial phase function as given in Eq. (2.46a). The first two moments of this phase function are

$$\langle\mu\rangle \equiv \langle\cos\theta\rangle = \alpha g \,, \tag{2.53a}$$

$$\langle\mu^2\rangle \equiv \langle\cos^2\theta\rangle = \alpha g^2 + \frac{2}{5}(1-\alpha) \,, \tag{2.53b}$$

where

$$g = \frac{N}{N+2} \,. \tag{2.53c}$$

It has been noted that the two phase functions, Eqs (2.50) and (2.52), cover most of the $\langle\mu\rangle$ and $\langle\mu^2\rangle$ values obtained by Mie scattering in the size parameter range 1–25. It may be mentioned that another similar phase function has also been studied by Bevilacqua et al. (1999). This consists of a sum of a high anisotropy phase function due to large scatterers and a low anisotropy phase function due to small scatterers.

Phase functions with such structures have actually been measured in other biological tissues as well (Flock et al., 1987; Jacques et al., 1987; Marchesini et al., 1989; van der Zee et al., 1993; Chicea and Chicea, 2006; Fernandez-Oliveras et al., 2012).

2.3.2.2 The Gegenbauer kernel phase function (GKPF)

A more general form of the HGPF may be obtained through the use of the generating function of the Gegenbauer polynomials (Reynolds and McCormick, 1980):

$$p_{gkpf}(\mu) = K(1 + g^2 - 2g\mu)^{-(1+\epsilon)} \,, \tag{2.54a}$$

where the normalization factor K is

$$K = 2\epsilon g(1-g^2)^{2\epsilon}\left[(1+g)^{2\epsilon} - (1-g)^{2\epsilon}\right]. \qquad (2.54b)$$

For $\epsilon = 1/2$, ϕ_{gkpf} is nothing but the HGPF. This phase function has been referred to as the Gegenbauer kernel phase function, as it can also be written as

$$p_{gkpf}(\mu) = K(1-g^2)^{-1}\sum_{n=0}^{\infty}\left(1+\frac{n}{\epsilon}\right)C_n^{(\epsilon)}(\mu)g^n\,, \quad |g|<1,\ \epsilon > -1/2\,, \qquad (2.55)$$

where C_n's are the Gegenbauer polynomials. Using orthogonality and recursion relations, the parameter g can be expressed as

$$g = \frac{\int_{-1}^{1}\mu\phi(\mu)(1-\mu^2)^{\epsilon-1/2}\,d\mu}{\int_{-1}^{1}\phi(\mu)(1-\mu^2)^{\epsilon-1/2}\,d\mu}\,, \qquad \epsilon > -1/2\,. \qquad (2.56)$$

For the HGPF ($\epsilon = 1/2$), the parameter g is identically equal to the mean cosine of the scattering angle $\langle\mu\rangle$. For $\epsilon \neq 1/2$, $\langle\mu\rangle$ may be obtained from the definition Eq. (2.15) to obtain

$$\langle\mu\rangle = \frac{\left[2g\epsilon L - (1+g^2)\right]}{2g(\epsilon-1)}\,, \qquad (2.57a)$$

where

$$L = \frac{(1+g)^{2\epsilon} + (1-g)^{2\epsilon}}{(1+g)^{2\epsilon} - (1-g)^{2\epsilon}}\,. \qquad (2.57b)$$

From the experimental data, g may be obtained from the equation

$$\frac{1+g^2-2g\mu_1}{1+g^2-2g\mu_2} = R\,, \qquad (2.58)$$

where μ_1 and μ_2 are two widely separated angles for which $p(\mu_1, g)$ and $p(\mu_2, g)$ are very different from each other. If $\mu_1 = -1$ and $\mu_2 = 1$, g may be calculated most easily. The value of ϵ may then be obtained by using the relation

$$\epsilon = -1 + \frac{\ln[p(\mu_2)/p(\mu_1)]}{\ln R}\,, \qquad (2.59)$$

which follows from Eqs (2.54a) and (2.58).

Sample computations illustrating the validity of this phase function and its comparison with the HGPF have been done in various contexts. It is generally concluded that, while large-diameter low-index particles can be reasonably well approximated by the HGPF, higher-index particles are seen to fit better by the GKPF. However, exceptions have been noted. For particles in blood (platelets, erythrocytes, and platelet aggregates) modeled as spheres with diameters in the range $3 < d < 240$ μm, and relative refractive index in the range $1.015 < n < 1.25$, the ϕ_{gkpf} may represent more closely the scattering pattern in comparison to ϕ_{hgpf} (Reynolds and McCormick, 1980). Further studies on RBCs (Yaroslavsky et al., 1997; Flock et al., 1987; Hammer et al., 1998) suggest that the GKPF can reasonably well represent single-scattering properties of RBCs. Roggan et al. (1999)

compared the HGPF and the GKPF with Mie phase function for a scatterer of $g = 0.9924$ (corresponding to RBC) in the near forward direction ($\theta = 0°$ to $20°$). It was concluded that the HGPF overestimates the RBC phase function by about 30 times in the very near forward direction. In contrast, the GKPF mimics the exact phase function without significant deviations. Liu (1994) also noted that the HGPF overestimates the exact phase function in the near forward direction for biological particles modeled as homogeneous spheres.

2.3.2.3 The delta-Eddington phase function

This is a particular case of the $\delta - M$ phase function. It consists in representing the forward peak of a highly anisotropic phase function by a Dirac-delta function (as in the delta-M phase function) and a two-term expansion of the rest of the phase function (see, e.g. Joseph et al., 1976; Crosbie and Davidson, 1985). One may thus write Eq. (2.22) as

$$p_{depf}(\mu) = 2\alpha\delta(1-\mu) + (1-\alpha)(a_0' + 3a_1' \cos\theta)\,, \tag{2.60}$$

where α is a weight factor ranging from 0 to 1 and determines the amount of light in the forward peak. From Eq. (2.24c),

$$a_0' = 1, \tag{2.61a}$$

and

$$a_1' = \frac{a_1 - \alpha}{1-\alpha}\,, \tag{2.61b}$$

where

$$a_1 = \langle\mu\rangle = \frac{1}{2}\int_{-1}^{1} \mu p(\mu)\, d\mu\,. \tag{2.61c}$$

The parameter α may be obtained from the relation

$$\frac{1}{2}\int P_2(\mu) p_{depf}(\mu)\, d\mu = \alpha\,. \tag{2.61d}$$

In the event that the phase function is approximated by the HGPF, one may write $\alpha = g^2$.

2.3.2.4 The Liu phase function (LPF)

Liu (1994) examined the HGPF for large spherical scatterers and noted that a moderately acceptable agreement with Mie theory exists only for mean cosine less than 0.8. The biomedical tissues generally have a mean cosine value which is greater than this. For such particles, most scattering is in the near forward direction. This prompted Liu (1998) to design a phase function by enforcing the constraint that his phase function must reproduce correctly the forward-scattering, where most of the scattering takes place. His phase function reads as

$$p_{lpf} = K\Big(1 + \epsilon\cos\theta\Big)^{n}\,, \tag{2.62a}$$

where ϵ and n are two independent parameters and K is a normalization factor given by

$$K = \frac{2\epsilon(n+1)}{[(1+\epsilon)^{n+1} - (1-\epsilon)^{n+1}]} \,. \tag{2.62b}$$

The parameters ϵ and n are the characteristic factor and the anisotropy index, respectively. A positive value of ϵ means the scattering is forward peak. The larger the value of ϵ, the stronger the forward peak. A value of $\epsilon = 0$ implies isotropic scattering and negative ϵ indicates strong backscattering. The anisotropy index is limited to non-negative even integers to ensure that the phase function remains non-negative throughout the θ domain. The larger the value of n, the stronger the scattering anisotropy. For $\epsilon = 1$, the Liu phase function is

$$p_{lpf} = \frac{n+1}{2^n}(1+\cos\theta)^n \,, \tag{2.63}$$

which is easily recognized to be the binomial phase function.

The parameters n and ϵ can be determined by solving the simultaneous equations

$$p_{lpf}(0) = K(1+\epsilon)^n \tag{2.64a}$$

and

$$\langle\mu\rangle = \frac{n+1}{n+2}\left[\frac{(1+\epsilon)^{n+1} + (1-\epsilon)^{n+1}}{(1+\epsilon)^{n+1} - (1-\epsilon)^{n+1}}\right] - \frac{1}{\epsilon(n+2)} \,. \tag{2.64b}$$

A comparison of the Liu phase function with the exact phase function for dielectric homogeneous sphere, the HGPF, and the CSPF for $x = 1.0$ is shown in Fig. 2.3. As expected, the LPF gives improvement at near forward-scattering angles. However, its agreement with the exact profile at other angles is poor in comparison to the HGPF and the CSPF. For larger particles as well, the LPF performs well only in the near forward direction. At larger angles, its performance is not good at all (Sharma et al., 1998). This is perhaps the reason why the Liu phase function has not really received much attention.

Sometimes only the value of the asymmetry parameter is known. In such cases, one may assume $\epsilon = 1$ and write the phase function as

$$p_{lpf}(\epsilon = 1) = K_s(1+\cos\theta)^n \tag{2.65a}$$

where the normalization factor is

$$K_s = \frac{n+1}{2^n} \,. \tag{2.65b}$$

Even this simplified phase function displays better agreement with Mie theory in comparison to the HGPF in the near forward direction for $g = 0.9$, 0.95, and 0.99 (the values commonly encountered in biomedical applications).

2.3.2.5 The Draine phase function (DPF)

A variety of cosmic dusts, such as interstellar dust, interplanetary dust, cometary dust, etc., exist in space. Studies relating to the light propagation in such regions of outer space give information on these dust particles. In fact, this is the only way to deduce information about the type of particles (size, shape, size distribution, refractive index, etc.) in these dusts. For these studies to take place, one requires knowledge of the phase function. Indeed, the most well-known phase function, the HGPF, was first introduced in the context of an astrophysical application. A linear combination of three HGPFs for zodiacal dust has been studied by Hong (1985). More recently, an improved analytic phase function has been suggested by Draine (2003) which reproduces the interstellar dust phase function well at various wavelengths. The modified phase function has been expressed as

$$p(\theta)_{dpf} = \frac{1-g^2}{(1+g^2-2g\cos\theta)^{3/2}} \frac{1+\alpha\cos^2\theta}{1+\alpha(1+2g^2)/3} . \tag{2.66}$$

At $\alpha = 0$, p_{dpf} reduces to the HGPF and to the CSPF for $\alpha = 1$. The parameters g and α are determined by requiring that p_{dpf} reproduces correctly the first moment $\langle\cos\theta\rangle$:

$$\langle\cos\theta\rangle = g\frac{1+\alpha(3+2g^2)/5}{1+\alpha(1+2g^2)/3} , \tag{2.67a}$$

and the second moment

$$\langle\cos^2\theta\rangle = \frac{1+2g^2+(3\alpha/35)(7+20g^2+8g^4)}{3+\alpha(1+2g^2)} , \tag{2.67b}$$

of the exact phase function. These restrictions lead to the following relations:

$$g = \left[(a^3+b^2)^{1/2}-b\right]^{1/3} - \left[(a^3+b^2)^{1/2}+b\right]^{1/3} + \frac{17}{9}\langle\cos\theta\rangle , \tag{2.68a}$$

if $a^3+b^2>0$ and

$$g = 2|a|^{1/2}\cos\left(\frac{\Psi}{3}\right) + \frac{17}{9}\langle\cos\theta\rangle , \quad \Psi = \arccos\left(\frac{-b}{|a|^{3/2}}\right) , \tag{2.68b}$$

if $a^3+b^2<0$. Here

$$a = \frac{7}{3}\langle\cos^2\theta\rangle - \frac{289}{81}\langle\cos\theta\rangle^2 , \tag{2.68c}$$

and

$$b = \frac{119}{18}\langle\cos\theta\rangle\langle\cos^2\theta\rangle - \frac{4913}{729}\langle\cos\theta\rangle^3 - \frac{7}{6}\langle\cos\theta\rangle . \tag{2.68d}$$

Using Eq. (2.68a), the parameter α is then found to be

$$\alpha = \frac{15(\langle\cos\theta\rangle - g)}{3(3+2g^2)g-5(1+2g^2)\langle\cos\theta\rangle} . \tag{2.68e}$$

This phase function leads to improved fit if $\alpha<1$. For values of $\langle\cos\theta\rangle$ and $\langle\cos^2\theta\rangle$ that lead to $\alpha>1$, g is obtained by setting $\alpha=1$. Then the relevant expression for g is

$$g = \left[(c^3 + d^2)^{1/2} - d\right]^{1/3} - \left[(c^3 + d^2)^{1/2} + d\right]^{1/3} + \frac{5}{9}\langle\cos\theta\rangle\,, \qquad (2.69a)$$

where

$$c = \frac{4}{3} - \frac{25}{81}\langle\cos\theta\rangle^2 \qquad (2.69b)$$

and

$$d = \frac{125}{729}\langle\cos\theta\rangle^3 + \frac{5}{9}\langle\cos\theta\rangle\,. \qquad (2.69c)$$

The formula given here differs from what has been given by Draine (2003) in their Appendix B. In place of 15 in the numerator in Eq. (2.68e), 25 appears in Draine (2003). This is perhaps a misprint in Draine (2003) which has been noted for the benefit of those who might use this function in their calculations.

The DPF constitutes significant improvement over HGPF for interstellar dust. This can be seen in Fig. 2.4, which shows relative rms error in some of the approximate phase functions. The relative rms error is defined as

$$h_{rel} = \left[\frac{1}{2}\int d\mu \left(\frac{p_{app}(\mu) - p_{ex}(\mu)}{\phi_{ex}(\mu)}\right)\right]^{1/2}. \qquad (2.70)$$

It is clear that the DPF is an improvement over the HGPF in almost the entire wavelength range of interest.

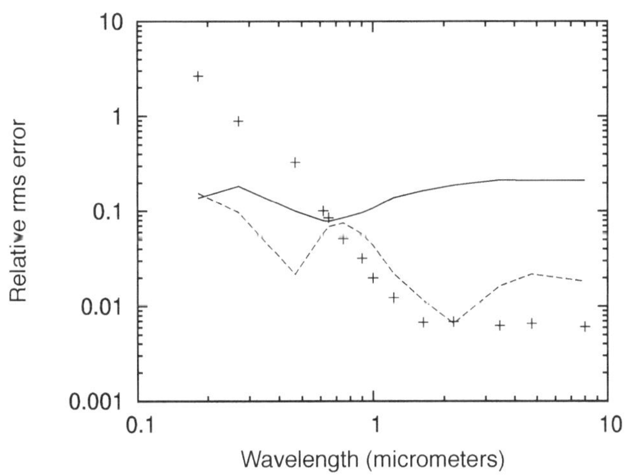

Fig. 2.4. The relative rms error as defined in Eq. (2.70) as a function of wavelength for a polydispersion governed by the inverse power law. Solid line: error in p_{hgpf}; dashed line: error in p_{dpf}; crosses: error in p_{rspf}.

2.3.2.6 Phase function for planetary regoliths

The Hapke's formula is a well known theoretical model for estimation of the reflection from a planetary surface (Hapke, 1981; Mishchenko et al., 2006). The model defines particle characteristics of a planetary surface by means of a phase function which has two parameters, b and c, and may be expressed as

$$\begin{aligned} p(\theta_p) &= \frac{(1-c)}{2}\left[\frac{(1-b^2)}{(1+2b\cos\theta_p+b^2)^{3/2}}\right] \\ &+ \frac{(1+c)}{2}\left[\frac{(1-b^2)}{(1-2b\cos\theta_p+b^2)^{3/2}}\right], \end{aligned} \tag{2.71}$$

where θ_p is the phase angle ($\theta_p = 180 - \theta$), and $0 \leq b \leq 1$ and c are parameters dependent on the material properties of the regolith. There is no restriction on c except that $p(\theta_p) \geq 0$ for all θ_p. It is an indicator of the relative strength of the backward peak with respect to the forward peak. A positive value of c indicates greater backscattering while a negative value of c implies greater forward-scattering. The parameter b describes the angular width of each lobe. If b is close to 1, the lobes are high and narrow and, if $b \leq 1$, the lobes are broader and small. An experimental study of the light-scattering by large irregular particles of various shapes (McGuire and Hapke, 1995) concludes that this phase function generally provides reasonably good description of the data while keeping the number of free parameters to the minimum necessary. The parameters b and c were studied for various types of particles and plotted against each other. The plot of these parameters fell on an L-shaped area, implying that this type of plot should be useful in estimating certain properties of particle type.

The shortcomings of the original Hapke model have been studied and modified from time to time by many authors (Mishchenko, 1994; Liang and Townshend, 1996; Mishchenko and Macke, 1997; Mishchenko et al., 1999).

2.3.3 Three-parameter phase functions (TPPF)

As was seen in section 3.1.1, the HGPF is capable of generating only the forward peak and not the backward rise in the phase function. This is because the HGPF is a monotonically decreasing phase function. The same holds for the GKPF. To parametrize phase functions which exhibit backscattering rise, a sum of the two HGPFs has been suggested by Irvine (1965):

$$p_{tppf}(\theta) = \alpha p_{hgpf}(\theta, g_1) + (1-\alpha) p_{hgpf}(\theta, g_2)\,. \tag{2.72}$$

This phase function allows introduction of peaks in the forward as well as the backward direction. The sign in front of the asymmetry parameter determines the direction of elongation. A positive sign results in the forward peak while a negative sign results in the backward rise. So, p_{ttpf} is a phase function capable of producing two peaks. For $-g_1 = g_2 = b$ and $\alpha = (1+c)/2$, this phase function reduces to Eq. (2.71). The subscript $tppf$ denotes a three-parameter phase function.

The parameters g_1, g_2, and α can be computed to fit three moments of the phase function (Kattawar, 1975). As in the HGPF, this phase function can also be expressed as an expansion

$$p_{ttpf}(\mu, g_1, g_2, \alpha) = \sum_{n=0}^{\infty} (\alpha g_1^n + (1-\alpha) g_2^n)(2n+1) P_n(\mu) \tag{2.73}$$

and

$$\langle P_n(\mu) \rangle = \alpha g_1^n + (1-\alpha) g_2^n \,. \tag{2.74}$$

It is then possible to set up the following equations:

$$\alpha g_1 + (1-\alpha) g_2 = g \,, \tag{2.75a}$$
$$\alpha g_1^2 + (1-\alpha) g_2^2 = h \,, \tag{2.75b}$$
$$\alpha g_1^3 + (1-\alpha) g_2^3 = t \,, \tag{2.75c}$$

for the first three moments of the phase function which can be solved to give

$$g_2 = \frac{t - hg - [(hg-t)^2 - 4(h-g^2)(tg-h^2)]^{1/2}}{2(h-g^2)} \,, \tag{2.76a}$$

$$g_1 = \frac{(gg_2 - h)}{g_2 - g} \,, \tag{2.76b}$$

$$\alpha = \frac{g - g_2}{g_1 - g_2} \,. \tag{2.76c}$$

Numerical tests showed that this phase function can be expected to yield excellent flux values. Accurate asymptotic radiance can be achieved if scattering albedo is greater than 0.9.

For the parameter set, $\alpha = 0.995$; $g_1 = 0.992$; $g_2 = -0.93$, this phase function has been found to reproduce correctly (Kienle et al., 2001) the experimentally observed phase function of white matter of a neonatal brain (van der Zee et al., 1993). For the oceanic medium, the following parametrization has been found (Haltrin, 2002) for g_2:

$$g_2 = -0.30614 + 1.0006 g_1 - 0.01826 g_1^2 - 0.03644 g_1^3 \,, \tag{2.77}$$

in the range $0.30664 < g \leq 1$, with α given by the relation

$$\alpha = \frac{g_2(1+g_2)}{(g_1+g_2)(1+g_2-g_1)} \,. \tag{2.78}$$

Sharma and Banerjee (2003) computed TPPF, HGPF, and Mie phase function for a biomedical soft tissue modeled as a turbid medium. The scatterers in the turbid medium were homogeneous spheres with a fractal size distribution. It was noted that neither phase function reproduces correctly the backscattering rise seen in experimentally observed scattering phase functions. Therefore, a simple redefinition of α:

$$\alpha = A + \frac{g_2(1+g_2)}{(g_1+g_2)(1+g_2-g_1)} \tag{2.79}$$

was employed which gave the desired backscattering rise by adjusting $A = 0.055$ in an empirical way.

Diffuse reflectance and fluence were computed using these phase functions in a Monte Carlo code (Jacques and Wang, 1995). Interestingly, despite significant difference in the phase functions, there was good agreement in reflectance (as a function of source-detector separation) and fluence (as a function of depth) graphs as long as the asymmetry parameter is the same for the phase functions. The TPPF can provide a simple, approximate analytic form for the phase function of a cylinder, too (Gillespie, 1992), although the width of the forward peak is not a very good match. There is relatively greater scattered intensity at the near forward and the intermediate angles for the case of the exact phase function than that in the TPPF.

2.3.4 Five-parameter phase function

A simple way of determining the first few Legendre expansion coefficients for the phase functions for an arbitrary particle of $x \leq 2$ has been given by Sharma et al. (1998). Their phase function reads as

$$\phi_{srspf} = b_0 + b_1 \cos\theta + b_2 \cos^2\theta + b_3 \cos^3\theta + b_4 \cos^4\theta, \tag{2.80}$$

where

$$b_0 = \phi_{ex}(\pi/2)\,, \tag{2.81a}$$

$$b_1 = \left.\frac{d\phi_{ex}}{d(\cos\theta)}\right|_{\theta=\pi/2}, \tag{2.81b}$$

$$b_2 = \frac{3}{8\pi}\Big[5 - 2\pi\big(\phi_{ex}(0) + \phi_{ex}(\pi) + 8\phi_{ex}(\pi/2)\big)\Big]\,, \tag{2.81c}$$

$$b_3 = \frac{\phi_{ex}(0) - \phi_{ex}(\pi)}{2} - a_1\,, \tag{2.81d}$$

$$b_4 = \frac{5}{8\pi}\Big[2\pi\big(\phi_{ex}(0) + \phi_{ex}(\pi) + 4\phi_{ex}(\pi/2)\big) - 3\Big]\,. \tag{2.81e}$$

Here, ϕ_{ex} refers to the exact phase function and the letters srs in the subscript refer to the initials of the authors of this paper (Sharma, Roy, and Somerford). The constraints in arriving at the expression are:

1. The ϕ_{srspf} is normalized according to the relation

$$2\pi \int_{-1}^{1} \phi_{srspf}(\cos\theta)\, d(\cos\theta) = 1\,. \tag{2.82}$$

2. The phase functions ϕ_{srspf} matches with ϕ_{ex} at $\theta = 0$, π, and $\pi/2$.
3. The slope of ϕ_{srspf} is identical with that of ϕ_{ex} at $\theta = \pi/2$.

Alternatively, Eq. (2.80) may also be expressed as

$$\begin{aligned}\phi_{srspf}(\cos\theta) = {} & a_0 P_0(\cos\theta) + a_1 P_1(\cos\theta) + a_2 P_2(\cos\theta) \\ & + a_3 P_3(\cos\theta) + a_4 P_4(\cos\theta).\end{aligned} \tag{2.83}$$

Comparing with the $\cos\theta$ series, the coefficients $a's$ may be expressed as

$$a_0 = b_0 + \frac{1}{3}b_3 + \frac{1}{5}b_5\,, \tag{2.84a}$$

$$a_1 = b_1 + \frac{3}{5}b_3\,, \tag{2.84b}$$

$$a_2 = \frac{2}{3}b_2 + \frac{4}{7}b_4\,, \tag{2.84c}$$

$$a_3 = \frac{2}{5}b_3\,, \tag{2.84d}$$

$$a_4 = \frac{8}{35}b_4\,. \tag{2.84e}$$

The phase function ϕ_{srspf} has been compared with the ϕ_{hgpf} and the ϕ_{ex} for several values of the size parameter and the refractive index m. It was found to yield better agreement with the exact phase function in comparison to the HGPF in the domain $x = \pi d/\lambda \leq 2.0$ and $m \leq 1.5$ (Sharma et al., 1998). The agreement was good even for a refractive index as high as 3.0 if $x \leq 1.0$. Figure 2.5 shows a comparison of ϕ_{srspf} with ϕ_{ex} for homogeneous spheres of $x = 0.1$, 0.5, 1.0, and $m = 1.5$.

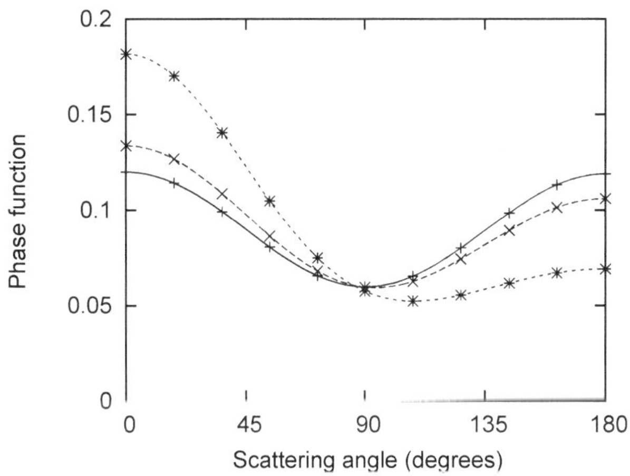

Fig. 2.5. A comparison of ϕ_{rspf} with ϕ_{ex} for $x = 0.1$, 0.5, 1.0, and $m = 1.5$. Solid line: ϕ_{ex} for $x = 0.1$; large dashed line: ϕ_{ex} for $x = 0.5$; small dashed line: ϕ_{ex} for $x = 1.0$. Corresponding ϕ_{rspf} values are plotted as points which virtually overlap with exact values.

Legendre coefficients computed using Eq. (2.84) agree well with those obtained from Eq. (2.8) (Sharma et al., 1999). This was demonstrated for two types of particles. The first type is absorbing spheres of $x = 0.7854$ and $m = 1.42 - i0.05$; the second is non-absorbing with $x = 1.48$ and $m = 1.818$. These are titanium dioxide pigments widely used in the paint industry to mix for the dispersion of dark pigments. The comparison is shown in Table 2.1. It can be seen that the Legendre coefficients computed using Eq. (2.84) agree very well with the exact Legendre coefficients.

Table 2.1. Comparison of Legendre coefficients from Eqs (2.8) and (2.84) for two monodispersions of Mie particles (Sharma et al. 1999).

	$x = 0.785$,	$m = 1.42 - i0.05$;	$x = 1.48$,	$m = 1.818$
a_j	Exact	From (2.84)	Exact	From (2.84)
a_0	7.958×10^{-2}	7.958×10^{-2}	7.958×10^{-2}	7.958×10^{-2}
a_1	2.801×10^{-2}	2.801×10^{-2}	1.361×10^{-1}	1.361×10^{-1}
a_2	4.165×10^{-2}	4.165×10^{-2}	7.233×10^{-2}	7.233×10^{-2}
a_3	5.486×10^{-3}	5.487×10^{-3}	1.662×10^{-2}	1.664×10^{-2}
a_4	3.129×10^{-4}	3.129×10^{-4}	3.211×10^{-3}	3.220×10^{-3}

A simple analytic expression can also be obtained for asymmetry parameters from Eq. (2.80):

$$\langle \mu \rangle_{rs} = \frac{2\pi}{15}\left[4b_1 + 3[\phi_{ex}(0) - \phi_{ex}(\pi)]\right] . \tag{2.85}$$

Equation (2.85) allows one to obtain the value of the asymmetry parameter in a straightforward way.

Sharma and Roy (2008) have examined how accurately the three phase functions, namely the ϕ_{hgpf}, the ϕ_{dpf}, and the ϕ_{srspf}, are able to emulate the exact phase function ϕ_{ex} for light propagation in interstellar dust grains. A power law size distribution was assumed for such grains with minimum and maximum sizes given by 0.02 μm and 0.25 μm respectively. Figure 2.4 depicts a comparison of relative *rms* errors for the HGPF, the DPF, and the SRSPF. The relative rms error is defined as in Eq. (2.70). It is clear that the errors in the ϕ_{hgpf} at various wavelengths are greater than 10% except in a region which has been identified by Draine (2003) as 0.47 $\mu\text{m} \leq \lambda \leq 0.97$ μm. The phase function ϕ_{dpf} constitutes a significant improvement over the ϕ_{hgpf}. Errors are less than 10% in the entire wavelength range $\lambda > 0.27$ μm for the ϕ_{dpf}. In contrast, the validity domain of ϕ_{srspf} is smaller in the sense that the errors are less than 10% only for $\lambda > 0.60$ μm. But the most important feature that is revealed by the comparison is that it is the SRSPF which gives the best fit in the entire wavelength region $\lambda > 0.7$ μm. In other words, the SRSPF is found to be the most appropriate phase function in the infrared region. At ultraviolet wavelengths, none of the phase functions is seen to produce a good fit.

The five-parameter phase function was studied for nonspherical particles, too (Sharma and Roy, 2000). Numerical comparison of exact phase functions for aligned spheroidal and infinitely long cylindrical particles shows excellent agreement with the five-term phase function for small parameter scatterers. This clearly demonstrates that ϕ_{srspf} is capable of reproducing the exact phase function truly and completely at all scattering angles for small nonspherical particles of arbitrary shape as long as the asymmetry parameter is less than about 0.6. As for a spherical scatterer, the HGPF differs significantly from the exact phase function for nonspherical particles, too.

2.3.5 Six-parameter phase function

The range of values of x over which the five-parameter phase function gives a good result can be increased a little by extending it to six parameters (Sharma et al., 1998). The additional constraint imposed to determine the sixth parameter is that the phase function also duplicates the asymmetry parameter of the phase function to be parametrized. With this additional constraint, one gets the following phase function:

$$\begin{aligned}\phi_{srspf}(\cos\theta) &= b'_0 + b'_1\cos\theta + b'_2\cos^2\theta + b'_3\cos^3\theta \\ &\quad + b'_4\cos^4\theta + b'_5\cos^5\theta\,, \end{aligned} \tag{2.86}$$

where $b'_0 = b_0,\ b'_1 = b_1,\ b'_2 = b_2,\ b'_4 = b_4$ with

$$b'_3 = \frac{g_{mie}}{8\pi} - \frac{10b_1}{3}\frac{5}{4}\left[\phi_{ex}(0) - \phi_{ex}(\pi)\right] \tag{2.87}$$

and

$$b'_5 = -\frac{g_{ex}}{8\pi} + \frac{7b_1}{3}\frac{7}{4}\left[\phi_{ex}(0) - \phi_{ex}(\pi)\right]. \tag{2.88}$$

The above phase function appears to work well even for $x = 2.0$ for $m \leq 3$. It has been shown to give some improvement over the five-parameter version. Oscillations appear in the phase function for $x \geq 2.0$ and ϕ_{srspf} starts showing deviations from the exact profile at these angles.

2.4 Analytic phase functions dependent on microphysical particle characteristics

The scattering phase functions discussed thus far in this article do not reflect the nature or the characteristics of the scatterer or scatterers. These phase functions were designed with the aim of arriving at simple analytic expressions for replicating a given curve with the help of some parameters related to some characteristics of the curve. In the HGPF, for example, this characteristic parameter is the asymmetry parameter g. However, the same value of g may be obtained for a variety of scattering phase functions. Clearly, the relationship between the characteristic input (g in this case) and the corresponding microphysical characteristics is ambiguous for such phase functions.

In contrast, the formulas with explicit dependence on the ensemble characteristics, such as size distribution and relative refractive index, are definitely preferable. The strength of such phase functions obviously lies in that they have a theoretical foundation and can be represented in terms of the microphysical properties of constituent particles and thus could possibly be employed profitably in an inverse problem.

2.4.1 Phase functions for small spherical particles

It was shown in section 3.4 that it is possible to parametrize small particle phase functions in terms of $\phi_{ex}(0)$, $\phi_{ex}(\pi/2)$, $\phi_{ex}(\pi)$, and $\langle\mu\rangle$. If one could express these quantities in terms of particle characteristics, the aim of expressing phase functions in terms of microphysical quantities of the scatterer stands achieved. To this end, a large number of spherical particle phase functions for several values of m and x were studied by Roy and Sharma (2008) for dielectric spheres. This led to the following power series expansions for $\phi_{mie}(0)$, $\phi_{mie}(\pi/2)$, $\phi_{mie}(\pi)$, and g:

$$\begin{aligned}\phi_{ex}(0) &\approx 0.1187 + 0.0045mx + \left(0.0356 - 0.0005p + 0.005p^2\right)x^2 \\ &+ \left(0.0032 + 0.008p\right)x^3 + \frac{(3m-3.02)(m+1)}{200m}x^4\,,\end{aligned} \tag{2.89a}$$

$$\phi_{ex}(\pi/2) \approx 0.0597 - 0.000qx^2 + 0.00125qx^3 - 0.0005\left(4m^2+1\right)x^4\,, \tag{2.89b}$$

$$\begin{aligned}\phi_{ex}(\pi) &\approx 0.1209 - 0.007mx - \left(0.038 + 0.0028p^2\right)x^2 \\ &- \frac{0.0112p^3}{m}x^3 - 0.00369mp^2x^4\,,\end{aligned} \tag{2.89c}$$

$$\begin{aligned}\langle\mu\rangle &\approx -0.01235px + \left(0.1621m - 0.0282p + 0.0561p^2\right)x^2 \\ &+ \left(\frac{0.0033}{m} - (0.126 + 0.06m)p\right)x^3 \\ &+ \left(\frac{0.0033}{m^2} + 0.088mp + 0.004p^4\right)\,,\end{aligned} \tag{2.89d}$$

where $p = (m-1)$ and $q = (2m^2-1)$. The above expressions are valid for $x \le 1.25$ and $m \le 2.0$.

Using the above parametrization, the single-particle phase function for a sphere can be expressed as

$$\phi_{sph}(\theta) = A_0 + A_1x + A_2x^2 + A_3x^3 + A_4x^4\,, \tag{2.90}$$

where $A_0, A_1, \ldots, A_4$ are functions of m and θ:

$$A_0 = b_0 + b_1\alpha_1 + b_2\alpha_2 + b_3\alpha_3 + b_4\alpha_4\,, \tag{2.91a}$$

$$A_1 = b_1\beta_1 + b_2\beta_2 + b_3\beta_3 + b_4\beta_4\,, \tag{2.91b}$$

$$A_2 = b_1\gamma_1 + b_2\gamma_2 + b_3\gamma_3 + b_4\gamma_4\,, \tag{2.91c}$$

$$A_3 = b_1\delta_1 + b_2\delta_2 + b_3\delta_3 + b_4\delta_4\,, \tag{2.91d}$$

$$A_4 = b_1\epsilon_1 + b_2\epsilon_2 + b_3\epsilon_3 + b_4\epsilon_4\,, \tag{2.91e}$$

with

$$
\begin{aligned}
b_0 &= \frac{15}{8\pi}\cos^2\theta\sin^2\theta\,, && (2.92a)\\
b_1 &= \frac{1}{4}\cos\theta\big(1+\cos\theta\big)\big(5\cos^2\theta-3\big)\,, && (2.92b)\\
b_2 &= \sin^2\theta\big(1-5\cos^2\theta\big)\,, && (2.92c)\\
b_3 &= -\frac{1}{4}\cos\theta\big(1-\cos\theta\big)\big(5\cos^2\theta-3\big)\,, && (2.92d)\\
b_4 &= b_0/\cos\theta\,, && (2.92e)
\end{aligned}
$$

and

$$
\begin{aligned}
\alpha_1 &= 0.1187,\ \beta_1 = 0.0045m,\ \gamma_1 = 0.005m^2 - 0.0105m + 0.0411,\\
\delta_1 &= 0.008m - 0.0048,\ \epsilon_1 = 0.015m - 0.0001 - 0.0151/m\,, && (2.93a)\\
\alpha_2 &= 0.0597,\ \beta_2 = 0,\ \gamma_2 = -0.0005q,\ \delta_2 = -0.00125q,\\
\epsilon_2 &= -0.0005\big(1+4m^2\big)\,, && (2.93b)\\
\alpha_3 &= 0.12090,\ \beta_3 = -0.007,\ \gamma_3 = -\big(0.0380 + 0.0028p^2\big)\,, && (2.93c)\\
\delta_3 &= -0.0112p^3/m,\ \gamma_3 = -0.00369mp^2, && (2.93d)\\
\alpha_4 &= 0,\ \beta_4 = -0.1235p,\ \gamma_4 = 0.0561m^2 + 0.0217m + 0.0843\,, && (2.93e)\\
\delta_4 &= \frac{0.0033}{m} - (0.126+0.06m)p,\ \epsilon_4\big(0.0033/m^2+0.088mp+0.004p^2\big)\,. && (2.93f)
\end{aligned}
$$

That these expressions are accurate representations of the asymmetry parameter and the phase function at three angular positions is illustrated via Figs 2.6 and 2.7. A comparison of plots of exact forward phase functions against x with those obtained using Eq. (2.89a) is shown in Fig. 2.6 for two values of refractive indices. These are $m = 1.5$ and $m = 2.0$. The corresponding comparison for the asymmetry parameter is depicted in Fig. 2.7. It is clear that the approximations are very accurate. The results from Eqs (2.89b) and (2.89c) are not shown here but have been found to the equally good. As a result, any phase function obtained using these expressions is also a true imitation of the original phase function (Roy and Sharma, 2008).

Explicit expressions for the small particle phase functions have also been obtained directly by expanding the Mie phase function in powers of the size parameter by many authors in the past (Penndorf, 1962; Chu and Churchill, 1960). More recently, Caldas and Semião (2001b) have re-examined the expansion of the Mie phase function for particles satisfying the condition $x \le 1/|m|$. It was shown that the phase function for such particles can be expressed as

$$
\phi_{cspfs}(\mu) = a_0P_0(\mu) + a_1P_1(\mu) + a_2P_2(\mu) + a_3P_3(\mu) + a_4P_4(\mu), \qquad (2.94)
$$

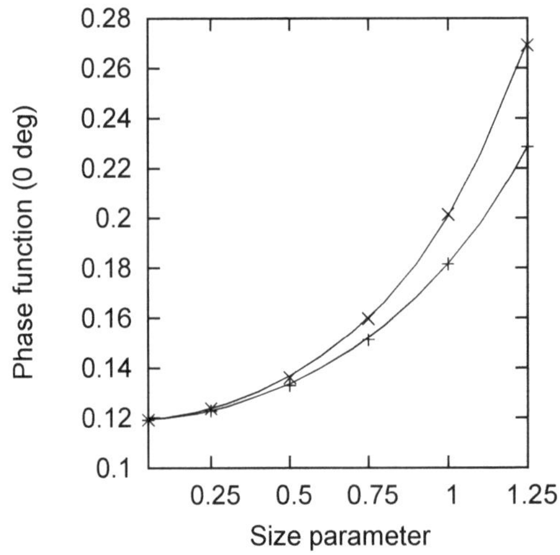

Fig. 2.6. Comparison of forward phase function from Eq. (2.89a) with exact forward phase function as a function of x for $m = 1.5$ and $m = 2.0$. The solid lines are exact results and points are obtained from Eq. (2.89a). Crosses are for $m = 2.0$ and pluses are $m = 1.5$.

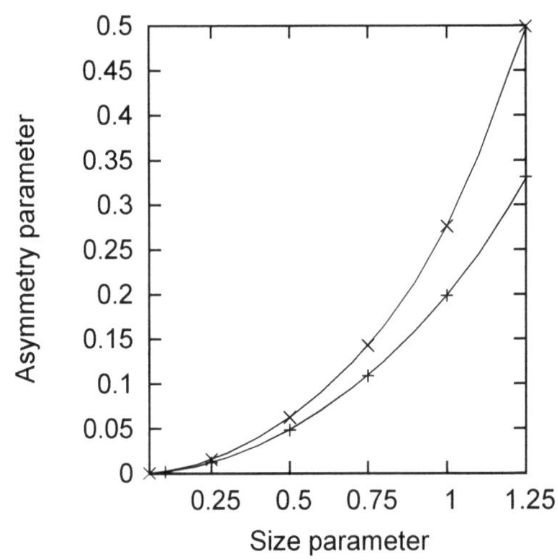

Fig. 2.7. Comparison of asymmetry parameter from Eq. (2.89d) with exact g as a function of x for $m = 1.5$ and $m = 2.0$. The solid lines are exact results and points are obtained from Eq. (2.89a). Crosses are for $m = 2.0$ and pluses are $m = 1.5$.

where

$$a_0 = 1, \tag{2.95a}$$

$$a_1 = \frac{12}{225}\frac{|P|^2x^6}{Q_s}\left[\left(5\mathrm{Re}\,(V') + 3\mathrm{Re}\,(S')\right) + \frac{1}{210}\left(35\mathrm{Re}\,(v'\bar{Q}') + 21\mathrm{Re}\,(S'\bar{q}')\right.\right.$$
$$\left.\left. + 10\mathrm{Re}\,(W') - 6\mathrm{Re}\,(T')\right)x^2 - \frac{2}{3}\left(5Im(V'\bar{P}) + 3Im(S'\bar{P})\right)x^3\right], \tag{2.95b}$$

$$a_2 = \frac{1}{2} + \frac{2}{55125}\frac{|P|^2x^8}{Q_s}\left[350\mathrm{Re}\,(V') + 245\mathrm{Re}\,(S'\bar{v}') + 40\mathrm{Re}\,(U') - 7|S'|^2\right], \tag{2.95c}$$

$$a_3 = \frac{24}{225}\frac{|P|^2x^6}{Q_s}\left[\mathrm{Re}\,(S') + \frac{1}{210}\left(7\mathrm{Re}\,(S'\bar{Q}') - 2\mathrm{Re}\,(T')\right)x^2 - \frac{2}{3}\mathrm{Im}\,(S'\bar{P})x^3\right], \tag{2.95d}$$

$$a_4 = \frac{4}{55125}\frac{|P|^2x^8}{Q_s}\left[28|S'|^2 + 15\mathrm{Re}\,(U')\right], \tag{2.95e}$$

with

$$P = \frac{m^2-1}{m^2+2}, \qquad Q = \frac{m^2-2}{m^2+2}, \tag{2.95f}$$

$$R = \frac{m^6 + 20m^4 - 200m^2 + 200}{(m^2+2)^2}, \qquad S = \frac{m^2-1}{2m^2+3}, \tag{2.95g}$$

$$T = \frac{m^2-1}{(2m^2+3)^2}, \qquad U = \frac{m^2-1}{3m^2+4}, \tag{2.95h}$$

$$V = m^2 - 1, \qquad W = (m^2-1)(2m^2-5)\,, \tag{2.95i}$$

and

$$Q' = Q,\ R' = 18R,\ S' = 5S/P,\ T' = 375T/P,\ U' = 28U/P\,, \tag{2.95j}$$

$$V' = V/P,\ w' = 5W/P\,. \tag{2.95k}$$

This phase function has proved to be very accurate within its range of applicability. The last letter ‘S’ in CSPFS stands for small particles. This also differentiates CSPFS from CSPF, which stands for Cornette and Shanks phase function.

2.4.2 Larger particles

An alternative phase function, valid for larger particles, has also been suggested by Caldas and Semião (2001a). The derivation employs the geometrical optics and the Fraunhofer diffraction to obtain a smooth and mathematically simple function that approximates accurately the particle phase function. This phase function reads as

$$\phi_{cspfl}(\mu) = a + b\frac{2-\mu}{3 - 2\mu + \sqrt{8(1-\mu)}} + c(1+\mu)^2\exp(-d^2(1-\mu^2))\,, \tag{2.96}$$

where

$$a = \frac{1}{2} - \frac{b}{4}(I_1 + 2), \qquad b = \frac{2(2\langle\mu\rangle - 1)}{2 - I_1 + 2I_2}, \tag{2.97a}$$

$$c = \frac{Pf - a - b}{4}, \qquad d^2 = \frac{2c}{2\langle\mu\rangle - I_2 b} - \frac{1}{2}, \tag{2.97b}$$

with

$$I_2 = \frac{1}{3} - \frac{1}{4}\ln(3), \qquad I_1 = \frac{5}{2}\ln(3) - 2\,, \tag{2.97c}$$

and

$$Pf = \frac{1}{4}x^2\frac{Q_e^2}{Q_s}\,. \tag{2.97d}$$

The letter l in subscript stands for large particles. Q_e and Q_s are extinction and scattering efficiencies of the scatterer and $\langle\mu\rangle$ is the asymmetry parameter.

This phase function was computed numerically and compared with the exact phase function. The exact phase functions were obtained by using the Mie theory for two types of particles: (i) fly ash ($m = 1.5 - i0.02$) and (ii) char particles ($m = 2.05 - i1.128$). The agreement between the Mie theory phase function profiles and CSPFL was indeed good when $x \geq 5$ and scatterers were absorbing. For a phase function of polydispersion characterized by mean size parameter $x = 5$, the maximum error was 30% with average error less than 5%. As x increases, the errors decrease. For $x = 100$, the maximum error is 10% and the average error becomes less than 5% for fly ash particles. For char particles, the respective errors are less than 5% and 2%.

2.4.3 Zhao phase function (ZPF)

An approximate scattering phase function which constitutes a good approximation for scattering by a particle of arbitrary shape of volume V and and projected area P at all angles and size parameters has been obtained by Zhao et al. (2006). The phase function is based on two well-known approximations, namely the anomalous diffraction approximation and the Rayleigh–Gans approximation. Theoretically, the validity domain of the Rayleigh–Gans approximation is $|m-1| \ll 1$ and $x|m-1| \leq 1$ and that of anomalous diffraction is $x \gg 1$ and $|m-1| \ll 1$ (see, e.g. van de Hulst, 1957; Sharma and Somerford, 2006). The ZPF may be expressed as

$$\phi_{zpf}(\theta) = e^{-c_2 x_{vp}^3}\phi(\theta)_{small} + \left[1 - e^{c_2 x_{vp}^3}\right]\phi(\theta)_{large}\,, \tag{2.98a}$$

where

$$x_{vp} = \frac{3kV}{4P} \tag{2.98b}$$

is the equivalent size parameter of the scatterer. By extensive trial and error, c_2 has been found to be $c_2 = 0.0128\,\mathrm{Im}\,(m)$. The phase functions $\phi(\theta)_{small}$ for small particles is based on the Rayleigh–Gans approximation:

$$\phi(\theta)_{small} = a_0\left[t|b_1|^2 + (1-t)(|b_2| + \gamma)^2\right](1 + \cos^2\theta)\,, \tag{2.99}$$

with

$$b_2 = \frac{1}{V}\int e^{i(m\mathbf{k}_i - \mathbf{k}_s).\mathbf{r}'}\,d\mathbf{r}'\,, \tag{2.100a}$$

where $\mathbf{k_i}$ and $\mathbf{k_s}$ are incident and scattered wave vectors,

$$t = e^{-c_1 x_{vp}^3}; \quad c_1 = 5 \text{ Re}\left[\frac{(m-1)}{8}\right], \tag{2.100b}$$

and

$$\gamma = \frac{x_{vp}^{9/2}}{(200 + x_{vp}^6)(1 + m^2 - 2m\cos\theta)^{3/4}}. \tag{2.100c}$$

The phase function in the large particle limit, $\phi(\theta)_{large}$, is based on the anomalous diffraction and has the form

$$\phi(\theta)_{large} = \frac{|f(\mathbf{k}_i, \mathbf{k}_s)|^2}{C_{sca}}, \tag{2.101a}$$

where $C_{sca} = \int |f(\mathbf{k}_i, \mathbf{k}_s|^2 d\Omega$ is the scattering cross-section and the scattering amplitude function $f(\mathbf{k}_i, \mathbf{k}_s)$ is expressed as

$$f(\mathbf{k}_i, \mathbf{k}_s) = -\frac{ik}{2\pi}\int e^{i(\mathbf{k}_i - \mathbf{k}_f).\mathbf{r}'} \times \frac{\partial}{\partial z'}\left[e^{ik\int_{-\infty}^{z'}(m-1)dz''}\right] d^3r'. \tag{2.101b}$$

In the above equations, $z'' = \mathbf{r}''.\hat{e}_z$, $\mathbf{r}'' = \mathbf{r} - \mathbf{r}'$, and $\mathbf{r}$ is the position vector at the observation point, $\hat{e}_{z'}$ is the unit vector along z'-axis, and the integration is overall space.

Numerical comparisons of this phase function with the exact theory and the HGPF for spherical particles in the size parameter range $0.5 \leq x \leq 500$ for $m = 1.2$ and $0.9 \leq x \leq 900$ for $m = 1.33 - i0.01$ show good agreement with exact profile at all angles and also constitute substantial improvement over the HGPF.

It may be noted that this phase function contains an integral which needs to be evaluated numerically (Eq. (2.103)). That is, this is not a simple analytic phase function in the strict sense. Analytic evaluation of the integration in Eq. (2.103) is, however, possible under certain conditions. Louedec and Urban (2012) have shown that a simplified form of the above expression,

$$\phi_{SRPF} = \left[x\frac{(1+\cos\theta)}{2}\frac{2J_1(x\sin\theta)}{x\sin\theta}\right]^2, \tag{2.102}$$

can be obtained using Ramsauer analytic solutions (Ramsauer, 1921). The subscript SRPF stands for the simplified Ramsauer phase function. This phase function gives good predictions but only at small angles. An interesting feature that has been noted from this phase function is that the full width at half maximum (FWHM) of the forward peak is related to x as

$$\Delta\theta_{FWHM} = \frac{2}{x}. \tag{2.103}$$

This behaviour was confirmed for various m values ($m = 1.05$, 1.33, and 1.60) by comparing it with the Mie calculations.

2.5 Densely packed particles

Densely packed particles are not independent scatterers, since the incident field is modified by the presence of all other particles. This happens when the interparticle distance becomes less than equal to three diameters (see, e.g. Bayvel and Jones, 1981). Also, the scattering becomes partly coherent. Thus, the concept of a single-scattering phase function requires to be modified for densely packed particles. The statistical model of Percus and Yevick (1958) has proved to be the most efficient to account for the dependent scattering effects. The adjusted phase function is then

$$\phi_{dpp}(m,k,a,\theta) = \phi(m,k,a,\theta)S(\theta)\,, \tag{2.104}$$

where the static structure factor, $S(\theta)$, is

$$S(\theta) = \frac{1}{1 - nC(p)}\,, \tag{2.105}$$

with

$$p = \frac{4\pi \sin(\theta/2)}{\lambda}\,. \tag{2.106}$$

The n is the number density of scattering particles and

$$\begin{aligned} C(p) &= 24\frac{f}{n}\left[\frac{\alpha+\beta+\delta}{u^2}\cos u - \frac{\alpha+2\beta+4\delta}{u^3}\sin u - \frac{2(\beta+6\delta)}{u^4}\right. \\ &\left. + \frac{2\beta}{u^4} + \frac{24\delta}{u^5}\sin\, u + \frac{24\delta}{u^6}(\cos u1)\right], \quad p \neq 0\,. \end{aligned} \tag{2.107a}$$

where

$$u = 2pa\,, \tag{2.107b}$$

$$\alpha = \frac{(1+2f)^2}{(1-f)^4}\,, \tag{2.107c}$$

$$\beta = -6f\frac{(1+f/2)^2}{(1-f)^4}\,, \tag{2.107d}$$

$$\delta = \alpha f/2\,. \tag{2.107e}$$

For forward-scattering ($p = 0$),

$$C(0) = \frac{24f}{n}\left(\frac{\alpha}{3} + \frac{\beta}{4} + \frac{\delta}{6}\right), \tag{2.108}$$

and $f = (4/3)\pi n a^3$ is the filling factor or the fraction of a volume occupied by the scatterers. For sparsely distributed scatterers $n = 0$, the structure factor is identically equal to unity. The results of a study of variation of $S(\theta)$ with θ for various f values show that, for small particles, the correlation effects are important at any scattering angle. For larger particles, however, these become important only for $\theta x \leq 7$ (Kokhanovsky, 2004).

Bascom and Cobbold (1995) have modified Eq. (2.107c) to

$$\alpha = \frac{(1+2(f-1))^{d-1}}{(1-f)^{d+1}}, \tag{2.109}$$

where d is a packing fraction dimension that describes the rate at which the empty space between scatterers diminishes as the number density increases. For spherical particles, $d = 3$. For this case, Eq. (2.109) reduces to Eq. (2.107c). The packing of sheet-like and rod-shaped scatterers is well described by $d = 2$ and $d = 1$, respectively. A biomedical tissue contains all these types of particles and the packing fraction may lie anywhere between 1 and 5 (Schmitt and Kumar, 1998).

2.6 Role of phase function in ray tracing by the Monte Carlo method

The Monte Carlo technique is one of the commonly used approaches to determining the distribution of light energy in a turbid medium (see, e.g. Jacques and Wang, 1995). To follow the path of a ray or a packet of photons, one needs to know the change in direction of the incident photon each time it encounters a scattering center. This change in direction is determined by using the scattering phase function which can be interpreted as a probability density function that defines the distribution of θ over the interval 0° to 180°. The quantity

$$\xi = \int_{-1}^{\mu} p(\mu')\, d\mu' \tag{2.110}$$

gives the probability that the scattering angles lies between $\cos^{-1}\mu$ and π. If one employs the exact phase function, the direction of scattered photons may be determined in the following way (Toublanc, 1996). Create a table of $p(\theta)$ versus scattering angle θ. The $p(\theta)$'s satisfy the normalization condition $\sum_{i=1}^{n} p_i = 1$, n being the number of θ values for which $p(\theta)$ has been tabulated. A random number ξ is generated with a uniform distribution between 0 and 1. It is then compared with the probability distribution to find θ, namely which angle corresponds to that number. Computationally, this may be achieved by implementing the following condition:

$$\sum_{i=1}^{v-1} p_i < \xi \le \sum_{i=1}^{v} p_i\,, \tag{2.111}$$

where the left sum is zero when $v = 0$.

If the HGPF is employed in place of the exact phase function, the integral on the right-hand side of Eq. (2.110) can be evaluated to yield a simple analytic expression for the scattered direction as

$$\cos\theta = \frac{1}{2g}\left[1 + g^2 - \left(\frac{1-g^2}{1-g+2g\xi}\right)^2\right]. \tag{2.112}$$

The GKPF also leads to an analytic expression for $\cos\theta$:

$$\cos\theta = \frac{1}{2g}\left[1 + g^2 - \frac{(1-g^2)^2}{[B\xi + (1-g)^3]^{3/2}}\right] \tag{2.113}$$

where $B = 2g(3+g^2)$. The delta-hyperbolic phase function gives:

$$\cos\theta = 1, \quad 0 \le \xi \le g, \tag{2.114a}$$

$$\cos\theta = 1 - 2\left(\frac{\xi - g}{1-g}\right), \quad g \le \xi \le 1, \tag{2.114b}$$

where g is the asymmetry of the phase function. For the transport phase function

$$\cos\theta = 1, \quad 0 \le \xi \le 1 - 2B, \tag{2.115a}$$

$$\cos\theta = \frac{1-\xi}{B} - 1, \quad 1 - 2B < \xi \le 1. \tag{2.115b}$$

Such an analytic evaluation does not seem possible for the modified Henyey–Greenstein phase function (CSPF) or for a Legendre polynomial expansion.

2.7 Distribution-specific analytic phase functions

In this section, we discuss some of the phase function, that have been designed for a collection of particles with a given size distribution.

2.7.1 Rayleigh–Gans phase function for modified gamma distribution

Kocifaj (2011) derived an analytic phase function in the framework of the Rayleigh–Gans approximation for a size distribution of an aerosol population of the form

$$f(r) = Aa^{\alpha}e^{-\beta a}; \quad \alpha \le 0, \ \beta > 0, \tag{2.116}$$

where a is the equivalent radius of the aerosol particles and the coefficient A is proportional to the volume concentration of aerosol particles. Particle sizes should be small and their refractive index should be close to the refractive index of the surrounding medium for the Rayleigh–Gans approximation to be applicable. Mathematically expressed, the validity domain of this approximation is $|m-1| < 1$; $x|m-1| \le 1$. The modified gamma size distribution is frequently used in atmospheric models because of its convenience in radiative transfer calculations (Shiffrin, 1971).

In the Rayleigh–Gans approximation, the intensity of scattered light by a particle of size a is (see, e.g. van de Hulst, 1957),

$$I_\lambda(\theta, a) = I_{0,\lambda}\left|\frac{3}{4\pi}\left(\frac{m^2-1}{m^2+1}\right)\right|^2 \times \frac{1+\cos^2\theta}{(1-\cos\theta)^2}\left(\frac{\sin q - q\cos q}{q}\right)^2, \tag{2.117}$$

where $q = (4\pi a/\lambda)\sin(\theta/2)$.

Scattering by a size-distributed population can be obtained by integrating the product of $I_\lambda(\theta, a)$ and $f(a)$ over all particle radii:

$$I_\lambda(\theta, \alpha, \beta) = \int_0^\infty I_\lambda(\theta, a) f(a)\, da\,. \tag{2.118}$$

Substituting Eqs (2.116) and (2.118) in Eq. (2.117), the relative scattered intensity can be expressed as

$$I_\lambda(\theta, \alpha, \beta) = \frac{1 + \cos^2\theta}{(1 - \cos\theta)^2}\left[L_\lambda((\alpha{+}1)(\alpha{+}2), \beta, \theta)) + \left(\frac{L_\lambda(0, \beta, \theta)}{L_\lambda(4, \beta, \theta}\right)^{\frac{\alpha+3}{2}} R_\lambda(\alpha, \beta, \theta)\right], \tag{2.119}$$

where

$$\begin{aligned} R_\lambda(\alpha, \beta, \theta) &= (\alpha + 1)(\alpha + 2)\cos\bigl[G_{\lambda,3}(\alpha, \beta, \theta)\bigr] \\ &\quad - 2(\alpha + 1)\sqrt{L_\lambda(4, \alpha, \theta)}\sin\bigl[G_{\lambda,2}(\alpha, \beta, \theta)\bigr] \\ &\quad - L_\lambda(4, \alpha, \theta)\cos\bigl[G_{\lambda,1}(\alpha, \beta, \theta)\bigr]\,, \end{aligned} \tag{2.120}$$

with

$$G_{\lambda,i}(\alpha, \beta, \theta) = (\alpha + i)\tan^{-1}\left(\frac{2}{\sqrt{L_\lambda(0, \beta, \theta)}}\right), \tag{2.121a}$$

$$L_\lambda(\xi, \beta, \theta) = \xi + \left(\frac{\lambda b}{4\pi\sin(\theta/2)}\right)^2. \tag{2.121b}$$

Numerical tests have shown that

$$L_\lambda(\xi, \beta, \theta) = \xi + \left(\frac{2|m| - 1}{m}\frac{\lambda\beta}{4\pi\sin(\theta/2)}\right)^2 \tag{2.122}$$

usually leads to more accurate results for $|m| < 2$ and hence may be used in place of Eq. (2.121b). The phase function can then be expressed as

$$p_{kpf}(\alpha, \beta, \theta, \lambda) = \frac{2I_\lambda(\alpha, \beta, \theta)}{P_\lambda(\alpha, \beta)}, \tag{2.123}$$

where

$$P_\lambda(\alpha, \beta) = \int_0^\pi I_\lambda(\alpha, \beta, \theta)\sin\theta\, d\theta\,, \tag{2.124}$$

which may be expressed as,

$$P_\lambda(\alpha, \beta) = \frac{\pi}{2}(1 + \gamma\alpha)^4\left(\frac{8\pi}{\lambda b}\right)^{4\gamma\left(1 - \frac{a}{180}\right)}, \tag{2.125}$$

by numerical experimentation. The relationship in Eq. (2.125) is accurate to within 5% at the intervals of interest [$\alpha = 1$–15, $\beta = 5$–$40\ \mu m^{-1}$, $\lambda = 0.4$–$0.8\ \mu m$].

The results from this phase function have been contrasted with results from rigorous Mie theory and the results from the HGPF. Non-absorbing sand-like particles with refractive index 1.61 were considered. The size distribution parameters were taken to be $\alpha = 0$ and $\beta = 20\ \mu\text{m}^{-1}$. The wavelength of the radiation was taken to be 400 nm in all computations. It was found that, while the Kocifaj phase function simulates the forward-scattering quite well, it underestimates side and backward scattering.

An analytic expression for the asymmetry parameter obtained is (Kocifaj, 2011)

$$\langle \mu \rangle \cong \frac{\cos^2 G(\alpha, \lambda\beta)}{1 + G^2(\alpha, \lambda\beta)}, \tag{2.126}$$

where $G(\alpha, \lambda\beta)$ is

$$G(\alpha, \lambda\beta) = \frac{10\gamma^2 + \lambda\beta}{8\pi} \frac{8}{1 + 5\alpha/(\gamma^2\pi^2)}, \tag{2.127}$$

with $\gamma = 0.577$ as Euler's constant. The asymmetry of the phase function as computed from Eq. (2.126) is accurate to within 1% for the most typical cases of α ranging from 0 to 15, β ranging from 5 to 40 μm^{-1}, and for the wavelength interval 0.4–0.8 μm^{-1}.

2.7.2 Junge size distribution

The phase functions discussed until now fail to reproduce the shape of the realistic marine phase functions. Fournier and Forand (1994) have derived a two-parameter approximate analytic phase function for an ensemble of particles which characterizes the marine phase functions with a high degree of accuracy. It is based on a Junge (hyperbolic) particle size distribution and assumes that each particle scatters according to the anomalous diffraction approximation. The validity domain of the anomalous diffraction approximation is $|m - 1| \ll 1$ and $x \gg 1$ (see, e.g. van de Hulst, 1957; Sharma and Somerford, 2006). The approximation is expected to be suitable for seawater particles which have a relative refractive index close to unity.

For a Junge size distribution, $N(r) \propto a^{-\alpha}$, where a is the volume-equivalent spherical radius of the particles. Oceanic particle size distributions typically have α values between 3 and 5. Fournier and Forand approximated the scattering phase function of a single particle with the relationship

$$\phi(x, \theta) = \frac{1 + (4x^2/3)}{(1 + (u^2x^2/3)}, \tag{2.128}$$

where $u = 2\sin(\theta/2)$. In its latest form, the marine phase function is given by (Fournier and Jonasz, 1999),

$$\begin{aligned} p_{ffpf}(\psi) = {} & \frac{1}{(1-\delta)^2\delta^\nu} \Big[\nu(1-\delta) - (1-\delta^\nu) + \big[\delta(1-\delta^\nu) \\ & - \nu(1-\delta)\big] \sin^{-2}\Big(\frac{\theta}{2}\Big)\Big] + \frac{1-\delta_{180}^\nu}{4(\delta_{180}-1)\delta_{180}^\nu}(3\cos^2\theta - 1), \end{aligned} \tag{2.129a}$$

where

$$\nu = \frac{3 - \alpha}{2}, \tag{2.129b}$$

and

$$\delta = \frac{4}{3(m-1)^2} \sin^2\left(\frac{\theta}{2}\right). \tag{2.129c}$$

The real refractive index of the particles is m, α is the slope parameter of the hyperbolic distribution, and δ_{180} is δ evaluated at $\theta = 180°$. The above equation can be integrated to obtain the backscatter fraction

$$B = \frac{b_b}{b} = 1 - \frac{1 - \delta_{90}^{\nu+1} - 0.5(1 - \delta_{90}^{\nu})}{(1 - \delta_{90})\delta_{90}^{\nu}}. \tag{2.130}$$

Huang et al. (2012) have assessed the errors of commonly used phase functions for single-component polydisperse seawater systems. The optimal factors corresponding to minimum fitting errors were also calculated. It was found that the HGPF as well as the two-term HGPF agree well with the theoretical ones for small particles, while the Fournier–Forand phase function can be used in the case of suspensions with large suspended particles. The accuracy of the one-term HGPF was found to be the worst.

It is interesting to note that a soft biomedical tissue can also be modeled as a turbid medium with scattering centres in the medium having a similar size distribution, namely the inverse power law. The power α in the size distribution and the relative refractive index of the tissue scatterers (Schmitt and Kumar, 1998; Wang, 2000) also have similar values. This suggests that the p_{ffpf} could be applicable for biomedical tissues as well. To the best of our knowledge, this possibility has not been explored until now.

2.7.3 Phase function for ice clouds

Ice clouds are composed of ice crystals of various sizes and shapes in different proportions. Experimental measurements and parametrization of scattered light intensity for such clouds have been done by Jourdon et al. (2003). Following this work, Kokhanovsky (2008) showed that the phase function for non-absorbing crystals in the cloud can be presented as a sum of two contributions:

$$p_{kpf}(\theta) = \frac{p^d(\theta) + p^g(\theta)}{2}, \tag{2.131}$$

where $p^d(\theta)$ and $p^g(\theta)$ are contributions due to the diffraction of light and geometrical optics component. All phase functions shown above satisfy the usual normalization condition

$$\frac{1}{2}\int_0^{\pi} p(\theta)\sin\theta\, d\theta = 1. \tag{2.132}$$

The geometrical optics component has been parametrized as

$$p^g(\theta) = \nu \exp(-\alpha\theta) + q \exp(-\beta\theta), \tag{2.133}$$

where

$$\nu = \frac{2}{s(\alpha) + bs(\beta)}, \quad q = \frac{2b}{s(\alpha) + bs(\beta)}, \tag{2.134a}$$

with $b \approx 0.023$, $\alpha = 4.3$, $\beta = 0.11$ and

$$s(\xi) = \frac{1 + \exp(-\pi\xi)}{1 + \xi^2} . \tag{2.134b}$$

For the diffraction part, the phase function can be written as

$$p^d(\theta) = \frac{4\langle J_1^2(ka\theta)\rangle}{\theta^2} , \tag{2.135}$$

where

$$\langle J_1^2(ka\theta)\rangle = \frac{\int_0^\infty daa^2 f(a) J_1^2(ka\theta)}{\int_0^\infty daa^2 f(a)} . \tag{2.136}$$

Equation (2.135) assumes that the random collection of irregularly shaped particles have the same angular structure of diffraction peaks as a collection of spherical particles (Kokhanovsky, 2006). This simple parametrization has been shown to yield good agreement with experimental observations for size distribution

$$f(a) = a^6 \exp(-6a/a_0) , \tag{2.137}$$

where a_0 is the mode radius. Calculations give $g = 0.756$, which is close to the measurements of g for ice clouds.

2.8 Concluding remarks

The basic aim of this article has been to make available in one place the information relating to approximate analytic scattering phase functions, developed over the years, for single-particle scattering as well as for a collection of scatterers in various contexts and applications. The need for approximate phase function crops up in various fields where the light-scattering technique is used as a tool either to compute the propagation in a turbid medium or to characterize the scattering medium. Since the technique spans very diverse fields and includes various branches of science, engineering, medicine, agriculture, and industry, a large amount of work has been done over the years; it is quite possible that progress in one field remains hidden to workers in other fields. This may even lead to duplication of work. For example, essentially the same phase functions have been discovered by Cornette and Shanks (1992) and Liu and Weng (2006) independently. Although the particle sizes or the particle size distribution or the optical properties in different problems may vary, the derivation of a phase function in one context can serve as a useful guide in a different context. As new applications of the light-scattering based on light propagation appear, it is hoped that this article will serve as a useful reference and guide for workers in the fields of scattering and radiative transfer.

References

Bai, L., Wu, Z., Tang, S., Li, M., Xie, P. and Wang, S. (2011) 'Study on phase function in Monte Carlo transmission characteristics of poly-disperse aerosol', *Opt. Engg.*, **50**, 16002-1–8.

Bascom, P.A.J. and Cobbold, R.S.C. (1995) 'On the fractal packing approach for understanding ultrasonic backscattering from blood', *J. Acoust. Soc. Am.*, **98**, 3040–9.

Bayvel, L.P. and Jones, A.R. (1981) *Electromagnetic Scattering and Its Applications*, London, Applied Science.

BenZvi, S.Y., Connolly, B.M., Matthews, J.A.J., Prouza, M., Visbal, E.F. and Westerhoff, S. (2007) 'Measurement of aerosol phase function at Pierre Auger Observatory', *Astroparticle Physics*, **28**, 312–20.

Bevilacqua, F. and Depeursinge, C. (1999) 'Monte Carlo study of diffuse reflectance at source detector separations close to one transport mean free path', *J. Opt. Soc. Am. A*, **16**, 2935–45.

Bevilacqua, F., Piguet, D., Marquet, P., Gross, J.D., Tromberg, B.J. and Depeursinge, C. (1999) 'In vivo local determination of tissue optical properties: Applications to human brain', *Appl. Opt.*, **38**, 4939–50.

Bianchi, S., Ferrara, A. and Giovanardi, C. (1996) 'Monte Carlo simulation of dusty spiral-galaxies: Extinction and polarization properties', *ApJ.*, **465**, 127.

Blasi, P., Le Saec, B. and Schlick, C. (1993) 'A rendering algorithm for discrete volume density objects', *Computer Graphics Forum*, **12**, 201–10.

Bohren, C.F. and Huffman, D.R. (1983) *Absorption and Scattering of Light by Small Particles*, New York, Wiley Interscience, New York.

Box, M.A. (1983) 'Power series expansion of the Mie scattering phase function', *Aust. J. Phys.*, **36**, 701–6.

Bucholtz, A. (1995) 'Rayleigh-scattering calculations for the terrestrial atmosphere', *Appl. Opt.*, **34**, 2765–73.

Bugaenko, O.I. (1976) 'Generalized spherical functions in Mie problem', *Izv. Acad. Sci. USSR. Atmos. Oceanic Phys. (English Translation)*, **12**, 366–70.

Caldas, M. and Semião, V. (2001a) 'A new approximate phase function for isolated particles and polydispersions', *J. Quant. Spectrosc. Radiat. Transfer*, **68**, 521–42.

Caldas, M. and Semião, V. (2001b) 'Radiative properties of small particles: Extension of the Penndorf model', *J. Opt. Soc. Am. A*, **18**, 831–8.

Chandrashekhar, S. (1950) *Radiative Transfer*, Oxford, Clarendon Press.

Chicea, D. and Chicea, L.M. (2006) 'On light scattering anisotropy of biological fluids (urine) characterization', *Rom. Journ. Phys.*, **52**, 383–8.

Chu, C.M. and Churchill, S.W. (1955) 'Representation of the angular distribution of radiation scattered by a spherical particles', *J. Opt. Soc. Am.*, **45**, 958–62.

Chu, C.M. and Churchill, S.W. (1960) 'Numerical solution of problems in multiple scattering of electromagnetic radiation', *J. Phys. Chem.*, **59**, 855–63.

Clarke, G.C., Chu, C.M. and Churchill, S.W. (1957) 'Angular distribution coefficients for radiation scattered by a spherical particles', *J. Opt. Soc. Am.*, **47**, 81–4.

Cornette, W.M. and Shanks, J.G. (1992) 'Physically reasonable analytic expression for the single-scattering phase function', *Appl. Opt.*, **31**, 3152–60.

Crosbie, A.L. and Davidson, G.W. (1985) 'Dirac-delta function approximations to the scattering phase function', *J. Quant. Spectrosc. Radiat. Transf.*, **33**, 391–409.

Dave, J.V. (1970) 'Coefficients of the Legendre and Fourier Series for the scattering functions of spherical particles', *Appl. Opt.*, **9**, 1888–96.

Davison, D. (1957) *Neutron Transport Theory*, Oxford, Clarendon Press, pp. 255–84.

de Rooij, W.A. and van der Stap, C.C.A.H. (1984) 'Expansion of Mie scattering matrices in generalized spherical functions', *Astron. Astrophys.*, **131**, 237–48.

Deirmendjian, A. (1969) *Electromagnetic Scattering on Spherical Polydispersions*, New York, Elsevier.

Devaux, C., Siewert, C.E. and Yener, Y. (1979) 'The effect of forward and backward scattering on the scattering albedo for a finite plane parallel atmosphere', *J. Quant. Spectrosc. Radiat. Transf.*, **21**, 505–9.

Ding, S., Xie, Y., Yang, P., Weng, F., Liu, Q., Baum, B. and Hu, Y. (2009) 'Estimates of radiation over cloud and dust aerosols: Optimized number of terms in phase function expansion', *J. Quant. Spectrosc. Radiat. Transf.*, **110**, 1190–8.

Domke, H. (1975) 'Fourier expansion of the phase matrix for Mie scattering', *Z Meteorologie*, **25**, 357–61.

Draine, B.T. (2003) 'Scattering by interstellar dust grains. 1: Optical and ultraviolet', *ApJ.*, **598**, 1017–25.

Edwards, J.M. and Slingo, A. (1996) 'Studies with a flexible new radiation code. I: Choosing a configuration for a large-scale model', *Q. J. R. Meteorol. Soc.*, **122**, 689–719.

Fernandez-Oliveras, A., Rubino, M. and Perez, M.M. (2012) 'Scattering anisotropy measurements in dental tissues and biomaterials', *J. Europ. Opt. Soc. Rap. Public.*, **7**, 12016-1–8.

Fishburne, E.S., Neer, M.E. and Sandri, G. (1976) *Voice Communication via Scattered Ultraviolet Radiation*, Report 274, Vol. 1, Princeton, NJ, Aeronautical Research Associates of Princeton.

Flock, S.T., Wilson, B.C. and Patterson, M.S. (1987) 'Total attenuation coefficient and scattering phase function of tissues and phantom materials at 633 nm', *Med. Phys.*, **14**, 826–34.

Fournier, G. and Forand, J.L. (1994) 'Analytic phase function for ocean water', *Proc. SPIE*, **2258**, 194–201.

Fournier, G. and Jonasz, M. (1999) 'Computer based underwater imaging analysis', *Proc. SPIE*, **3761**, 62–7.

Fowler, B.W. (1983) 'Expansion of Mie-theory phase functions in series of Legendre polynomials', *J. Opt. Soc. Am.*, **73**, 19–22.

Fried, D., Glena, R.D., Featherstone, J.D.B. and Seka, W. (1995) 'Nature of light scattering in dental enamel and dentin at visible and near infrared wavelengths', *Appl. Opt.*, **7**, 1278–85.

Fu, Q., Thorsen, T.J., Su, J., Ge, J.M. and Huang, J.P. (2009) 'Test of Mie-based single-scattering properties of non-spherical dust aerosols in radiative flux calculations', *J. Quant. Spectrosc. Radtiat. Transf.*, **110**, 1640–53.

Gillespie, P. (1992) *An Analytic Phase Function for Cylindrical Particles*, Report ADA255 750 of Atmospheric Science Laboratory, White Sands Missile range.

Haltrin, V.I. (1988) 'Exact solution of the characteristic equation for transfer in the anisotropically scattering and absorbing medium', *Appl. Opt.*, **27**, 599–602.

Haltrin, V.I. (1997) 'Theoretical and empirical phase functions for Monte Carlo calculations of light scattering in sea water', *Proc. 4th Int. Conf. on Remote Sensing for Marine and Coastal Environments: Technology and Applications*, Vol. 1, pp. 509–18.

Haltrin, V.I. (2002) 'One term two parameter phase function Henyey–Greenstein phase function for light scattering in seawater', *Appl. Opt.*, **41**, 1022–8.

Hammer, M., Schweitzer, D., Michel, B., Thamm, E. and Kolb, A. (1998) 'Single scattering by red blood cells', *Appl. Opt.*, **37**, 7410–18.

Hansen, J.E. (1969) 'Exact and approximate solutions for multiple scattering by cloudy and hazy planetary atmospheres', *J. Atmos. Sci.*, **26**, 478–87.

Hapke, B.W. (1981) 'Bidirectional reflectance spectroscopy', *I. Theory. J. Geophys. Res.*, **86**, 3039–54.

Henyey, L.G. and Greenstein, J.L. (1941) 'Diffuse radiation in the Galaxy', *ApJ.*, **93**, 70–83.

Hong, S.S. (1985) 'Henyey–Greenstein representation of the volume scattering phase function for zodiacal dust', *Astron. Astrophys.*, **146**, 67–75.

Hu, Y.-X., Wielicki, B., Lin, B., Gibson, G., Tsay, S.-C., Stamnes, K. and Wong, T. (2000) 'δ-fit: A fast and accurate treatment of particle scattering phase functions with weighted singular-value decomposition least-squares fitting', *J. Quant. Spectrosc. Radiat. Transf.*, **65**, 681–90.

Huang, Y., Liu, G., Jin, W., Ding, K., Li, H., Cao, F. and Wang, X. (2012) 'Comparison study of several underwater light scattering phase functions', *Optoelectronics Letts.*, **8**, 0233–6.

Irvine, W.M. (1965) 'Multiple scattering by large particles', *ApJ.*, **142**, 1563–75.

Iwabuchi, H. and Suziki, T. (2009) 'Fast and accurate radiance calculations using truncation approximation for anisotropic phase functions', *J. Qnat. Spectrosc. Radiat. Transf.*, **110**, 1926–39.

Jacques, S.L. and Wang, L.-H. (1995) 'Monte Carlo modelling of light transport in tissues', in Welch, A.J. and van Gemert, M.J.C. (eds), *Optical Thermal Response of Laser Irradiated Tissue*, New York, Plenum, pp. 73–100.

Jacques, S.L., Alter, C.A. and Prahl, S.A. (1987) 'Angular dependence of He Ne laser light scattering by human dermis', *Laser Life Sci.*, **1**, 309–33.

Jonasz, M. and Fournier, G.R. (2007) *Light Scattering by Particles in Water*, USA, Elsevier.

Joseph, J.H., Wiscombe, W.J. and Weinman, J.S. (1976) 'The Delta–Eddington approximation for radiative flux transfer', *J. Atmos. Sci.*, **33**, 2452–9.

Jourdan, O., Oshchepkov, S., Gayet, J.-F., Sheherbakov, V. and Isaka, H. (2003) 'Statistical analysis of cloud light scattering and microphysical properties obtained from airborne measurements', *J. Geophys. Res.*, **D5**, 4155.

Kagiwada, H. and Kalaba, R. (1967) *Multiple Anisotropic Scattering in Slabs with Axially Symmetric Fields*, Memo. RM-5245-PR, Santa Monica, Rand Corp.

Kamiuto, K. (1987) 'Study of the Henyey–Greenstein approximation to scattering phase function', *J. Quant. Spectrosc. Radiat. Transf.*, **37**, 411–13.

Kanhert, F.M. (2003) 'Numerical methods in electromagnetic scattering theory', *J. Quant. Spectrosc. Radiat. Transf.*, **775**, 79–80.

Kaper, H.G., Shultis, J.K. and Veninga, J.G. (1970) 'Numerical evaluation of the slab albedo problem solution in one-speed anisotropic transport theory', *J. Comp. Phys.*, **6**, 288–313.

Kattawar, G.W. (1975) 'A three-parameter analytic phase function for multiple scattering calculations', *J. Quant. Spectrosc. Radiat. Transf.*, **15**, 839–49.

Kienle, A., Forster, S.K. and Hibst, R. (2001) 'Influence of phase function on determination of the optical properties of biological tissues by spatially resolved reflectance', *Opt. Lett.*, **26**, 1571–3.

Kim, W. and Majumdar, A. (2006) 'Phonon scattering cross section of polydispersed spherical nanoparticles', *J. Appl. Phys.*, **99**, 084306-1–7.

Kocifaj, M. (2011) 'Approximate analytical scattering phase function dependent on microphysical characteristics of dust particles', *Appl. Opt.*, **50**, 2493–9.

Kokhanovsky, A.A. (2004) *Optics of Light Scattering Media, Problems and Solutions*, 3rd edn, Chichester, UK, Springer–Praxis.

Kokhanovsky, A.A. (2006) *Cloud Physics*, Berlin, Springer.

Kokhanovsky, A.A. (2008) 'Phase functions of mixed clouds', *Atmos. Res.*, **89**, 218–21.

Liang, S. and Townshend, J.R.G. (2006) 'A modified Hapke model for soil bidirectional reflectance', *Remote Sens. Environ.*, **55**, 1–10.

Liou, K.N. (2002) *An Introduction to Atmospheric Radiation*, San Diego, Academic Press.

Liu, P. (1994) 'A new phase function approximating to Mie scattering for radiative transport equation', *Phys. Med. Biol.*, **39**, 1025–36.

Liu, Q. and Weng, F. (2006) 'Combined Henyey–Greenstein and Rayleigh phase function', *Appl. Opt.*, **45**, 7475–9.

Louedec, K. and Urban, M. (2012) 'Ramsauer approach for light scattering on nonabsorbing spherical particles and application to the Henyey–Greenstein phase function', *Appl. Opt.*, **51**, 7842–52.

Marchesini, R., Bertoni, A., Andreola, S., Melloni, E. and Sichirollo, A.E. (1989) 'Extinction and absorption coefficients and scattering phase functions of human tissues in vitro', *Appl. Opt.*, **28**, 2318–24.

McCormick, N.J. and Sanchez, R. (1981) 'Inverse problem transport calculations for anisotropic scattering coefficients', *J. Math. Phys.*, **22**, 199–208.

McGuire A.F. and Hapke, B.W. (1995) 'An experimental study of light scattering by large, irregular particles', *Icarus*, **113**, 134–55.

Mengüc, M.P. and Subramanian, S. (1990) 'A step-function approximation for the experimental determination of the effective scattering phase function of particles', *J. Quant. Spectrosc. Radiat. Transf.*, **43**, 253–65.

Mishchenko, M. (1994) 'Asymmetry parameters of the phase function for densely packed scattering grains', *J. Quant. Spectrosc. Radiat. Transf.*, **52**, 95–110.

Mishchenko, M. and Travis, L.D. (1998) 'Capabilities and limitations of a current FORTRAN implementation of the T-matrix method for randomly oriented rotationally symmetric scatterers', *J. Quant. Spectrosc. Radiat. Transf.*, **60**, 309–24.

Mishchenko, M.I. and Macke, A. (1997) 'Asymmetry parameters of the phase function for isolated and densely packed spherical particles with multiple internal inclusions in the geometric optics limit', *J. Quant. Spectrosc. Radiat. Transf.*, **57**, 767–94.

Mishchenko, M.I., Dlugach, J.M., Yanovitskij, E.G. and Zakharova, N.D. (1999) 'Bidirectional reflectance of flat, optically thick particle layers: An efficient radiative transfer solution and applications to snow and soil surfaces', *J. Quant. Spectrosc. Radiat. Transf.*, **63**, 409–32.

Mishchenko, M.I., Travis, L.D. and Lacis, A.A. (2002) *Scattering, Absorption and Emission of Light by Small Particles*, Cambridge, Cambridge University Press.

Mishchenko, M.I., Travis, L.D. and Lacis, A.A. (2006) *Light Scattering of Light by Particles: Radiative Transfer and Coherent Backscattering*, Cambridge, UK, Cambridge University Press.

Mobley, D., Sundman, L.K. and Boss, E. (2002) 'Phase function effects on oceanic light fields', *Appl. Opt.*, **41**, 1035–50.

Nakajima, T. and Asano, S. (1977) 'On the peak truncation approximation of phase function for radiative transfer', *Sci. Rep. Tohoku Univ., Ser. 5, Geophysics*, **24**, 89–102.

Nishita, T., Miyawaki, Y. and Nakamao, E. (1987) 'A shading model for atmospheric scattering considering luminous intensity distribution of light sources', *Computer Graphics*, **21**, 303–10.

Pegoraro, V., Schott, M. and Parker, S.G. (2010) 'A closed-form solution to single scattering for general phase functions and light distributions', *Eurographics Symposium on Rendering*, **29**, 1365–74.

Penndorf, R.B. (1957) 'Tables of refractive index for standard air and the Rayleigh scattering coefficient for the spectral region between 0.2 and 20 μm and their application to atmospheric optics', *J. Opt. Soc. Am.*, **47**, 176–82.

Penndorf, R.B. (1962) 'Scattering and extinction coefficients for small absorbing and nonabsorbing aerosols', *J. Opt. Soc. Am.*, **52**, 896–904.

Percus, J.K. and Yevick, G.J. (1958) 'Analysis of classical statistical mechanics by means of collective coordinates', *Phys. Rev.*, **110**, 1–13.

Petzold, J. (1972) *Volume Scattering Functions for Selected Ocean Waters*, Tech. Rep. SIO 72–78, Scripps Institution of Oceanography, San Diego, CA.

Pharr, M. and Humphreys, G. (2004) *Physically Based Rendering: From Theory to Implementation*, San Francisco, CA, Morgan Kauffman Publications.

Pomraning, G.C. (1988) 'On the Henyey–Greenstein approximation to scattering phase function', *J. Quant. Spectrosc. Radiat. Transf.*, **39**, 109–13.

Potter, J.F. (1970) 'The delta function approximation in radiative transfer theory', *J. Atmos. Sci.*, **27**, 943–9.

Prasher, R.S. (2004) 'Thermal transport cross section and phase function of longitudinal phonons for scattering by nanoparticles and microparticles', *J. Appl. Phys.*, **96**, 5202–11.

Ramsauer, C.W. (1921) 'Uber den Wirkungsquerschnitt der Gasmolekule gegenuber langsamen Electronen', *Ann. Physik*, **64**, 513–40.

Reynolds, L.O. and McCormick, N.J. (1980) 'Approximate two parameter phase function for light scattering', *J. Opt. Soc. Am.*, **70**, 1206–12.

Riewe, F. and Green, A.E.S. (1978) 'Ultraviolet aureole around a source at a finite distance', *Appl. Opt.*, **17**, 1923–9.

Rinzema, K., Hoenders, B.J., Ferwerda, H.A. and Ten Bosch, J.J. (1993) 'Low-degree polynomial phase-functions with high g-value', *Phys. Med. Biol.*, **38**, 1343–50.

Roggan, A., Friebel, M., Dorschel, K., Hahn, A. and Muller, G. (1999) 'Optical properties of circulating human blood in the wavelength range 400 = 2500 nm', *J. Biomed. Opt.*, **4**, 36–46.

Roy, A.K. and Sharma, S.K. (2008) 'A simple parametrized phase function for small non-absorbing spheres', *J. Quant. Spectrosc. Radiat. Transf.*, **109**, 2804–12.

Rozanov, V.V. and Lyapustin, A.I. (2010) 'Similarity of radiative transfer equation: Error analysis of phase function truncation techniques', *J. Quant. Spectrosc. Radiat. Transf.*, **111**, 1964–79.

Schmitt, J.M. and Kumar, G. (1998) 'Optical scattering properties of soft tissue: A discrete particle model', *Appl. Opt.*, **37**, 2788–97.

Sekera, Z. (1952) *Legendre Series of the Scattering Functions for Spherical Particles*, Rep. No. 5, Contr. No. AF19(122)-239, Department of Meteorology, University of California, Los Angeles, CA, ASTIA No. AD-3870.

Sharma, S.K. and Banerjee, S. (2003) 'Role of approximate phase functions in Monte Carlo simulation of light propagation in tissues', *J. Opt. A: Pure Appl. Opt.*, **5**, 294–302.

Sharma, S.K. and Banerjee, S. (2012) 'On the validity of anomalous diffraction approximation in light scattering studies of soft biomedical tissues', *J. Mod. Opt.*, **59**, 701–9.

Sharma, S.K. and Roy, A.K. (2000) 'New approximate phase functions: Test for nonspherical particles', *J. Quant. Spectrosc. Radiat. Transf.*, **64**, 327–37.

Sharma, S.K. and Roy, A.K. (2008) 'On the validity of phase functions used in calculations of electromagnetic wave scattering by interstellar dust', *ApJ. (Supplement Series)*, **177**, 546–50.

Sharma, S.K. and Somerford, D.J. (2006) Light Scattering by Optically Soft Particles: Theory and Applications, Chichester, UK, Springer–Praxis.

Sharma, S.K., Roy, A.K. and Somerford, D.J. (1998) 'New approximate phase functions for scattering of unpolarized light by dielectric particles', *J. Quant. Spectrosc. Radiat. Transf.*, **60**, 1001–10.

Sharma, S.K., Somerford, D.J. and Shah, H.S. (1999) 'A new method to calculate Legendre coefficients for use in multiple scattering problems', *J. Opt. (India)*, **28**, 123–31.

Shiffrin, K.S. (1971) *Theoretical and Applied Problems of Light Scattering*, Minsk, Nauka i Technika, pp. 228–44.

Siegel, R. and Howell, J.R. (1981) *Thermal Radiation Heat Transfer*, 4th edn, Washington, Hemisphere Publishing Corporation.

Siewert, C.E. and Williams, M.M.R. (1977) 'The effect of anisotropic scattering on the critical slab problem in neutron transport theory using a synthetic kernel', *J. Phys. D: Appl. Phys.*, **10**, 2031–40.

Toublanc, D. (1996) 'Henyey–Greenstein and Mie phase functions in Monte Carlo radiative transfer computations', *Appl. Opt.*, **35**, 3270–4.

Turcu, I. (2004) 'Effective phase function for light scattered by disperse systems: The small angle approximation', *J. Opt. A: Pure Appl. Opt.*, **6**, 537–43.

Turcu, I. (2006) 'Effective phase function for light scattered by blood', *Appl. Opt.*, **45**, 639–47.

van de Hulst, H.C. (1957) *Light Scattering by Small Particles*, New York, Wiley.

van der Zee, P., Essenpreis, M. and Delpy, D.T. (1993) 'Optical properties of brain tissue', *Proc. SPIE*, **1888**, 454–65.

Wang, R.K. (2000) 'Modelling optical properties of soft tissues by fractal distribution of scatterers', *J. Mod. Opt.*, **47**, 103–20.

Wiscombe, W.J. (1977) 'The delta-M method: Rapid yet accurate radiative flux calculations for strongly asymmetric phase functions', *J. Atmos. Sci.*, **34**, 1408–22.

Wiscombe, W.J. (1980) 'Improved Mie scattering algorithms', *Appl. Opt.*, **19**, 1505–9.

Wriedt, T. (1998) 'A review of elastic light scattering theories', *Particle and Particle Systems Characterization*, **15**, 67–74.

Yaroslavsky, A.N., Yaroslavsky, I.V., Goldbach, T. and Schwarzmaier, H.-J. (1997) 'Different phase function approximations to determine optical properties of blood: A comparison', *SPIE*, **2982**, 324–30.

Ying, C.F. and Truell, R. (1956) 'Scattering of a plane longitudinal wave by a spherical obstacle in an isotropically elastic solid', *J. Appl. Phys.*, **27**, 1086.

Yoon, G. (1988) 'Absorption and scattering of laser light in biological mediamathematical modeling for determining optical properties', Ph.D. thesis, University of Texas, Austin.

Zhao, J., Shi, G., Che, H. and Cheng, G. (2006) 'Approximations of the scattering phase functions of particles', *Adv. Atmos. Sci.*, **23**, 802–8.

Zhao, Z. and Sun, X. (2010) 'Error analysis of using Henyey–Greenstein in Monte Carlo radiative transfer simulations', *Proceedings of Progress in Electromagnetics Research Symposium*, Xi'an, China, pp. 1449–52.

Zuckerman, N. and Lukes, J.R. (2008) 'Acoustic phonon scattering from particles embedded in an isotropic medium: A molecular dynamics study', *Phys. Rev. B*, **77**, 094302-1–20.

3 Scattering of electromagnetic plane waves in radially inhomogeneous media: ray theory, exact solutions and connections with potential scattering theory

John A. Adam

3.1 Complementary levels of description in light scattering

This chapter addresses several related topics: the existence of direct transmission (or zero-order) bows in radially inhomogeneous spheres for some specific refractive index profiles; a method for obtaining analytic solutions of the radial electromagnetic wave equations (one for each polarization) in both spherical and cylindrical geometry; scalar plane wave scattering by a transparent sphere, and its connection with the scattering matrix of potential scattering theory. This connection is well illustrated in a series of recent papers by Lock (2008a, 2008b, 2008c; see section 3.3). Finally, the vector problem (namely the Mie solution of electromagnetic scattering) is summarized. Two appendices respectively examine details of the ray path integral (section 3.2) and several properties of the S-matrix (section 3.5) in a simplified scattering problem.

Geometrical optics and wave (or physical) optics are two very different but valuable and complementary approaches to describing a wide variety of optical phenomena, the rainbow being probably the most well-known example. However, there is a broad 'middle ground' between these two descriptions: the '*semiclassical*' regime. Thus, there are essentially three domains within which scattering phenomena may be described: the scattering of waves by objects which are (i) small, (ii) comparable with, and (iii) large in size compared to the wavelength of the incident (plane wave) radiation. There may be considerable overlap of region (ii) with the others, depending on the problem of interest, but, basically, the wave-theoretic principles in region (i) tell us why the sky is blue (amongst many other things!). At the other extreme, the 'classical' domain (iii) enables us in particular to be able to describe the basic features of the rainbow in terms of ray optics. The wave-particle duality so fundamental in quantum mechanics is relevant to region (ii) because the more subtle features exhibited by such phenomena involve both these aspects of description and explanation. Indeed, it is useful to relate (somewhat loosely) the regimes (i)–(iii) above to three domains, as stated by Grandy (2000):

(1) *The classical domain*: geometrical optics; particle and particle/ray-like trajectories.
(2) *The wave domain*: physical optics; acoustic and electromagnetic waves; quantum mechanics.

(3) *The semiclassical domain*: 'the vast intermediate region between the above two, containing many interesting physical phenomena.'

Geometrical optics is associated with 'real' rays, but their analytic continuation to complex values of some associated parameters enables the concept of 'complex rays' to be used, often in connection with surface or 'evanescent' rays travelling along a boundary while penetrating the less dense medium in an exponentially damped manner. However, complex rays can also be used to describe the phenomenon of *diffraction*: the penetration of light into regions that are forbidden to the real rays of geometrical optics. In fact, the primary bow light/shadow transition region is associated physically with the confluence of a pair of geometrical rays and their transformation into complex rays; mathematically this corresponds to a pair of real saddle points merging into a complex saddle point. Diffraction into the shadow side of a rainbow occurs by virtue of rays striking the surface tangentially. On the other hand, rays that that just miss grazing the sphere may 'tunnel' into the interior. This phenomenon is well known in quantum mechanics; specifically, tunneling through a classically forbidden potential barrier. It occurs in the 'above-edge' region of semiclassical scattering that permits rays just outside the sphere to interact with it and contribute to the radiation field (Nussenzveig, 1992; Grandy, 2000; Lock, 2012, private communication).

As shown by Nussenzveig in a series of very elegant but technical papers (Nussenzveig, 1969a, 1969b, 1979; see also Nussenzveig, 1965), scattering of scalar waves by a transparent sphere is in many respects isomorphic to the problem of scattering of particles by a spherical potential well. In quantum mechanics, as will be shown in Appendix 2, the bound states of a potential well correspond to poles in the elements of a certain matrix, the *scattering matrix*, on the negative real energy axis, whereas *resonances* of the well correspond to poles that are just below the positive real energy axis of the second Riemann sheet associated with those matrix elements. The closer these poles are to the real axis, the more the resonances behave like very long-lived bound states, or 'almost bound' states of the system. In very simplistic terms, if a particle with a resonance energy is 'shot' at the well from far enough away, it is captured by the well for a considerable time, and acts like a bound particle, but eventually it escapes from the well (this, for example, is a crude description of the mechanism of α-decay from a nucleus, though that is a decay phenomenon, not a scattering one). The reciprocal of the half-width of the resonance is a measure of the lifetime of the resonance particle in the well.

3.2 Scattering by a transparent sphere: ray description

In the following discussion, i refers to the angle of incidence for the incoming ray, r is the radial distance within a sphere of radius a (which may be taken to be unity), and $D(i)$ is the deviation undergone by the ray from its original direction. Below, the subscripts 0 and 1 will be used to distinguish the respective deviations of the exiting ray for the direct transmission (or zero-order) and the primary bow. For $p - 1$ internal reflections in a spherical droplet of *constant* refractive index $n > 1$, straightforward geometrical optics reveals that the deviation from its original direction of a ray incident from infinity upon the sphere at angle of incidence i is, in radians ($i \in [0, \pi/2]$),

$$D_{p-1}(i) = (p-1)\,\pi + 2i - 2p \arcsin\left(\frac{\sin i}{n}\right) . \tag{3.1}$$

In general, an extremum of this angle exists at $i = i_c$, where

$$i_c = \arccos\left[\frac{n^2-1}{p^2-1}\right]^{1/2} , \quad p > 1 . \tag{3.2}$$

Naturally, for real optical phenomena such as rainbows, n is such that i_c exists. A primary bow corresponds to $p = 2$, a secondary bow to $p = 3$, and so forth. That a zero-order (or direct transmission bow) corresponding to $p = 1$ cannot exist for constant n is readily shown from Eq. (3.1). Nevertheless, it has been established that such relative extrema (for zero and higher-order bows) can exist for radially inhomogeneous spheres (for more details, see Adam and Laven, 2007; Adam, 2011). In fact, multiple zero-order and primary bows may exist depending on the refractive index profile. A well-known result is that the curvature of the ray path is towards regions of higher refractive index n. This is a consequence of Snel's law of refraction generalized to continuously varying media. Thus, within the sphere, if $n'(r) < 0$, an incoming ray bends towards the origin; if $n'(r) > 0$, it bends away from it. From Fig. 3.1, it can be seen that, for direct transmission in the former case,

$$i + 2\Theta(i) + (i - |D_0(i)|) = \pi \Rightarrow |D_0(i)| = 2i - \pi + 2\Theta(i) . \tag{3.3}$$

In this equation, $2\Theta(i)$ is the angle through which the radius vector turns from the point at which the ray enters the sphere to its point of exit. It is readily noted that, for one internal reflection (corresponding to a primary bow),

$$|D_1(i)| = 2i - \pi + 4\Theta(i) . \tag{3.4}$$

In what follows, the absolute value notation will be dropped. The deviation formulae can be extended to higher-order bows in an obvious fashion. The quantity $\Theta(i)$

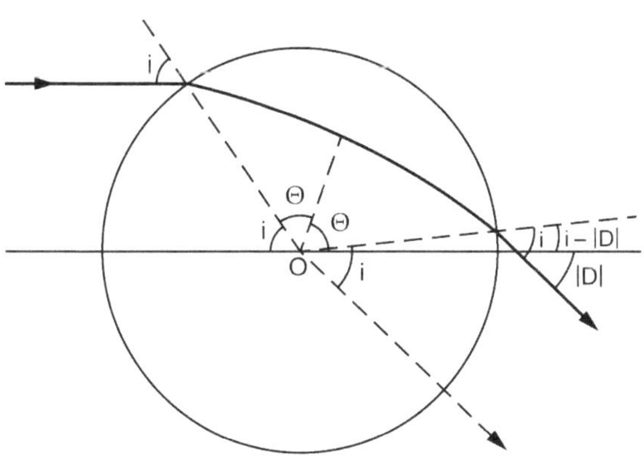

Fig. 3.1. The ray path for direct transmission through a radially inhomogeneous sphere sketched for $n'(r) < 0$.

is an improper definite integral to be defined in section 3.2.1. Analytic expressions for $\Theta(i)$ are difficult to obtain except for a few specific $n(r)$ profiles; several examples are indicated below. For a constant refractive index, $\Theta(i)$ is a standard integral resulting in the inverse secant function, and can be readily evaluated. Specifically,

$$D_0(i) = 2i - 2\tilde{r}(i) \quad \text{and} \tag{3.5a}$$
$$D_1(i) = 2i + \pi - 4\tilde{r}(i) \tag{3.5b}$$

where $\tilde{r}(i)$ is the angle of refraction inside the sphere. Of course, these results are readily determined from elementary geometry and are the $p = 1$ and $p = 2$ cases referred to earlier. As already noted, there can be no 'zero-order rainbow' for the direct transmission of sunlight in uniform spheres, only primary and secondary bows (ignoring theoretically possible but practically almost unobservable higher-order bows).

In Fig. 3.2, the dashed curve D_h represents the deviation $D_1(i)$ through a homogeneous sphere of *constant* refractive index $n = 4/3$. The other graphs represent the deviations corresponding to a zero bow and a primary bow for the particular (but arbitrary) choice of refractive index

$$n_1(r) = 1.3 - 0.2\cos\left\{[1.9\,(r - 0.85)]^2\right\}. \tag{3.6}$$

Note that both $D_0(i)$ and $D_1(i)$ exhibit fairly broad double extrema in this case. It is interesting to note that the relative maximum for D_1 is much less pronounced than that for D_0. Further discussion of such extrema can be found in Adam (2011). Note that, mathematically at least, a primary 'rainbow' is, amongst other things (Adam, 2002a, 2002b): (1) a concentration of light rays corresponding to an extremum of the deviation or scattering angle (this extremum is identified as the Descartes' or rainbow ray); (2) a caustic, separating a two-ray region from a zero-ray (or shadow) region; (3) an integral superposition of waves over a (locally) cubic wave-front (the *Airy approximation*); (4) a coalescence of two real saddle points;

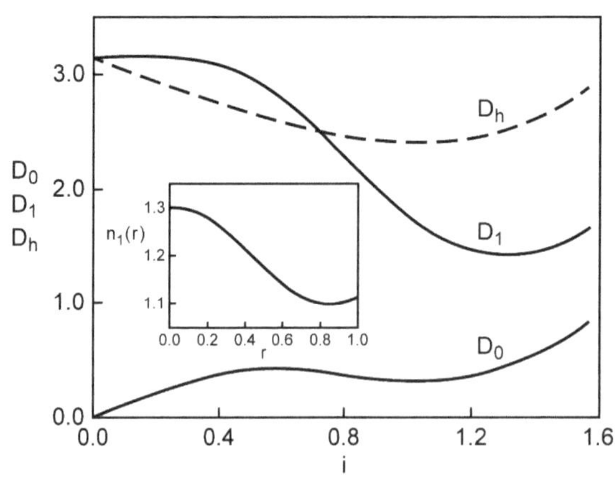

Fig. 3.2. Deviation functions for both a homogeneous (D_h) and inhomogeneous spheres (D_0 and D_1) for the profile $n_1(r)$ (inset).

(5) a result of scattering by a square-well potential; (6) an example of 'Regge-pole dominance', and (7) a *fold diffraction catastrophe.*

3.2.1 The ray path integral

In a spherically symmetric medium with refractive index $n(r)$, each ray path satisfies the following equation (Born and Wolf, 1999):

$$rn(r)\sin\varphi = \text{constant}\,, \tag{3.7}$$

where φ is the angle between the radius vector $\bar{r}$ and the tangent to the ray at that point. This expression may be thought of as the optical analog of the conservation of angular momentum for a particle moving under the action of a central force. The result, known as *Bouguer's formula* (for Pierre Bouguer, 1698–1758), implies that all the ray paths $r(\theta)$ are curves lying in planes through the origin (θ is the polar angle). Elementary differential geometry establishes that

$$\sin\varphi = \frac{r(\theta)}{\sqrt{r^2(\theta) + (dr/d\theta)^2}}\,. \tag{3.8}$$

From this the angular deviation of a ray $\Theta(i)$ within the sphere can be determined and subsequently the total angle of deviation $D(i)$ through which an incoming ray at angle of incidence i is rotated. From this, the formula for $\Theta(i)$ is found to be

$$\Theta(i) = \sin i \int_{r_c(i)}^{1} \frac{dr}{r\sqrt{r^2 n^2(r) - \sin^2 i}}\,. \tag{3.9}$$

The lower limit $r_c(i)$ is the point at which the integrand is singular and is therefore the solution of Eq. (3.10) below in which (for a *unit* sphere) $\sin i$ is the *impact parameter.* The quantity $r_c(i)$ is the radial point of closest approach to the center of the sphere, sometimes called the *turning point.* The value of is determined implicitly from the following expression:

$$\eta\big(r_c(i)\big) \equiv r_c(i) n\big(r_c(i)\big) = \sin i\,. \tag{3.10}$$

The nature of $\eta(r) = rn(r)$ will be very significant in what follows; in particular, $r_c(i)$ will be unique if $\eta(r)$ is a monotonic function. The integral in Eq. (3.9) can be evaluated analytically in certain special cases. Consider first the (somewhat unphysical and singular) power law profile $n(r) = n(R)(r/R)^m$ where m can be of either sign (Brockman and Alexopoulos, 1977; see also Alexopoulos, 1971a, 1971b, 1972, 1974; Uslenghi, 1964, 1969a, 1979b; Uslenghi and Weston, 1970). By a judicious change of variable, this can be reduced to the standard result for a constant refractive index. For the choice of a 'shifted hyperbolic' profile of the form $n(r) = (ar + b)^{-1}$, the integral in Eq. (3.9) can be evaluated in terms of elementary transcendental functions (Adam and Laven, 2007). The complexity of these integrals increases rapidly with even relatively simple expressions for $n(r)$. In the case of a linear profile, Eq. (3.3) can be evaluated in terms of incomplete elliptic integrals of the first and third kinds (Vetrano et al., 2005). A parabolic profile of the form $n(r) = a - br^2$ also yields a result in terms of a purely imaginary elliptic integral of the third kind (Vetrano et al., 2005). Several specific profiles were also studied by Gould and Burman (1964, and references therein) and Burman (1965).

Whether the ray path integral is evaluated analytically or numerically, it contributes to the *direct problem* of geometrical optics, namely (for direct transmission) the total angular deviation $2\Theta(i)$ of the ray inside the sphere for a given profile $n(r)$. Coupled with the refraction at the (in general discontinuous) boundary entrance and exit points, this naturally yields the total deviation of an incoming ray as a function of its angle of incidence. The corresponding *inverse problem* is to determine the profile $n(r)$ from knowledge of the quantity $\Theta(i)$. This is generally more difficult to accomplish. Another reason for pursuing the inverse problem is that it would be valuable to find at least some sufficient conditions under which inhomogeneous spheres can exhibit bows of any order, but especially of zero order (particularly with regard to industrial techniques such as rainbow refractometry, for example; see references in Adam (2011)). By choosing a generic profile for $D_0(i)$ or $D_1(i)$, it should be possible in principle to examine the implications on $n(r)$ for such profiles. From a strict mathematical point of view, inverse problems in general are notorious for their lack of solution uniqueness. In practical terms, it is not significant in this context, and we shall address the topic no further here.

3.3 Analysis of specific profiles

We now examine two specific (and possibly singular) refractive index profiles for the *unit* sphere, generalizing somewhat that considered in Uslenghi (1969a; see also Uslenghi and Weston, 1970). Before so doing, we introduce some new notation. Electromagnetic waves possess two different polarizations: the transverse electric (*TE*) and transverse magnetic (*TM*) modes. Spherical *TE* modes have a magnetic field component in the direction of propagation, in this case that is in the radial direction, and spherical *TM* modes have an electric field in the radial direction.

The first profile to be considered is

$$n(r) = n_1 r^{1/b-1}\left(2 - r^{2/b}\right)^{1/2}, \quad n_1 = n(1) > 1\,. \tag{3.11}$$

Note that, if $b = 1$ and $n_1 = 1$, this profile corresponds to the classic Luneberg lens (Luneberg, 1964). Using the result in Eq. (3.5a) $D_0(i) = 2i - \pi + 2\Theta$, and substituting for $n(r)$ in the Θ-integral, after some algebra, the deviation angle can be shown to be

$$D_0(i) = \pi(b-1) + 2i - b \arcsin\left(\frac{\sin i}{n_1}\right). \tag{3.12}$$

For a zero-order bow to exist for some critical angle of incidence $i_c \in [0, \pi/2]$, it is necessary and sufficient that $D_0'(i_c) = 0$. This is the case if

$$\cos i_c = 2\left(\frac{n_1^2 - 1}{b^2 - 4}\right)^{1/2}, \tag{3.13}$$

which implies that $b \geq 2n_1$ if we restrict ourselves to the least potentially singular case of $b > 0$. We have therefore established that a zero bow can exist, unless

$n_1 = 1$, whence Eq. (3.12) is a linear function of incidence angle i. It is interesting to note that the *TE* wave equation has an exact solution for this choice of profile, finite for $0 \leq r \leq 1$, namely

$$S_l(r) = r^{l+1} \exp\left(-\frac{bkr^{2/b}}{2}\right) \times {}_1F_1\left(\frac{1}{2} + \frac{b}{2}\left(l + \frac{1}{2} - k\right); 1 + b\left(l + \frac{1}{2}\right); bkr^{2/b}\right). \tag{3.14}$$

Here, ${}_1F_1$ refers to the confluent hypergeometric function. The *TM* equation cannot be expressed in terms of well-known functions, though it can be written in terms of generalized hypergeometric functions and solved by power series expansions in special cases. In a recent series of papers, Lock (2008a, 2008b, 2008c) analyzed the scattering of plane electromagnetic waves by a modified *Luneberg lens*. This 'lens' is a dielectric sphere of radius a with a radially varying refractive index, specifically

$$n(r) = \frac{1}{f}\left[1 + f^2 - \left(\frac{r}{a}\right)^2\right]^{1/2}. \tag{3.15}$$

Here, f is a parameter determining the focal length of the lens. If $0 < f < 1$, the focus is inside the sphere (i.e. the focal length $< a$); for $f = 1$, it is on the surface and, for $f > 1$, the focal point is outside the sphere. Note that, in contrast to the refractive index profiles in Eqs (3.11) and (3.16) below, for the profile in Eq. (3.15), $n(a) = 1$. Lock also found the existence of a transmission bow for this profile; indeed, this will occur for $f > 1$, whereas, for $f = 1$, this bow evolves into an orbiting ray and, if $0 < f < 1$, this ray in turn evolves into a family of morphology-dependent resonances. In a wave-theoretic approach to this problem, Lock (2008b) studied the related radial 'Schrödinger' equation for the *TE* mode using the effective potential approach, discussed in section 3.4 below.

When a family of rays has a near-grazing incidence on a dielectric sphere, the so-called 'far zone' consists of (i) an illuminated region containing rays refracted into the sphere and making $p - 1$ internal reflections (where $p \geq 1$) before exiting the sphere, and (ii) a shadow zone into which no rays enter. Lock (1988) also showed that the asymptotic form of the Airy theory bow far into the illuminated region becomes the interference pattern of two supernumerary rays (with slightly different optical path lengths through the sphere).

The other choice for refractive index profile discussed here is

$$n(r) = \frac{2n_1 r^{1/c-1}}{1 + r^{2/c}}, \quad n_1 = n(1). \tag{3.16}$$

Detailed algebraic manipulation indicates that, in this case,

$$D_0(i) = \pi(c - 1) + 2i. \tag{3.17}$$

Obviously, $D_0'(i) \neq 0$ for any value of i—that is, there is no zero-order bow for this profile! Both *TE* and *TM* modes have finite solutions for $0 \leq r \leq 1$, expressible in terms of the hypergeometric functions ${}_2F_1$, but we do not state them here. For the special case of $c = 1$ and $n_1 = 1$, this profile corresponds to the classic *Maxwell fish-eye* lens (Leonhardt and Philbin, 2010). The reader is also referred to the related analyses of Tai (1956, 1958, 1963). For a discussion of what are now called Luneberg lenses, see Luneberg (1964) and for their generalization, see Uslenghi (1969b). Other singular profiles are discussed by Alexopoulos (1972, 1974).

3.4 The generation of exact solutions for radially inhomogeneous media

For the theory of electromagnetic scattering, in particular for radially symmetric media, the electric field vector E must satisfy the scattering boundary conditions and the vector wave equation, where $k = \omega/c$ is the wavenumber in free space

$$\nabla \times \nabla \times E - k^2 n^2(r) E = \mathbf{0} \, . \tag{3.18}$$

We concentrate on spherical symmetry initially. By expanding$E$in terms of vector spherical harmonics, the following radial equations are obtained for the TE respectively TM (Johnson, 1993),

$$\frac{d^2 S_l(r)}{dr^2} + \left[k^2 n^2(r) - \frac{l(l+1)}{r^2}\right] S_l(r) = 0 \, ; \tag{3.19}$$

$$\frac{d^2 T_l(r)}{dr^2} - \frac{2n'(r)}{n(r)} \frac{dT_l(r)}{dr} + \left[k^2 n^2(r) - \frac{l(l+1)}{r^2}\right] T_l(r) = 0 \, . \tag{3.20}$$

If we make the substitution $T_l(r) = n(r) u_l(r)$ in the latter equation, it reduces to the form

$$\frac{d^2 u_l(r)}{dr^2} + \left[k^2 n^2(r) - n(r) \frac{d^2}{dr^2} \left(\frac{1}{n(r)}\right) - \frac{l(l+1)}{r^2}\right] u_l(r) = 0 \, . \tag{3.21}$$

Each of these equations can be reworked into a time-independent Schrödinger equation form (Schiff, 1968), with $\psi(r)$ now being a generic dependent variable for the two modal equations above. Thus

$$\frac{d^2 \psi(r)}{dr^2} + \left[k^2 - V(r) - \frac{l(l+1)}{r^2}\right] \psi(r) = 0 \tag{3.22}$$

or equivalently, as indicated earlier,

$$\frac{d^2 \psi(r)}{dr^2} + \left[k^2 - V(r) - \frac{\lambda^2 - 1/4}{r^2}\right] \psi(r) = 0 \, , \tag{3.23}$$

where $k^2 = E$ is the energy of the 'particle', and $\lambda = l + 1/2$. The 'scattering potential' is now

$$V(r) = k^2 \left[1 - n^2(r)\right] \tag{3.24}$$

for the TE mode, and (by eliminating the first derivative term in Eq. (3.20)):

$$V(r) = k^2 \left[1 - n^2(r) + k^{-2} n(r) \frac{d^2}{dr^2} \left(n(r)\right)^{-1}\right] \tag{3.25}$$

for the TM mode. Thus, for scattering by a dielectric sphere, the corresponding potential has finite range. Note that, for constant refractive index, these two equations are identical in form.

3.4.1 A summary of the method

In Eq. (3.21), let $U(r) = n^2(r)$; then that equation takes the form

$$\frac{d^2u_l(r)}{dr^2} + \left[k^2U + \frac{U''}{2U} - \frac{3}{4}\left(\frac{U'}{U}\right)^2 - \frac{l\,(l+1)}{r^2}\right] u_l(r) \equiv \tag{3.26a}$$

$$\frac{d^2u_l(r)}{dr^2} + \left[k^2U + s(U) - \frac{l\,(l+1)}{r^2}\right] u_l(r) = 0\,. \tag{3.26b}$$

Given any known solution, $w(t)$ expressed in terms of the standard functions (e.g. Bessel functions, Whittaker functions, hypergeometric functions) of the differential equation

$$\frac{d^2w}{dt^2} + p(t)\frac{dw}{dt} + q(t)w = 0\,,$$

we suppose that

$$w(t) = u(t)\,\exp\left(\int g(t)\,dt\right), \quad t = t(r)\,. \tag{3.27}$$

On eliminating the coefficient of du/dr, we find that

$$\frac{d^2u}{dr^2} + \left[s\big(t'(r)\big) + \big(t'(r)\big)^2\left(q - \frac{1}{4}p^2 - \frac{1}{2}\frac{dp}{dt}\right)\right] u = 0\,, \tag{3.28a}$$

where the prime refers to a derivative with respect to r and

$$s\left(t'(r)\right) = \frac{1}{2}\frac{d^3u}{dr^3}\left(\frac{du}{dr}\right)^{-1} - \frac{3}{4}\frac{d^2u}{dr^2}\left(\frac{du}{dr}\right)^{-2}\,. \tag{3.28b}$$

The last expression is known as the Schwarzian derivative of $t'(r)$. The differential Eq. (3.28a) possesses a solution of the form

$$u\left(t(r)\right) = w\left(t(r)\right)\left(t'(r)\right)^{-1/2}\exp\left[\frac{1}{2}\int pt'(r)\,dr\right]. \tag{3.29}$$

If, for given functions $p(t)$ and $q(t)$, we find both a transformation $t = t(r)$ and a function $U(r) = n^2(r)$ that is independent of the centrifugal parameter l, such that

$$k^2U(r) + s(U) - \frac{l(l+1)}{r^2} = s\big(t'(r)\big) + \big(t'(r)\big)^2\left(q - \frac{1}{4}p^2 - \frac{1}{2}\frac{dp}{dt}\right), \tag{3.30}$$

then the equation

$$\frac{d^2u}{dr^2} + \left[k^2U(r) + s(U) - \frac{l\,(l+1)}{r^2}\right] u = 0\,,$$

and consequently Eq. (3.21), possesses the solution in Eq. (3.29). Note that, in the notation of Sharaf (1969), $s(U)$ is written as $D(U, r)$. The hypergeometric equation

$$t\,(1-t)\frac{d^2w}{dt^2} + \left[\gamma - (\alpha+\beta+1)t\right]\frac{dw}{dt} - \alpha\beta w = 0\,, \tag{3.31}$$

possesses the linearly independent solutions (provided $\gamma \neq 1$)

$$w_1 = {}_2F_1(\alpha,\beta,\gamma;t) \quad \text{and} \quad w_2 = t^{1-\gamma}{}_2F_1(A-\beta, A-\alpha, 2-\gamma; t)\,, \tag{3.32a}$$

$$\text{where} \quad A = \alpha+\beta-\gamma+1\,, \tag{3.32b}$$

so the coefficients p and q are therefore given by

$$p = \frac{\gamma}{t} - \frac{A}{1-t};\; q = -\frac{\alpha\beta}{t(1-t)}\,, \quad \text{and also}$$

$$q - \frac{1}{4}p^2 - \frac{1}{2}\frac{dp}{dt} = \frac{\gamma(1-\gamma/2)}{2t^2} + \frac{A(1-A/2)}{2(1-t)^2} + \frac{\gamma A/2 - \alpha\beta}{t(1-t)}\,. \tag{3.33}$$

For Whittaker's equation (a self-adjoint version of the confluent hypergeometric equation)

$$\frac{d^2W}{dt^2} + \left(-\frac{1}{4} + \frac{\lambda}{t} + \frac{1/4-\mu^2}{t^2}\right)W = 0\,, \quad \text{it is clear that}$$

$$p = 0;\; q = -\frac{1}{4} + \frac{\lambda}{t} + \frac{1/4-\mu^2}{t^2}, \quad \text{and hence} \quad q - \frac{1}{4}p^2 - \frac{1}{2}\frac{dp}{dt} = q\,. \tag{3.34}$$

$$\text{For Bessel's equation,} \quad t^2\frac{d^2w}{dt^2} + t\frac{dw}{dt} + \left(t^2-\nu^2\right)w = 0\,,$$

$$p = \frac{1}{t},\; q = \frac{1-\nu^2}{t^2};\; q - \frac{1}{4}p^2 - \frac{1}{2}\frac{dp}{dt} = 1 + \frac{1/4-\nu^2}{t^2}\,. \tag{3.35}$$

For any one of the three standard equations, the expression $q(t)-(p^2(t)/4)-p'(t)/2$ can be written in the general form $Lf(t)+Mg(t)+Nh(t)$, where $f(t)$, $g(t)$, and $h(t)$ are independent functions of t, and L, M, and N are algebraic functions of the constant parameters of the equations, and are therefore also independent of t. Using this new expression in Eq. (3.30), we obtain

$$k^2u(r) + s(u) - \frac{l(l+1)}{r^2} = s\big(t'(r)\big) + \big(t'(r)\big)^2(Lf+Mg+Nh)\,, \tag{3.36}$$

This equation contains five independent functions of t: $s(t')$, $(t')^2f(t)$, $(t')^2g(t)$, $(t')^2h(t)$, and r^{-2}. These may be reduced to match the number of 'disposable' parameters (such as a, b, c in the hypergeometric equation) by introducing the as yet undetermined multipliers λ, μ, ν, P, Q, and R such that

$$s(t') = -(t')^2(\lambda f + \mu g + \nu h) \tag{3.37a}$$

and

$$r^{-2} = (t')^2(Pf+Qg+Rh) \equiv (t')^2F(t)\,. \tag{3.37b}$$

Then, from Eqs (3.37a) and (3.37b) (see Westcott, 1968), recalling that $U(r) = n^2(r)$, we obtain the expression

$$n^2(r) = (t')^2\left[K_1^2f(t) + K_2^2g(t) + K_3^2h(t)\right]\,, \tag{3.38}$$

where $K_1^2 = k^{-2}\left[L - \lambda + Pl(l+1)\right]$, $K_2^2 = k^{-2}\left[M - \mu + Ql(l+1)\right]$, and $K_3^2 = k^{-2}\left[N - \nu + Rl(l+1)\right]$ are arbitrary constants. These expressions give L, M, and N (and hence a, b, and c) in terms of the set $\{\lambda,\ \mu,\ \nu,\ P,\ Q,\ R,\ K_1^2,\ K_2^2,\ K_3^2,\ l,\ k^2\}$. Eventually, it is found that the refractive index profile $n(r)$ is independent of k and l, though the solution will not be. Further details can be found in the papers by Heading (1965), Sharaf (1969a, 1969b, 1970), and Westcott (1968a, 1969). Westcott also published a short summary of refractive index profiles and *TE*/*TM* solutions corresponding to them in terms of the above three differential equations for both spherically and cylindrically stratified media (1968b, 1968c). The core of the Heading–Sharaf–Westcott 'algorithm' presented is to find $t(r)$ satisfying Eq. (3.37) simultaneously for the respective functions $f(t)$, $g(t)$, and $h(t)$ corresponding to the Bessel, Whittaker, and hypergeometric equations above. Below, we examine a subset of the much wider class of both profiles and solutions that can be generated by this algorithm.

3.4.2 Specific profiles

3.4.2.1 Spherically stratified isotropic media

In what follows, $k(r) = kn(r) = \omega n(r)/c_0$, where the free space wavenumber k is expressed in terms of the angular wave frequency ω and the speed of light in vacuo, c_0. The governing ordinary differential equations for the radial variation of the Hertz vector are given by $u(r)/r$, where, in Westcott's (1968b, 1968c) notation,

$$\frac{d^2u}{dr^2} + \left[k_{\mathit{eff}}^2(r) - \frac{l(l+1)}{r^2}\right]u = 0\,. \tag{3.39}$$

For fields of the 'magnetic' type, $k_{\mathit{eff}}^2 = k^2(r)$, and, for fields of the 'electric' type,

$$k_{\mathit{eff}}^2 = k^2(r) - \frac{d^2}{dr^2}\left\{\frac{1}{k(r)}\right\}. \tag{3.40}$$

(a) **Profiles based on Bessel's equation:** $k(r) = ar^b$. A solution for fields of the 'electric type' (*TM* mode) is given by

$$u(r) \propto r^{1/2} Z_v\left(\frac{a}{1+b}r^{b+1}\right), \quad v^2 = \frac{b}{1+b} + \left\{\frac{2l+1}{2(1+b)}\right\}^2, \tag{3.41}$$

where Z_v is any solution of Bessel's equation of order ν. For fields of the 'magnetic type' (*TE* mode), the order of the Bessel function is $(2l+1)/2(b+1)$.

(b) **Profiles based on Whittaker's equation:** (i) $k(r) = \dfrac{a}{r\ln\beta r}$.

For the *TM* mode,

$$u(r) \propto r^{1/2} W_{\pm p,m}\left\{\pm(2l+1)\ln\beta r\right\},\ p = -(2l+1)^{-1},\ m = \left(\frac{1}{4} - a^2\right)^{1/2}. \tag{3.42}$$

For the *TE* mode, the corresponding solutions are

$$u(r) \propto (r \ln \beta r)^{1/2} Z_\nu \left\{ \pm \left(l + \frac{1}{2} \right) \ln \beta r \right\}, \quad \nu = \left(\frac{1}{4} - a^2 \right)^{1/2}. \tag{3.43}$$

(ii) $k(r) = \dfrac{a}{r (\ln \beta r)^{1/2}}$. For the *TM* mode

$$u(r) \propto r^{1/2} W_{\pm p,0} \left\{ \pm (2l+1) \ln \beta r \right\}, \quad p = \left(a^2 - \frac{1}{2} \right) (2l+1)^{-1}. \tag{3.44}$$

For the *TE* mode, the corresponding solutions are

$$u(r) \propto r^{1/2} W_{\pm p,1/2} \left\{ \pm (2l+1) \ln \beta r \right\}, \quad p = a^2 (2l+1)^{-1}. \tag{3.45}$$

Both profiles are singular at the origin; for real r, the singularity at $\beta r = 1$ is removed if β is a complex number.

(c) **Profiles based on the hypergeometric equation:** (i) $k(r) = \dfrac{a_0}{r (1 + \beta r^\alpha)}$. The special case of $\alpha = 1$ has been noted above.

(ii) $k(r) = \dfrac{a_0 r^{(\alpha/2)-1}}{1 + \beta r^\alpha}$. For $\alpha = 2$ the Maxwell fish-eye lens profile is recovered. The singularity at the origin is a removable one if $|\alpha| \geq 2$.

(iii) $k(r) = \dfrac{a_0}{r(1 + \beta r^\alpha)^{1/2}}$. In this instance, the singularity at the origin is a removable one if $\alpha \leq -2$.

Each of these three profiles has independent solutions of the form

$$u(r) \propto r^{[1+(c-1)\alpha]/2} (1 + \beta r^\alpha)^{(a-b+1)/2}_{2} F_1 (a, c-b; c; -\beta r^\alpha) \quad \text{and} \tag{3.46a}$$

$$u(r) \propto r^{[1-(c-1)\alpha]/2} (1 + \beta r^\alpha)^{(a-b+1)/2}_{2} F_1 (1+a-c, 1-b; 2-c; -\beta r^\alpha). \tag{3.46b}$$

Naturally, the constants a, b, and c assume different forms for each profile. Westcott (1968b, 1968c) introduces the parameters noted above, namely

$$L = \frac{c}{2}\left(1 - \frac{c}{2}\right), \quad M = \frac{A}{2}\left(1 - \frac{A}{2}\right) \quad \text{and} \quad N = \frac{cA}{2} - ab, \quad \text{where } A = a + b - c + 1.$$

Since L, M, and N are defined in terms of the original profile parameters, it follows that they can be solved to provide the constants a, b, and c. However, it does not appear to be noted in the literature that there are *two* possible solutions for c (c_1 and c_2, which may be complex conjugates if $L > 1/4$); similarly, there are two solutions for A (and hence $a + b - c_{1,2}$) in terms of M, so, when these are solved for a and b using N, there exists the possibility for multiple representations of these solutions, though duplication may occur. Presumably, for specific parameter profiles, such solutions can be 'whittled down' to a smaller number of independent cases.

For profile (i),

$$L = \left[a_0^2 - \frac{1}{4}(2l+1)^2\right]\alpha^{-2} + \frac{1}{4}, \tag{3.47a}$$

$$M = -\frac{3}{4} - \alpha^{-1} - \frac{1}{4}(2l+1)\alpha^{-2}, \tag{3.47b}$$

$$N = -\frac{1}{2}(2l+1)^2\alpha^{-2} - \frac{1}{2} - \alpha^{-1}. \tag{3.47c}$$

For profile (ii),

$$L = \frac{1}{2}\alpha^{-1} - \frac{1}{4}(2l+1)\alpha^{-2}, \tag{3.48a}$$

$$M = -\frac{1}{2}\alpha^{-1} - \frac{1}{4}(2l+1)\alpha^{-2}, \tag{3.48b}$$

$$N = \left[a_0^2\beta^{-1} - \frac{1}{2}(2l+1)^2\right]\alpha^{-2}. \tag{3.48c}$$

Finally, for profile (iii),

$$L = \left[a_0^2 - \frac{1}{4}(2l+1)^2\right]\alpha^{-2} + \frac{1}{4}, \tag{3.49a}$$

$$M = -\frac{1}{2}\alpha^{-1} - \frac{1}{4}(2l+1)\alpha^{-2}, \tag{3.49b}$$

$$N = \left[a_0^2 - \frac{1}{2}(2l+1)^2\right]\alpha^{-2} - \frac{1}{2}\alpha^{-1}. \tag{3.49c}$$

For magnetic-type fields (TE modes), the corresponding solutions are somewhat simpler, and may be expressed as

$$u(r) \propto r^{[1+(c-1)\alpha]/2}\left(1+\beta r^{\alpha}\right)^{A/2}_{2}F_1\left(a,b;c;-\beta r^{\alpha}\right) \tag{3.50a}$$

and

$$u(r) \propto r^{[1-(c-1)\alpha]/2}\left(1+\beta r^{\alpha}\right)^{A/2}_{2}F_1\left(1+a-c,1+b-c;2-c;-\beta r^{\alpha}\right), \tag{3.50b}$$

where now

$$L = \left[a_0^2 - \frac{1}{4}(2l+1)^2\right]\alpha^{-2} + \frac{1}{4}, \tag{3.51a}$$

$$M = a_0^2\alpha^{-2}, \tag{3.51b}$$

$$N = 2a_0^2\alpha^{-2} \tag{3.51c}$$

for profile (i); for profile (ii),

$$L = -\frac{1}{4}(2l+1)^2\alpha^{-2} + \frac{1}{4}, \tag{3.52a}$$

$$M = N = -a_0^2\beta^{-1}\alpha^{-2}, \tag{3.52b}$$

and, for profile (iii),

$$L = \left[a_0^2 - \frac{1}{4}(2l+1)^2\right]\alpha^{-2} + \frac{1}{4}, \tag{3.53a}$$

$$M = 0, \tag{3.53b}$$

$$N = a_0^2\alpha^{-2}. \tag{3.53c}$$

3.4.2.2 Cylindrically stratified isotropic media

In the case of decoupled (non-obliquely propagating) electromagnetic waves in isotropic cylindrical media (the field quantities being independent of the cylindrical axis z), the governing differential equations are similar to those for the spherical case. The E-parallel field E_z and the H-parallel field H_z are respectively expressed by the relations (with subscript N used to avoid confusion with the refractive index profile n):

$$E_z = \sum_{N=0}^{\infty} b_N P_N(r)\cos N\theta, \tag{3.54a}$$

$$H_z = \sum_{N=0}^{\infty} d_N T_N(r)\cos N\theta, \tag{3.54b}$$

where the coefficients b_N and d_N are constants. In terms of Westcott's $f_N(r)$ and $g_N(r)$ functions (here represented in each case by $u(r)$), defined by $P_N(r) = r^{-1/2} f_N(r)$ and $T_N(r) = r^{-1/2} g_N(r)$, each equation takes the form

$$\frac{d^2u}{dr^2} + \left[k_{\mathit{eff}}^2(r) - \frac{N^2 - \frac{1}{4}}{r^2}\right]u = 0. \tag{3.55}$$

For fields of the magnetic type, $k_{\mathit{eff}}^2 = k^2(r)$, and for fields of the electric type (*TE*/*TM* modes, respectively)

$$k_{\mathit{eff}}^2 = k^2(r) - \frac{k(r)}{r}\frac{d}{dr}\left\{r\frac{d}{dr}\left(\frac{1}{k(r)}\right)\right\}. \tag{3.56}$$

If we replace N by $l + 1/2$ in the above equation for magnetic type, we recover the form given for the corresponding spherically stratified case discussed above, so this type needs no further analysis.

(a) **Profiles based on Bessel's equation:** $k(r) = ar^b$. A solution for fields of the electric type (*TM*) is given by

$$u(r) \propto r^{1/2} Z_v\left(\frac{a}{1+b}r^{b+1}\right), \quad v^2 = \frac{N^2 + b^2}{(b+1)^2}, \tag{3.57}$$

where Z_v is any solution of Bessel's equation of order ν.

(b) **Profiles based on Whittaker's equation:** (i) $k(r) = \dfrac{a}{r \ln \beta r}$.

For the TM mode,

$$u(r) \propto r^{1/2} W_{\pm p,m} \left\{ \pm 2(1+N^2)^{1/2} \ln \beta r \right\}, \; p = -(1+N^2)^{-1/2}, \; m = \left(\frac{1}{4} - a^2 \right)^{1/2}. \tag{3.58}$$

(ii) $k(r) = \dfrac{a}{r \left(\ln \beta r \right)^{1/2}}$.

For the TM mode,

$$u(r) \propto r^{1/2} W_{\pm p,0} \left\{ \pm 2(1+N^2)^{1/2} \ln \beta r \right\}, \quad p = \frac{1}{2} \left(a^2 - 1 \right) \left(1 + N^2 \right)^{-1/2} \tag{3.59}$$

Both the above profiles are singular at the origin; for real r the singularity at $\beta r = 1$ is removed if β is a complex number.

(c) **Profiles based on the hypergeometric equation:** (i) $k(r) = \dfrac{a_0}{r \left(1 + \beta r^{\alpha} \right)}$, (ii) $k(r) = \dfrac{a_0 r^{(\alpha/2)-1}}{1 + \beta r^{\alpha}}$, and (iii) $k(r) = \dfrac{a_0}{r \left(1 + \beta r^{\alpha} \right)^{1/2}}$; each of these three profiles has independent solutions of the form

$$u(r) \propto r^{[1+(c-1)\alpha]/2} \left(1 + \beta r^{\alpha} \right)^{(a-b+1)/2}_{2} F_1 \left(a, c-b; c; -\beta r^{\alpha} \right) \quad \text{and} \tag{3.60a}$$

$$u(r) \propto r^{[1-(c-1)\alpha]/2} \left(1 + \beta r^{\alpha} \right)^{(a-b+1)/2}_{2} F_1 \left(1+a-c, 1-b; 2-c; -\beta r^{\alpha} \right). \tag{3.60b}$$

Again, the constants a, b, and c assume different forms for each profile. As with the spherical case, the parameters $L = \dfrac{c}{2}\left(1 - \dfrac{c}{2}\right)$, $M = \dfrac{A}{2}\left(1 - \dfrac{A}{2}\right)$, and $N = \dfrac{cA}{2} - ab$, where $A = a + b - c + 1$. Similar potential solution duplication is possible for the reasons cited above. For profile (i),

$$L = \left[a_0^2 - 1 - N^2 \right] \alpha^{-2} + \frac{1}{4}, \tag{3.61a}$$

$$M = -\frac{3}{4} - 2\alpha^{-1} - 2(1+N^2)\alpha^{-2}, \tag{3.61b}$$

$$N = -2(1+N^2)\alpha^{-2} - \frac{1}{2} - 2\alpha^{-1}. \tag{3.61c}$$

For profile (ii),

$$L = \alpha^{-1} - (1+N^2)\alpha^{-2}, \tag{3.62a}$$

$$M = -\alpha^{-1} - (1+N^2)\alpha^{-2}, \tag{3.62b}$$

$$N = \left[a_0^2 \beta^{-1} - 2(1+N^2) \right] \alpha^{-2}. \tag{3.62c}$$

Finally, for profile (iii),

$$L = \left[a_0^2 - (1+N^2) \right] \alpha^{-2} + \frac{1}{4}, \tag{3.63a}$$

$$M = -\alpha^{-1} - (1+N^2)\alpha^{-2}, \tag{3.63b}$$

$$N = \left[a_0^2 - 2(1+N^2) \right] \alpha^{-2} - \alpha^{-1}. \tag{3.63c}$$

Other (less comprehensive) exact solutions may be found in the papers by Burman (1965), Gould and Burman (1964), and Yeh and Kapielian (1965).

3.4.3 The non-existence of bound state solutions

We examine one property of Eqs (3.19) and (3.21) in more detail. Although they are formally identical to the radial Schrödinger equation, there are important differences for both the scalar and the vector problems. Pure 'bound state' solutions, that is real, regular and square-integrable solutions corresponding to $k^2 < 0$ ($\operatorname{Im} k > 0$), do not in general exist in the 'non-QM case'. To see this, assume that $S_l(r)$ is a square-integrable solution of Eq. (3.19). On multiplying by $\bar{S}_l(r)$ (the complex conjugate of $S_l(r)$) and integrating by parts, we obtain

$$\bar{S}_l(r)S_l'(r)\Big|_0^\infty - \int_0^\infty \left[|S_l'(r)|^2 + \left\{\frac{l\,(l+1)}{r^2} - k^2n^2(r)\right\}|S_l(r)|^2\,dr\right] = 0\,. \quad (3.64)$$

The integrated term vanishes because, to be square-integrable, $S(r)$ must vanish at infinity, and we have noted already that near the origin, $S_l(r) \sim r^{l+1}$. Hence

$$\int_0^\infty \left[\left|S_l'(r)\right|^2 + \frac{l(l+1)}{r^2}\left|S_l(r)\right|^2\right]\,dr = \int_0^\infty k^2n^2(r)\,|S_l(r)|^2\,dr\,. \quad (3.65)$$

Clearly, this cannot be satisfied for $k^2 < 0$ unless $n^2(r) < 0$ in some interval or set of intervals. This actually 'opens the door' for some insight into properties of exotic 'metamaterials' for which the refractive index may be purely imaginary (Pendry, 2000). Regarding the second of the two potentials in Eq. (3.21), if we write $T_l(r) = U_l(r)n(r)$, then, from Eq. (3.21), $U_l(r)$ satisfies the equation

$$\frac{d^2U_l(r)}{dr^2} + \left[k^2n^2(r) - n(r)\frac{d^2}{dr^2}\left[\frac{1}{n(r)}\right] - \frac{l\,(l+1)}{r^2}\right]U_l(r) = 0\,. \quad (3.66)$$

A similar procedure to that above yields the less useful form

$$\int_0^\infty \left[\left|U_l'(r)\right|^2 + \left\{\frac{l(l+1)}{r^2} + n(r)\frac{d^2}{dr^2}\left(\frac{1}{n(r)}\right)\right\}\left|S_l(r)\right|^2\right]\,dr$$
$$= \int_0^\infty k^2n^2(r)\left|U_l(r)\right|^2 dr\,. \quad (3.67)$$

Clearly, this expression places some conditions on the concavity of $n^{-1}(r)$, but, with the Liouville transformation (Eftimiu, 1982, 1985), $r \mapsto s$: $s = \int_0^r n^2(t)\,dt$, and $U_l \mapsto W_l$: $W_l(s) = m(s)W_l(s)$, where $m(s) = n(r(s))$, it follows that

$$\frac{d^2U_l(r)}{dr^2} = m^2\,(s)\left[m\,(s)\,\frac{d^2W_l\,(s)}{ds^2} - W_l\,(s)\,\frac{d^2m\,(s)}{ds^2}\right] \quad (3.68)$$

and

$$\frac{d^2}{dr^2}\left(\frac{1}{n(r)}\right) = -m^2\,(s)\,\frac{d^2m\,(s)}{ds^2}\,. \quad (3.69a)$$

Therefore, Eq. (3.66) simplifies to the form

$$\frac{d^2 W_l(s)}{ds^2} + \left[\frac{k^2}{m^2(s)} - \frac{l(l+1)}{m^4(s)\, r^2(s)}\right] W_l(s) = 0\,. \tag{3.69b}$$

The transformation $r \mapsto s$ is monotonic (and linear for $r > 1$), and $s \sim r$ in the neighborhood of the origin, so the previous analysis carries over, and we can conclude that, for $n^2 > 0$, no bound states are possible.

3.5 Scalar wave scattering by a transparent sphere

The essential mathematical problem for scalar waves can be thought of either in terms of classical mathematical physics, such as the scattering of sound waves, or in wave-mechanical terms, such as the non-relativistic scattering of particles by a square potential well (or barrier) of radius a and depth (or height) V_0 (Nussenzveig, 1965, 1977, 1992; Grandy, 2000). In either case, we can consider a scalar plane wave impinging in the direction $\theta = 0$on a sphere of radius a. In what follows, a boldface letter refers to a vector quantity so, here, $r = \langle |r|, \theta, \varphi\rangle$ (or $\langle r, \theta, \varphi\rangle$) denotes a position vector in space (using a spherical coordinate system). Suppose that we had started with the 'classical wave equation' with dependent variable $\tilde{\psi}(r,t) = \psi(r)e^{-i\omega t}$. For the scalar electromagnetic problem, the angular frequency ω, wavenumber k, and (constant) refractive index n are related by $\omega = kc/n$, c being the speed of light *in vacuo*. Then, for a penetrable (= 'transparent' = 'dielectric') sphere, the spatial part of the wave function $\psi(r)$ satisfies the scalar Helmholtz equation

$$\nabla^2\psi + k^2 n^2 \psi = 0\,, \quad r < a\,, \tag{3.70a}$$

$$\nabla^2\psi + k^2\psi = 0\,, \quad r > a\,. \tag{3.70b}$$

Again, k is the wavenumber and $n > 1$ is the (for now, constant) refractive index of the sphere. We can expand the wave function $\psi(r)$ as

$$\psi(r) = \sum_{l=0}^{\infty} B_l(k)\, u_l(r) r^{-1} Y_l^m(\theta,\varphi) \equiv \sum_{l=0}^{\infty} A_l(k)\, u_l(r) r^{-1} P_l(\cos\theta)\,, \tag{3.71}$$

the coefficients A_l and B_l are related by a multiplicative normalization constant that need not concern us here. The reason that the spherical harmonics $Y_l^m(\theta,\varphi)$ reduce to the Legendre polynomials in the above expression is because the cylindrical symmetry imposed on the system by the incident radiation renders it axially symmetric (i.e. independent of the azimuthal angle φ). The equation satisfied by $u_l(r)$ is exactly provided by Eq. (3.22), where, from Eq. (3.24), the potential $V(r)$ is now k-dependent—that is,

$$V(r) = k^2\left(1 - n^2\right)\,, \quad r < a \tag{3.72a}$$

$$V(r) = 0\,, \quad r > a\,. \tag{3.72b}$$

Since $n > 1$ within the sphere, this potential corresponds to that of a spherical potential well of depth $V_0 = k^2(n^2 - 1)$. This leads very naturally to a discussion of the effective potential, wherein the potential $V(r)$ is combined with the 'centrifugal barrier' term $l(l+1)/r^2$.

3.5.1 Morphology-dependent resonances: the effective potential $U_l(r)$ (constant n)

A rather detailed study of the radial wave equations was carried out by Johnson (1993), specifically for the *Mie solution* of electromagnetic theory. A crucial part of his analysis was the use of the effective potential for the *TE* mode of the Mie solution but, without any loss of generality, we may still refer to the scalar problem here. This potential is defined as

$$U_l(r) = V(r) + \frac{l(l+1)}{r^2} = k^2(1-n^2) + \frac{l(l+1)}{r^2}, \quad r \leq a, \tag{3.73a}$$

$$= \frac{l(l+1)}{r^2} \approx \frac{\lambda^2}{r^2}, \quad r > a. \tag{3.73b}$$

For large enough values of l, $[l(l+1)]^{1/2} \approx l + 1/2$. It is clear that $U_l(r)$ has a discontinuity at $r = a$ because of the 'addition' of a potential well to the centrifugal barrier. Thus there arises a 'spike' corresponding to a barrier surrounding a well (see Fig. 3.3), and this suggests the possible existence of resonances, particularly between the top of the former and bottom of the latter, where there are three turning points (where the energy k^2 is equal to $U_l(r)$). Such resonances are called 'shape resonances' (or sometimes 'morphology-dependent resonances'); they are quasi-bound states in the potential well that escape by tunneling through the centrifugal barrier. The widths of these resonances depend on where they are located; the smaller the number of nodes of the radial wave function within the well, the deeper that state lies in the well. This in turn determines the width (and lifetime) of the state, because the tunneling amplitude is 'exponentially sensitive' to the barrier height and width (Nussenzveig, 1992). Since the latter decreases rapidly with the depth of the well, the smaller is the barrier transmissivity and the lowest-node resonances become very narrow for large values of $\beta = ka$. The lifetime of the resonance (determined by the rate of tunneling through the barrier) is inversely proportional to the width of the resonance, so these deep states have the longest lifetimes.

Note that, as k^2 is reduced, the bottom B of the potential rises (and, for some value of k, the energy will coincide with the bottom of the well (Johnson, 1993)); however, at the top of the well, $U_l(a) = \lambda^2/a^2$ is independent of k^2 but, if k^2 is increased, it will eventually coincide with the top of the well (T). Consider a value of k^2 between the top and the bottom of the well: within this range, there will be three radial turning points, the middle one obviously occurring at $r = a$ and the largest at $r = b$ for which $U_l(b) = \lambda^2/b^2$. The smallest of the three (r_{min}) is found by solving the equation

$$k^2 = \frac{\lambda^2}{r_{\text{min}}^2} - \left(n^2 - 1\right) k^2 \tag{3.74}$$

to obtain, in terms of the impact parameter

$$b(\lambda) = \lambda/k, \quad r_{\text{min}} = \frac{\lambda}{nk} \equiv \frac{b}{n}. \tag{3.75}$$

By applying Snel's law for given b, it is readily shown that the distance of nearest approach of the equivalent ray to the center of the sphere is just r_{min}; indeed, there

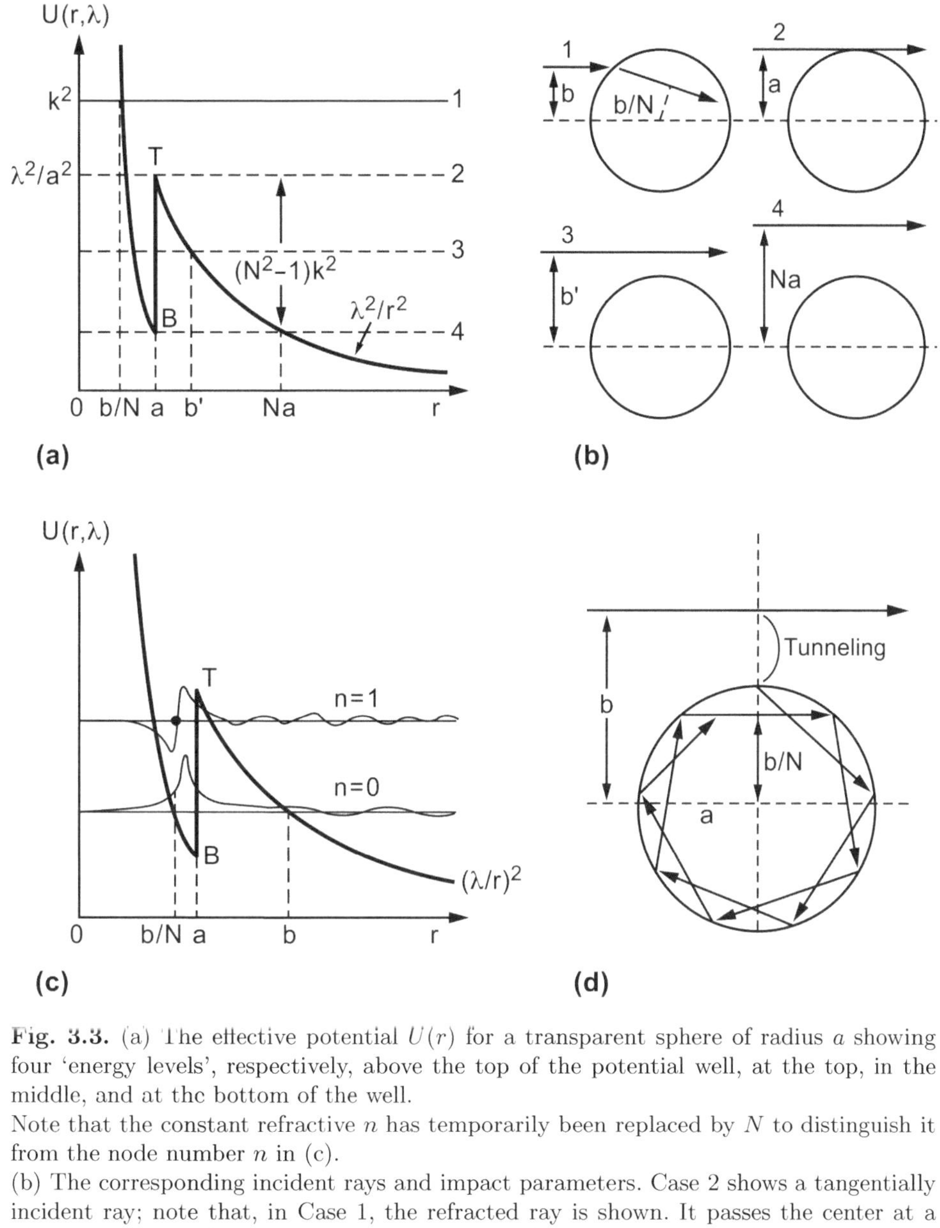

Fig. 3.3. (a) The effective potential $U(r)$ for a transparent sphere of radius a showing four 'energy levels', respectively, above the top of the potential well, at the top, in the middle, and at the bottom of the well.

Note that the constant refractive n has temporarily been replaced by N to distinguish it from the node number n in (c).

(b) The corresponding incident rays and impact parameters. Case 2 shows a tangentially incident ray; note that, in Case 1, the refracted ray is shown. It passes the center at a distance of $l = b/N$; that this is the case is readily shown from simple geometry: from Snel's law of refraction $\sin i = N \sin r = b/a$, and since $l = a \sin r$, the result follows directly.

(c) Similar to (a), but with resonant wave functions shown, corresponding to node numbers $n = 0$ and $n = 1$ (the latter possessing a single node).

(d) The 'tunneling' phenomenon illustrated for an impact parameter $b > a$, being multiply reflected after tunneling, between the surface $r = a$ and the caustic surface $r = b/N$ (the inner turning point).

are in general many nearly-total internal reflections (because of internal incidence beyond the critical angle for total internal reflection) within the sphere between $r = b/n$ and $r = a$. This is analogous to orbiting in a ray picture; the very low leakage of these states allows the resonance amplitude and energy to build up significantly during a large resonance lifetime which in turn can lead to nonlinear optical effects. In acoustics, these are called 'whispering gallery modes'.

The energy at the bottom of the well (i.e. $\lim_{r\to a^-} U_l(r)$) corresponding to the turning point at $r = a$ is determined by the impact parameter inequalities $a < b < na$ or, in terms of $\lambda = kb$,

$$U_l\left(a^-\right) = \left(\frac{\lambda}{na}\right)^2 < k^2 < \left(\frac{\lambda}{a}\right)^2 = U_l\left(a^+\right) . \tag{3.76}$$

This is the energy range between the top and bottom of the well (and in which the resonances occur). To cross the 'forbidden region', $a < r < b$ requires tunneling through the centrifugal barrier and, near the resonance energies, the usual oscillatory/exponential matching procedures lead to very large ratios of internal to external amplitudes (see Fig. 3.3(c)); these resonances correspond to 'quasi-bound' states of electromagnetic radiation (that would be bound in the limit of zero leakage).

(To avoid confusion of the node number n with the refractive index in Fig. 3.3, the latter has temporarily been written as N.) Mathematically, the resonances are complex eigenfrequencies associated with the poles λ_n of the scattering function (or S-matrix element to be discussed in section 3.6) $\mathcal{S}_l(\lambda, k)$ in the first quadrant of the complex λ-plane; these are known as *Regge poles* (for real k). Corresponding to the energy interval $[U_l(a^-), U_l(a^+)]$, the real parts of these poles lie in the interval $(\beta, n\beta)$ (or equivalently (ka, nka)); this corresponds to the tunneling region. The imaginary parts of the poles are directly related to resonance widths (and therefore lifetimes). As the node number n decreases, $\operatorname{Re}\lambda_n$ increases and $\operatorname{Im}\lambda_n$ decreases very rapidly (reflecting the exponential behavior of the barrier transmissivity). As β increases, the poles λ_n trace out Regge trajectories, and $\operatorname{Im}\lambda_n$ tend exponentially to zero. When $\operatorname{Re}\lambda_n$ passes close to a 'physical' value, $\lambda = l + 1/2$, it is associated with a resonance in the lth partial wave; the larger the value of β, the sharper the resonance becomes for a given node number n.

3.6 Connection with the scattering matrix

Consider first, for simplicity, a scalar plane wave incident upon an *impenetrable* sphere of radius a. The solution of the Helmholtz Eq. (3.70b) (outside the sphere is) (Grandy, 2000)

$$\psi_k(r,\theta) = \frac{1}{2}\sum_{l=0}^{\infty}(2l+1)i^l\left[h_l^{(2)}(kr) + \mathcal{S}_l(\beta)h_l^{(1)}(kr)\right]P_l(\cos\theta)\,, \tag{3.77}$$

where $h_l^{(1)}(kr)$ and $h_l^{(2)}(kr)$ are spherical Hankel functions of the first and second kind, respectively, and

$$\mathcal{S}_l(\beta) = -\frac{h_l^{(2)}(\beta)}{h_l^{(1)}(\beta)}; \quad \beta \equiv ka = \frac{2\pi a}{\lambda}. \tag{3.78}$$

The quantity $\mathcal{S}_l(\beta)$ is the element (for a given l-value) of the scattering or S-matrix. For 'elastic' (or non-absorptive) scattering, $\mathcal{S}_l(\beta)$ is a phase factor, and a very important one—it completely determines the nature of scattering in a potential field. As $|\bar{r}| = r \to \infty$,

$$h_l^{(1)}(kr) \sim (-i)^{l+1}\frac{e^{ikr}}{kr}; \tag{3.79a}$$

$$h_l^{(2)}(kr) \sim i^{l+1}\frac{e^{-ikr}}{kr}, \tag{3.79b}$$

hence inside the summation we have the term

$$\frac{(-1)^{l+1}}{kr}\mathcal{S}_l(\beta)\left[e^{ikr} + \frac{(-1)^{l+1}e^{-ikr}}{\mathcal{S}_l(\beta)}\right]. \tag{3.80}$$

So-called 'bound states' (of interest in quantum mechanics) are characterized by a pure imaginary wavenumber $k = ik_i$, $k_i > 0$ corresponding to energy $E = k^2 < 0$. In order for such a solution to be square-integrable in (a, ∞), it is necessary that the second term vanish in Eq. (3.80) above. Formally, this will be the case if $\beta = ka$ is a pole of $\mathcal{S}_l(\beta)$. This is the essential significance of the poles of the S-matrix in what follows.

For a spherical square well or barrier, corresponding to a transparent sphere with constant refractive index n, the form of the scattering matrix elements for scalar waves is more complicated than Eq. (3.78). In fact (Nussenzveig, 1992; see also Sanz et al. 1981), in terms of spherical Bessel (j_l) and spherical Hankel functions,

$$\mathcal{S}_l(\beta) = -\frac{\beta j_l(\alpha)h_l^{\prime(2)}(\beta) - \alpha j_l'(\alpha)h_l^{(2)}(\beta)}{\beta j_l(\alpha)h_l^{\prime(1)}(\beta) - \alpha j_l'(\alpha)h_l^{(1)}(\beta)}. \tag{3.81}$$

The vanishing of the denominator of this expression—defining the poles—is an expression of the matching at the finite boundary of the potential of the regular internal solution with the appropriate external solution of the Schrödinger equation. Using the notation of Nussenzveig (1992), the expression in Eq. (3.81) is equivalent to

$$\mathcal{S}_l(\beta) = -\frac{h_l^{(2)}(\beta)}{h_l^{(1)}(\beta)}\left[\frac{\ln' h_l^{(2)}(\beta) - n\ln' j_l(\alpha)}{\ln' h_l^{(1)}(\beta) - n\ln' j_l(\alpha)}\right] \tag{3.82}$$

where $\ln'$ represents the logarithmic derivative operator, j_l is a spherical Bessel function. The 'size parameter' $\beta = ka$ plays the role of a dimensionless external wavenumber, and $\alpha = n\beta$ is the corresponding *internal* wavenumber. Not surprisingly, $\mathcal{S}_l(\beta)$ may be equivalently expressed in terms of cylindrical Bessel and Hankel functions of half-integer order. Note that, for $l = 0$, the S-matrix element takes the simpler form (Nussenzveig, 1959)

$$\mathcal{S}_0(\beta) = e^{-2i\beta} \frac{\alpha \cot \alpha + i\beta}{\alpha \cot \alpha - i\beta}. \tag{3.83}$$

The lth 'partial wave' in the series solution in Eq. (3.71) (or Eq. (3.77)) is associated with an *impact parameter* $p_l = (l + 1/2)/k$—That is, only rays 'hitting' the sphere ($p_l \leq a$) are significantly scattered, and the number of terms that must be retained in the series to get an accurate result is of order β. Unfortunately, for visible light scattered by water droplets in the atmosphere, β is approximately several thousand and the partial wave series converges very slowly. This is certainly a non-trivial problem! In the next section, we examine the resolution of this difficulty for both the scalar and the vector wave problems.

3.7 The vector problem: the Mie solution of electromagnetic scattering theory

Having made considerable reference to the scalar problem, and its connection with the potential scattering theory, we now turn to the vector problem which, for electromagnetic waves, possesses two polarizations (the *TE* and *TM* modes); each radial equation can be examined in turn as a scalar problem. Mie theory is based on the solution of Maxwell's equations of electromagnetic theory for a monochromatic plane wave from infinity incident upon a homogeneous isotropic sphere of radius a. The surrounding medium is transparent (as the sphere may be), homogeneous, and isotropic. The incident wave induces forced oscillations of both free and bound charges in synchrony with the applied field, and this induces a secondary electric and magnetic field, each of which has components inside and outside the sphere (van de Hulst, 1981; Born and Wolf, 1999).

In this section, reference will be made to the intensity functions i_1, i_2, the Mie coefficients a_l, b_l, and the angular functions π_l, τ_l. The intensity functions are proportional to the square of the magnitude of two incoherent, plane-polarized components scattered by a single particle; they are related to the scattering amplitudes S_1 and S_2 in the notation of Nussenzveig (1979). The function $i_1(\beta, n, \theta)$ is associated with the electric oscillations perpendicular to the plane of scattering (sometimes called horizontally polarized) and $i_2(\beta, n, \theta)$ is associated with the electric oscillations parallel to the plane of scattering (vertically polarized). The scattered spherical wave is composed of many partial waves, the amplitudes of which depend on $a_l(\beta, n)$ and $b_l(\beta, n)$. In physical terms, these may be interpreted as the lth electrical and magnetic multipole waves, respectively. The first set is that part of the solution for which the radial component of the magnetic vector in the incident wave is zero; in the second set, the corresponding radial component of the electric vector is zero. A given partial wave can be thought of as coming from an electric or a magnetic multipole field, the first wave coming from a dipole field, the second from a quadrupole, and so on. The angular functions $\pi_l(\theta)$ and $\tau_l(\theta)$ are, as their name implies, independent of size (β) and refractive index (n).

For a point P located a distance r from the origin of coordinates, at polar angle θ and azimuthal angle φ, the scattered intensities I_θ and I_φ are, respectively,

$$I_\theta = \left(\frac{i_2}{kr}\right)^2 \cos^2\varphi \tag{3.84a}$$

and

$$I_\varphi = \left(\frac{i_1}{kr}\right)^2 \sin^2\varphi\,, \tag{3.84b}$$

where $i_j = |S_j|^2$, $j = 1, 2$ and the amplitude functions S_j are given by

$$S_1 = \sum_{l=1}^{\infty} \frac{2l+1}{l(l+1)} \left[a_l \pi_l(\cos\theta) + b_l \tau_l(\cos\theta)\right]\,, \quad \text{and} \tag{3.85a}$$

$$S_2 = \sum_{l=1}^{\infty} \frac{2l+1}{l(l+1)} \left[a_l \tau_l(\cos\theta) + b_l \pi_l(\cos\theta)\right]\,. \tag{3.85b}$$

l is the order of the induced electric or magnetic multipole. The Legendre functions $\pi_l(\cos\theta)$ and $\tau_l(\cos\theta)$ are defined in terms of the associated Legendre functions of the first kind, $P_l^1(\cos\theta)$ as

$$\pi_l(\cos\theta) = \frac{P_l^1(\cos\theta)}{\sin\theta} \tag{3.86a}$$

and

$$\tau_l(\cos\theta) = \frac{d}{d\theta} P_l^1(\cos\theta)\,. \tag{3.86b}$$

The scattering coefficients a_l and b_l are defined in terms of the previously encountered Riccati–Bessel functions of the first and second kinds, respectively. The coefficients a_l and b_l may be written in terms of the Riccati–Hankel function of the first kind, $\zeta_l^{(1)}(z) = zh_l^{(1)}(z) = \psi_l(z) + i\xi_l(z)$, namely

$$a_l = \frac{\psi_l(\beta)\psi_l'(\alpha) - n\psi_l(\alpha)\psi_l'(\beta)}{\zeta_l^{(1)}(\beta)\psi_l'(\alpha) - n\psi_l(\alpha)\zeta_l^{(1)\prime}(\beta)} \quad \text{and} \tag{3.87a}$$

$$b_l = \frac{\psi_l(\alpha)\psi_l'(\beta) - n\psi_l(\beta)\psi_l'(\alpha)}{\zeta_l^{(1)\prime}(\beta)\psi_l(\alpha) - n\psi_l'(\alpha)\zeta_l^{(1)}(\beta)}\,. \tag{3.87b}$$

For future reference, the Riccati–Hankel function of the second kind is defined by $\zeta_l^{(2)}(z) = zh_l^{(2)}(z) = \psi_l(z) - i\xi_l(z)$. The dimensionless size parameters $\beta = ka$ and $\alpha = n\beta$ are again used in Eqs (3.87a) and (3.87b). These expressions can be simplified by the introduction of phase shift angles and this results in considerable simplification if the refractive index is real (van de Hulst, 1981). There it is demonstrated that the Mie formulae lead, for large values of β, to a principle for localizing rays and separating diffracted, refracted, and reflected light (in the sense of geometrical optics). The principle asserts that the term of order l in the partial wave expansion corresponds approximately to a ray of distance $(l + 1/2)/k$ from the center of the particle (this is just the impact parameter). When $\beta \gg 1$, the expansions for the S_j ($j = 1, 2$) may be truncated at $l + 1/2 \approx \beta$ (in practice, $l_{\max} \sim \beta + 4\beta^{1/3} + 2$; see Nussenzveig (1969a, 1992) and Wang and van de Hulst (1998)). The remaining sum is separated into two parts: a diffracted light field component independent of the nature of the particle, and reflected and refracted rays dependent on the particle (see also Ungut et al., 1981).

From Eqs (3.87a) and (3.87b) above, we can define the new quantities (Grandy, 2000)

$$P_l^e \equiv \psi_l(\beta)\psi_l'(\alpha) - n\psi_l(\alpha)\psi_l'(\beta)\,, \tag{3.88a}$$

$$Q_l^e \equiv \xi_l(\beta)\psi_l'(\alpha) - n\psi_l(\alpha)\xi_l'(\beta)\,, \tag{3.88b}$$

$$P_l^m \equiv \psi_l(\alpha)\psi_l'(\beta) - n\psi_l(\beta)\psi_l'(\alpha)\,, \tag{3.88c}$$

$$Q_l^m \equiv \xi_l'(\beta)\psi_l(\alpha) - n\psi_l'(\alpha)\xi_l(\beta)\,. \tag{3.88d}$$

These quantities are real if n is real. Then, the external coefficients (in particular) may be written as

$$a_l = \frac{P_l^e}{P_l^e + iQ_l^e}\,, \tag{3.89a}$$

$$b_l = \frac{P_l^m}{P_l^m + iQ_l^m}\,. \tag{3.89b}$$

Furthermore, we may define implicitly (for real n) the real *phase shifts* δ_l as follows:

$$\tan\delta_l^e \equiv \frac{P_l^e}{Q_l^e} \tag{3.90a}$$

and

$$\tan\delta_l^m \equiv \frac{P_l^m}{Q_l^m}\,. \tag{3.90b}$$

Hence

$$a_l = \frac{1}{2}\left[1 - \exp\left(2i\delta_l^e\right)\right]\,, \tag{3.91a}$$

$$b_l = \frac{1}{2}\left[1 - \exp\left(2i\delta_l^m\right)\right]\,. \tag{3.91b}$$

Also, it is readily shown that

$$a_l = \frac{\left(P_l^e\right)^2}{\left(P_l^e\right)^2 + \left(Q_l^e\right)^2} - i\frac{P_l^e Q_l^e}{\left(P_l^e\right)^2 + \left(Q_l^e\right)^2}\,, \tag{3.92}$$

from which it follows that, for no absorption (i.e. elastic scattering)

$$\text{Re}\,(a_l) = |a_l|^2 = \sin^2\delta_l^e \in [0, 1]\,, \tag{3.93a}$$

and

$$\text{Im}\,(a_l) = \frac{1}{2}\sin 2\delta_l^e \in \left[-\frac{1}{2}, \frac{1}{2}\right]\,. \tag{3.93b}$$

A similar set of equations can be deduced for b_l. It is interesting to note that the locus of a_l and b_l in the complex δ_l-plane is a circle of radius with center at $(1/2, 0)$. By direct comparison with the scalar problem, the vector problem can be characterized by (for real n) the unitary matrix

$$\mathcal{S}_l = \begin{pmatrix} \mathcal{S}_l^e & 0 \\ 0 & \mathcal{S}_l^m \end{pmatrix}. \tag{3.94}$$

If we now write

$$a_l = \frac{1}{2}\left[1 - \mathcal{S}_l{}^e(k)\right], \tag{3.95a}$$

$$b_l = \frac{1}{2}\left[1 - \mathcal{S}_l{}^m(k)\right], \tag{3.95b}$$

substitution into Eqs (3.91a) and (3.91b) yields the expressions in terms of α and β

$$\mathcal{S}_l^e(k) = -\frac{\zeta_l^{(2)}(\beta)}{\zeta_l^{(1)}(\beta)}\left[\frac{\ln'\zeta_l^{(2)}(\beta) - n^{-1}\ln'\psi_l(\alpha)}{\ln'\zeta_l^{(1)}(\beta) - n^{-1}\ln'\psi_l(\alpha)}\right], \tag{3.96a}$$

$$\mathcal{S}_l^m(k) = -\frac{\zeta_l^{(2)}(\beta)}{\zeta_l^{(1)}(\beta)}\left[\frac{\ln'\zeta_l^{(2)}(\beta) - n\ln'\psi_l(\alpha)}{\ln'\zeta_l^{(1)}(\beta) - n\ln'\psi_l(\alpha)}\right]. \tag{3.96b}$$

In these expressions, the notation $\ln' f(z) = d(\ln f(z))/dz$ has been used. As we have seen, $\mathrm{Re}\,(a_l)$ reaches its maximum value (unity) when $Q_l^e = 0$ (for the *TM* modes) and, similarly, a maximum occurs for $\mathrm{Re}\,(b_l)$ when $Q_l^m = 0$(*TE* modes). These conditions correspond to Johnson's (1993) condition for resonance and, as Grandy (2000) shows in some detail, they are also equivalent to the poles of the Mie coefficients a_l and b_l in the complex β-plane, which are *in turn* equivalent to the poles of the scattering matrix elements $\mathcal{S}_l^m(\lambda, \beta)$ and $\mathcal{S}_l^e(\lambda, \beta)$ in the complex λ-plane. A valuable examination of the formal analogies between Mie theory and time-independent quantum scattering by a radial potential for both transparent and absorbing 'particles' has been carried by Gousbet (2004).

3.8 Conclusion

This article attempts to categorize and summarize some of the many and various connections that exist between exact analytic representations in ray and wave theory, and correspondences with potential scattering theory. By identifying these related areas in the broader field of mathematical physics, it is hoped that the reader will recognize how each of the levels of description can inform the others, resulting in a greater appreciation for the whole. By examining the complementary approaches of wave and potential scattering theory, the resulting radial equations (for scalar and vector wave equations) can be regarded as time-independent Schrödinger-like equations. Consequently, it is possible to exploit some of the mathematical techniques in potential scattering theory because every refractive index profile $n(r)$ defines a (wavenumber-dependent) scattering potential $V(k; r)$ for the problem. This is significantly different from the case of time-independent potential scattering in quantum mechanics because it ensures that there are no bound states of the system. The close correspondence between the resonant modes in scattering by a potential of the 'well-barrier' type and the behavior of electromagnetic 'rays' in a transparent (or dielectric) sphere is discussed in some detail.

Note: Portions of this review appeared in Adam (2013). Some of the material in section 3.2 and Appendix 1 has been adapted from Adam (2011).

Appendix 1: Properties of $\eta(r)$ and interpretation of the ray path integral

A careful analysis of the integral In Eq. (3.9) for $\Theta(i)$ in the neighborhood of the singularity yields two possibilities depending on whether or not $\eta(r)$ is a monotone increasing function:

(i) *Monotonic case*: If $\eta'(r_c) \neq 0$, then, in the neighborhood of $r = r_c$, the integral for Θ has the dominant behavior $(r - r_c)^{1/2}$ which tends to zero as $r \to r_c^+$.
(ii) *Non-monotonic case*: If $\eta'(r_c) = 0$, then, in the neighborhood of $r = r_c$, the integral for Θ has the dominant behavior $\ln |r - r_c|$ which tends to $-\infty$ as $r \to r_c^+$.

To see this, we expand the quantity $r^2 n^2(r)$ about the point $r = r_c$. The radicand then takes the form

$$\begin{aligned} r^2 n^2(r) - K^2 &= r_c^2 n^2(r_c) - K^2 + \frac{d}{dr}\left[r^2 n^2(r)\right]_{r_c}(r - r_c) \\ &+ \frac{1}{2}\frac{d^2}{dr^2}\left[r^2 n^2(r)\right]_{r_c}(r - r_c)^2 + O\left((r - r_c)^3\right). \quad \text{(A1.1)} \end{aligned}$$

Simplifying (and neglecting extraneous multiplicative and additive constants), we find that, if $(d[r^2 n^2(r)]/dr)_{r_c} > 0$, then the integral in Eq. (3.9) has the functional form (Adam, 2011)

$$I \propto \int (r - r_c)^{-1/2} dr \propto (r - r_c)^{1/2} \to 0 \quad \text{(A1.2)}$$

as $r \to r_c^+$. If, on the other hand, $(d[r^2 n^2(r)]/dr)_{r_c} = 0$, then

$$I \propto \int |r - r_c|^{-1} dr \propto \ln |r - r_c| \to -\infty \quad \text{(A1.3)}$$

as $r \to r_c^+$.

Generic $\eta(r)$ profiles for these two cases are illustrated schematically in Figs 3.4 and 3.5. In the monotonic case, the radius of closest approach for a given angle of incidence is denoted by r_i in Fig. 3.4; the distance of the ray trajectory from the center of the sphere is indicated on the r-axis. This is also indicated in the non-monotonic case in Fig. 3.5. To interpret this figure, it is best to consider rays with angles of incidence increasing away from zero. The radius (point) of closest

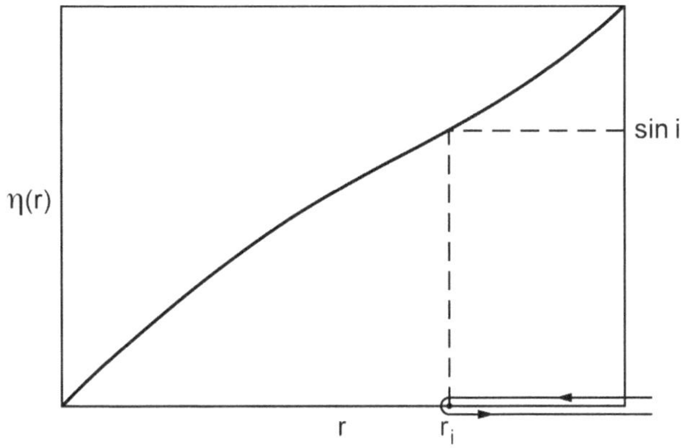

Fig. 3.4. $\eta(r) = rn(r)$ for the monotonic case. The point of closest approach is $r = r_c$.

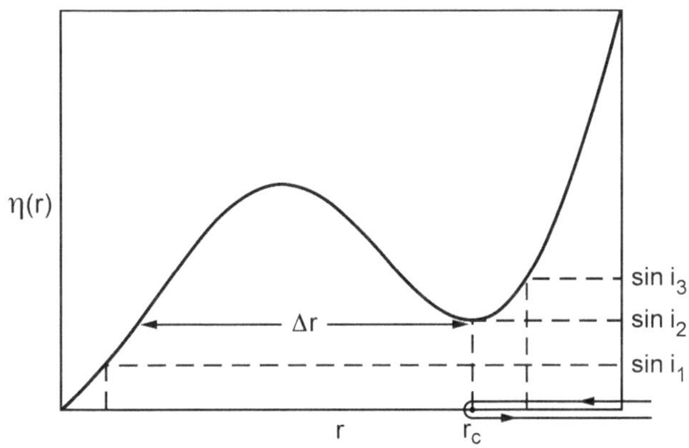

Fig. 3.5. $\eta(r) = rn(r)$ for the non-monotonic case. The point of closest approach for $i > i_2$ is $r = r_c$, and a zone of width Δr exists into which no ray penetrates.

approach increases in a continuous manner until $i = i_2$ as shown. At that stage, the point of closest approach increases discontinuously by an amount Δr to $r = r_c$, thereafter increasing continuously once again. This behavior corresponds to a spherical 'zone' of thickness Δr into which *no rays* can penetrate. The situation is reversible: starting with $i = \pi/2$ and reducing it yields the same zonal gap.

In scattering theory, the logarithmic singularity (ii) above is associated with the phenomenon of *orbiting*. An extremum of $\eta(r)$ arises at $r = r_c$ when

$$n'(r_c) = -\frac{n(r_c)}{r_c} < 0\,, \tag{A1.4}$$

meaning that the refractive index profile $n(r)$ either possesses a local minimum at $r = r_m > r_c$ or it tends monotonically to a constant value as r increases to one (see Adam, 2011). Of course, unlike the case of classical and/or atomic or molecular scattering, $n(r)$ and its corresponding potential $V(r)$ is in general piecewise continuous.

Appendix 2: Poles and resonances on the k-plane and E-plane

For algebraic simplicity, we consider the (simple) poles of the S-matrix for the one-dimensional scalar problem (Baym, 1969; Burke, 1977). In this approach, the analysis is based on a slightly different formulation of the governing time-independent 'Schrödinger' equation, namely

$$\frac{1}{2}\frac{d^2u(x)}{dx^2} + [k^2 - V(x)]u(x) = 0\,. \tag{A2.1}$$

For a square well of depth $V_0 > 0$ (i.e. $V(r) = -V_0,\ |x| < a/2$ and is zero elsewhere), the incident 'wave' is represented by

$$u\,(x) = Ae^{ikx}\,, \quad x < -a/2\,, \tag{A2.2}$$

and a transmitted wave

$$u\,(x) = Ae^{ik(x-a)}S(E)\,, \quad x > a/2\,. \tag{A2.3}$$

The transmission coefficient $S(E)$ is the one-dimensional scattering matrix in this problem. It can be shown that (Baym, 1969)

$$S(E) = \left\{\cos Ka - \frac{i}{2}\left(\frac{k}{K} + \frac{K}{k}\right)\sin Ka\right\}^{-1}\,, \tag{A2.4}$$

where now $k = \sqrt{2E}$ and $K = \sqrt{2(E+V_0)}$. Note the similarity of this expression with the denominator of the S-matrix in Eq. (3.36). The transmissivity of the well is defined as

$$T(E) = |S(E)|^2 = \left\{1 + \frac{V_0^2\sin^2 Ka}{4E(E+V_0)}\right\}^{-1}\,. \tag{A2.5}$$

This expression has maxima equal to one whenever $\sin Ka = 0$, namely when $Ka = n\pi$, $n = 1, 2, 3, \ldots$ Equivalently, $E = n^2\pi^2/2a^2 - V_0 > 0$. These maxima correspond to resonances—perfect transmission—in this system. The well contains an integral number of half wavelengths when this condition is satisfied.

We examine $S(E)$ as an analytic function of the energy E in what follows. For $E > 0$, $0 < T(E) \leq 1$.

Therefore, poles of $T(E)$ (and $S(E)$) can only occur when $-V_0 < E < 0$. In fact $S(E)$ has a pole whenever

$$\cos Ka - \frac{i}{2}\left(\frac{k}{K} + \frac{K}{k}\right)\sin Ka = 0\,, \tag{A2.6}$$

namely when

$$\cot Ka = \frac{1}{2}\left(\frac{K}{k} - \frac{k}{K}\right)\,. \tag{A2.7}$$

Furthermore, from the identity $2\cot 2\theta = (\cot\theta - \tan\theta)$, the solutions of Eq. (A2.7) can be recast in terms of odd and even parity bound state solutions, namely

$$K \cot\left(\frac{Ka}{2}\right) = ik\,, \tag{A2.8a}$$

and

$$K \tan\left(\frac{Ka}{2}\right) = -ik\,. \tag{A2.8b}$$

(Again, notice the similarity of Eq. (A2.8a) with $\alpha\cot\alpha = i\beta$ from Eq. (3.36).) Suppose now that a resonance occurs at$E = E_r \equiv k_r^2/2 > 0$.In the *vicinity* of such a value of the resonance energy, we may expand the expression $\left(\frac{k}{K} + \frac{K}{k}\right)\tan Ka$ as

$$\left(\frac{k}{K} + \frac{K}{k}\right)\tan Ka = \frac{d}{dE}\left[\left(\frac{k}{K} + \frac{K}{k}\right)\tan Ka\right]_{E_r}(E - E_r) + O\left(E - E_r\right)^2\,. \tag{A2.9}$$

To first order in $(E - E_r)$, on simplifying, we find that

$$\left(\frac{k}{K} + \frac{K}{k}\right)\tan Ka \approx a\left[\frac{dK}{dE}\left(\frac{k}{K} + \frac{K}{k}\right)\right]_{E_r}(E - E_r) \equiv \frac{4}{\Gamma}\left(E - E_r\right)\,. \tag{A2.10}$$

We can rewriting Eq. (A2.4) as

$$\begin{aligned} S(E) &= \sec Ka\left\{1 - \frac{i}{2}\left(\frac{k}{K} + \frac{K}{k}\right)\tan Ka\right\}^{-1} \approx \sec Ka\left\{1 - i\frac{2}{\Gamma}\left(E - E_r\right)\right\}^{-1} \\ &= \sec Ka\left(\frac{i\Gamma/2}{E - E_r + i\Gamma/2}\right) \approx \left(\frac{i\Gamma/2}{E - E_r + i\Gamma/2}\right)\,. \end{aligned} \tag{A2.11}$$

To this order of approximation, then, the pole of $S(E)$ lies in the fourth quadrant of the complex E-plane. There is a branch cut along the real axis, $E > 0$, since, if $E = |E|e^{i\theta}$, and $E^{1/2} = |E|^{1/2}e^{i\theta/2}$, in the limit $\theta \to 2\pi^-$, $\sqrt{E} = -|E|^{1/2}$, and for $E < 0$, $k = i|2E|^{1/2}$. As can be seen from the term $\exp(ikx)$ in Eq. (A2.3), therefore, $E < 0$ corresponds to a decaying transmitted wave, and Eq. (A2.1) then defines the conditions for the bound states to exist within the potential well. These conditions are exactly the Eqs (A2.8a) and (A2.8b) above.

Similarly, for the more general three-dimensional case, we would expect that, near a resonance, $\mathcal{S}_l(E)$ also has a pole in the fourth quadrant. This pole is in the analytic continuation of $\mathcal{S}_l(E)$ from above to below the positive real axis, and lies on the second Riemann sheet of $\mathcal{S}_l(E)$. The bound states of the well correspond to poles of $\mathcal{S}_l(E)$ on the negative real energy axis. The closer the resonances are to the real axis, the 'stronger' they become—that is, the more they behave like very long lived bound states (Baym, 1969).

Finally, a nice connection can be made to the phase shift from Eq. (A2.5). Retaining E as the independent variable, we can write

$$S(E) = e^{i\delta(E)}|T(E)|^{1/2}\,. \tag{A2.12}$$

For notational convenience, we write Eq. (A2.4) as $S(E) = [A(E) - iB(E)]^{-1}$, with obvious choices for A and B. Then it follows that

$$\tan \delta(E) = \frac{B(E)}{A(E)} = \frac{1}{2}\left(\frac{k}{K} + \frac{K}{k}\right) \tan Ka \approx \frac{2}{\Gamma}(E - E_r) \tag{A2.13}$$

on using Eq. (A2.10). Hence

$$\delta(E) \approx \arctan\left[\frac{2}{\Gamma}(E - E_r)\right] . \tag{A2.14}$$

Note also that

$$\frac{d\delta(E)}{dE} = \frac{2\Gamma}{\Gamma^2 + 4(E - E_r)^2} , \tag{A2.15}$$

and this derivative has a maximum value when $E = E_r$, that is at a resonance, so $\delta(E)$ varies rapidly there.

References

Adam, J.A. (2002a) 'Like a bridge over colored water: A mathematical review of *The Rainbow Bridge: Rainbows in Art, Myth and Science* (by R. Lee and A. Fraser)', *Notices of the American Mathematical Society*, **49**, 1360–71.

Adam, J.A. (2002b) 'The mathematical physics of rainbows and glories', *Phys. Reports.*, **356**, 229–365.

Adam, J.A. (2011) 'Zero-order bows in radially inhomogeneous spheres: Direct and inverse problems', *Appl. Opt.*, **50**, F50–9.

Adam, J.A. (2013) "Rainbows' in homogeneous and radially inhomogeneous spheres: connections with ray, wave and potential scattering theory', in Bourama, T. (ed.), *Advances in Interdisciplinary Mathematical Research: Applications to Engineering, Physical and Life Sciences*, vol. 37, Ch. 3, Springer.

Adam, J.A. and Laven, P. (2007) 'On rainbows from inhomogeneous transparent spheres: A ray-theoretic approach', *Appl. Opt.*, **46**, 922–29.

Alexopoulos, N.G. (1971a) 'High-frequency backscattering from a perfectly conducting sphere coated with a radially inhomogeneous dielectric', *Radio Sci.*, **10**, 893–901.

Alexopoulos, N.G. (1971b) 'Scattering from inhomogeneous lenses with singular points', *J.O.S.A.*, **61**, 876–8.

Alexopoulos, N.G. (1972) 'Scattering from inhomogeneous cylindrically symmetric lenses with a line infinity in the index of refraction', *J.O.S.A.*, **62,** 1088–94.

Alexopoulos, N.G. (1974) 'On the refractive index of media with poles or zeros in the index of refraction', *IEEE Trans. Antennas Propagat.*, **AP-22**, 242–51.

Baym, G. (1969) *Lectures on Quantum Mechanics*, New York, W.A. Benjamin.

Born, M. and Wolf, E. (1999) Principles of Optics: Electromagnetic Theory of Propagation, Interference and Diffraction of Light, Cambridge, Cambridge University Press.

Brockman, C.L. and Alexopoulos, N.G. (1977) 'Geometrical optics of inhomogeneous particles: Glory ray and the rainbow revisited', *Appl. Opt.*, **16**, 166–74.

Burke, P.G. (1977) *Potential Scattering in Atomic Physics*, New York and London, Plenum Press.

Burman, R.R. (1965) 'Some electromagnetic wave functions for propagation in cylindrically stratified media', *IEEE Trans. Antennas Propagat.*, **AP-13**, 646–7.

Eftimiu, C. (1982) 'Direct and inverse scattering by a sphere of variable index of refraction', *J. Math. Phys.*, **23**, 2140–6.
Eftimiu, C. (1985) 'Inverse electromagnetic scattering for radially inhomogeneous dielectric spheres', in *Inverse Methods in Electromagnetic Imaging: Proceedings of the NATO Advanced Research Workshop, Bad Windsheim, West Germany, September 18–24, 1983*, Part 1 (A85-48926 24-70), Dordrecht, D. Reidel Publishing Co., pp. 157–76.
Frisk, G.V. and DeSanto, J.A. (1970) 'Scattering by spherically symmetric inhomogeneities', *J. Acoust. Soc. Am.*, **47**, 172–80.
Gould, R.N. and Burman, R.J. (1964) 'Some electromagnetic wave functions for propagation in stratified media', *J. Atmos. Terrest. Phys.*, **26**, 335–40.
Gousbet, G. (2004) 'Cross-sections in Lorenz–Mie theory and quantum scattering: Formal analogies', *Opt. Comm.*, **231**, 9–15.
Grandy, W.T., Jr (2000) *Scattering of Waves from Large Spheres*, Cambridge, Cambridge University Press.
Heading, J. (1965) 'Refractive index profiles based on the hypergeometric equation and the confluent hypergeometric equation', *Proc. Camb. Phil. Soc.*, **61**, 897–913.
Johnson, B.R. (1993) 'Theory of morphology-dependent resonances: Shape resonances and width formulas', *J. Opt. Soc. Am.*, **A10**, 343–52.
Leonhardt, U. and Philbin, T. (2010) *Geometry and Light: The Science of Invisibility*, New York, Dover Publications.
Lock, J.A. (1988) 'Cooperative effects among partial waves in Mie scattering', *J. Opt. Soc. Am.*, **A5**, 2032–44.
Lock, J.A. (2008a) 'Scattering of an electromagnetic plane wave by a Luneburg lens: I. Ray theory', *J. Opt. Soc. Am.*, **A25**, 2971–9.
Lock, J.A. (2008b) 'Scattering of an electromagnetic plane wave by a Luneburg lens: II. Wave theory', *J. Opt. Soc. Am.*, **A25**, 2980–90.
Lock, J.A. (2008c) 'Scattering of an electromagnetic plane wave by a Luneburg lens: III. Finely stratified sphere model', *J. Opt. Soc. Am.*, **A25**, 2991–3000.
Luneberg, R.K. (1964) *The Mathematical Theory of Optics*, Berkeley and Los Angeles, University of California Press.
Nussenzveig, H.M. (1959) 'The poles of the S-matrix of a rectangular potential well or barrier', *Nucl. Phys.*, **11**, 499–521.
Nussenzveig, H.M. (1965) 'High-frequency scattering by an impenetrable sphere', *Ann. Phys.*, **34**, 23–95.
Nussenzveig, H.M. (1969a) 'High-frequency scattering by a transparent sphere: I. Direct reflection and transmission', *J. Math. Phys.*, **10**, 82–124.
Nussenzveig, H. M. (1969b) 'High-frequency scattering by a transparent sphere: II. Theory of the rainbow and the glory', *J. Math. Phys.*, **10**, 125–78.
Nussenzveig, H.M. (1977) 'The theory of the rainbow', *Sci. Am.*, **236**, 116–27.
Nussenzveig, H.M. (1979) 'Complex angular momentum theory of the rainbow and the glory', *J. Opt. Soc. Am.*, **69**, 1068–79.
Nussenzveig, H.M. (1992) *Diffraction Effects in Semiclassical Scattering*, Cambridge, Cambridge University Press.
Pendry, J.B. (2000) 'Negative refraction makes a perfect lens', *Phys. Rev. Lett.*, **85**, 3966–9.
Rosu, H. and Reyes, M. (1994) 'Electromagnetic modes of Maxwell fisheye lens', *Il Nuovo Cim.*, **16D**, 517–22.
Sanz, P., Sanudo, J. and Sesma, J. (1981) 'Trajectories of the S-matrix poles in the complex wavenumber plane', *J. Math. Phys.*, **22**, 2594–7.
Schiff, L.I. (1968) *Quantum Mechanics*, 3rd edn, New York, McGraw-Hill.

Sharaf, A.L. (1969a) 'Exact solutions for fields of electric type in spherically stratified isotropic media', *Proc. Camb. Phil. Soc.*, **66**, 119–27.
Sharaf, A.L. (1969b) 'Some refractive index profiles for spherically and planar stratified media', *J. Lond. Math. Soc.*, **2(1)**, 193–202.
Sharaf, A.L. (1970) 'Exact solutions for electromagnetic fields in cylindrically stratified isotropic media', *J. Lond. Math. Soc.*, **2(1)**, 92–8.
Tai, C.T. (1956) 'The electromagnetic theory of the spherical Luneberg lens', *Appl. Sci. Res.*, **B7**, 113–30.
Tai, C.T. (1958) 'Maxwell fish-eye treated by Maxwell equations', *Nature*, **182**, 1600–1.
Tai, C.T. (1963) 'Some wave functions and potential functions pertaining to spherically stratified media', *J. Res. Nat. Bur. Stand.*, **67D**, 199–202.
Ungut, A., Grehan, G. and Gouesbet, G. (1981) 'Comparisons between geometrical optics and Lorenz–Mie theory', *Appl. Opt.*, **20**, 2911–18.
Uslenghi, P.L.E. (1964) 'A note on dielectric lenses', *Can. J. Phys.*, **42**, 2359–64.
Uslenghi, P.L.E. (1969a) 'Electromagnetic and optical behavior of two classes of dielectric lenses', *IEEE Trans. Ant. Prop.*, **17**, 235–6.
Uslenghi, P.L.E. (1969b) 'On the generalized Luneburg lenses', *IEEE Trans. Ant. Prop.*, **17**, 644–5.
Uslenghi, P.L.E. and Weston, V.H. (1970) 'High-frequency scattering from a metal-like dielectric lens', *Appl. Sci. Res.*, **23**, 147–63.
van de Hulst, H.C. (1981) *Light Scattering by Small Particles*, New York, Dover.
Vetrano, M.R., van Beeck, J.P. and Riethmuller, M. (2005) 'Generalization of the rainbow Airy theory to nonuniform spheres', *Opt. Lett.*, **30**, 658–60.
Wang, R.T. and van de Hulst, H.C. (1991) 'Rainbows: Mie computations and the Airy approximation', *Appl. Opt.*, **30**, 106–17.
Westcott, B.S. (1968a) 'Electromagnetic wave propagation in cylindrically stratified isotropic media', *Electron. Lett.*, **4**, 323–4.
Westcott, B.S. (1968b) 'Exact solutions for electromagnetic wave propagation in spherically stratified isotropic media', *Proc. Camb. Phil. Soc.*, **64**, 227–35.
Westcott, B.S. (1969) 'Exact solutions for transverse electromagnetic wave propagation in a cylindrically stratified axially magnetized plasma', *Proc. Camb. Phil. Soc.*, **66**, 129–43.
Yeh, C. and Kapielian, Z.A. (1965) 'Scattering from a cylinder coated with an inhomogeneous dielectric sheath', *Can. J. Phys.*, **41**, 143–51.

Part II

Remote Sensing

4 Spectral dependence of MODIS cloud droplet effective radius retrievals for marine boundary layer clouds

Zhibo Zhang, Steven Platnick, Andrew S. Ackerman, and Hyoun-Myoung Cho

4.1 Introduction

Low-level warm marine boundary layer (MBL) clouds cover large regions of Earth's surface. They have a significant role in Earth's radiative energy balance (Klein and Hartmann, 1993) and hydrological cycle. Despite the fundamental role of low-level warm water clouds in climate, our understanding of these clouds is still limited. In particular, connections between their properties (e.g. cloud fraction, cloud water path, and cloud droplet size) and environmental factors such as aerosol loading and meteorological conditions continue to be uncertain or unknown. Modeling these clouds in climate models remains a challenging problem. As a result, the influence of aerosols on these clouds in the past and future, and the potential impacts of these clouds on global warming remain open questions leading to substantial uncertainty in climate projections. To improve our understanding of these clouds, we need continuous observations of cloud properties on both a global scale and over a long enough timescale for climate studies. At present, satellite-based remote sensing is the only means of providing such observations.

The cloud droplet effective radius (r_e) is one of the most important cloud parameters that are routinely monitored from space. The r_e is defined as (Hansen and Travis, 1974)

$$r_e(z) = \frac{\int_0^\infty r^3 n(r,z)\,dr}{\int_0^\infty r^2 n(r,z)\,dr}, \tag{4.1}$$

where $n(r,z)$ is the cloud droplet size distribution (DSD) at altitude z in cloud. The cloud droplet effective radius determines the optical thickness of cloud for a given amount of water (Twomey, 1974). It also has a significant influence on precipitation formation processes (Lebsock et al., 2008; Kubar et al., 2009). Therefore, it is a key microphysical parameter required to estimate radiative effects of clouds, study aerosol–cloud-precipitation interactions, and validate cloud parameterization in global climate models.

Many remote sensing methods exist to infer cloud r_e from various types of satellite instruments (e.g. Prabhakara et al., 1988; Nakajima and King, 1990; Austin et al., 2009). Of particular interest in this review are cloud r_e retrievals from MODIS (or Moderate Resolution Imaging Spectroradiometer). MODIS is a key instrument aboard the Terra and Aqua satellites. Terra's orbit around Earth is timed so that it passes from north to south across the equator in the morning, while Aqua passes south to north over the equator in the afternoon. Terra and Aqua MODIS are viewing the entire Earth's surface every one to two days, acquiring data in 36 spectral bands. A variety of cloud parameters, from cloud fraction to cloud top height to cloud optical thickness (τ) and r_e, can be inferred from MODIS multi-spectral observations (Platnick et al., 2003). The operational MODIS cloud retrieval algorithm, as described in detail in the next section, is based on the so-called bi-spectral solar reflective method (referred to as bi-spectral method hereafter), which utilizes cloud reflectance measurements from two spectral bands to retrieve cloud τ and r_e, simultaneously (Nakajima and King, 1990). One measurement is usually made in the visible or near-infrared spectral region (e.g. 0.86 μm), where water absorption is negligible and therefore cloud reflection is mainly determined by τ, and the other in the shortwave infrared (SWIR) (e.g. 2.1 μm or 3.7 μm), where water is significantly absorptive and cloud reflectance primarily decreases with increasing cloud droplet size. The τ and r_e retrievals based on the bi-spectral method are widely used for validating climate models (Kay et al., 2012; Pincus et al., 2012), studying aerosol–cloud interactions (Quaas and Boucher, 2005; Quaas et al., 2009) and facilitating other cloud remote sensing techniques (Lebsock and L'Ecuyer, 2011).

Unlike cloud optical properties (such as τ) that can be spectrally dependent, cloud r_e, a physical cloud parameter, should be independent of the observation method or what spectral band is used. For example, the MODIS instrument has three SWIR bands, centered at 1.6 μm, 2.1 μm, and 3.7 μm, respectively. In the MODIS operational cloud product (MOD06) Collection 5 (C5) processing algorithm, the combination of the 0.86 and 2.1 μm bands are used for τ and r_e retrievals over open ocean (Platnick et al., 2003). Hereafter, we will refer to the r_e retrieval based on the 2.1-μm band observation as $r_{e,2.1}$. Besides $r_{e,2.1}$, the MODIS also provides two other r_e retrievals, one based on the 1.6-μm band and the other based on the 3.7-μm band observations (hereafter referred to as $r_{e,1.6}$ and $r_{e,3.7}$). One might expect these three r_e retrievals to be in close agreement. However, several studies have found substantial differences between them (Nakajima et al., 2010a; Seethala and Horváth, 2010; Zhang and Platnick, 2011). In particular, it is found that:

- MODIS $r_{e,3.7}$ retrievals for marine warm clouds are generally smaller than $r_{e,2.1}$ (and $r_{e,1.6}$);
- Geographically, the differences between $r_{e,3.7}$ and $r_{e,2.1}$ based on MODIS Level 3 monthly mean product show strong dependence on cloud regimes, small variations (close to zero) over the costal stratocumulus regimes, and large variations over the cumulus cloud regimes;
- At pixel level, the difference between $r_{e,3.7}$ and $r_{e,2.1}$ correlates with cloud τ, r_e, and the degree of sub-pixel inhomogeneity.

A detailed analysis of the differences between $r_{e,3.7}$ and $r_{e,2.1}$ MODIS retrievals will be given in section 4.3. $r_{e,1.6}$ is not considered in this study, mainly because the 1.6-μm band on Aqua MODIS has nonfunctional or noisy detectors that lead to the striping issue (Wang et al., 2006). As a result, it is difficult to make pixel-to-pixel comparisons between $r_{e,1.6}$ and other effective radius retrievals.

The spectral dependence of MODIS r_e raises many questions. The most important one, for all the users of MODIS products, is whether the difference between $r_{e,3.7}$ and $r_{e,2.1}$ is an artifact due to, say, algorithm issues and inherent limitations of the retrieval method, or something meaningful that contains information about the cloud. Several lines of evidence suggest that it is unlikely to be due to technical issues (e.g. code bugs, ancillary data issues, etc.) of the operational MODIS cloud retrieval algorithm. First, several studies based on independent retrieval algorithms also found substantial differences between $r_{e,3.7}$ and $r_{e,2.1}$ that are similar to the operational MODIS cloud product (Nakajima et al., 2010a; Minnis et al., 2011). Second, both theoretical and numerical studies indicate that $r_{e,3.7}$ and $r_{e,2.1}$ are expected to differ significantly under certain circumstances owing to their difference sensitivities to, for example, cloud vertical structure (Platnick, 2000) , 3D radiative effects (Zhang and Platnick, 2011; Zhang et al., 2012), and the presence of drizzle drops in the cloud (Nakajima and King, 1990; Zhang, 2013).

The question is then: what has caused the differences between $r_{e,3.7}$ and $r_{e,2.1}$? A number of recent studies have attempted to address this question from different perspectives (Platnick, 2000; Nakajima et al., 2010a, 2010b; Seethala and Horváth, 2010; Zhang et al., 2010, 2012; Zinner et al., 2010; Zhang and Platnick, 2011; Zhang, 2013). Several mechanisms have been proposed to explain the above-mentioned spectral dependence of MODIS r_e retrieval, which can be divided into two categories based on their underlying physics and consequent implications. In the first category are those mechanisms related to the fact that $r_{e,3.7}$ and $r_{e,2.1}$ have different sensitivities to cloud vertical structure and the presence of large drizzle drops in warm water clouds, with implications that the spectral difference actually carries useful information about the cloud that can be used for remote sensing and model validation. In the other category are those mechanisms related to the inherent limitations of the bi-spectral method, such as, the lack of consideration of 3D radiative effects and sub-pixel inhomogeneity in the retrieval method. In such cases, the spectral difference implies significant retrieval uncertainties should be cautioned when using MODIS r_e retrieval products. Overall, these studies reveal that the spectral difference of MODIS r_e retrievals is a complicated issue that has complex causes.

This chapter provides an overview of the current understanding of the spectral dependence of MODIS r_e retrievals. In section 4.2, we briefly revisit the operational MODIS cloud r_e retrieval algorithm to set the stage for later discussion. In section 4.3, the MODIS $r_{e,3.7}$ and $r_{e,2.1}$ retrievals for MBL clouds are compared to illustrate the spectral dependence of MODIS cloud r_e retrieval. Section 4.4 introduces several potential mechanisms that may cause or contribute to the spectral dependence of MODIS r_e retrieval. The relative role of these mechanisms in different cloud regimes is discussed in section 4.5. Current outstanding issues and an overview for future work are given in section 4.6.

4.2 Operational MODIS r_e retrieval algorithm

In order to set the stage for later discussion in this section, we briefly describe the operational MODIS r_e and τ retrieval algorithm based on the bi-spectral method. The retrieval makes use of a pair of measurements of cloud reflectivity, one from a visible or near-infrared (VNIR) MODIS band (0.86-μm band over ocean) and the other from a shortwave infrared (SWIR) band (e.g. 2.1 μm or 3.7 μm band) (Nakajima and King, 1990; Platnick et al., 2003). The visible band measurement provides the information for τ retrieval because water absorption in the VNIR region is almost negligible and, as a result, cloud reflectivity is mainly determined by τ. The SWIR measurement provides the information for r_e retrieval because significant water absorption in SWIR makes cloud reflectance decrease with increasing cloud r_e. In the operational algorithm, this method is implemented by using the so-called look-up-table (LUT), as shown in Fig. 4.1. The LUT contains pre-computed cloud reflectivities at visible and SWIR bands for various combinations of r_e and τ under different Sun–satellite viewing geometries and surface reflectances. As illustrated in Fig. 4.1, in practice. r_e and τ are retrieved by projecting the observed reflectivities onto the LUT. Once r_e and τ are retrieved, the liquid water path (LWP) of the cloud can be easily derived from the equation $LWP = \frac{2}{3}\rho\tau r_e$ assuming that the cloud layer is vertically homogenous and the extinction coefficient Q_e of a cloud droplet is about 2. The MODIS cloud r_e and τ retrievals have been shown to agree reasonably well with other satellite products (e.g. Zhang et al., 2009; Minnis et al., 2011; Pincus et al., 2012; Stubenrauch et al., 2012; Walther and Heidinger, 2012) and *in situ* measurements (e.g. Painemal and Zuidema, 2011; King et al., 2012).

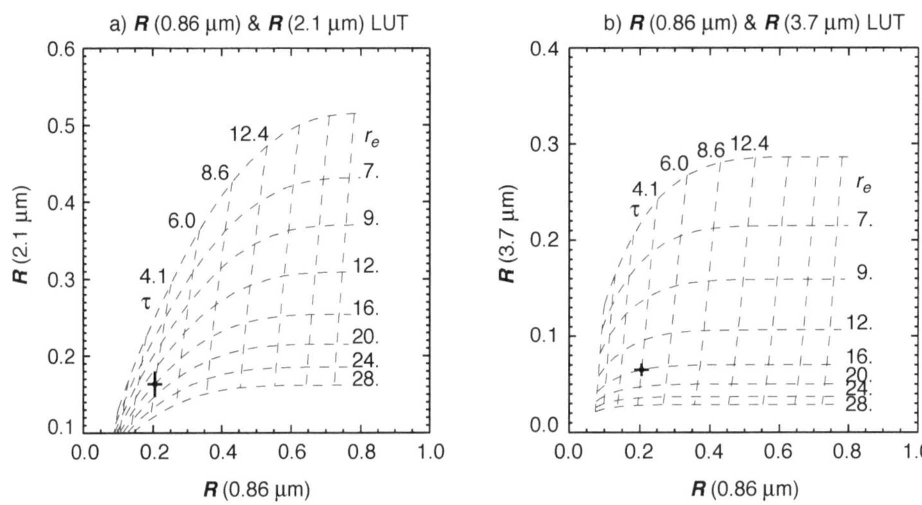

Fig. 4.1. Examples of the look-up-table of cloud bi-directional reflection function as functions of cloud optical thickness and effective radius, based on the combination of (a) 0.86 and 2.1-μm bands, and (b) 0.86 and 3.7-μm bands.

It is important to note that the bi-spectral method is based on several important assumptions about the cloud (or the cloudy pixel):

1. Within the pixel, the cloud is vertically homogeneous. (referred to as the vertical homogenous assumption).
2. Within the pixel, the cloud is horizontally homogenous. The cloud reflectance of the pixel of interest is independent of the properties of surrounding pixels. (referred to as the plane-parallel and homogeneous cloud assumption).
3. Within the pixel, the sizes of cloud droplets follow certain analytical distributions, namely the following monomodal Gamma distribution (King et al., 1998):

$$n(r) = Nr^{\frac{1-3v_e}{v_e}} \exp\left(-\frac{1}{v_e}\frac{r}{r_e}\right), \tag{4.2}$$

   where $n(r)$ is DSD, N is a constant, and v_e is the effective variance assumed to be 0.1 in the operational MODIS algorithm (monomodal Gamma DSD assumption hereafter).

These assumptions are necessary because, at most, two independent pieces of information can be extracted from a pair of cloud reflectivities. Since the information content has been used for r_e and τ retrievals, other aspects of cloud would have to be assumed. These assumptions are justified for some clouds, particularly, non-precipitating stratocumulus clouds (Martin et al., 1994; Di Girolamo et al., 2010). However, for other clouds, they can be problematic. For example, it is known that trade wind cumulus clouds can be far from plane-parallel (Liang et al., 2009; Di Girolamo et al., 2010). The vertical homogeneous assumption and the monomodal Gamma DSD often break down when precipitation begins to form within the cloud. The warm rain processes, such as collision-coalescence, could broaden the DSD and sometimes give rise to a second mode, the so-called drizzle or rain mode, leading to a *bi-modal* DSD (Berry, 1967; Berry and Reinhardt, 1974; Pruppacher and Klett, 1997). In addition, the collision-coalescence processes make drizzle drops grow bigger as they fall from cloud top towards cloud base, inducing vertical structures within the cloud (Pruppacher and Klett, 1997). When these conditions occur and the above-mentioned assumptions break down, MODIS r_e and τ retrievals face substantial uncertainties. It will be shown later in section 4.4 that the mechanisms that cause significant differences between $r_{e,3.7}$ and $r_{e,2.1}$ are more or less connected to breakdown of these fundamental assumptions about cloud made in MODIS retrieval.

4.3 Spectral dependence of MODIS r_e retrievals for MBL clouds

As stated from the beginning, we focus only on the warm liquid-phase clouds over ocean. Since such clouds reside mostly in the MBL, we will refer to them as MBL clouds for simplicity. We attempt to identify a MODIS pixel as MBL cloud based on the following criteria: (i) the pixel is labeled as 'confident cloudy' by the 1-km MODIS cloud mask product (MOD35); (ii) over ocean; (iii) labeled as 'liquid water' by the MODIS 1-km 'Cloud_Phase_Optical_Properties' data set within MOD06; (iv) cloud top temperature warmer than 273K. These conditions are expected to remove most of the situations that may complicate the analyses, such as thin cirrus overlapping lower clouds.

4.3.1 Geographical pattern

The monthly mean $r_{e,2.1}$ and $r_{e,3.7}$ for MBL clouds based on May 2007 MODIS/Aqua data are shown in Fig. 4.2a and 4.2b, and their difference $\Delta r_{e,3.7-2.1} = r_{e,3.7} - r_{e,2.1}$ in Fig. 4.2c respectively. Several well-known coastal stratocumulus regions, such as off the coasts of California, Peru, and Namibia/Angola, are clearly seen in the figure. One can also see from Fig. 4.2 that the transition from the coastal stratocumulus cloud regimes to the offshore cumulus cloud regimes is quite sharp. Accompanying this transition, $r_{e,2.1}$ increases substantially from $8 \sim 10$ μm near the coast to as large as $20 \sim 25$ μm far offshore. However, $r_{e,3.7}$ is significantly smaller than its counterparts. It is easily seen that $\Delta r_{e,3.7-2.1}$ shows an obvious dependence on cloud regime. For example, over coastal stratocumulus cloud regions, $\Delta r_{e,3.7-2.1}$ is close to zero, or even slightly positive. However, over the broken cumulus cloud regions, where water cloud fraction is small, $r_{e,3.7}$ is seen to be smaller than $r_{e,2.1}$ by as much as 5–10 μm on average.

4.3.2 Correlation with key cloud parameters

In the rest of this section, we explore the correlations between MODIS r_e retrieval differences and several key factors. In doing so, we attempt to identify regimes where the r_e retrieval differences can be attributed more to cloud physics, such as drizzle, than to retrieval uncertainties and artifacts caused by, for example, 3D radiative effects.

Figure 4.3a shows the joint histogram between $r_{e,3.7}$ and $r_{e,2.1}$ based on about 1.5 billion Level 2 marine water cloud pixels collected during May 2007 by Aqua MODIS from 60S to 60N. Evidently, the density of points is highest along the one-to-one line, attesting that pixel-level $r_{e,3.7}$ and $r_{e,2.1}$ modes agree reasonably well. It is interesting to see that the bias between $r_{e,3.7}$ and $r_{e,2.1}$ is quite small when r_e is smaller than about $12 \sim 13$ μm. However, when r_e is larger than about 15 μm, the histogram distribution is clearly weighted toward the $r_{e,2.1}$ side and the deviation from the one-to-one line increases with increasing $r_{e,2.1}$. Figure 4.3b presents the same story, but from a different perspective. It shows the joint histogram between $\Delta r_{e,3.7-2.1}$ and $r_{e,2.1}$. Note that, in order to reduce the data sampling impact on the shape of the joint histogram, the histogram of $\Delta r_{e,3.7-2.1}$ at $r_{e,2.1}$ bin has been normalized with respect to its peak value at that bin. Therefore, the red color in Fig. 4.3b corresponds to the most frequent $\Delta r_{e,3.7-2.1}$ at a given $r_{e,2.1}$ bin. The gray dotted line in Fig. 4.3b shows the PDF of $r_{e,2.1}$ derived from one month of Level 2 data. Interestingly, the bin-normalized histogram of $\Delta r_{e,3.7-2.1}$ vs. $r_{e,2.1}$ not only shows an increasing spread in $\Delta r_{e,3.7-2.1}$ with increasing $r_{e,2.1}$, but also a clear systematic transition, in terms of the most likely observed $\Delta r_{e,3.7-2.1}$ for a given $r_{e,2.1}$ (i.e. the red area), from near-zero values when $r_{e,2.1} < 15$ μm to larger negative values when $r_{e,2.1} > 15$ μm. The fact that this threshold-like behavior

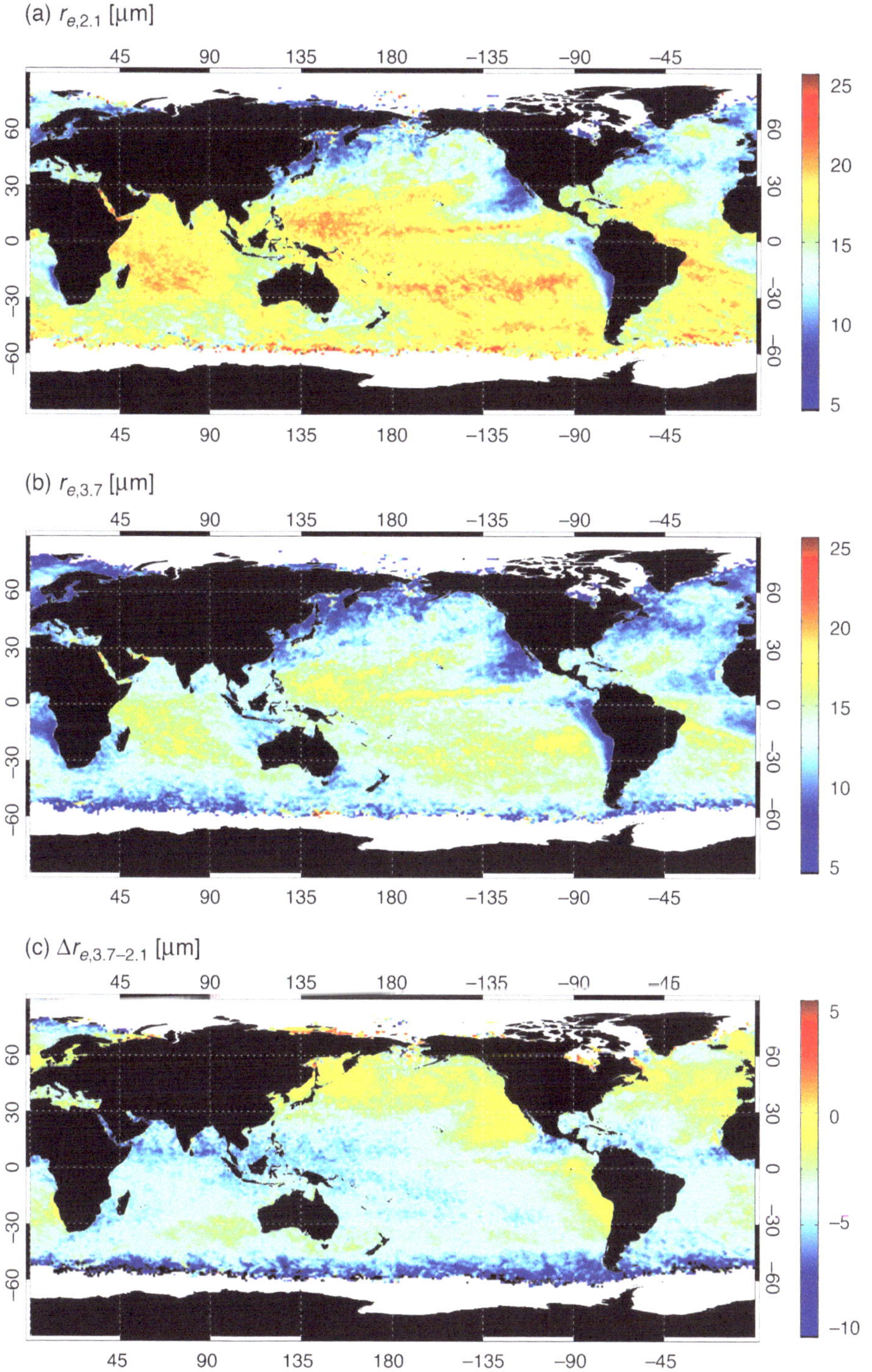

Fig. 4.2. Monthly mean MODIS (a) $r_{e,2.1}$, (b) $r_{e,3.7}$, and (c) $\Delta r_{e,3.7-2.1}$ retrievals for warm (cloud top temperature larger than 273 K) liquid-phase clouds over ocean.

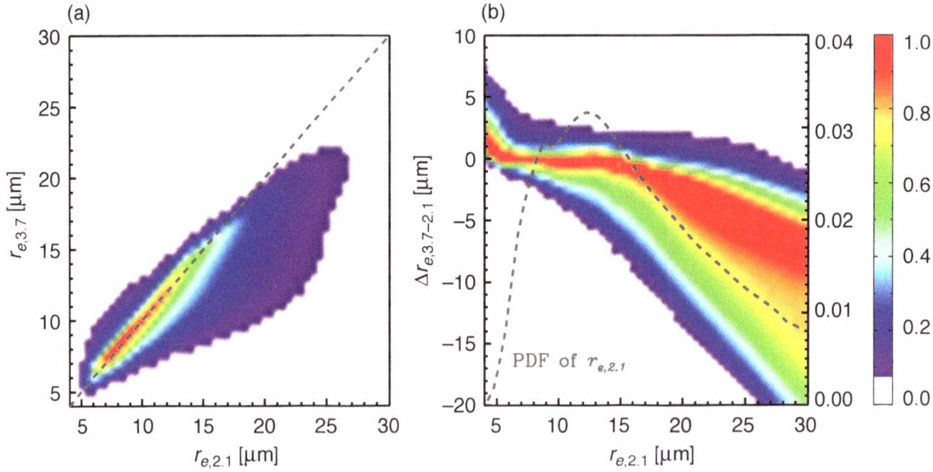

Fig. 4.3. The joint histogram of pixel level (a) $r_{e,2.1}$, vs. $r_{e,3.7}$, (b) $\Delta r_e 3.7 - 2.1$ vs. $r_{e,2.1}$, which is bin-normalized with respect to the PDF of $r_{e,2.1}$ (gray dashed line).

takes place at $r_e \sim 15$ μm is particularly interesting because $r_e \sim 15$ μm has been suggested to be the threshold for the collision-coalescence process to take place in marine water clouds (Gerber, 1996). However, further studies are needed to determine whether this is simply a coincidence or due to more fundamental physical reasons.

Figure 4.4 shows the bin-normalized joint histograms between $\Delta r_{e,3.7-2.1}$ and cloud optical thickness (τ). One can easily note that, when clouds are optically thin (e.g. $\tau < 5$), $\Delta r_{e,3.7-2.1}$ varies quite remarkably from -15 μm to 10 μm. However, when the cloud becomes sufficiently thick ($\tau > 5$), the statistics of $\Delta r_{e,3.7-2.1}$ become stable and show little dependence on τ. The large variability of $\Delta r_{e,3.7-2.1}$ for thin clouds in Fig. 4.4 is likely a result of the combined effects of random retrieval uncertainties (see section 4.4.1), 3D radiative effects (section 4.4.5), and the so-called plane-parallel r_e retrieval bias (section 4.4.4). For thin clouds, the signal is comparable or smaller than, the noise caused by instrument uncertainties, ancillary data uncertainties, and discretization and interpolation of the LUT. As a result, the uncertainty associated with the MODIS r_e retrievals for thin clouds is large. In addition, thin and broken clouds often have significant horizontal heterogeneity, providing favorable conditions for 3D radiative effects and plane-parallel r_e retrieval bias. Caution must therefore be taken when interpreting the meaning of $\Delta r_{e,3.7-2.1}$ for clouds with $\tau < 5$ because many factors other than cloud physics, such as retrievals errors and artifacts, all play a significant role in this regime.

When analyzing the potential connection between cloud horizontal inhomogeneity and r_e retrieval failure, we will use a so-called sub-pixel inhomogeneity index (H_σ) from MODIS. It is defined as (Liang et al., 2009)

$$H_\sigma = \frac{\mathrm{stdev}[R_i(0.86\ \mu\mathrm{m},\ 250\ \mathrm{m})]}{\mathrm{mean}[R_i(0.86\ \mu\mathrm{m},\ 250\ \mathrm{m})]}\,, \tag{4.3}$$

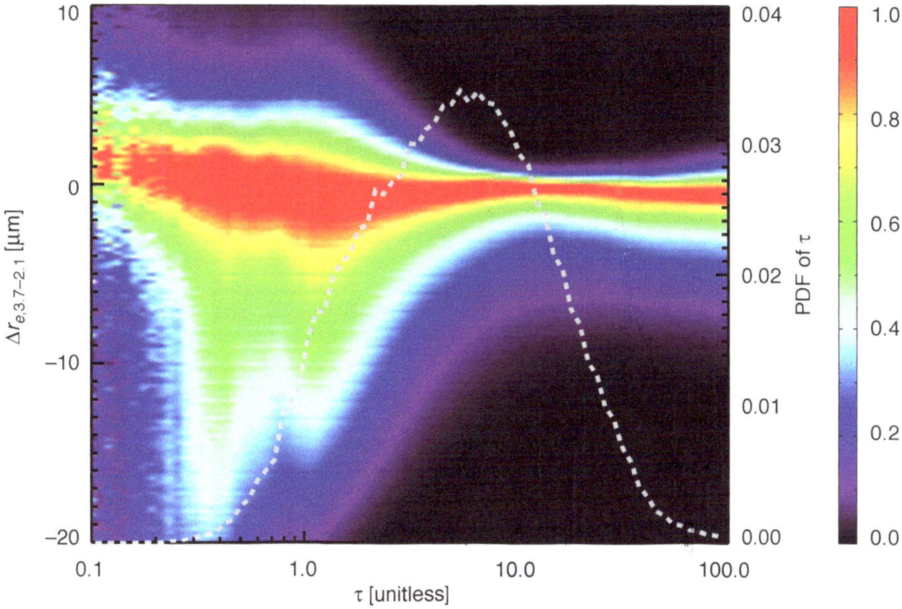

Fig. 4.4. Joint histogram of $\Delta r_{e,3.7-2.1}$ and cloud optical thickness (τ), bin-normalized by the PDF of τ.

where stdev[R_i(0.86 μm, 250 m)] and mean[R_i(0.86 μm, 250 m)] indicate the standard deviation and mean of the measured reflectances, respectively, for the principle 16 250-m-resolution sub-pixels within the 1-km MODIS pixel retrieval footprint. Thus, H_σ has a spatial resolution (i.e. 1 km) consistent with the cloud property retrieval and increases with pixel inhomogeneity. Recent studies found that the H_σ index derived from high-resolution (250-m) MODIS cloud reflectance measurement provides an objective and quantitative measurement of the horizontal inhomogeneity of 1-km MODIS pixel (Di Girolamo et al., 2010; Zhang and Platnick, 2011; Zhang et al., 2012). The H_σ index will be reported in the MOD06 product in the coming Collection 6. The dependence of $\Delta r_{e,3.7-2.1}$ on cloud horizontal inhomogeneity is shown in Fig. 4.5. Figure 4.5a and 4.5b show the bin-normalized joint histograms of the sub-pixel cloud inhomogeneity (H_σ) defined vs. $r_{e,2.1}$ and $r_{e,3.7}$, respectively. Figure 4.5c shows the bin-normalized joint histograms of the H_σ vs. $\Delta r_{e,3.7-2.1}$. Optically thin clouds ($\tau < 5$) are excluded from this figure for the above-mentioned reason, but results are similar if we include thin pixels (not shown). The most compelling feature in Fig. 4.5a is the sharp transition of $r_{e,2.1}$ at H_σ around 0.3 $\sim$ 0.5. When H_σ is smaller than 0.3, the most likely $r_{e,2.1}$ (i.e. red area in Fig. 4.5a) for a given H_σ stays relatively constant, within 10 $\sim$ 15 μm. However, when H_σ exceeds about 0.3, the most likely $r_{e,2.1}$ value increases dramatically with H_σ. Interestingly, this is not the case in Fig. 4.5b, where the most likely value of $r_{e,3.7}$ shows only weak dependence on the sub-pixel inhomogeneity. It is therefore not surprising to see in Fig. 4.5c the most likely values of $\Delta r_{e,3.7-2.1}$

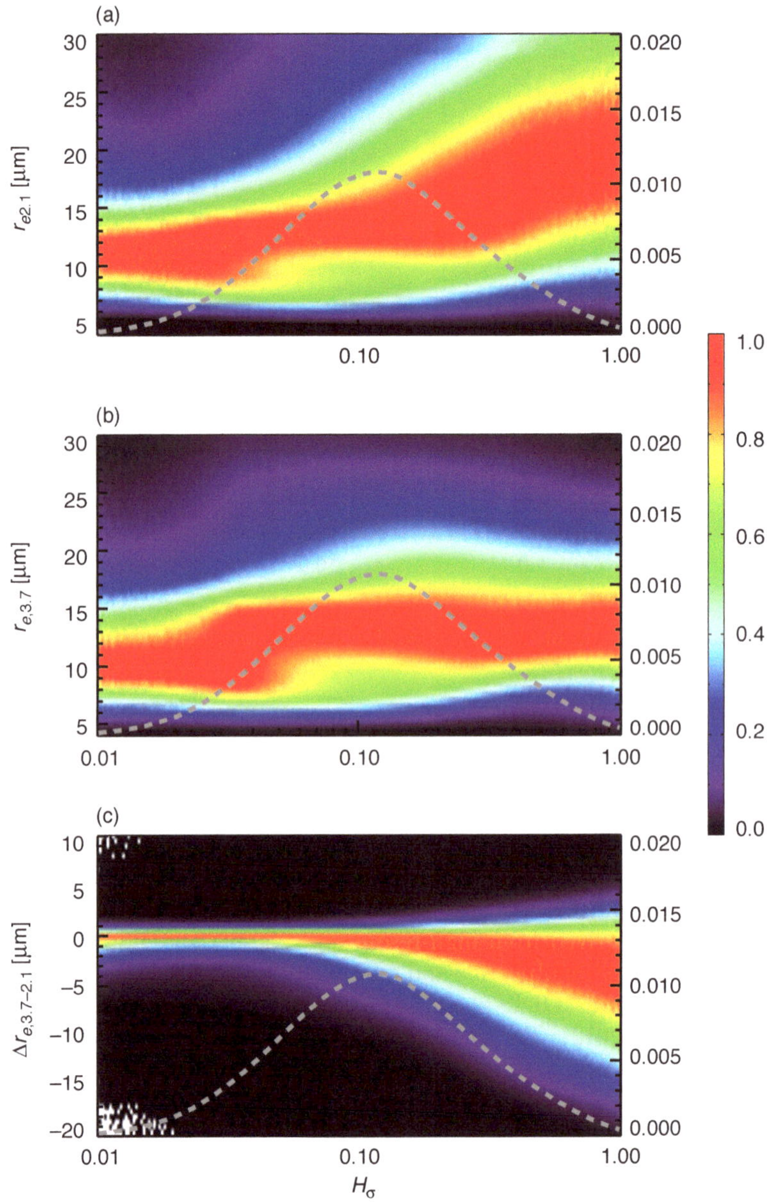

Fig. 4.5. Joint histograms of (a) $r_{e,2.1}$ vs. sub-pixel inhomogeneity index (H_σ), (b) $r_{e,3.7}$ vs. H_σ, and (c) $\Delta r_{e,3.7-2.1}$ vs. H_σ. All histograms in this figure are bin-normalized with respect to the PDF of H_σ. Note that thin clouds with $\tau < 5$ have been excluded in this figure, but results are similar if we include thin pixels (not shown).

shifting from near-zero to the negative side when H_σ exceeds about 0.3. The potential reasons for the dependence of $\Delta r_{e,3.7-2.1}$ on cloud inhomogeneity H_σ will be discussed later in sections 4.4.4 and 4.4.5.

In addition to r_e, τ, and H_σ, we have also investigated the dependence of $\Delta r_{e,3.7-2.1}$ on other factors such as cloud top temperature, solar zenith angle, satellite viewing angle, scattering angle, surface reflectance, etc., none of which shows an impact on $\Delta r_{e,3.7-2.1}$ as dramatic and clear as r_e, τ, and H_σ.

4.4 Potential reasons for the spectral difference

The fact that $\Delta r_{e,3.7-2.1}$ is correlated with several key cloud parameters, including r_e, τ, and H_σ, indicates that the spectral dependence of the MODIS r_e retrieval is a complex issue likely caused by multiple mechanisms. Indeed, several hypotheses have been proposed to explain the causes of $\Delta r_{e,3.7-2.1}$ and its behaviors described in the previous section. This section provides an overview of these hypotheses. It is helpful to begin with a classification. The existing hypotheses can be grouped into two categories based on their underlying mechanisms and consequent implications. In one category are those related to the inherent limitations of the bi-spectral method, such as retrieval uncertainties (section 4.4.1), the sub-pixel inhomogeneity (section 4.4.4), and lack of consideration of 3D radiative effects (section 4.4.5) in the retrieval method. In such cases, $\Delta r_{e,3.7-2.1}$ carries little, if any, information about the microphysical property of the pixel, but is rather a retrieval artifact indicating significant uncertainties in either $r_{e,2.1}$ or $r_{e,3.7}$, or both. In the other category are those related to the fact that $r_{e,3.7}$ and $r_{e,2.1}$ have different sensitivities to cloud vertical structure (section 4.4.2) and the presence of large drizzle drops (section 4.4.3) in warm water clouds. The implication of this is that the spectral r_e difference actually contains useful information about the cloud that can be used and actually has been used for remote sensing and model validation.

4.4.1 Random error

As shown in Fig. 4.4, $\Delta r_{e,3.7-2.1}$ for clouds with $\tau < 5$ varies more widely than that for thicker clouds. The large variation of $\Delta r_{e,3.7-2.1}$ for thin clouds is likely caused by random retrieval uncertainties. There are many sources of uncertainty in the MODIS retrieval, such as instrument uncertainties, ancillary data uncertainties, and discretization and interpolation of the LUT. These sources are usually uncorrelated, leading to random errors in the observed cloud reflectances. When the cloud is thin, the signal from the cloud is comparable to, or even smaller than, the retrieval uncertainties, which can result in large errors in $r_{e,2.1}$ and $r_{e,3.7}$, and a highly variable $\Delta r_{e,3.7-2.1}$. An example to illustrate the impact of random uncertainties on MODIS τ and r_e retrievals is given in Fig. 4.6. In the example, one point in the MODIS operational LUT, with $\tau = 4.1$ and $r_e = 16$ μm, (i.e. the center of the cross in Fig. 4.1), is chosen for the purpose of illustration. Uncorrelated random errors, which are assumed to follow the normal distribution with standard deviation $\sigma = 10\%$, are added to the reflectances of this point at all three bands to mimic the above-mentioned retrieval uncertainties. The magnitude (one standard deviation) of error is indicated by the size of the crosses in Fig. 4.1. One million such samples are generated and then used to obtained τ and r_e retrieval samples. The probability density functions (PDF) of τ retrievals are shown in Fig. 4.6a.

The two τ retrievals, one based on 0.86-μm and 2.1-μm band LUT and the other based 0.86-μm and 3.7-μm band LUT, both vary closely around the true value of $\tau = 4.1$ and they agree with each other very well. The PDFs of $r_{e,2.1}$, $r_{e,3.7}$, and $\Delta r_{e,3.7-2.1}$ are shown in Fig. 4.6b. Interestingly, $r_{e,2.1}$ has a narrower PDF than $r_{e,3.7}$, even though the same magnitude of error (i.e. 10% in reflectance) is given to the two bands. This difference is mainly due to the fact that the 0.86-μm and 2.1-μm band LUT ($LUT_{0.86\&2.1}$ hereafter) is generally more condensed in the r_e direction than the 0.86-μm and 3.7-μm band LUT ($LUT_{0.86\&3.7}$) (i.e. $|\partial \ln R(2.1\ \mu\text{m})/\partial r_e| < |\partial \ln R(3.7\ \mu\text{m})/\partial r_e|$). As a result, the same magnitude of reflectance error (e.g. 10%) leads to different r_e error. The fact that $LUT_{0.86\&2.1}$ is generally less orthogonal than $LUT_{0.86\&3.7}$ also plays a role. The orthogonality of LUT is determined by the extent to which the cloud reflectance in the SWIR band is independent from that in the 0.86-μm band. Because is generally more orthogonal than (i.e. $|\partial R(2.1\ \mu\text{m})/\partial R(0.86\ \mu\text{m})| > |\partial R(3.7\ \mu\text{m})/\partial R(0.86\ \mu\text{m})|$), error in the 0.86-$\mu$m band has less impact on $r_{e,3.7}$ than $r_{e,2.1}$, which contributes to the PDF difference between $r_{e,2.1}$ and $r_{e,3.7}$ in Fig. 4.6b. This effect of LUT orthogonality is demonstrated in Fig. 4.6c and 4.6d, which shows the PDFs of τ retrievals (Fig. 4.6c), $r_{e,2.1}$, $r_{e,3.7}$, and $\Delta r_{e,3.7-2.1}$ (Fig. 4.6d) when we only add random errors to $R(0.86\ \mu\text{m})$ but *not* to $R(2.1\ \mu\text{m})$ and $R(3.7\ \mu\text{m})$. Due to the LUT orthogonality difference, the error, which is solely in the $R(0.86\ \mu\text{m})$, causes only a small error in $r_{e,3.7}$ but causes a significant error in $r_{e,2.1}$ retrievals. This difference in error between the two bands leads to $\Delta r_{e,3.7-2.1}$ varying between $-5\ \mu$m and $5\ \mu$m. It is worth mentioning here that the orthogonality of LUT also plays an important role in the so-called plane-parallel bias that will be discussed in section 4.4.4. As a result, the widths of the $r_{e,2.1}$, $r_{e,3.7}$, and the $\Delta r_{e,3.7-2.1}$ PDFs in Fig. 4.6b are quite wide, which suggests that retrieval uncertainties caused by instrument uncertainties, ancillary data uncertainties, etc. probably play an important role in causing the large variation of $\Delta r_{e,3.7-2.1}$ for thin clouds in Fig. 4.4.

As shown in Fig. 4.6, random error in reflectance tends to result in random error in $r_{e,2.1}$ and $r_{e,3.7}$, which in turn leads to a $\Delta r_{e,3.7-2.1}$ with a mean value close to zero. Therefore, the random retrieval uncertainties cannot explain why $r_{e,3.7}$ is systematically smaller than $r_{e,2.1}$, especially when cloud is thick (i.e. high signal-to-noise ratio).

4.4.2 Vertical cloud structure

Several studies have shown that $r_{e,2.1}$ and $r_{e,3.7}$ retrieval have different sensitivities to the in-cloud vertical structure of cloud droplet microphysics, leading to spectral difference in r_e retrieval (Platnick, 2000; Kokhanovsky, 2004; Nakajima et al., 2010a; Zhang et al., 2010). The concept of 'weighting function' developed by Platnick (2000) provides a convenient framework to assess the sensitivity of r_e retrieval based on the bi-spectral method to the vertical structure of cloud. One of the two weighting functions introduced in Platnick (2000) is based on the maximum penetration depth of photon into cloud. This weighting function, $w_m(\tau, \tau_c)$, is defined as

$$w_m(\tau, \tau_c) = \frac{1}{R(\tau_c)} \frac{dR(\tau)}{d\tau}\,, \tag{4.4}$$

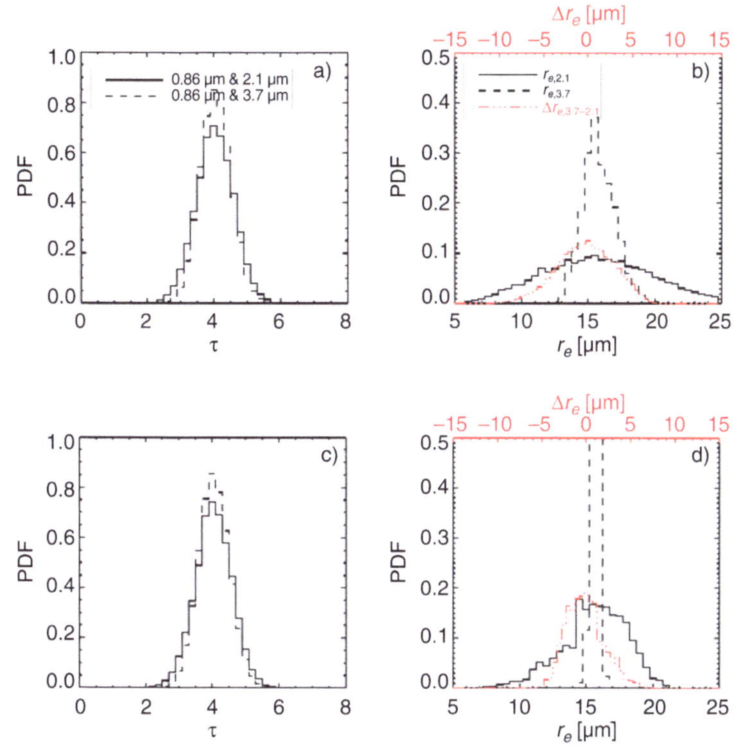

Fig. 4.6. An example to illustrate the impact of random error in cloud reflectance on τ and r_e retrievals. In (a) and (b), random error is assigned to both the 0.86-μm band and the SWIR band (2.1-μm or 3.7-μm band). In (c) and (d), random error is only assigned to the SWIR band. Impacts on τ retrievals are shown in (a) and (c). Impacts on r_e retrievals and Δr_e are shown in (b) and (d). See text for details.

where τ_c is the total cloud optical thickness, τ is the optical depth of a level in cloud from cloud top, $R(\tau_c)$ is the cloud reflectance, and $dR(\tau) = R(\tau + d\tau) - R(\tau)$ represents the fraction of all reflected photons that penetrate to a maximum optical depth between τ and $\tau + d\tau$. As shown in Platnick (2000), the shape of $w_m(\tau, \tau_c)$ can be used to quantitatively interpret the impact of the in-cloud vertical structure on MODIS retrievals. Additionally, it is shown that the value predicted from the equation

$$r_e^* = \int_0^{\tau_c} r_e(\tau) w_m(\tau, \tau_c) \, d\tau \,, \tag{4.5}$$

where r_c^*, is the expected r_e retrieval value based on the vertical weighting, agrees to within ± 1 μm with the value from full radiative transfer simulation. Note that the vertical structure of cloud has little impact on cloud optical thickness retrieval.

Figure 4.7 shows an example of $w_m(\tau, \tau_c)$ for $r_{e2.1}$ (black line) and $r_{e3.7}$ (red line) for a cloud layer with adiabatic cloud vertical structure, as shown by the dashed $r_e(\tau)$ line. The $w_m(\tau, \tau_c)$ for $r_{e3.7}$ peaks at cloud top, while $w_m(\tau, \tau_c)$ for $r_{e2.1}$ peaks lower in the cloud. This difference and Eq. (4.5) suggest that $r_{e3.7}$ retrieval

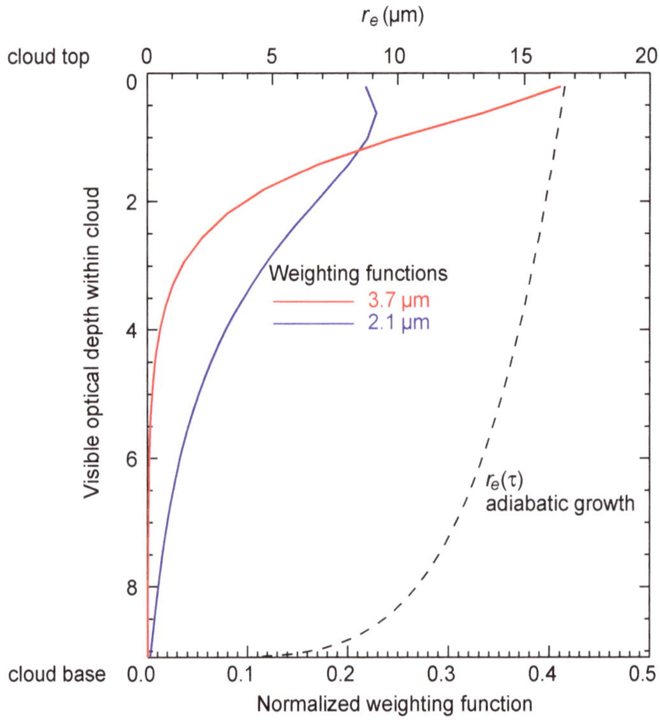

Fig. 4.7. Normalized weighting function $w_m(\tau, \tau_c)$ for $r_{e2.1}$ (blue line) and $r_{e3.7}$ (red line) as a function of cloud optical depth for a cloud with adiabatic vertical structure. Dashed line shows the vertical variation of r_e as a function of cloud optical depth.

has more sensitivity to the cloud microphysics at the uppermost part of cloud while $r_{e2.1}$ retrieval penetrates deeper into the cloud with maximum sensitivity at about $\tau \sim 2$ or 3. For the case in Fig. 4.7, $r_{e2.1}$ and $r_{e3.7}$ predicted from $w_m(\tau, \tau_c)$ based on Eq. (4.5) are 15.2 μm and 16.1 μm. Both agree well with full retrievals.

It is important to note that the general shape of the weighting function is *not* sensitive to the detailed cloud vertical structure. Zhang et al. (2010) demonstrated that the weighting functions for two cases, one with $r_e(\tau)$ increasing and the other decreasing from cloud top toward cloud base, are quite similar to each other. The invariance of the weighting function shape, according to Eq. (4.5), suggests that the retrieved r_e^* is largely determined by cloud vertical structure $r_e(\tau)$. And so is $\Delta r_{e,3.7-2.1}$ as far as cloud vertical structure is concerned. As a consequence, $r_{e,3.7} > r_{e,2.1}$ ($\Delta r_{e,3.7-2.1} > 0$) according to Eq. (4.5) when the cloud droplet size increases from cloud base toward cloud top, and vice versa when cloud droplet size decreases from cloud base toward cloud top. Various microphysical processes can affect cloud vertical structure, such as adiabatic growth, entrainment mixing, collision-coalescence, and sedimentation. It is therefore important to understand the influence of these processes on the vertical structure of cloud microphysics and their consequent impact on $\Delta r_{e,3.7-2.1}$.

In a classic adiabatic growth model (Brenguier et al., 2000), cloud r_e increases monotonically from cloud base to cloud top. In such a case, $r_{e,3.7}$ should be larger than $r_{e,2.1}$, as shown in the example in Fig. 4.7. Indeed, some recent studies, such as Painemal and Zuidema (2011) and Zhang and Platnick (2011), have found some positive $\Delta r_{e,3.7-2.1}$, which is qualitatively consistent with the classic adiabatic growth model as shown in Fig. 4.7. However, as mentioned in section 4.3, only a small fraction of MODIS MBL cloud observations have positive $\Delta r_{e,3.7-2.1}$. Moreover, because the adiabatic condensation growth becomes less efficient as the droplet grows larger, the droplet growth rate dr_e/dz remains relative small over the upper portion of the water cloud (Brenguier et al., 2000). As a result, the magnitude of $\Delta r_{e,3.7-2.1}$ induced by adiabatic cloud vertical structure is expected to be small, generally smaller than 2 μm (Platnick, 2000).

Cloud top entrainment could lead to a decreasing-with-height r_e structure at cloud top. Several recent studies speculated that this cloud top entrainment-induced structure plays an important role in causing smaller $r_{e,3.7}$ retrieval than $r_{e,2.1}$ (Breon and Doutriaux-Boucher, 2005; Seethala and Horváth, 2010). It should be pointed out that the decreasing-with-height r_e structure at cloud top assumed in these studies seems to indicate a homogenous mixing process (Baker et al., 1980). On the contrary, observational studies have actually found more inhomogeneous mixing cases than homogenous mixing cases (Gerber et al., 2005). A couple of recent studies analyzed *in situ* measurements of MBL clouds off the coast of Peru made during the VOCALS-REx (or American Monsoon Systems (VAMOS) Ocean-Cloud-Atmosphere-Land Study Regional Experiment) campaign (Painemal and Zuidema, 2011; King et al., 2012). They did not observe the decreasing-with-height r_e structure at cloud top. In addition, the strong correlation of $\Delta r_{e,3.7-2.1}$ with $r_{e,2.1}$ and the threshold behavior of $\Delta r_{e,3.7-2.1}$ at $r_{e,2.1} \sim 15$ μm (Fig. 4.3) do not easily fit into the cloud top entrainment structure argument.

Warm rain processes (e.g. collision-coalesce) give rise to embryo drizzle drops at cloud top, and also make them grow bigger as they fall from the cloud top, potentially leading to a decreasing-with-height r_e structure from cloud top toward cloud base (Berry, 1967; Berry and Reinhardt, 1974; Pruppacher and Klett, 1997). Therefore, cloud vertical structure induced by warm rain processes has been argued in many studies to be the primary reason for the large negative $\Delta r_{e,3.7-2.1}$ (Chang and Li, 2002, 2003; Chen et al., 2007; Nakajima et al., 2010a, 2010b; Kokhanovsky and Rozanov, 2011). Some behaviors of $\Delta r_{e,3.7-2.1}$ shown in section 4.3 seem in favor of this argument. For example, it is seen in Fig. 4.3b that $\Delta r_{e,3.7-2.1}$ is close to zero when $r_{e,2.1} < 15$ μm and decreases quickly with $r_{e,2.1}$ when $r_{e,2.1} >$ 15 μm. This threshold behavior could be explained as a result of increasing drizzle probability when cloud r_e becomes larger than 15 μm (Gerber, 1996; Nakajima et al., 2010a, 2010b). Although there is increasing evidence indicating that warm rain processes play an important role in causing the large negative $\Delta r_{e,3.7-2.1}$, there are also studies suggesting otherwise. For example, a couple of case studies based on large-eddy simulation of MBL clouds and radiative transfer simulations found only a negligible impact of drizzle on both r_e retrievals and thereby $\Delta r_{e,3.7-2.1}$ (Zinner et al., 2010; Zhang et al., 2012). Although it is difficult to draw any statistical conclusions from these case studies, they demonstrated a new path toward better understanding the impact of drizzle on MODIS r_e retrieval that is worthy of further exploration.

Finally, it should be noted that some $\Delta r_{e,3.7-2.1}$ behaviors, the correlation of $\Delta r_{e,3.7-2.1}$ with cloud horizontal inhomogeneity index H_σ shown in section 4.3 in particular, cannot be easily explained by cloud vertical structure. This indicates that other factors also have a role.

4.4.3 Cloud droplet size distribution

As mentioned in section 4.2, in the operational MODIS retrieval algorithm, cloud DSD follows the monomodal Gamma distribution with fixed effective variance $v_e = 0.1$ (i.e. Eq. (4.2)). Several studies have shown that, when cloud microphysics deviates from what is assumed in the MODIS retrieval algorithm, the retrieved r_e also deviates from the true r_e. Invalid microphysics assumptions affect $r_{e,2.1}$ and $r_{e,3.7}$ retrievals to a different extent, leading to spectral differences (Chang and Li, 2001; Minnis et al., 2004; Zhang, 2013).

The monomodal Gamma distribution assumption for cloud DSD is especially problematic when MBL clouds are precipitating, because warm rain processes (e.g. collision-coalescence) could broaden cloud DSD and even give rise to a second mode, the so-called drizzle or rain mode, creating *bi-modal* DSD (Berry, 1967; Berry and Reinhardt, 1974; Pruppacher and Klett, 1997). In such a case, the MODIS retrieval process is to find an r_e^* (the superscript * is to indicate that it is a retrieved value under monomodal DSD assumption) that satisfies the following equation:

$$R_{\lambda,LUT}\left(r_e^*\right)\big|_{v_e} = R_\lambda[n(r)]\,, \tag{4.6}$$

where R_λ is the observed SWIR band cloud reflectance that is a function of the true *bi-modal* cloud DSD $n(r)$ and $R_{\lambda,LUT}$is the SWIR band cloud reflectance in the LUT which is pre-computed based on the monomodal DSD assumption. Additionally, λ denotes the wavelength of the SWIR band used for r_e retrieval (e.g. 2.1 μm or 3.7 μm).

Recently, Zhang (2013) developed a semi-analytical model to explain and predict MODIS r_e retrieval results from Eq. (4.6). In a numerical test, the r_e retrievals predicted by this model agree well with numerical solutions based on radiative transfer simulations (Zhang, 2013). The model is based on the fact that dependence of R_λ on cloud microphysics arises mainly from the dependence of cloud single-scattering albedo ω_λ on cloud microphysics (Zhang et al., 2009). Figure 4.8a shows the $\omega_{2.1\ \mu\mathrm{m}}$ (solid blue line) and $\omega_{3.7\ \mu\mathrm{m}}$ (solid red line) of individual cloud droplets computed using a MIE code (Wiscombe, 1979) as a function of droplet radius over the range $r \in [1\ \mu\mathrm{m},\ 1000\ \mu\mathrm{m}]$. Figure 4.8b shows the so-called bulk scattering albedo $\overline{\omega}_{2.1\ \mu\mathrm{m},LUT}$ and $\overline{\omega}_{3.7\ \mu\mathrm{m},LUT}$ in the MODIS LUT, which are averaged over monomodal Gamma DSD, for $r_e \in [5\ \mu\mathrm{m},\ 30\ \mu\mathrm{m}]$. Zhang (2013) expanded $\omega_{2.1\ \mu\mathrm{m}}$ and $\omega_{3.7\ \mu\mathrm{m}}$ into polynomials of r:

$$\omega_\lambda(r) = \sum_{i=0}^{N_\lambda} k_{i,\lambda} r^i\,, \tag{4.7}$$

where $k_{i,\lambda}$ are fitting constants given in Table 4.1, and $\overline{\omega}_{2.1\ \mu\mathrm{m},LUT}$ and $\overline{\omega}_{3.7\ \mu\mathrm{m},LUT}$ into a linear function of r_e:

$$\overline{\omega}_\lambda(r_e) \approx \overline{k}_{0,\lambda} + \overline{k}_{1,\lambda} r_e\,, \tag{4.8}$$

Table 4.1. Fitting coefficients in Eq. (4.7) and the coefficients (values in parentheses) for each nonlinear term in Eq. (4.9). The unit of droplet radius in Eq. (4.7) is micron.

	k_0 (r_0)	k_1 ($k_1/\overline{k}_1$)	k_2 ($k_2/\overline{k}_1$)	k_3 ($k_3/\overline{k}_1$)	k_4 ($k_4/\overline{k}_1$)	k_5 ($k_5/\overline{k}_1$)
2.1 μm	9.952E-01	−1.736E-03	2.442E-06	−1.187E-09		
	(1.17 μm)	(9.266E-01)	(−1.303E-03)	(6.335E-07)		
3.7 μm	9.623E-01	−6.443E-03	3.348E-05	−7.606E-08	7.757E-11	−2.907E-14
	(1.74 μm)	(9.314E-01)	(−4.840E-03)	(1.099E-05)	(−1.121E-08)	(4.201E-12)

where fitting constants $\overline{k}_{0,\lambda}$ and $\overline{k}_{1,\lambda}$ are given in Table 4.2. Without going into detail, Zhang (2013) arrived at the following equation to explain and predict MODIS r_e retrieval results from Eq. (4.6):

$$r^*_{e,\lambda} = r_{0,\lambda} + \frac{k_{1,\lambda}}{\overline{k}_{1,\lambda}} r_e + \sum_{i=2}^{N_\lambda} \frac{k_{i,\lambda}}{\overline{k}_{1,\lambda}} \frac{\langle r^{i+2} \rangle}{\langle r^2 \rangle}, \tag{4.9}$$

where $r^*_{e,\lambda}$ is the predicted MODIS r_e retrieval, $\langle r^i \rangle = \int_0^\infty r^i n(r)\, dr$ is the ith moment of the true DSD, and $r_{0,\lambda} = (k_{0,\lambda} - \overline{k}_{0,\lambda})/\overline{k}_\lambda$ is a constant. With the coefficients known, r^*_e can be easily solved from the above equation once the $\langle r^{i+2} \rangle / \langle r^2 \rangle$ terms are derived from any given DSD.

Table 4.2. Fitting coefficients in Eq. (4.8). The unit of droplet radius in Eq. (4.8) is micron.

	$\overline{k}_0$	$\overline{k}_1$
2.1 μm	9.974E-01	−1.874E-03
3.7 μm	9.744E-01	−6.918E-03

It can be seen from Table 4.1 that $k_{1,\lambda}/\overline{k}_\lambda$ is close to unity, which suggests that r^*_e should be close to the true value of r_e, if the higher-order terms, namely $(k_{i,\lambda} \langle r^{i+2} \rangle) / (\overline{k}_\lambda \langle r^2 \rangle)$, in Eq. (4.9) are small. The magnitude of higher-order terms is determined by two competing factors. On one hand, the $\langle r^{i+2} \rangle / \langle r^2 \rangle$ term, which is in the same order of magnitude as r^i_e, increases algorithmically with i. On the other hand, as seen in Table 4.1, the $k_{i,\lambda}/\overline{k}_{1,\lambda}$ term decreases with polynomial order i. Provided that r_e ranges generally from a few to a few tens of microns, the decrease of $k_{i,\lambda}/\overline{k}_{1,\lambda}$ terms can be expected to be dominant, leading to a decreasing impact of nonlinear terms on r^*_e with increasing order i. Therefore, the most important nonlinear term is the second-order term ($i = 2$). Because $k_{2,\lambda}/\overline{k}_{1,\lambda} < 0$ (see Table 4.1) and $\langle r^4 \rangle / \langle r^2 \rangle > 0$, the sign of this term is negative, implying that r^*_e tends to underestimate the true value when this term is significant. Physically, this term is a result of the nonlinear relationship between $\overline{\omega}_\lambda$ and r. As seen in Fig. 4.8a, ω_λ decreases monotonically with r, and the rate becomes slower as ω_λ approaches

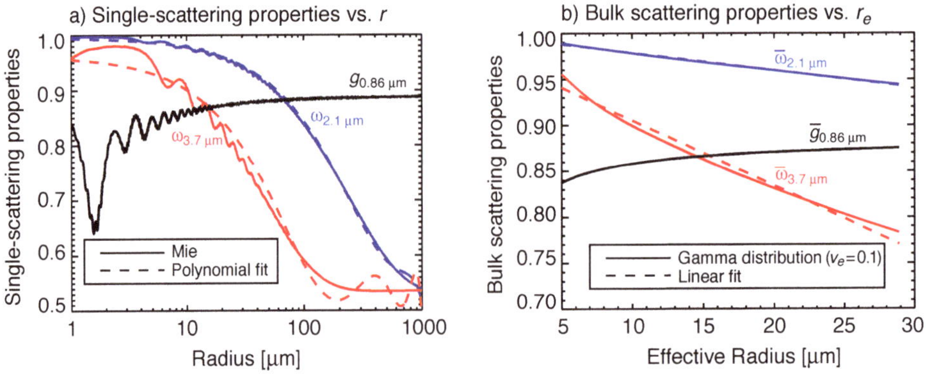

Fig. 4.8. (a) The single-scattering albedo of cloud droplet in the 2.1-μm (blue) and 3.7-μm (red) band, and the asymmetry factor in the 0.86-μm band (black) plotted as a function of cloud droplet radius. The dashed lines indicate polynomial fitting to the single-scattering albedos based on Eq. (4.7). (b) The bulk single-scattering albedo in the 2.1-μm (blue) and 3.7-μm (red) band, and the asymmetry factor in the 0.86-μm band (black) plotted as a function of cloud droplet effective radius. The dashed lines indicate polynomial fitting to the single-scattering albedos based on Eq. (4.8). In the computation, monomodal Gamma size distribution with $v_e = 0.1$ is assumed.

the asymptotic value of 0.5. Thus, the curvature of the relationship between ω_λ and r results in the negative value of the $k_{2,\lambda}/\overline{k}_{1,\lambda}$ term. The magnitude of this term is determined by the ratio $\langle r^4 \rangle / \langle r^2 \rangle$ which can be shown to be equivalent to $r_e^2(1 + v_e)$. This relationship reveals that the impact of the second-order term depends on r_e and to a much lesser extent on v_e, namely the width of the cloud DSD. This relationship also implies that, in order for the nonlinear term to significantly affect $\overline{\omega}_\lambda$, both r_e and v_e have to be sufficiently large. Interestingly, such conditions happen to be favored in precipitating clouds, as they tend to have larger r_e and precipitating processes (e.g. collision-coalescence) tend to broaden the DSD (i.e. increase v_e) (Berry, 1967; Berry and Reinhardt, 1974; Pruppacher and Klett, 1997).

Focusing now on the spectral dependence of Eq. (4.9), it is seen from Table 4.1 that the magnitude of $k_{i,2.1\mu m}/\overline{k}_{1,2.1\mu m}$ is substantially smaller than $k_{i,3.7\mu m}/\overline{k}_{1,3.7\mu m}$. It indicates a stronger impact of the nonlinear terms on r_e retrieval in the 3.7-μm band than the 2.1-μm band. For the $r^*_{e,2.1\mu m}$ retrieval, only the second-order nonlinear term is significant. But, for the $r^*_{e,3.7\mu m}$ retrieval, higher-order terms need also to be included, although the second-order term is still the dominant nonlinear term. As discussed above, the second-order nonlinear term tends to result in underestimated r_e^* retrieval. This underestimation is expected to be stronger in the 3.7-μm band than in the 2.1-μm band, resulting in negative $r^*_{e,3.7\mu m} - r^*_{e,2.1\mu m}$. This theoretical prediction is consistent with the numerical results in Minnis et al. (2004) and Nakajima et al. (2010a). It was found in these studies that increasing rain mode in a bi-modal DSD leads to underestimated $r_e$retrieval and the underestimation is more severe in the 3.7-μm band than the 2.1-μm band.

In summary, Zhang (2013) illustrated that r_e retrieval-based monomodal DSD assumption tends to underestimate the r_e of bi-modal DSD, due to the nonlinear relationship between cloud droplet single-scattering albedo and cloud droplet size. The underestimation is stronger for $r_{e,3.7\mu m}$ than for $r_{e,2.1\mu m}$, leading to a negative $\Delta r_{e,3.7-2.1}$.

4.4.4 Plane-parallel r_e bias

As mentioned in section 4.2, one of the fundamental assumptions made in the MODIS retrieval algorithm is that a cloudy 1 km × 1 km pixel is horizontally homogenous. However, it is known that clouds can have significant horizontal variability at smaller scales, leading to sub-pixel inhomogeneity (e.g. Davis et al., 1994; Marshak et al., 1995). When the (horizontally) homogeneous pixel assumption breaks down, both MODIS τ and r_e retrievals face challenging issues. For τ retrieval, for example, sub-pixel inhomogeneity can cause the well-known plane-parallel albedo bias. That is, the τ retrieval based on the average reflectance of a heterogeneous cloudy pixel tends to be smaller than the average of the sub-pixel scale τ (e.g. Cahalan et al., 1994; Barker, 1996; Oreopoulos and Davies, 1998). Sub-pixel inhomogeneity also affects r_e retrieval in various ways, one of which is the so-called plane-parallel r_e bias (Zhang and Platnick, 2011; Zhang et al., 2012). Plane-parallel r_e bias tends to result in overestimation of r_e and it affects the $r_{e2.1}$ retrieval more than $r_{e3.7}$, therefore leading to spectral difference (negative $\Delta r_{e,3.7-2.1}$) (Zhang and Platnick, 2011; Zhang et al., 2012).

The plane-parallel r_e bias is defined as the impact of *small-scale variability in* τ on r_e retrievals that use area-averaged reflectance (Zhang and Platnick, 2011; Zhang et al., 2012). This bias is illustrated using two idealized examples in Fig. 4.9, which shows forward calculations of reflectances in 0.86-μm, 2.1-μm and 3.7-μm bands. In Fig. 4.9a and 4.9b, we assume that half of a MODIS pixel overlying a black surface is covered by a cloud with $\tau_1 = 2.8$ and $r_e = 8$ μm and the other half is covered by a cloud with $\tau_2 = 30.8$ and $r_e = 8$ μm. Panels c and d assume the same optical thickness but use $r_e = 18$ μm. Focusing on the τ retrieval, the figure illustrates the well-known 'plane-parallel albedo bias' (Cahalan et al., 1994): the retrieved τ based on the mean reflectance of inhomogeneous pixels tends to be smaller than the mean of the sub-pixel τ. In this example the value of retrieved from the averaged reflectance ($\tau = 10.8$) is substantially smaller than the average value ($\overline{\tau} = 16.8$).

This problem is more acute for retrievals of r_e because the reflectance used to infer r_e also depends on τ over much of the range of plausible values, as has been shown in section 4.4.1. If the reflectance at non-absorbing and absorbing wavelengths depended only on τ and r_e, respectively (i.e. if the LUT was orthogonal), reflectance at absorbing wavelengths would be uniform in our example and particle size could be retrieved perfectly. As Fig. 4.9 demonstrates, however, the LUT is not orthogonal. The nonlinearity leads to a simultaneous underestimation of τ (plane-parallel-albedo bias) and overestimation of r_e (plane-parallel r_e bias). The area over which this is true is larger in the less-absorbing band, which explains why $r_{e,3.7}$ is smaller than $r_{e,2.1}$. The impact becomes more pronounced as r_e increases: in Fig. 4.9c and 4.9d, the true $r_e = 18$ μm while $r_{e,2.1}$ and $r_{e,3.7}$ retrieved from averaged reflectances are 24 μm and 18.1 μm, respectively, resulting in a $\Delta r_{e,3.7-2.1}$ around -6 μm.

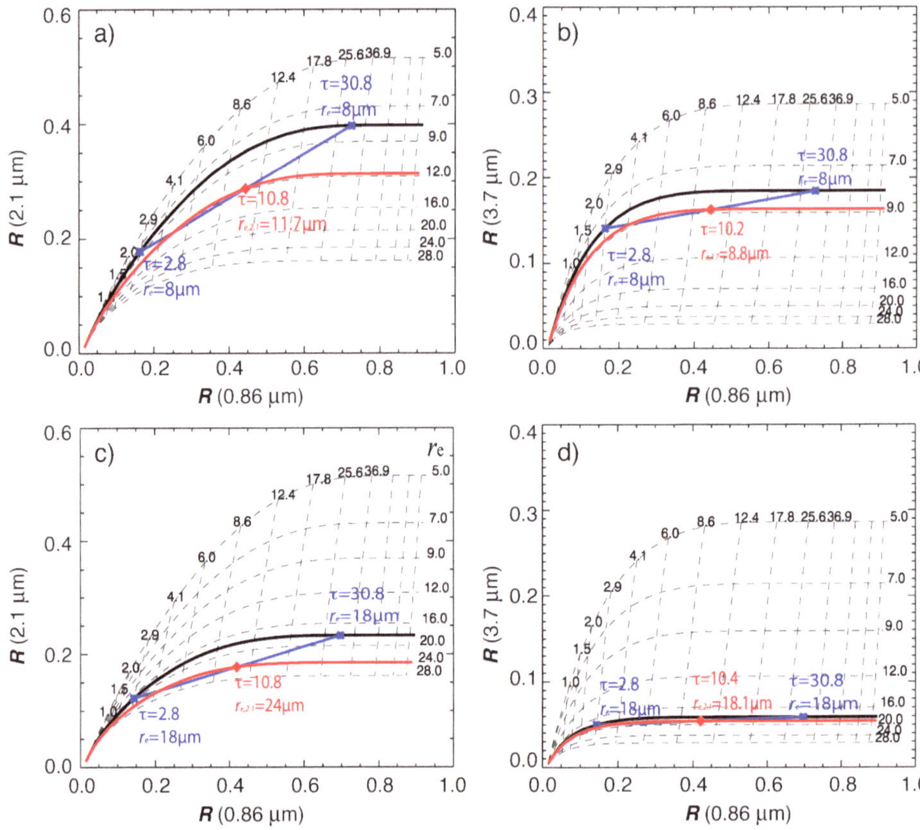

Fig. 4.9. Two theoretical cases to illustrate the nonlinearity effect in r_e retrievals resulting from sub-pixel cloud inhomogeneity. Numbers on top of the Nakajima–King look-up-table (LUT) curves correspond to values of τ contour lines in the LUT and the numbers on the right of the curves correspond to values of r_e contour lines in the LUT.

Recently, Zhang et al. (2012) developed a numerical MODIS retrieval simulator based on the combination of an LES model with bin microphysics (Ackerman et al., 2004; Fridlind and Ackerman, 2010) and radiative transfer models (both 1D and 3D). In the case study based on this simulator, they confirmed that the r_e plane-parallel bias, as conceptually illustrated in Fig. 4.9, indeed play an important role in causing the spectral difference between $r_{e,2.1}$ and $r_{e,3.7}$, especially for those pixels with high sub-pixel inhomogeneity.

As shown earlier in Fig. 4.5a, the statistics of MODIS $r_{e,2.1}$ show a strong dependence on the sub-pixel inhomogeneity index, H_σ. When a cloud is relatively homogeneous (i.e. $H_\sigma < 0.3$) the peak of the PDF of MODIS $r_{e,2.1}$ is around $12 \sim 14$ μm. When H_σ becomes larger than about 0.3, the peak increases quickly with H_σ, by up to about 20 μm when H_σ is about unity. Interestingly, MODIS $r_{e,3.7}$ shows little dependence in Fig. 4.5b, leading to significant negative $\Delta r_{e,3.7-2.1}$ when $H_\sigma > 0.3$. The behaviors of $r_{e,2.1}$, $r_{e,3.7}$ and $\Delta r_{e,3.7-2.1}$ in Fig. 4.5 are consistent with the concept of r_e plane-parallel bias as shown Fig. 4.9, therefore suggesting

r_e plane-parallel bias is an important factor causing the $\Delta r_{e,3.7-2.1}$. On the other hand, Fig. 4.5 also suggests that the impact of r_e plane-parallel bias is limited to those highly inhomogeneous pixels, with H_σ larger than about 0.3.

4.4.5 3D radiative transfer effect

As shown in a couple of recent studies (Zhang and Platnick, 2011; Zhang et al., 2012), 3D radiative transfer effects generally influence $r_{e,2.1}$ and $r_{e,3.7}$ retrievals to different extents. This makes 3D radiative transfer a potentially important factor in causing the spectral dependence of MODIS r_e retrieval as described in section 4.3.

As mentioned in section 4.2, the MODIS retrieval algorithm assumes the so-called plane-parallel cloud model. However, clouds in reality have significant horizontal structures and variabilities at various scales, from thousands of kilometers to a few meters (Cahalan et al., 1994; Davis et al., 1994; Wood and Hartmann, 2006; Wood et al., 2008). Horizontal cloud variability gives rise to horizontal photon transport (Davis and Marshak, 2010), which makes the observed reflectance dependent on not only the property of pixels within the field of view, but also the properties of surroundings (Marshak et al., 1995). In such circumstances, cloud properties retrieved under the plane-parallel cloud assumption are subject to significant errors. This is the so-called 3D radiative transfer effect. Take the so-called 'shadowing effect', for example (Várnai and Marshak, 2002; Marshak et al., 2006). Consider, for example, a cloudy pixel that is in the shadow of an adjacent pixel. The darkening of the pixel would be fallaciously interpreted under the plane-parallel cloud assumption, leading to underestimated τ and overestimated r_e retrievals (Várnai and Marshak, 2002; Marshak et al., 2006; Zhang and Platnick, 2011; Zhang et al., 2012).

It was shown in a couple of recent studies (Zhang and Platnick, 2011; Zhang et al., 2012) that 3D radiative transfer effects generally have a stronger impact on the $r_{e,2.1}$ retrieval than on the $r_{e,3.7}$ retrieval. For example, when the shadowing effect leads to an overestimation of r_e, the magnitude of overestimation is larger for $r_{e,2.1}$ than for $r_{e,3.7}$. The reason for this is two-fold. First, 3D radiative transfer effect results from horizontal photon transport and the strength of droplet absorption in the 3.7-μm band exerts a stronger limit on horizontal photon transport than the weaker 2.1-μm band absorption. Secondly, as shown in sections 4.4.1 and 4.4.4, the LUT for the 0.86-μm and 2.1-μm combination is less orthogonal than that for the 0.86-μm and 3.7-μm combination. As a consequence, the $r_{e,2.1}$ retrieval is more sensitive to 3D radiative transfer effects on the 0.86-μm band reflectance than $r_{e,3.7}$ retrieval.

Recently, Zhang et al. (2012) studied the impact of 3D radiative effects, as well as other factors, on MODIS cloud property retrievals using an LES model with a binned microphysical scheme (Ackerman et al., 2004; Fridlind and Ackerman, 2010) coupled with a 3D radiative transfer model (Cahalan et al., 2005; Pincus and Evans, 2009). Figure 4.10 shows one of the LES cases in this study, which is based on an idealized case study (Stevens et al., 2001) from the Atlantic Trade Wind Experiment (ATEX), with an average cloud droplet concentration (weighted by liquid water mixing ratio) of about 30 cm^{-3}. For more details about this LES case and the configurations of radiative transfer models, readers are referred to Zhang et al. (2012). Figure 4.10a provides a planar view of the cloud τ derived from the LES for this case. Figure 4.10b shows the cloud extinction coefficient along the vertical cross-section indicated by the red dashed line in Fig. 4.10a. In radiative transfer simulations, solar zenith angle is assumed to be 20° and viewing zenith angle to be 0°. The 1D and 3D simulations of cloud reflectances along the cross-section for the 0.86-μm, 2.1-μm and 3.7-μm bands are shown in Figs. 4.10c, 4.10d and 4.10e, respectively. Although 1D and 3D simulations are in general agreement, the influence of horizontal photon transport can be clearly seen in a few regions, especially in Fig. 4.10c. For example, at $x = 7$ km the 0.86-μm cloud reflectance based on 1D simulation is substantially larger than 3D simulation. But, on both sides of this reflectance peak (i.e. $x = 6.5$ km and $x = 7.5$ km), 3D reflectance is larger than 1D. These differences between 1D and 3D simulations are the result of horizontal photon transport due to multiple scattering (Várnai and Davies, 1999). Interestingly, this horizontal photon transport effect is much reduced in the 2.1-μm and 3.7-μm simulations. This is because the absorption in the SWIR bands limits multiple scattering and therefore reduces the horizontal transport of photons.

Figure 4.11 shows the impacts of 3D radiative transfer effects on $r_{e,2.1}$ and $r_{e,3.7}$ for the case in Fig. 4.10. In the figure, we use the difference between the τ retrieval based on 3D radiative transfer simulation (referred to as '3D τ') and that based on 1D radiative transfer simulation (referred to as '1D τ') as an index of the 3D radiative transfer effects. By definition, 3D $\tau >$ 1D τ in the case of illumination, and 3D $\tau <$ 1D τ in the case of shadowing. As expected, the r_e retrievals based on 3D radiative transfer simulation ('3D r_e') appear smaller than those based on 1D radiative transfer simulation ('1D r_e') in the case of illuminating effect (i.e. 3D $r_e -$ 1D $r_e < 0$), and vice versa in the case of shadowing effect (i.e. 3D $r_e -$1D $r_e > 0$). What is interesting in Fig. 4.11 is that the impact of 3D radiative effects on $r_{e,2.1}$ is generally stronger than on $r_{e,3.7}$, for the reasons discussed earlier in this section. Another important point to note in Fig. 4.11 is that illuminating and shadowing effects naturally come in pairs and tend to cancel each other out (Marshak et al., 2006). Therefore, 3D radiative effects seem unlikely to cause large systematic bias in r_e retrieval. Moreover, since 3D radiative effects are a result of horizontal photon transport, its impact on MODIS retrievals is limited to pixels with large horizontal inhomogeneity. In addition, when a cloud pixel has strong horizontal inhomogeneity, both 3D radiative effects and the plane-parallel r_e bias discussed in section 4.4.4 may play a role at the same time, creating either positive or negative $\Delta r_{e,3.7-2.1}$ (Zhang et al., 2012).

4.5 Discussion

In the previous section, we reviewed several mechanisms that could cause significant difference between MODIS $r_{e,2.1}$ and $r_{e,3.7}$ retrievals for MBL clouds. But the most fundamental reason is that MBL clouds in reality do *not* always satisfy the ideal cloud model (e.g. plane-parallel and homogenous) on which MODIS is based. When the cloud deviates from the ideal model, the deviation affects the $r_{e,2.1}$ and $r_{e,3.7}$ retrievals to different extents or in different ways, leading to significant difference between the two. From this perspective, spectral dependence is an inherent and

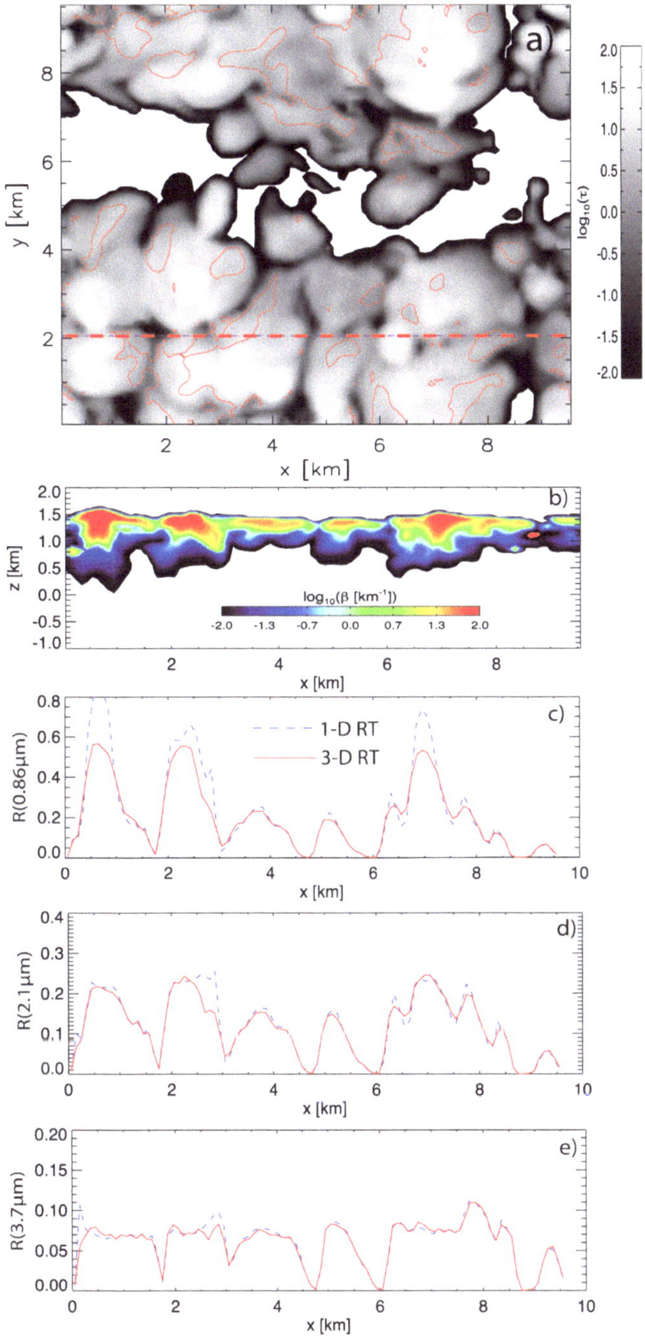

Fig. 4.10. (a) Planar view of the cloud τ of the ATEX clean case at 6 h of simulation time. (b) The cross-section of cloud extinction coefficient (β) along $y = 2$ km in (a). Cloud bi-directional reflectance along the cross section is shown for the (c) 0.86-μm, (d) 2.1-μm, and (e) 3.7-μm MODIS bands simulated using 1D (blue) and 3D (red) radiative transfer models.

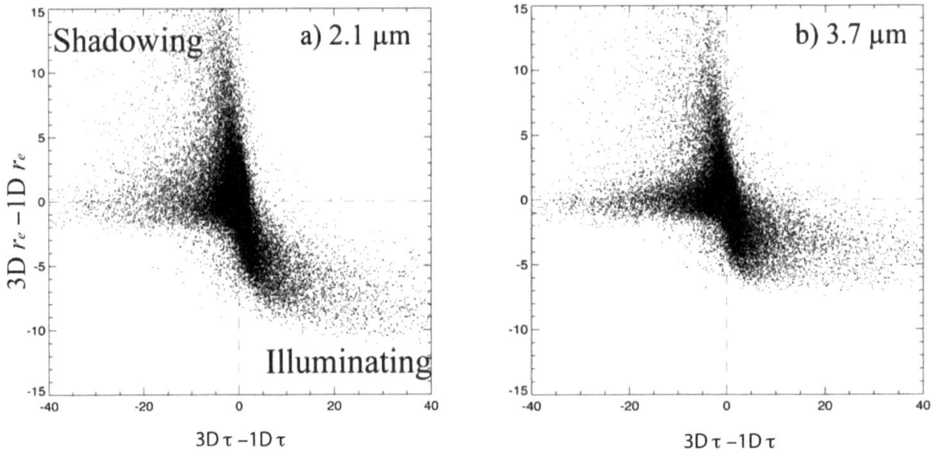

Fig. 4.11. The difference between 3D and 1D τ retrieval vs. the difference between (a) 3D $r_{e,2.1}$ and 1D $r_{e,2.1}$ at LES resolution, (b) 3D $r_{e,3.7}$ and 1D $r_{e,3.7}$ at LES resolution.

inevitable feature of the r_e retrieval based on the bi-spectral method. On one hand, it reflects the limitations of the method but, on the other, it indicates that the operational MODIS retrieval results are generally consistent with our expectations.

4.5.1 Which one is better?

If spectral dependence is inevitable in MODIS r_e retrieval, then a natural question arises: which one, $r_{e,2.1}$ or $r_{e,3.7}$, is better or more correct? Unfortunately, there is no simple answer to this question. As shown in section 4.4.1, $r_{e,3.7}$ is less affected by random retrieval uncertainties, if uncertainties associated with 2.1-μm and 3.7-μm band observations are of similar magnitude. Also, in comparison with $r_{e,2.1}$, $r_{e,3.7}$ is generally less affected by the plane-parallel r_e and 3D radiative effects as discussed in sections 4.4.4 and 4.4.5. In these cases, $r_{e,3.7}$ is expected to perform better than $r_{e,2.1}$. However, as discussed in section 4.4.2, from the perspective of vertical weighting, $r_{e,2.1}$ and $r_{e,3.7}$ simply reflect different parts of the cloud. Simply put, there is no right or wrong. Moreover, as discussed in section 4.4.3, when the true cloud DSD is bi-modal, both $r_{e,2.1}$ and $r_{e,3.7}$ tend to underestimate the true r_e, but the underestimation is more severe in the $r_{e,3.7}$ than in the $r_{e,2.1}$. Therefore, whether to use $r_{e,2.1}$ or $r_{e,3.7}$ depends on the intended application and the geographical location of interest. For example, several studies have attempted to retrieve the vertical profile of MBL clouds utilizing the combination of $r_{e,2.1}$ and $r_{e,3.7}$ retrievals from MODIS (Chang and Li, 2002, 2003; Kokhanovsky and Rozanov, 2011). For these applications, both $r_{e,2.1}$ and $r_{e,3.7}$ are needed and their difference serves as a useful signal rather than a data issue. As shown in Fig. 4.2, $r_{e,2.1}$ and $r_{e,3.7}$ agree quite well over the coastal stratocumulus regions. So, if those regions are of interest, it perhaps makes little difference which retrieval is used. In contrast, over broken cloud regions where clouds have significant horizontal heterogeneity, $r_{e,3.7}$ would be less affected by plane-parallel r_e bias and 3D radiative effect and therefore might be a better choice than $r_{e,2.1}$.

It should be noted here that $r_{e,3.7}$ is reported as having a respective difference to successful $r_{e,2.1}$ retrieval in the Collection 5 of MODIS cloud product. Thus, in Collection 5, the sampling of $r_{e,3.7}$ is biased by the success of $r_{e,2.1}$. In addition, although retrieved in Level 2 products, $r_{e,3.7}$ is not aggregated to Level 3 products. In the current Collection 6 of MODIS cloud product, these sampling biases will be removed and $r_{e,3.7}$ will be sampled independently and aggregated to Level 3 data. These changes will facilitate the use of $r_{e,3.7}$.

4.5.2 Cloud regime classification

Summarizing the discussions in the previous section, one can see that none of the hypotheses can uniquely explain all of the $\Delta r_{e,3.7-2.1}$ features shown in section 4.3. This is surprising because all these mechanisms are more or less entangled together in the retrieval. For example, when a cloud pixel has strong horizontal inhomogeneity, both the plane-parallel r_e bias (section 4.4.4) and 3D radiative effects (section 4.4.5) may play a role at the same time. When warm rain processes begin to develop in MBL clouds, it can lead to both cloud vertical structure and bi-model DSD, in which case both the vertical weighting effect (section 4.4.2) and the DSD sensitivity effect (section 4.4.3) will influence MODIS r_e retrievals. Thus, for a better understanding, we need to untangle these mechanisms and sort out their relative importance in different cloud regimes. By 'cloud regimes', we mean groups of MBL clouds with different characteristics in terms of optical (e.g. τ), microphysical (e.g. r_e), and macrophysical (e.g. horizontal properties). An attempt of such cloud regime classification and analysis has been made in Fig. 4.12. It shows a color contour of the mean value of $\Delta r_{e,3.7-2.1}$ projected on sub-pixel heterogeneity index H_σ and $r_{e,2.1}$ for MBL clouds with $\tau > 5$. The black contour lines indicate the joint frequency histogram of H_σ and $r_{e,2.1}$ based on one month of MODIS observation. Based on the combination of $r_{e,2.1}$ and H_σ, MBL cloud pixels observed by MODIS can be classified into the following three regimes:

- *Regime 1*: MBL cloud pixels in this regime have low horizontal heterogeneity ($H_\sigma < 0.3$) and relatively small effective radius ($r_{e,2.1} < 20$ μm). Low H_σ index suggests that the plane-parallel r_e bias and 3D radiative effects should be small in this regime. Small $r_{e,2.1}$ suggests that the probability of warm rain precipitation is relatively small because the critical value of r_e for collision-coalescence to occur is around 15 μm (Gerber, 1996; Rosenfeld et al., 2012). As a result, the vertical weighting effect and the DSD sensitivity effect can be expected to be small. It now becomes clear that MBL cloud pixels in this regime, with low horizontal heterogeneity and low possibility of precipitation, satisfy those assumptions made in MODIS cloud retrieval mentioned in section 4.2. For this reason, it is not surprising to see good agreement between $r_{e,2.1}$ and $r_{e,3.7}$ in this regime ($\Delta r_{e,3.7-2.1}$ is within ± 2 μm), although a small difference, probably due to random retrieval uncertainties, may still exist.
- *Regime 2*: MBL cloud pixels in this regime are characteristic with high horizontal heterogeneity ($H_\sigma > 0.3$). Retrieval uncertainties, plane-parallel r_e bias, and 3D radiative effects may all play a role in this regime. Note that these effects have different impacts on r_e retrievals and therefore $\Delta r_{e,3.7-2.1}$. Retrieval uncertainties, as discussed in section 4.4.1, are most likely to cause random

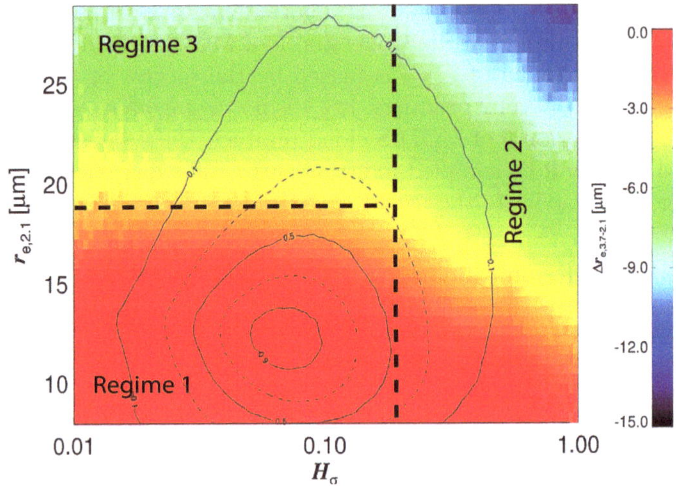

Fig. 4.12. The color contour of monthly mean $\Delta r_{e,3.7-2.1}$ for clouds with $\tau > 5$ on the space specified by $r_{e,2.1}$ and H_σ. The black lines indicate the relative frequency of each grid box, specified by certain combinations of $r_{e,2.1}$ and H_σ (unity corresponds to the most frequently observed combination of $r_{e,2.1}$ and H_σ). See text for details on cloud regime classification and implications for MODIS effective radius retrievals.

errors, rather than bias, in $\Delta r_{e,3.7-2.1}$. Similarly, as discussed in section 4.4.5, 3D radiative effects can cause either positive or negative $\Delta r_{e,3.7-2.1}$, depending on the nature (e.g. illuminating or shadow) of the 3D effect. In contrast, the plane-parallel r_e bias, as discussed in section 4.4.4, tends to cause negative $\Delta r_{e,3.7-2.1}$.

- *Regime 3*: MBL cloud pixels in this regime have low horizontal heterogeneity ($H_\sigma < 0.3$) and relatively large effective radius ($r_{e,2.1} > 20$ μm). Large effective radius suggests that MBL clouds in this regime are possibly precipitating. The warm rain process can give rise to the vertical weighting effect and the DSD sensitivity effect, as discussed in section 4.4.2 and 4.4.3 respectively. Because both effects tend to make $r_{e,2.1}$ larger than $r_{e,3.7}$, thus the bias of this regime is large and negative, as seen in Fig. 4.12.

4.6 Outlook of future work

Recently, the spectral dependence of MODIS r_e retrievals for MBL clouds has received significant and increasing attention. A number of recent studies reviewed in this chapter have significantly improved our understanding of the nature and potential causes of this issue. Nevertheless, many outstanding questions remain and future work is needed particularly in the following areas: (i) *More quantitative understanding*: Together ,those hypotheses reviewed in section 4.4 provide a qualitative explanation for the $r_{e,2.1}$ and $r_{e,3.7}$ differences. However, a more quantitative understanding is still lacking. It remains unclear whether the threshold behavior of $\Delta r_{e,3.7-2.1}$ at $r_{e,2.1} \sim 15$ μm in Fig. 4.3 is a coincidence or a result of warm rain process. In order to answer questions like this, hypotheses must be evaluated in a more quantitative way in future work. **(ii)** *Independent measurements*: For more objective and quantitative evaluation of MODIS $r_{e,2.1}$ and $r_{e,3.7}$ retrievals, independent measurements are needed. For example, several studies have investigated the impact of 3D radiative effects on cloud masking (Zhao and Di Girolamo, 2006), cloud optical thickness, and cloud effective radius retrievals (Marshak et al., 2006) using the high spatial resolution ASTER (Advanced Spaceborne Thermal Emission and Reflection Radiometer) observations. The ASTER observation may prove useful in future analyses to understand the relative importance of the impacts of 3D radiative effects and plane-parallel r_e retrieval bias on cloud property retrievals. Several studies have demonstrated that high temporal-spatial resolution cloud DSD measurements from air-borne *in situ* instruments, collocated with MODIS observation, are very helpful for understanding the errors and uncertainties in MODIS retrievals. Other remote sensing measurements, in particular those from A-Train satellite sensors (e.g. CloudSat, POLDER, and CALIPSO), would add unique and valuable perspectives on MBL cloud microphysics that are worthy of exploration in future work. **(iii)** *Combining observations with numerical models*: Every remote sensing method has its limitations. Even though a combination of various methods, such as those available from A-Train sensors, provides a complementary perspective, there will always be gaps that cannot be covered by observations. As shown in a number of recent studies, a combination of LES models with satellite retrieval simulators provides a flexible and powerful tool for understanding the impact of various factors on MODIS cloud property retrievals (e.g. Zinner et al., 2010; Zhang et al., 2012). This new avenue of research should be further explored in future work. **(iv)** *Implication studies*: MODIS cloud products are popularly used in climate change studies, aerosol indirect effect studies, and climate model validations. The implications of the spectral dependence of MODIS r_e retrievals for these 'down-stream' applications should be investigated in future work. **(v)** *Novel use of $r_{e,2.1}$ and $r_{e,3.7}$*: As discussed in sections 4.4.2 and 4.4.3, in certain circumstances, the difference between $r_{e,2.1}$ and $r_{e,3.7}$ actually contains useful information about MBL clouds. Actually, a number of studies have attempted to retrieve the vertical profile of precipitating MBL clouds utilizing the combination of effective radius retrievals from different spectral bands (Chang and Li, 2002, 2003; Kokhanovsky and Rozanov, 2011). Further work is needed along these lines to fully explore the potential of MODIS observations.

Acknowledgments

The authors acknowledge funding from NASA through grant NNX11AI98G managed by Dr. Richard Eckmann. Z.Z.'s research is also supported by the start-up funding from UMBC. Computational support was provided by the NASA Advanced Supercompting Division.

References

Ackerman, A.S., Kirkpatrick, M.P., Stevens, D.E. and Toon, O.B. (2004) 'The impact of humidity above stratiform clouds on indirect aerosol climate forcing', *Nature*, **432**, 1014–17.

Austin, R.T., Heymsfield, A.J. and Stephens, G.L. (2009) 'Retrieval of ice cloud microphysical parameters using the CloudSat millimeter-wave radar and temperature', *J. Geophys. Res.*, **114**, D00A23, doi:10.1029/2008JD010049.

Baker, M.B., Corbin, R.G. and Latham, J. (1980) 'The influence of entrainment on the evolution of cloud droplet spectra: I. A model of inhomogeneous mixing', *Quarterly Journal of the Royal Meteorological Society*, **106**, 581–98.

Barker, H.W. (1996) 'A parameterization for computing grid-averaged solar fluxes for inhomogeneous marine boundary layer clouds: Part I: Methodology and homogeneous biases', *J. Atmos. Sci.*, **53**, 2289–303.

Berry, E.X. (1967) 'Cloud droplet growth by collection', *J. Atmos. Sci.*, **24**, 688–701.

Berry, E.X. and Reinhardt, R.L. (1974) 'An analysis of cloud drop growth by collection: Part I: Double distributions', *Journal of Atmospheric Sciences*, **31**, 1814–24.

Brenguier, J.-L., Pawlowska, H., Schüller, L., Preusker, R. and co-authors (2000) 'Radiative properties of boundary layer clouds: Droplet effective radius versus number concentration', *J. Atmos. Sci.*, **57**, 803–21.

Breon, F. and Doutriaux-Boucher, M. (2005) 'A comparison of cloud droplet radii measured from space', *IEEE Transactions on Geoscience and Remote Sensing*, **43**, 1796.

Cahalan, R.F., Oreopoulos, L., Marshak, A., Evans, K.F., Davis, A.B. and co-authors (2005) 'THE I3RC: Bringing together the most advanced radiative transfer tools for cloudy atmospheres', *Bull. Amer. Meteor. Soc.*, **86**, 1275–93.

Cahalan, R.F., Ridgway, W., Wiscombe, W.J., Bell, T.L. and Snider, J.B. (1994) 'The albedo of fractal stratocumulus clouds', *J. Atmos. Sci.*, **51**, 2434–55.

Chang, F.L. and Li, Z. (2001) 'The effect of droplet size distribution on the determination of cloud droplet effective radius', in Proceedings of the 11th ARM Science Team Meeting, Atlanta, GA.

Chang, F.L. and Li, Z. (2002) 'Estimating the vertical variation of cloud droplet effective radius using multispectral near-infrared satellite measurements', *J. Geophys. Res.*, **107**, 4257.

Chang, F.L. and Li, Z. (2003) 'Retrieving vertical profiles of water-cloud droplet effective radius: Algorithm modification and preliminary application', *J. Geophys. Res.*, **108**, 4763.

Chen, R., Chang, F.L., Li, Z., Ferraro, R. and Weng, F. (2007) 'Impact of the vertical variation of cloud droplet size on the estimation of cloud liquid water path and rain detection', *J. Atmos. Sci.*, **64**, 3843–53.

Davis, A., Marshak, A., Wiscombe, W. and Cahalan, R. (1994) 'Multifractal characterizations of nonstationarity and intermittency in geophysical fields: Observed, retrieved, or simulated', *Journal of Geophysical Research—All Series*, **99**, 8055.

Davis, A.B. and Marshak, A. (2010) 'Solar radiation transport in the cloudy atmosphere: A 3D perspective on observations and climate impacts', *Reports on Progress in Physics*, **73**, 026801.

Di Girolamo, L., Liang, L. and Platnick, S. (2010) 'A global view of one-dimensional solar radiative transfer through oceanic water clouds', *Geophys. Res. Lett.*, **37**, L18809.

Fridlind, A.M. and Ackerman, A.S. (20100 'Estimating the sensitivity of radiative impacts of shallow, broken marine clouds to boundary layer aerosol size distribution parameter uncertainties for evaluation of satellite retrieval requirements', *J. Atmos. Oceanic Technol.*, **28**, 530–8.

Gerber, H. (1996) 'Microphysics of marine stratocumulus clouds with two drizzle modes', *J. Atmos. Sci.*, **53**, 1649–62.

Gerber, H., Frick, G., Malinowski, S.P., Brenguier, J.L. and Burnet, F. (2005) 'Holes and entrainment in stratocumulus', *J. Atmos. Sci.*, **62**, 443–59.

Hansen, J.E. and Travis, L.D. (1974) 'Light scattering in planetary atmospheres', *Space Science Reviews*, **16**, 527–610.

Kay, J., Hillman, B.R., Klein, S.A., Zhang, Y., Medeiros, B. and co-authors (2012) 'Exposing global cloud biases in the Community Atmosphere Model (CAM) using satellite observations and their corresponding instrument simulators', *J. Climate*, **25**, 5190–207.

King, M.D., Tsay, S.-C., Platnick, S., Wang, M. and Liou, K.-N. (1998) 'Cloud retrieval algorithms for MODIS: Optical thickness, effective particle radius, and thermodynamic phase', NASA (ed.), available online at *http://modis-atmos.gsfc.nasa.gov/reference_atbd.html.*

King, N., Bower, K., Crosier, J. and Crawford, I. (2012) 'Evaluating MODIS cloud retrievals with in situ observations from VOCALS-REx', *Atmospheric Chemistry & Physics Discussions*, **12**, 23679–729.

Klein, S.A. and Hartmann, D.L. (1993) 'The seasonal cycle of low stratiform clouds', *J. Climate*, **6**, 1587–606.

Kokhanovsky, A. (20040 'The depth of sunlight penetration in cloud fields for remote sensing', *IEEE Geoscience and Remote Sensing Letters*, **1**, 242–5.

Kokhanovsky, A. and Rozanov, V.V. (2011) 'Droplet vertical sizing in warm clouds using passive optical measurements from a satellite', *Atmos. Meas. Tech. Discuss.*, **4**, 5597–629.

Kubar, T.L., Hartmann, D.L. and Wood, R. (2009) 'Understanding the importance of microphysics and macrophysics for warm rain in marine low clouds: Part I: Satellite observations', *J. Atmos. Sci.*, **66**, 2953–72.

Lebsock, M.D. and L'Ecuyer, T.S. (2011) 'The retrieval of warm rain from CloudSat', *J. Geophys. Res.*, **116**, D20209.

Lebsock, M.D., Stephens, G.L. and Kummerow, C. (2008) 'Multisensor satellite observations of aerosol effects on warm clouds', *J. Geophys. Res.*, **113,** D15205, doi:10.1029/2008JD009876.

Liang, L., Di Girolamo, L. and Platnick, S. (2009) 'View-angle consistency in reflectance, optical thickness and spherical albedo of marine water-clouds over the northeastern Pacific through MISR–MODIS fusion', *Geophys. Res. Lett.*, **36**, L09811.

Marshak, A., Davis, A., Wiscombe, W. and Cahalan, R. (1995) 'Radiative smoothing in fractal clouds', *J. Geophys. Res.*, **100**, 26247–61.

Marshak, A., Platnick, S., Várnai, T., Wen, G. and Cahalan, R.F. (2006) 'Impact of three-dimensional radiative effects on satellite retrievals of cloud droplet sizes', *J. Geophys. Res.*, **111**, D09207, doi:10.1029/2005JD006686.

Martin, G.M., Johnson, D.W. and Spice, A. (1994) 'The measurement and parameterization of effective radius of droplets in warm stratocumulus clouds', *J. Atmos. Sci.*, **51**, 1823–42.

Minnis, P., Arduini, R.F., Young, D.F., Ayers, J.K., Albracht, B. and co-authors (2004) 'An examination of the impact of drizzle drops on satellite retrieved effective particle sizes', in Proceedings of the 14th International Conference on Clouds and Precipitation, Bologna, Italy.

Minnis, P., Sun-Mack, S., Chen, Y., Khaiyer, M.M., Yi, Y. and co-authors (2011) 'CERES Edition-2 cloud property retrievals using TRMM VIRS and terra and aqua MODIS data: Part II: Examples of average results and comparisons with other data', *IEEE Trans. Geosci. Remote Sens.*, **49**, 4401–30.

Nakajima, T. and King, M.D. (1990) 'Determination of the optical thickness and effective particle radius of clouds from reflected solar radiation measurements: Part I: Theory', *J. Atmos. Sci.*, **47**, 1878–93.

Nakajima, T.Y., Suzuki, K. and Stephens, G.L. (2010a) 'Droplet growth in warm water clouds observed by the A-Train: Part I: Sensitivity analysis of the MODIS-derived cloud droplet sizes', *J. Atmos. Sci.*, **67**, 1884–96.

Nakajima, T.Y., Suzuki, K. and Stephens, G.L. (2010b) 'Droplet growth in warm water clouds observed by the A-Train: Part II: A multisensor view', *J. Atmos. Sci.*, **67**, 1897–907.

Oreopoulos, L. and Davies, R. (1998) 'Plane parallel albedo biases from satellite observations: Part I: Dependence on resolution and other factors', *J. Climate*, **11**, 919–32.

Painemal, D. and Zuidema, P. (2011) 'Assessment of MODIS cloud effective radius and optical thickness retrievals over the Southeast Pacific with VOCALS-Rex in-situ measurements', *J. Geophys. Res.* **116**, D24206, doi:10.1029/2011JD016155.

Pincus, R. and Evans, K.F. (2009) 'Computational cost and accuracy in calculating three-dimensional radiative transfer: Results for new implementations of Monte Carlo and SHDOM', *J. Atmos. Sci.*, **66**, 3131–46.

Pincus, R., Platnick, S., Ackerman, G.S. and Hofmann, R.J.P. (2012) 'Reconciling simulated and observed views of clouds: MODIS, ISCCP, and the limits of instrument simulators', *J. Climate*, **25**, 4699–720.

Platnick, S. (2000) 'Vertical photon transport in cloud remote sensing problems', *J. Geophys. Res.*, **105**, 22919–35.

Platnick, S., King, M.D., Ackerman, S.A., Menzel, W.P., Baum, B.A. and co-authors (2003) 'The MODIS cloud products: Algorithms and examples from Terra', *IEEE Trans. Geosci. Remote Sens.*, **41**, 459–73.

Prabhakara, C., Fraser, R.S., Dalu, G., Wu, M.-L.C., Curran, R.J. and co-authors (1988) 'Thin cirrus clouds: Seasonal distribution over oceans deduced from Nimbus-4 IRIS', *J. Appl. Meteor.* **27**, 379–99.

Pruppacher, H. and Klett, J. (1997) *Microphysics of Clouds and Precipitation*, Dordrecht, The Netherlands, Kluwer Academic Publishers.

Quaas, J. and Boucher, O. (2005) 'Constraining the first aerosol indirect radiative forcing in the LMDZ GCM using POLDER and MODIS satellite data', *Geophys. Res. Lett.*, **32**, L17814.

Quaas, J., Ming, Y., Menon, S., Takemura, T., Wang, M. and co-authors (2009) 'Aerosol indirect effects: General circulation model intercomparison and evaluation with satellite data', *Atmos. Chem. Phys. Discuss.*, **9**, 12731–79.

Rosenfeld, D., Wang, H. and Rasch, P.J. (2012) 'The roles of cloud drop effective radius and LWP in determining rain properties in marine stratocumulus', *Geophys. Res. Lett.*, **39**, L13801.

Seethala, C. and Horváth, Á. (2010) 'Global assessment of AMSR-E and MODIS cloud liquid water path retrievals in warm oceanic clouds', *J. Geophys. Res.*, **115**, D13202.

Stevens, B., Ackerman, A.S., Albrecht, B.A., Brown, A.R., Chlond, A. and co-authors (2001) 'Simulations of trade wind cumuli under a strong inversion', *J. Atmos. Sci.*, **58**, 1870–91.

Stubenrauch, C., Rossow, W., Kinne, S., Ackerman, S., Cesana, G. and co-authors (2012) 'ASSESSMENT OF GLOBAL CLOUD DATASETS FROM SATELLITES: Project and database initiated by the GEWEX Radiation Panel', *BAMS*, doi:10.1175/BAMS-D-12-00117.1.

Twomey, S. (1974) 'Pollution and the planetary albedo', *Atmospheric Environment*, **41**, 120–5.

Várnai, T. and Davies, R. (1999) 'Effects of cloud heterogeneities on shortwave radiation: Comparison of cloud-top variability and internal heterogeneity', *J. Atmos. Sci.*, **56**, 4206–24.

Várnai, T. and Marshak, A. (2002) 'Observations of three-dimensional radiative effects that influence MODIS cloud optical thickness retrievals', *J. Atmos. Sci.*, **59**, 1607–18.

Walther, A. and Heidinger, A.K. (2012) 'Implementation of the Daytime Cloud Optical and Microphysical Properties algorithm (DCOMP) in PATMOS-x', *Journal of Applied Meteorology and Climatology*, **51**, 1371–90.

Wang, L., Qu, J.J., Xiong, X., Hao, X., Xie, Y. and co-authors (2006) 'A new method for retrieving band 6 of Aqua MODIS', *Geoscience and Remote Sensing Letters, IEEE*, **3**, 267–70.

Wiscombe, W.J. (1979) 'MIE scattering calculations, advances in technique and fast, vector-shaped computer codes', *NCAR Tech. Note*, **TN-140+STR**, 66.

Wood, R. and Hartmann, D.L. (2006) 'Spatial variability of liquid water path in marine low cloud: The importance of mesoscale cellular convection', *J. Climate*, **19**, 1748–64.

Wood, R., Comstock, K.K., Bretherton, C.S., Cornish, C., Tomlinson, J. and co-authors (2008) 'Open cellular structure in marine stratocumulus sheets', *J. Geophys. Res.*, **113**, D12207.

Zhang, Z. (2013) 'On the sensitivity of cloud effective radius retrieval based on spectral method to bi-modal droplet size distribution: A semi-analytical model', *J. Quant. Spectrosc. Radiat. Transfer*, **129**, 79–88.

Zhang, Z. and Platnick, S. (2011) 'An assessment of differences between cloud effective particle radius retrievals for marine water clouds from three MODIS spectral bands', *Journal of Geophysical Research—Atmospheres*, **116**, doi:10.1029/2011JD016216.

Zhang, Z., Ackerman, A., Feingold, G., Platnick, S., Pincus, R. and co-authors (2012) 'Effects of cloud horizontal inhomogeneity and drizzle on remote sensing of cloud droplet effective radius: Case studies based on large-eddy simulations', *J. Geophys. Res.*, doi:10.1029/2012JD017655.

Zhang, Z., Platnick, S., Yang, P., Heidinger, A.K. and Comstock, J.M. (2010) 'Effects of ice particle size vertical inhomogeneity on the passive remote sensing of ice clouds', *J. Geophys. Res.*, **115**, D17203.

Zhang, Z., Yang, P., Kattawar, G., Riedi, J., C.-Labonnote, L. and co-authors (2009) 'Influence of ice particle model on satellite ice cloud retrieval: Lessons learned from MODIS and POLDER cloud product comparison', *Atmospheric Chemistry and Physics*, **9**, 7115–29.

Zhao, G. and Di Girolamo, L. (2006) 'Cloud fraction errors for trade wind cumuli from EOS-Terra instruments', *Geophys. Res. Lett.*, **33**, L20802.

Zinner, T., Wind, G., Platnick, S. and Ackerman, A.S. (2010) 'Testing remote sensing on artificial observations: Impact of drizzle and 3D cloud structure on effective radius retrievals', *Atmos. Chem. Phys.*, **10**, 1221–59.

5 Remote sensing of above cloud aerosols

Kirk Knobelspiesse, Brian Cairns, Hiren Jethva, Meloë Kacenelenbogen,
Michal Segal-Rosenheimer, and Omar Torres

5.1 Introduction

The direct and indirect radiative effects of aerosols suspended in the atmosphere above clouds (ACA) are a highly uncertain component of both regional and global climate. Much of this uncertainty is observational in nature most orbital remote sensing algorithms were not designed to simultaneously retrieve aerosol and cloud optical properties in the same vertical column. Thus the climate modeling community has limited data to inform model development efforts, which encapsulate the current understanding of climate. Furthermore, field measurements have identified regions with consistent ACA, and regional simulations show that the radiative forcing may be significant. For this reason, there has been a recent push to develop the ability to determine ACA distribution, optical properties and cloud interactions, while also providing a means to validate models. Several algorithms have been created that utilize existing instruments for these purposes. However, the observational data sources, algorithm characteristics, geophysical assumptions and retrieved products from these methods are highly variable. This chapter is a review of these techniques, their uncertainties, and the associated validation efforts. We will also discuss the future of ACA remote sensing, both with regards to new instruments and the potential for new algorithms.

5.2 Above cloud aerosols (ACA), and their role in climate

Despite years of research, the role of atmospheric aerosols in the global climate remains highly uncertain. This is because aerosols have diverse chemical and optical properties, sources, and sinks, and persist in the atmosphere at timescales short enough that their global distribution is highly varied. Furthermore, there are a number of ways that aerosols can interact with climate, such as *direct* scattering or absorption of radiation, *indirect* effects on clouds (and their interactions with radiation and meteorology, e.g. Twomey, 1972; Albrecht, 1989), and *semi-direct* effects on atmospheric temperature profiles relevant to clouds (Hansen et al., 1997).

For these reasons, the global climate effects of aerosols, as expressed in general circulation models (GCM), are highly uncertain. In the 4th Assessment Report (AR4) of the Intergovernmental Panel on Climate Change (IPCC), the global mean *direct* aerosol radiative forcing (RF, the impact on the Earth–atmosphere energy balance) has been identified as negative (-0.5 ± 0.4 W/m^2), with a *medium-low* level of scientific understanding (LOSU) (Fraser et al., 1985). *Indirect* aerosol RF was also identified as negative (-0.7 with 95% uncertainty between -1.1 and 0.4 W/m^2), with a *low* LOSU. These are the largest negative forcing elements identified by the AR4, and oppose the positive RF of long-lived greenhouse gases such as carbon dioxide and methane (but have radically different spatial and temporal distributions than greenhouse gases). Since the role of atmospheric aerosols in the global climate are so uncertain, it is difficult to make appropriate policy decisions regarding aerosols.

Much of the RF uncertainty in GCM's is due to the lack of knowledge about aerosols and their climate interactions, and this is because of observational limitations. This is especially the case for above cloud aerosols (ACA). Until recently, most satellite retrievals of aerosol optical properties were limited to cloud-free scenes (Remer, 2009). This means that estimates of aerosol global *direct* RF had to be determined using clear-sky observations and simulations (such as Haywood et al., 1997; Bellouin et al., 2005), or guided by focused observation in key locations. Another factor complicating accurate global ACA RF estimates is that ACA have a variety of (often difficult to determine) optical properties relevant to RF, and because there are several ways that ACA can interact with the climate. Systematic observations of ACA have been generally unexplored until recently (Yu and Zhang, 2013).

In this section, we briefly describe ACA impacts on climate. In the next section, we will discuss the observations of ACA that are currently being made by sensors in orbit, including a summary in Table 5.2. Section 5.4 will describe aircraft-based measurements of ACA. These measurements are useful because they can serve as validation tools for orbital data, tests of new instrument and algorithm designs, and a means to understand a particular location in detail. Sections 5.3 and 5.4 will be the primary focus of this paper. The future for ACA observations will be illustrated in section 5.5. Finally, concluding remarks will be made in section 5.6.

Part of the difficulty of understanding the role of aerosols in climate is due to the variety of ways in which that they interact with climate, and the dependence of these interactions on both the aerosol microphysical properties and the characteristics of the surrounding scene. The conceptually simplest type of interaction are the so-called *direct effects*, where the aerosols absorb and scatter solar radiation. More complicated are the *indirect effects*, where aerosols modify cloud optical properties by acting as condensation nuclei, and *semi-direct effects*, where the direct absorption of aerosols modifies the atmospheric temperature vertical profile and thus cloud formation. Many reviews of aerosol effects on climate exist (Haywood and Boucher, 2000; Lohmann and Feichter, 2005; Yu et al., 2006; Forster et al., 2007), so we provide only a brief summary of the role that ACA, specifically, play in climate.

5.2.1 Direct effects

The scientific community has been aware of *direct effects* of aerosols since at least the late 1960s (McCormick and Ludwig, 1967; Charlson and Pilat, 1969; Atwater,

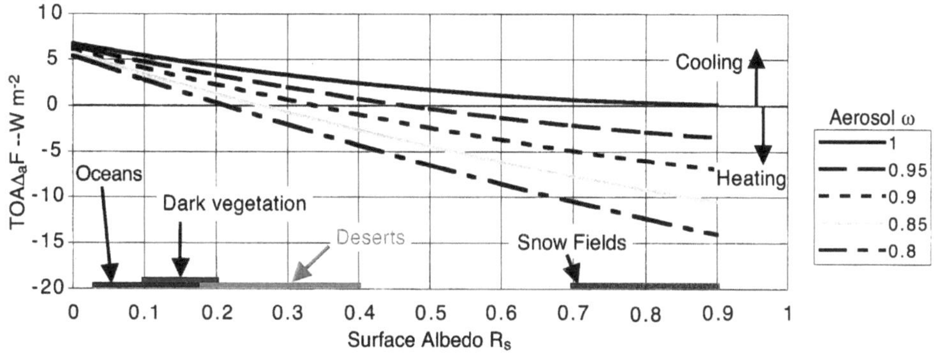

Fig. 5.1. (adapted from Russell et al., 2002). Aerosol-induced change in top-of-atmosphere upwelling flux for aerosols with different single-scattering albedo (ω) values. The simulations that were the basis for this figure assume an aerosol optical depth (AOD) of 0.1 and atmospheric transmission (T) of 0.76. Cloud albedos are roughly in the range of 0.4 to 0.95 (Kokhanovsky et al., 2006), meaning that most aerosols that are cooling over the ocean would heat over a cloud.

1970) and soon after began to estimate their impacts on climate (Mitchell, 1971; Coakley et al., 1983). The aerosol direct radiative effect (DRE, the RF due to aerosol direct effects) is negative (cooling) when primarily scattering aerosols increase the planetary albedo (reflectance). It is positive (warming) when absorbing aerosols reduce the planetary albedo. Both of these terms refer to the effect on the global energy balance, and impacts at specific locations may be different. For example, a positive forcing due to absorbing aerosols may warm the atmosphere surrounding the aerosols, but cool the surface. In any case, the magnitude of these impacts and the threshold between negative and positive DRE are governed not just by aerosol optical properties, but also by the brightness of the surface underneath (Fraser and Kaufman, 1985). Figure 5.1 (adapted from Russell et al., 2002) is an example of this relationship, where the Top of Atmosphere (TOA) DRE has been computed for a variety of surface albedos (Rs) and aerosol single-scattering albedos (SSAs, the ratio of the scattering to total aerosol extinction). In this case, the negative of the DRE is shown. For a moderately absorbing aerosol with SSA = 0.90, the DRE is cooling over most surfaces (such as oceans, vegetation, and desert) but warming over bright snow surfaces. Clouds as an underlying surface (the case of ACA) are not included in this figure. However, low-altitude cloud types such as marine stratocumulus can have albedos between roughly 0.4 and 0.95 (Kokhanovsky et al., 2006), over which the total aerosol direct effect can be either cooling or warming. Thus, ACA have the potential to have both positive and negative DRE, and this relationship is based on a variety of factors including aerosol and cloud optical properties, both of which are difficult to determine for ACA scenes with current remote sensing techniques.

Because of the difficulties of orbital remote sensing, systematic observations of aerosols came later than the first simulations. Furthermore, remote sensing observations have typically been confined to clear-sky (cloudless) conditions (examples to the contrary are discussed later), and this limitation cascades to estimates of climate interactions. For example, Bellouin et al. (2005) used the Moderate Resolution Imaging Spectrometer (MODIS) to estimate the total DRE due to aerosols. Since the standard MODIS aerosol products are only generated in clear-sky conditions, the all-sky (global) DRE was estimated by multiplying the clear-sky DRE by the clear-sky fraction. This, in effect, assumes the DRE due to ACA is negligible, or at least omitted from the analysis.

As described in Yu et al. (2006), ACA have an unresolved impact on climate. Regional assessments, such as in the tropical Indian Ocean (Podgorny and Ramanathan, 2001) or for the southern Atlantic Ocean near Africa (Keil and Haywood, 2003; Wilcox, 2012; Peters et al., 2011) indicate that aerosols have the potential to have a strong, if localized, effect. Systematic measurements on a global basis (for which orbital platforms are well suited) are needed if these uncertainties are to be resolved.

5.2.2 Indirect and semi-direct effects

Aerosols can also have a climate impact by modifying clouds. This was originally discussed by Twomey (1977), who noted that aerosols act to increase the reflectance of clouds by providing more condensation nuclei for droplet formation, thus increasing the total surface area of droplets in a cloud (assuming constant liquid water path). Albrecht (1989) found that this reduction in cloud droplet size also reduced the amount of drizzle, leading to longer cloud lifetimes and a net increase in cloud cover. Both of these effects have a negative forcing, but they are highly sensitive to the cloud type, meteorological conditions, and aerosol properties (Rosenfeld and Lensky, 1998). Presumably, some portion of ACA fall into clouds and have an indirect climate effect.

Aerosol semi-direct effects, whereby aerosols modify the atmospheric temperature profile, can also be important. The climate response to a DRE is sensitive to the location of that forcing in the atmosphere (Hansen et al., 1997). Furthermore, warming due absorbing aerosols can reduce cloud cover by reducing relative humidity and by suppressing convection (Ackerman et al., 2000; Johnson et al., 2004). These are processes with considerable variability and complicated feedbacks (Koren et al., 2008) and, like indirect effects, require systematic observation to be fully understood (Kaufman et al., 2005). For a detailed review of semi-direct aerosol assessments, see Koch and Del Genio (2010).

5.3 Orbital observations of ACA

5.3.1 Passive ultraviolet (UV) observations

5.3.1.1 Physical basis

The presence of absorbing aerosols in cloud-free scenes as well as above clouds has been qualitatively detected with satellite UV observations since the introduction of the Total Ozone Mapping Spectrometer (TOMS) UV aerosol index (AI) (Herman et al., 1997a; Torres et al., 1998). The TOMS UV AI is the difference between measured and calculated 340 nm to 380 nm radiance ratios. The calculated radiances are obtained assuming a purely molecular atmosphere bounded at the bottom by a Lambertian surface (Torres et al., 1998). When absorbing aerosols are present above clouds, the measured UV AI differs from the cloud-only case, which is due to the wavelength-dependent reduction of reflectance at the top of the atmosphere as a result of aerosol absorption. For an opaque underlying cloud layer, it is possible to estimate the aerosol and cloud optical depths by an inversion of the observed radiances. Such inversion must accurately account for cloud reflectance. A retrieval algorithm to simultaneously estimate cloud and aerosol optical depth (COD, AOD) from Ozone Monitoring Instrument (OMI) UV measurements has recently been developed (Torres et al., 2012). It uses input radiances at 354 nm and 388 nm.

5.3.1.2 Sources of uncertainty

Assumptions of aerosol and cloud particle size distribution, as well as aerosol SSA and aerosol–cloud vertical distribution, are needed. For cloud and aerosol optical depths of 10 and 0.5, respectively, the uncertainty of retrieved parameters has been estimated. AOD errors between −26% and 54% are associated with combined uncertainties of ±0.03 and ±2 km in SSA and aerosol–cloud separation, respectively. In most cases, errors in retrieved COD are smaller than ±20% (Torres et al., 2012). The geographic distribution of OMI-derived AOD above clouds qualitatively compares to MODIS retrievals (Jethva et al., 2013) and Polarization and Directionality of the Earth's Reflectances (POLDER) (Waquet et al., 2009; see comparison in Torres et al., 2012) results. Direct comparisons between OMI-determined above cloud aerosol optical depth (ACAOD) in the UV and POLDER ACAOD at 865 nm found a linear relationship, with a coefficient of determination value of 0.67. Figure 5.2 depicts retrieved fields of aerosol and COD on September 10, 2005 over the south-eastern Atlantic Ocean.

The color ratio (CR) technique applied to the OMI and MODIS observations (described in the next section and primarily based on Jethva et al. (2013)) requires assuming the vertical profiles of aerosols and clouds. For the aerosol events studied in Torres et al. (2012) and Jethva et al. (2013), and the OMI/MODIS results shown in Fig. 5.2 here, this assumption was based on climatologies of Cloud, Aerosol, Lidar with Orthogonal Polarization (CALIOP)-observed vertical profiles of clouds and aerosol over the south-eastern Atlantic Ocean. The marine stratocumulus cloud deck of about 500 m depth is often observed at about 1 km height. The aerosol layer is generally lofted in the atmosphere with peak concentration at about 3–4 km height. Due to the very narrow overlap of the CALIOP Lidar track on the OMI and MODIS broad swaths, the exact detailed profiles of clouds and aerosols measured by CALIOP cannot be applied to the full swath of OMI and MODIS. For this reason, the vertical distribution of aerosols was represented as a generalized Gaussian profile with peak concentration at 3 km and a full-width-half-maximum of 1 km. The 500-m-thick cloud deck was placed between 1 km and 1.5 km. In a gross sense, both aerosol and cloud profiles are generally in good agreement with the CALIOP observations over this region during the biomass-burning season (July through September) (Torres et al., 2012; Jethva et al., 2013).

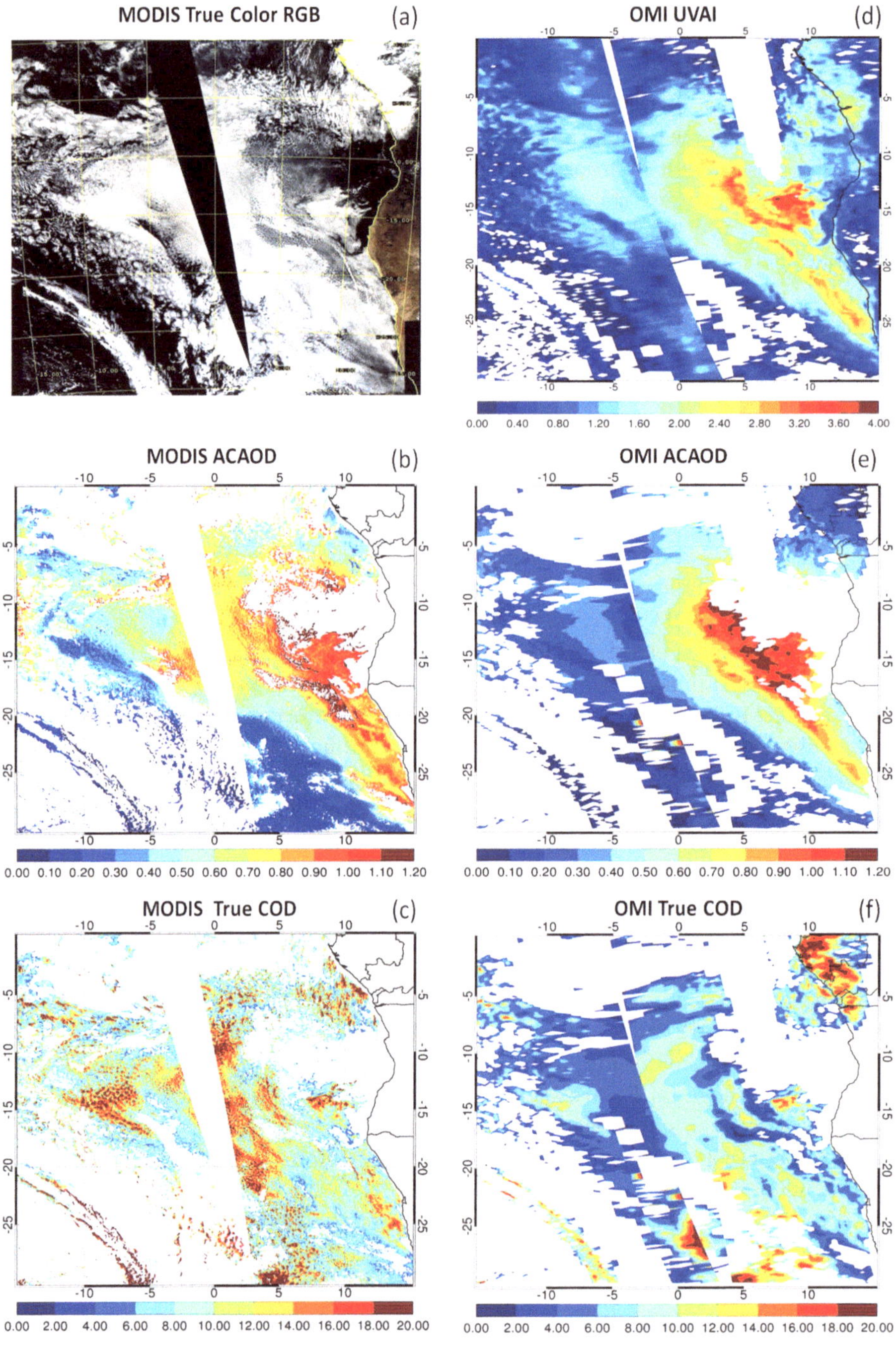

Fig. 5.2. An above cloud smoke aerosols event observed by OMI and MODIS on September 10, 2005, over the south-eastern Atlantic Ocean. (a) MODIS true color RGB image, (b) ACAOD retrieved from MODIS at 5-km resolution, (c) COD retrieved by MODIS, (d) OMI UV aerosol index, (e) ACAOD retrieved from OMI's near-UV measurements, and (f) COD retrieved from OMI.

5.3.2 Passive visible (VIS) near-infrared (NIR) observations

5.3.2.1 Physical basis

The spectral dependence of aerosol absorption approximately follows the power law, which relates the aerosol absorption optical depth to wavelength with the Absorption Ångström Exponent, α_a, defined from $\tau_a(\lambda) = \tau_0\lambda^{-\alpha_a}$, where τ_a is the aerosol absorption optical depth at wavelength λ and τ_o is a reference optical depth, typically defined at $\lambda = 1$ μm. Light-absorbing aerosols such as carbonaceous particles generated from biomass-burning activities and wind-blown mineral dust exhibit a strong extinction spectral dependence from the UV to shortwave infrared (SWIR) spectral regions (Russell et al., 2010). Water clouds, on the other hand, exhibit near-neutral wavelength dependence of extinction in this part of the spectrum. Therefore, we would expect absorbing ACA to have stronger effects at shorter wavelengths than at longer wavelengths. This produces a strong 'color ratio', by which the spectral ratio TOA observations by satellites such as MODIS are sensitive to ACA.

Based on the 'color ratio' principle, Jethva et al. (2013) developed a technique to detect and derive the ACAOD and underlying COD from the MODIS spectral reflectance measurements. This application is an extension to the visible spectral region of the near UV retrieval technique of Torres et al. (2012) described in section 5.3.1. The presence of absorbing aerosols above clouds reduces the TOA radiance—an effect known as 'cloud darkening', which can be seen by eye in 'true color' MODIS images. Under such situations, the TOA reflectance measured by MODIS show a sharp decline in the TOA reflectance at shorter wavelengths such that the magnitude of the CR (the observed 470/860-nm reflectance ratio) reduces significantly from the reference ratio for clouds, typically near unity. Under some conditions (such as specific amounts of aerosol absorption or cloud–aerosol separation), the reductions in CR and VIS/SWIR reflectance are a strong function of AOD and COD. This forms the basis of the ACA retrieval from MODIS, in which the observed CR and reflectance are fitted to pre-calculated ACA fields to simultaneously determine AOD and COD. Figure 5.2 shows the example of an above cloud smoke aerosols event observed on September 10, 2005, over the south-eastern Atlantic Ocean. Among the passive techniques to quantify ACA, this is the only one that is currently configured to simultaneously retrieve the COD underlying the aerosol layer. This is important because the accurate quantification of the above cloud DRE requires accurate estimation of both ACAOD as well as COD. Furthermore, neglect of ACA in cloud retrievals from passive sensors such as MODIS would likely produce a negative bias in the cloud retrieval, particularly for the COD (Haywood et al., 2004; Wilcox et al., 2009; Coddington et al., 2010; Torres et al., 2012; Jethva et al., 2013).

Although the CR technique is currently implemented to make use of the MODIS TOA measurements at two channels (470 nm in the VIS and 870 nm in the SWIR), it can be applied to the other satellite measurements that have channels in this portion of the spectrum. For instance, the Multi-angle Imaging SpectroRadiometer (MISR) has four spectral bands centered at 446, 558, 672, and 867 nm and observes the scene at multiple viewing angles. The recently launched Visible-Infrared

Imager Radiometer Suite (VIIRS) on the Suomi National Polar-orbiting Partnership (NPP) spacecraft has a total of eight spectral bands in the wavelength range 412–865 nm.

5.3.2.2 Sources of uncertainty

The CR method requires assumptions about aerosol and cloud size distribution, aerosol SSA, and relative placements of aerosol layer and cloud. For the example retrieval shown in Fig. 5.2, these parameters were obtained from the nearest Aerosol Robotic Network (AERONET) station (for aerosol properties) during the event. Vertical distributions of aerosols and clouds were based on climatological estimates from CALIOP. In addition to spatial extrapolation errors, these parameters are also uncertain. For instance, the aerosol SSA retrieved by AERONET is known to be uncertain within ± 0.03 for AOD larger than 0.4 at 440 nm, and less certain for smaller AOD (Dubovik et al., 2000). The CR technique for ACA retrieval is undergoing continuous expansion.

The uncertainty in the ACAOD/COD retrieval was also estimated given the perturbation in the assumed state, namely aerosol SSA and size. The approach completely relies on radiative transfer (RT) simulations, in which the reference or observed state was retrieved with the retrieval Lookup Table (LUT) generated from the perturbed states. The difference between the reference and retrieved AOD/COD was then considered as error associated with the given perturbation. An over-estimated aerosol SSA by $+0.03$ at COD of 10 produces a positive error (over-estimation in ACAOD) of about 46% and 68% at actual AODs of 0.5 and 1.0, respectively, which reduce to 32% and 42% at COD equals 20. An underestimated SSA by -0.03, on the other hand, yields negative errors of less than -12% at nearly all values of CODs. The error associated with the uncertain aerosol layer height is found be significantly lower than that obtained with uncertain SSA. The estimated error in ACAOD was less than 15% for an underestimated height by -1 km and less than -4% for an over-estimated height by $+1$ km (Jethva et al., 2013). Finally, a case study comparing OMI- and MODIS-derived ACAOD finds that most match-ups are within the expected uncertainties (-10% to 50%).

Difficulty determining cloud properties can also be a source of error in this and all passive ACA retrieval algorithms. ACA uncertainty increases for low COD due to the difficulty separating cloud and aerosol signals. For MODIS, a COD of 5 is the lower limit for successful ACA retrieval. This method also assumes cloud and aerosol fields are uniform within each pixel. For 1-km-scale MODIS pixels, this can be assessed by examining the variability of neighboring pixels (Jethva et al., 2013). The coarser spatial resolution of OMI (13×24 km^2) will make ACA retrievals from that instrument more vulnerable to sub-pixel heterogeneity. Finally, this technique uses the modified Gamma distribution to describe droplet size. While the uncertainty of this assumption needs to be assessed, it is commonly used and probably not a large contributor to overall uncertainty.

5.3.2.3 Ongoing developments

The CR technique for ACA retrieval is being expanded. In particular, the longer SWIR channels, where aerosol attenuation is expected to be minimal, are being tested for the retrieval. An inter-comparison of the ACAOD retrieved from different A-Train sensors will be made to check the inter-consistency between different types of retrievals. Unlike ample availability of direct AOD measurements made in the clear-sky conditions by ground-based sunphotometers, the ACA measurements are sparse. Therefore, it is hard to directly validate the satellite-based ACA retrieval. However, a comparative analysis can be performed in which retrievals from different sensors/platforms are inter-compared. Field experiments with a focus on obtaining ACAOD measurements using airborne sunphotometry and Lidar during satellite overpass times is a prime requirement to validate the CR and other ACA retrieval methods.

5.3.3 Passive hyperspectral observations

5.3.3.1 Physical basis

Orbital spectrometers can be used to infer the presence of ACA if the spectral range includes UV, VIS, and SWIR wavelengths. The Scanning Imaging Absorption Spectrometer for Atmospheric Chartography (SCIAMACHY), an instrument on the European Space Agency's (ESA) polar-orbiting Environmental Satellite (Envisat), is sensitive between 240 nm and 2380 nm, and is therefore appropriate for these purposes (Bovensmann et al., 1999). de Graaf et al. (2012) used SCIAMACHY to directly determine the aerosol DRE by building a LUT of cloud (and aerosol-free) reflectances. SWIR observations, which are minimally affected by aerosols, were matched to the LUT to determine the COD, cloud droplet effective radius, and surface albedo underneath an ACA layer. Cloud top pressure and effective fraction are determined by observations in the Oxygen-A band. The difference between simulated and observed reflectance at UV and VIS wavelengths can therefore be attributed to aerosols, providing for a retrieval of aerosol DRE that sidesteps the uncertainties associated with aerosol optical property retrieval.

5.3.3.2 Sources of uncertainty

The average value of the DRE for ACA in the south-eastern Atlantic Ocean determined by SCIAMACHY in de Graaf et al. (2012) was 23 ± 8 W/m^2. The uncertainty of this, and other SCIMACHY DRE results, is due to several factors.

1. Cloud determination: The process of determining the cloud and surface optical properties by matching SWIR observations to the LUT is sensitive to the uncertainty of SCIAMACHY observations in the SWIR, accuracy of cloud simulations comprising the LUT, and to a lesser extent LUT interpolation. An assessment of cloud-free scenes from SCIAMACHY found the differences between measured and simulated spectra are well within the SCIAMACHY reflectance accuracy of 3%. The spectrally integrated apparent aerosol effect for scenes without ACA shows an average bias of -5 W/m^2, which is corrected in the final product. The standard deviation of the measured and simulated apparent aerosol effect differences is 7 W/m^2, which is incorporated into the total uncertainty.

2. Reflectance anisotropy: SCIAMACHY reflectances represent a single viewing and solar illumination geometry, which cannot resolve the potential anisotropic difference between an ACA scene and that of the cloud-only scenes simulated in the LUT. The effect of this 'anisotropy factor' is dependent upon the solar zenith angle (it is small for values less than 60°), the ACA (dependence in linear and increases with ACA load), and is higher in the UV. Simulation-based assessments of uncertainty are roughly 1–2 W/m^2.
3. Residual SWIR aerosol absorption: Aerosol absorption in the SWIR affects the cloud property determination, leading to a subsequent underestimation of aerosol DRE. Simulations in de Graaf et al. (2012) estimate that the uncertainty due to this effect is about 1 W/m^2. Presumably, this uncertainty becomes larger with greater ACAOD.
4. Cloud top pressure and effective fraction determination: Cloud top pressure and effective fraction are initially determined using the Oxygen-A band, which is then used to select the appropriate subset of the LUT for comparison to SWIR observations. While ACA loading can affect this aspect of the retrieval, the impact is small. Simulations show that the large ACAOD of 1.5 only creates a DRE bias of 1.1 W/m^2. Uncertainty due to this component of the system is therefore not considered in the total error budget.

Ultimately, the total uncertainty for ACA DRE is estimated by de Graaf et al. (2012) to be 8 W/m^2.

5.3.4 Passive polarimetric observations

5.3.4.1 Physical basis

The pre-eminent passive polarimetric instrument currently in orbit is the POLDER on the Polarization and Anisotropy of Reflectances for Atmospheric Sciences Coupled with Observations from a Lidar (PARASOL) spacecraft of the French Centre National D'Études Spatiales (CNES) (Tanré et al., 2011). PARASOL was launched in December 2004 into the so-called 'A-Train' orbit. That constellation also includes the Aura, Cloud-Aerosol Lidar and Infrared Pathfinder Satellite Observations (CALIPSO), and Aqua spacecraft, carrying the OMI, CALIOP, and MODIS sensors, (see sections 5.3.1, 5.3.2, and 5.3.5, and Anderson et al. (2005)). Earlier versions of POLDER were launched in 1996 and 2003 on the unfortunately short-lived Advanced Earth Observing System (ADEOS) I and II spacecraft, and each collected data for less than a year. Fortunately, the POLDER on PARASOL is still functioning, although PARASOL was removed from the A-Train constellation in 2011 with the depletion of the fuel required to maintain orbit position.

POLDER observes the total and linearly polarized reflectance of each pixel in nine channels at up to 16 viewing angles, between 55° forward and aft of the satellite track. Narrowband channels are centered between 443 nm and 1020 nm. The 490, 670, and 870-nm channels are sensitive to polarization by use of a filter wheel, and are accurate to 1% to 2% in the Degree of Linear Polarization (DoLP, the ratio of linearly polarized to total reflectance) (Deschamps et al., 1994). The information contained within these measurements is used to retrieve AOD and some aerosol optical and microphysical parameters for cloud-free scenes over land (Herman et al., 1997b; Leroy et al., 1997; Nadal and Bréon, 1999; Deuzé et al., 2001) and water (Bréon and Deschamps, 1993; Deuzé et al., 2000; Herman et al., 2005; Harmel and Chami, 2011). POLDER has also been used to determine the optical properties of clouds (Bréon and Goloub, 1998; Goloub et al., 2000; Parol et al., 2004), although these algorithms assume aerosols are not present above a cloud. This issue has been considered, for stratospheric aerosols (Lafrance and Herman, 1998), and more recently for case studies of ACA by Waquet et al. (2009, 2013).

The physical principle of polarimetric remote sensing of ACA is described schematically in Fig. 5.3. Simulations of the reflected Q component of the Stokes polarization vector (see Hansen and Travis (1974) for a review of polarized light

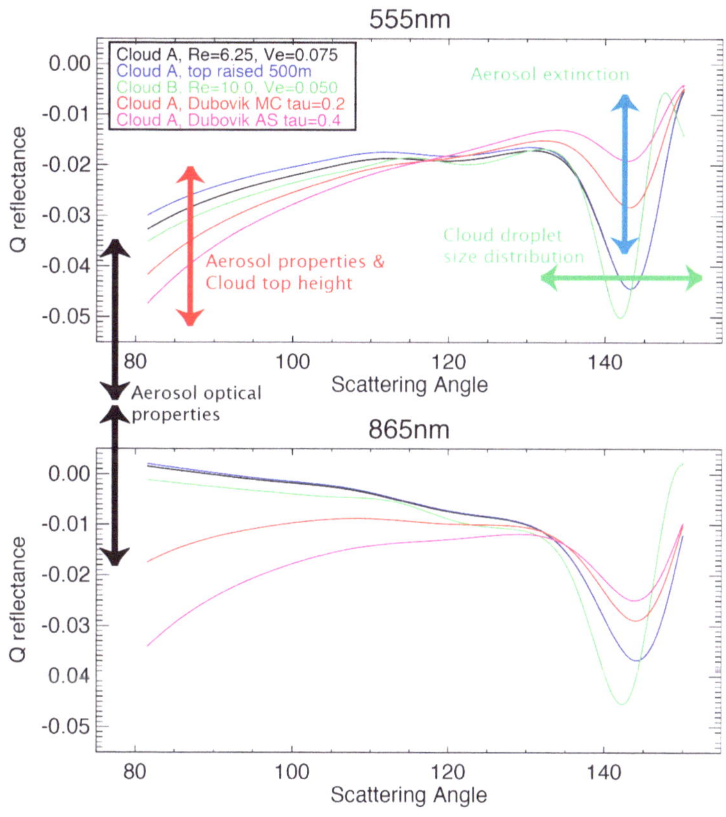

Fig. 5.3. Simulated multi-angle, multi-spectral, polarimetric observations (Q Stokes element in the scattering plane) for various types of clouds (black, blue, and green) and ACA (red and magenta). Simulated aerosols are the 'Mexico City' (MC) and 'African Savannah' (AS) classes from Dubovik et al. (2002), where 'tau' indicates the AOD at 550 nm. Arrows indicate how variability in aerosol and cloud parameters impact reflectance. Cloud droplet size distribution determines the angular location of the cloud-bow (green arrow), aerosol extinction determines the amplitude of the cloud-bow (blue arrow), and side scattering reflectances are sensitive to aerosol optical properties and cloud top altitude (red arrow). Furthermore, aerosols have different impacts at different wavelengths (black arrow). The distinct manifestation of these effects in the observations indicates the retrievability of the parameters that create them.

scattering), observed at TOA, are shown for various cloud and ACA scenarios. Polarized scattering from the cloud-bow (analogous to a rainbow but representing scattering from cloud, rather than rain, droplets) is quite distinct. The scattering angle of the peak indicates the cloud droplet effective radius. The presence of side lobes (weakly expressed here at about 125°) indicate the width of the cloud droplet size distribution (Alexandrov et al., 2012) while the amplitude of the main cloud-bow indicates ACA attenuation. Reflectance at side scattering angles (less than roughly 110°) expresses both aerosol optical properties and the quantity of Rayleigh scattering above the cloud (and thus the cloud top height). Polarized reflectances are insensitive to COD greater than 3, meaning that parameter does not need to be constrained when determining aerosol optical properties (Goloub et al., 2000). Given sufficient angular sampling, spectral channels, and external information about cloud top height, an algorithm can retrieve the cloud droplet size distribution, ACAOD, and microphysical properties (size distribution and complex refractive index) of ACA. Retrievals for scenes with COD less than 3 would need additional constraints for the COD and possibly the ground surface reflectance, although this has not been tested.

POLDER's 16 viewing angles are not sufficient to resolve the scattering angle of the cloud-bow peak in a single pixel. Nevertheless, these data can still be used to determine ACA optical properties. Waquet et al. (2009) was the first attempt to quantify ACA from POLDER data. In this algorithm, ACA pixels were identified by comparing the results of two methods to determine cloud top height. Determination of cloud top height from the differential absorption in the Oxygen-A band (Vanbauce et al., 1998) is mostly insensitive to ACA, while the Rayleigh cloud top pressure (determined from short wavelength channels at low scattering angles; Goloub et al., 1994; Parol et al., 1999) is affected by ACA (see Fig. 5.3). Differences between the values indicate the presence of ACA. In these pixels, 670-nm and 865-nm polarized observations in the scattering angle range of 90° to 110° were matched to scaled single-scattering simulations of the atmosphere to determine ACAOD and the aerosol effective radius (scaling is an approximation that simulates multiple scattering). Restricting the retrieval to this subset of the POLDER observations lessens sensitivity to cloud properties and the use of a single, rather than multiple, scattering RT calculation. The consequence is less access to information contained in the full POLDER set of observations, highlighting the delicate balance required to maximize the information content of a set of observations while minimizing sensitivity to errors or ancillary parameters that must be constrained.

A more recent and sophisticated approach is in Waquet et al. (2013), which describes two algorithms. The 'research' algorithm aggregates data within a 200 km by 200 km pixel to provide adequate scattering angle sampling of the cloud-bow. An optimal estimation (using multiple scattering RT computations) of both cloud and aerosol optical properties can then be performed. Cloud droplet effective radiance and variance, ACAOD, aerosol fine size mode effective radius, and the extinction Ångström Exponent are all retrieved with this algorithm, along with the cloud top height determination from the standard POLDER Oxygen-A band retrieval (Goloub et al., 1994; Parol et al., 2004). Because of sensitivity to nonspherical aerosol extinction in the cloud-bow, this method can also determine the fraction of ACAOD due to nonspherical particles. Other aerosol, microphysical parameters, such as complex refractive index, the ratio of fine to coarse size mode aerosols and the coarse size mode particle effective radius can only be estimated with limited accuracy. The primary drawback of this technique is the large spatial extent of these 'super-pixels', meaning that spatial variability of aerosol and cloud properties could impact retrievals. The 'operational' POLDER algorithm uses a single pixel, which has up to 16 viewing angles. This means that cloud optical properties cannot be estimated, and that a retrieval is valid only if COD is greater than 5. This algorithm is based upon what was presented in Waquet et al. (2009), with several key differences. It uses more accurate RT simulations, compares measurements of both the cloud-bow and side scattering (rather than side scattering alone), and uses realistic nonspherical aerosol models to simulate dust.

5.3.4.2 Sources of uncertainty

5.3.4.2.1 Research algorithm

The POLDER research algorithm is sensitive to the homogeneity of aerosol and cloud properties within the 200 × 200 km^2 'hyper-pixel'. This homogeneity is assessed by performing the operational aerosol retrieval and the Oxygen-A band cloud height determination on the individual pixels comprising the hyper-pixel region. Significant variability of either aerosol or cloud properties is grounds to flag the hyper-pixel as potentially inaccurate, and therefore should be removed from the data-processing stream.

Since the research algorithm uses the optimal estimation technique, individual uncertainty estimates can be determined for each retrieval (this assessment does not include the impact of model uncertainties such as those described above). Waquet et al. (2013) assesses several cases of ACA retrievals. As shown in Table 5.1, SSA uncertainty decreases as ACAOD increases. The nonspherical particle fraction can also be determined for large quantities of nonspherical aerosols. Otherwise, the ACAOD, Ångström Exponent, aerosol fine size mode effective radius, and cloud droplet size distribution can all be determined. Except for cloud size distribution, accuracies improve with aerosol loading.

Table 5.1. Uncertainty assessment for POLDER research algorithm case studies (reformatted from Waquet et al. (2013)). Uncertainties are expressed in percentage, except SSA and F, which are absolute values. In this table, ACAOD; above cloud aerosol optical depth at 865 nm, α, Ångström exponent; $r_{eff,a,f}$, aerosol effective radius in microns, fine size mode; m_i, aerosol imaginary refractive index, SSA; single-scattering albedo at 865 nm; $r_{eff,c}$; cloud droplet effective radius in microns, $v_{eff,c}$, cloud droplet effective variance, f fraction nonspherical aerosol particles $*$ indicates a case study with nonspherical aerosols modeled with spheroids, – parameters that either could not be uniquely retrieved or were fixed in the optimal estimation.

ACAOD	ΔACAOD%	$\Delta\alpha$%	$\Delta r_{eff,a,f}$%	Δm_i%	ΔSSA	Δf	$\Delta r_{eff,c}$%	$\Delta v_{eff,c}$%
0.28	3.6	10.6	1.5	27.1	0.045	–	3.5	16.7
0.063	6.3	26.7	4.1	–	–	–	1.4	11.1
1.05*	6.7	143.9	15.0	5.0	0.002	0.07	11.9	32.7

Like most other retrieval algorithms, aerosol, and cloud layers are assumed to be plane parallel for the research algorithm. Waquet et al. (2013) assessed the impact of this approximation for observations in the cloud-bow, and found between 4% and 8% polarized radiance errors (depending on wavelength and viewing geometry).

5.3.4.2.2 Operational algorithm

The operational algorithm retrieves the ACAOD and an aerosol type from a LUT of seven aerosol models. Six of these models are of small spherical aerosols, while the seventh is nonspherical dust. Only two of the POLDER polarization sensitive channels (670 nm and 865 nm) channels are used. This is intended to limit the sensitivity to cloud and aerosol vertical distribution (prescribed in the LUT), which is greatest for the 490-nm channel. Uncertainty estimates are more limited than the research algorithm, but differences between ACAOD and the AOD in cloud-free adjacent scenes are less than 0.2 at 865 nm.

5.3.4.3 Ongoing developments

The POLDER instrument on PARASOL has been in orbit since 2004, so its remaining functional lifetime is limited. While instrument concepts currently exist at both NASA and CNES, no instrument with multi-angle polarimetric sensitivity will be launched in the near term. The Aerosol Polarimetry Sensor (APS) on the NASA Glory spacecraft (Mishchenko et al., 2007) was intended to measure aerosol and cloud optical properties, but was unfortunately lost during launch in 2011. APS would have had nine spectral channels in the VIS, NIR, and SWIR, all with polarimetric sensitivity roughly an order of magnitude more accurate than POLDER. It would have had up to 255 different angular views of each pixel, meaning that a far greater quantity of information would have been available (with the consequence of a single pixel wide field of view). Because of this, retrievals similar to, or more sophisticated than, the POLDER research algorithm could have been performed in a single pixel (Hasekamp, 2010).

The current successor to POLDER is the Multi-directional, Multi-polarization and Multispectral Instrument (3MI) proposed as a sensor on the EUMETSAT Polar System – Second Generation (EPS-SG) (Marbach et al., 2013; Bréon et al., 2011). 3MI will have characteristics similar to POLDER, but with more spectral polarization information (nine polarimetrically sensitive channels in the VIS, NIR, and SWIR) and a finer spatial resolution. Since the number of viewing angles is unchanged, observation of AAC will probably still require the creation of 'hyperpixels' as in the research retrieval algorithm in Waquet et al. (2013).

The NASA Earth Science Decadal Survey calls for polarimetrically sensitive instruments on board both the Pre-Aerosol, Cloud, Ecosystems (PACE) (Del Castillo, 2012), and Aerosol, Cloud, Ecosystems (ACE) missions (National Research Council, 2007). While the final form of these instruments has not yet been finalized, several polarimeter concepts exist. These include the APS design, the Airborne Multiangle SpectroPolarimetric Imager (AirMSPI) (Diner et al., 2013), and the HyperAngular Rainbow Polarimeter (HARP) that was recently selected for development as a 3U ($30 \times 10 \times 10$ cm) Cubesat as part of the NASA In-space Validation of Earth Science Technologies (InVEST) Program. Sensitivity to ACA of all these designs will depend upon the particular observational configuration. Specifically, the number of available viewing angles, number and range of spectral channels, spatial resolution, coverage, and radiometric and polarimetric accuracy contribute to the success of ACA remote sensing.

5.3.5 Active Lidar observations

5.3.5.1 Physical basis

The ACA AOD retrievals using passive satellite sensors described in previous sections are limited research products for which there is a lack of global experimental validation. In addition, these sensors often require information about cloud and aerosol vertical distribution to provide an accurate measure of aerosol loading. This is typically provided by the CALIOP or from differential absorption measurements in the Oxygen-A band (Waquet et al., 2013).

CALIOP is a three-channel elastic backscatter Lidar flying on board the CALIPSO spacecraft. CALIOP/CALIPSO has been part of the A-Train satellite constellation since April 2006, and is the only active space-borne sensor providing routine aerosol backscatter and inferred extinction profiles in both cloud-free and cloudy conditions (Winker et al., 2009). CALIOP measures high-resolution (1/3 km in the horizontal and 30 m in the vertical in the low and middle troposphere) profiles of the attenuated backscatter by aerosols and clouds at visible (532 nm) and near-infrared (1064 nm) wavelengths, along with polarized backscatter in the visible channel (Winker et al., 2009). These data are distributed as part of the level 1 CALIOP products. The level 2 products are derived from the level 1 products using a succession of complex algorithms (e.g. Winker et al., 2009). The level 2 retrieval scheme is composed of a feature detection scheme (Vaughan et al., 2009) that classifies features according to layer type (i.e. cloud vs. aerosol) (Liu et al., 2010) and sub-type (e.g. aerosol species) (Omar et al., 2009), and, finally, an extinction retrieval algorithm (Cattrall et al., 2005; Young and Vaughan, 2009) that estimates aerosol backscatter and extinction coefficient profiles and the total column AOD based on modeled values of the extinction-to-backscatter ratio (Sa) inferred for each aerosol layer.

5.3.5.2 Sources of uncertainty

We list the most important uncertainties in the CALIOP AOD retrieval above clouds:

1. CALIOP calibration: The accuracy of the CALIOP level 1 products (and, by extension, many of the level 2 products) critically depends on the accuracy of the calibration of the attenuated backscatter profiles. The night-time CALIOP 532-nm parallel attenuated backscatter measurement is calibrated by determining the ratio between the measured signal and the total backscatter estimated from an atmospheric scattering model (Powell et al., 2009; Hostetler et al., 2006; Russell et al., 1979) across an altitude range of 30–34 km, where aerosol loading is assumed to be low and there is still sufficient molecular backscatter to produce a robust signal. Despite some aerosol contamination of the 532-nm channel

calibration over the tropics (Vernier et al., 2009), the error bars on the 532-nm calibration coefficients over the North American continent should generally be within ±3% (Rogers et al., 2011). CALIOP 532-nm calibration coefficients are much more accurate than the current 1064-nm calibration coefficients. The latter is expected to change by as much as 25% in the version 4 CALIOP data release (Vaughan et al., 2012).

2. CALIOP detection threshold: CALIOP attempts to retrieve aerosol extinction coefficients (and hence AOD) only in those regions where the CALIOP layer detection and classification schemes identify the presence of aerosol layers. As a consequence, aerosols with backscatter intensities below the CALIOP layer detection threshold will not contribute to the CALIOP AOD estimates. When such faint aerosol layers are present, the CALIOP retrieved AOD is likely to be underestimated. The minimum day or nighttime CALIOP backscatter threshold to detect aerosols at an altitude of ~2–3 km is around ~2–4 × 10^{-4}/km/sr (when using the highest horizontal averaging of 80 km) (Winker et al., 2009). Consequently, if we assume a Lidar extinction-to-backscatter ratio of 50 sr, the minimum detectable extinction coefficient is on the order of 0.01–0.02 per km (corresponding to a lowest detectable AOD of 0.02–0.04 in a homogenous 2-km planetary boundary layer).
3. CALIOP extinction-to-backscatter Lidar ratio: Classification errors can occur in the aerosol subtyping algorithm (Omar et al., 2009), leading to an incorrect assumption about the appropriate extinction-to-backscatter ratio (Sa). A comparison study between the airborne High Spectral Resolution Lidar (HSRL) (Hair et al., 2008) and the CALIOP aerosol classification by Burton et al. (2013) shows relatively poor agreement for polluted dust (35%) and smoke (i.e. 13% of CALIOP agrees with the HSRL results), compared to marine (62%), polluted continental (54%), and desert dust (80%).
4. CALIOP Signal-to-Noise Ratio: One of the sources of CALIOP misclassification of the aerosol layer is low CALIOP signal-to-noise ratio (SNR), especially in daytime. CALIOP's SNR is lower than for typical ground-based or airborne Lidars because the instrument is far from the atmosphere, the laser pulse energy is limited by the available electrical power, and the footprint is moving across Earth's surface at nearly 7 km/sec. The CALIOP SNR can also be decreased in the presence of clouds overlying the ACA, a thick aerosol layer overlying clouds (i.e. the signal is attenuated), or due to sunlight reflected from clouds (i.e. the noise is increased).
5. Another reason for a potential CALIOP misclassification of the aerosol comes from the fact that CALIOP uses loading-dependent Lidar measurements and information that is only indirectly related to aerosol type (volume depolarization, attenuated backscatter, aerosol location, height, and surface type), rather than the exclusively intensive aerosol properties used by the HSRL aerosol classification (Burton et al., 2012). An erroneous Sa could also be assumed even when the aerosol type is correctly defined. For example, Schuster et al. (2012) used a coincident CALIOP–AERONET data set of 147 AERONET stations to find that the Lidar ratio for the CALIOP global dust model (40 sr) often underestimates the local Lidar ratio. The prelaunch goal of the CALIPSO mission was to retrieve aerosol extinction coefficients accurate to within ±40% (Winker

et al., 2003). The CALIOP AOD fractional error is similar to the Lidar ratio fractional error for small AOD values (Winker et al., 2009) but, as the AOD increases, the AOD fractional error will quickly become much higher than the Lidar ratio fractional error.

5.3.5.3 Validation and assessment

The advantages of Lidar-derived properties near and above clouds are a high vertical and temporal resolution paired with a narrow source of illuminating radiation, which limits cloud adjacency effects (3D cloud radiative effects) and cloud contamination of data products (Zhang et al., 2005; Wen et al., 2007; Várnai and Marshak, 2009). In addition to inferring the aerosol height in passive satellite retrievals of ACA (see previous sections), CALIOP observations are used to quantify frequency of occurrence and zonal or global seasonal mean of ACAOD as well as vertical separation between the aerosol and underlying clouds (Yu and Zhang, 2013; Devasthale and Thomas, 2011; Costantino and Bréon, 2010).

The studies mentioned above would greatly benefit from the quantification of CALIOP ACAOD biases and uncertainties (Jethva et al., 2011; Yu and Zhang, 2013). Validating products such as the standard CALIOP ACA occurrence, ACA altitude, geometrical thickness, and optical depth is challenged by the lack of suitable validation data sets. To the best of our knowledge, the existing peer-reviewed evaluations of CALIOP aerosol detection and optical properties retrievals have been largely restricted to cloud-free conditions (Kim et al., 2008; Pappalardo et al., 2010; Omar et al., 2009; Kacenelenbogen et al., 2011; Rogers et al., 2011; Burton et al., 2013; Winker et al., 2013). For this reason, we investigate this briefly in section 5.4.3.

5.3.5.4 Ongoing developments

In addition to the standard CALIOP retrieval, Hu et al. (2007) and Chand et al. (2008) introduced alternative CALIOP retrieval methods that use liquid water clouds as targets of known reflectivity under the aerosol layer. This allows AOD and Angström Exponent to be deduced directly from aerosol effects on light transmission. Because such 'constrained retrievals' do not require knowledge of a Lidar ratio (as is the case for the CALIOP standard retrieval), they are more accurate (Young and Vaughan, 2009). The first technique is based on measurements of water cloud depolarization ratio (DR) (Hu et al., 2007) at 532 nm and retrieves ACAOD regardless of the nature of the overlying material (but requires opaque underlying water clouds). If the annual mean global low cloud coverage is approximately 40% (e.g. Rossow and Schiffer, 1999), the clouds that arc opaque to the Lidar signal often reside over the ocean and are mostly limited to the tropical region (Devasthale and Thomas, 2011). This leads to DR-retrieved CALIOP ACA observations that are substantially limited in spatial coverage. The second technique is based on measurements of cloud color ratio (CCR) (Chand et al., 2008) (ratio of the signal from the cloud at 1064 nm to that at 532 nm). This method does not require opaque clouds. It responds to an increase in the water cloud layer integrated attenuated CCR compared to the value for unobstructed clouds, and is primarily sensitive to fine-mode aerosols. The CCR method can only detect aerosols that have an appreciable difference in extinction for the two wavelengths. One reason for favoring the DR compared to the CCR method is that, although the DR method requires more limited opaque water cloud coverage, the calibrations required to generate the 532-nm depolarization ratios are much more accurate than the current 1064-nm CALIPSO calibration. Chand et al. (2008) show that, during August 2006 and offshore from the west coast of Africa (region of high biomass-burning influence), the median random error in the respective DR and CCR-derived AOD is 0.12 and 0.09 during the day.

5.4 Validation with in situ and suborbital observations

5.4.1 *In situ* observations from field campaigns

Suborbital observations can improve our capability to detect and quantify aerosol properties (radiative and microphysical) above clouds. Campaigns such as SAFARI (Southern African Fire–Atmosphere Research Initiative; Swap et al., 2003) and TARFOX (Tropospheric Aerosol Radiative Forcing Observational eXperiment; Russell et al., 1999) studied the aerosol radiative effect in both cloud-laden and clear-sky conditions. Using *in situ* (air sample-derived SSA and size distribution) and satellite (Advanced Very High Resolution Radiometer (AVHRR) AOD product; Nagaraja Rao et al., 1989; Holben et al., 1992) observations from TARFOX, Bergstrom and Russell (1999) estimated the annual average aerosol radiative flux change under all-sky conditions (using the International Satellite Cloud Climatology Project (ISCCP) cloudiness data set) due to all natural and anthropogenic aerosol components for measurements over the North Atlantic to be -0.8 to -1.1 W/m^2. Their clear-sky estimations were in the range of -1.7 to -5.1 W/m^2, due to seasonal variability. During SAFARI, Keil and Haywood (2003) used *in situ* aerosol and cloud measurements on board the Met Office C-130 Hercules aircraft, and combined them with RT calculations to learn about the influence of clouds on direct aerosol forcing and to quantify the overall sign and magnitude of the biomass-burning aerosol TOA solar RF over oceanic regions off the South African coast. The instruments used in their analysis included a Passive Cavity Aerosol Spectrometer probe (PCASP) to obtain size distribution and particle number concentration, a nephelometer to measure the aerosol scattering coefficients, a Particle Soot Absorption Photometer (PSAP) to obtain aerosol absorption coefficients, a Fast Forward Scattering Spectrometer Probe (FFSSP) to measure cloud drop size distribution, and a liquid water content (LWC) probe. Their results indicated negative clear-sky aerosol forcing of -13 W/m^2 which becomes $+11.5$ W/m^2 when the cloud layer below the aerosol is included in their RT model.

5.4.2 Airborne sunphotometers

Sunphotometers are versatile instruments that are sensitive to aerosols and are relatively simple to operate and calibrate. They are available on airborne platforms such as the 14-channel Ames Airborne Tracking Sunphotometer (AATS-14; Russell

Table 5.2. Basic properties of orbital remote sensing instruments that observe ACA. In this table, ACAOD, above cloud aerosol optical depth; $r_{eff,a,f}$, aerosol effective radius in microns, fine size mode; $r_{eff,c}$, cloud droplet effective radius in microns; $v_{eff,c}$; cloud droplet effective variance; α, Ångström Exponent; f, fraction nonspherical aerosol particles; m_r, aerosol real refractive index; m_i, aerosol imaginary refractive index; Z_c, cloud top height; DRE, Direct Radiative Effect; R, reflectance; R_p, polarized reflectance. All wavelengths are in nanometers.

Instrument	Input data	Algorithm	Assumptions	Products	References
OMI	R(354), R(388), 13×24 km^2 resolution	LUT search	All aerosol and cloud microphysical properties, aerosol and cloud vertical distribution, homogeneity within each pixel	COD, ACAOD	Torres et al. (2012)
MODIS	R(470), R(870), 1×1 km^2 resolution	LUT search	All aerosol and cloud microphysical properties, aerosol and cloud vertical distribution, homogeneity within each pixel	COD, ACAOD	Jethva et al. (2013)
SCIMACHY	R(240) to R(2380), 60×30 km^2 resolution	Simulation–observation difference (no aerosol properties retrieved)	Cloud property determination assumes no aerosol extinction in SWIR, identical viewing angle for observations and simulations	COD, $r_{eff,c}$, Aerosol DRE	de Graaf et al. (2007, 2012)
POLDER research algorithm	Rp(490), Rp(670), Rp(865) from 200×200 km^2 pixel, in 0.5° scattering angle bins	Optimal estimation	Uniform properties within 200×200 km^2 pixel, aerosol vertical distribution, homogeneity within each pixel	ACAOD, $r_{eff,a,f}$, $r_{eff,c}$, $v_{eff,c}$, α, f, Less sensitivity: m_r, m_i	Waquet et al. (2009, 2013)
POLDER operational algorithm	Rp(670), Rp(865) for up to 16 viewing angles between 55° forward and aft of satellite track, 6×6 km^2 resolution	LUT search	COD > 5, LUT limitations, aerosol vertical distribution (weakly sensitivite), homogeneity within each pixel	ACAOD, α	Waquet et al. (2009, 2013)
CALIOP	Attenuated backscatter profiles at 532 and 1064 nm, polarized backscatter at 532 nm, 0.33 km horizontal, 30 m vertical resolution	Feature detection, classification and extinction retrieval scheme	LUT limitations for aerosol type determination, assumed Lidar ratio	Attenuated backscatter profile, cloud and aerosol type, ACAOD, COD, Z_c	Winker et al. (2009); Vaughan et al. (2009); Liu et al. (2010); Young and Vaughan (2009)

et al., 2005) and the Spectrometer for Sky-Scanning, Sun-Tracking Atmospheric Research (4STAR; Dunagan et al., 2013). AATS-14 has been deployed in many field campaigns, including ARCTAS (Arctic Research of the Composition of the Troposphere from Aircraft and Satellites), to validate satellite AOD retrievals (e.g. MODIS) on days dominated by smoke and haze pollution, under clear or cloudy atmospheric conditions (Livingston et al., 2014). Although ACA scenarios were not observed in the ARCTAS case study, measurements of heavy smoke plumes were classified as clouds in the MODIS retrieval algorithm, which underestimated AOD under these conditions. This example amplifies the necessity of airborne platforms to validate satellite observations under challenging conditions such as heavy-aerosol loads in clear and cloudy conditions. The newly developed sunphotometer 4STAR was recently deployed on board the US Department of Energy's (DOE) Gulfstream-I (G-1) aircraft in the TCAP (Two Column Aerosol Project) campaign, which was designed for closure radiation studies of aerosol and clouds, and to evaluate new remote sensing retrieval algorithms of AOD in the presence of clouds, among other goals. The 4STAR acquired clear-sky AOD and sky radiance measurements that mimic the AERONET sky measurements to derive aerosol intensive properties such as SSA and size distribution following the method by Dubovik et al. (2002). On several days during the summer TCAP phase (July 2012), elevated smoke layers were detected by the HSRL on board the NASA B-200 aircraft, which flew coordinated legs above the G-1 in some instances. Figure 5.4 is a MODIS false color image for the July 9, 2012, G-1 flight path. High thin ice clouds are in blue, and thick ice clouds are in magenta. Brown–yellow patches above the ocean are smoke layers. The G-1 flight path is the solid cyan line. As seen in the figure, although the flight path was not intersecting ACA instances, the instrument flew through a variety of mixed smoke and high cloud scenes, and was able to capture Ångstrom Exponent changes for atmospheric columns measured above an aircraft altitude of

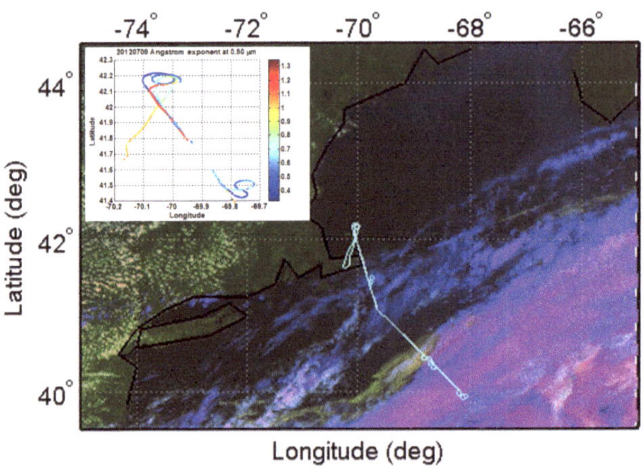

Fig. 5.4. False color MODIS-AQUA image from July 9, 2012 (image time: 18:20 UT), during the TCAP summer phase. Flight proceeded between 17:00 and 20:30 UT. The aircraft flight path is shown as asolid cyan line. Blue and magenta patches are thin and thick ice clouds (respectively), and the green–blue patch above the ocean is smoke. Figure insert shows the 4STAR Ångström Exponent values for columns above the aircraft.

3 km (see insert). This, again, amplifies the ability of airborne sunphotometers as a valuable validation tool for the detection of AOD and spectral dependence of ACA. However, direct comparisons with ACA retrievals from orbital instrumentation have yet to occur, which underscores the need for analysis of previous campaigns and organization of future campaigns specifically to observe ACA.

5.4.3 Active sensors

To assess the ability of CALIOP to detect ACA, Kacenelenbogen et al. (2014) used measurements acquired by HSRL (Hair et al., 2008). The data in this study were collected on over 800 flight hours from 10 field missions between 2006 and 2009, many of which included CALIOP validation flights. In comparison to CALIOP, the HSRL technique directly retrieves the vertical profiles of aerosol extinction coefficients and extinction-to-backscatter ratios, without requiring ancillary aerosol measurements or assumptions about aerosol type (Hair et al., 2008). Each CALIOP aerosol profile product was filtered with specific quality control criteria and co-located to the nearest (i.e. within 30 min and/or 5 km) HSRL profile. The cloud top height was defined by HSRL for both the HSRL and the CALIOP ACAOD calculations. In addition, co-located HSRL and CALIOP profiles were selected when showing no aerosol or cloud above the HSRL airplane according to CALIOP. Figure 5.5 shows the location of the co-located CALIOP and HSRL ACAOD retrievals (red) compared to all HSRL ACAOD retrievals along the CALIOP tracks (whether CALIOP retrieves an ACAOD or not) (blue).

Figure 5.5 shows a higher number of HSRL ACAOD retrievals ($N = 668$) compared to co-located CALIOP and HSRL ACAOD retrievals ($N = 151$). This leads to CALIOP detecting ACA in $\sim$23% of the cases in which it is observed by HSRL ($N = 151$ compared to 668). With the exception of a few cases over Alaska, most of the coincident CALIOP-HSRL ACA cases ($N = 151$) are found over the Eastern, Central, and South Central United States. According to the HSRL classification scheme (Burton et al., 2012), these ACA cases are mostly composed of

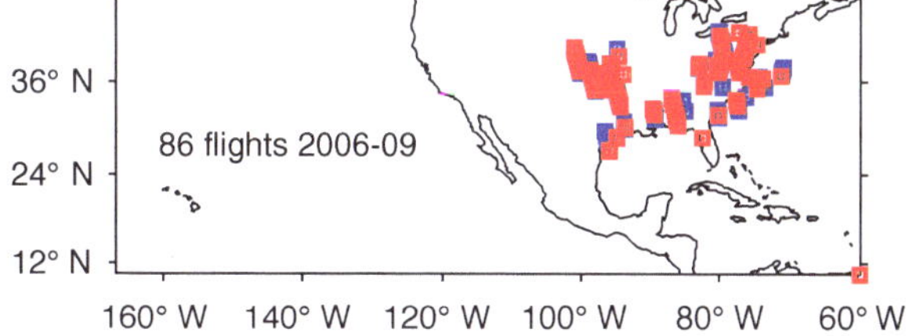

Fig. 5.5. Location of the co-located CALIOP and HSRL ACAOD retrievals (red, $N =$ 151) compared to all HSRL ACAOD retrievals along the CALIOP track (blue, $N = 668$).

urban (~46%), dusty mix (~27%), and biomass-burning smoke (~13%). A majority of the ACAOD values lie below 0.1 with an average AOD of ~ 0.04 ± 0.05, which is reasonable when compared to a global MODIS background AOD of about 0.13 over ocean and 0.19 over land (Remer et al., 2008) and a global average mid-visible remote sensing composite AOD near 0.13 (Kinne, 2009). The cloud top height underneath the ACA cases resides in average around 1.9 ± 1.02 km.

CALIOP shows essentially no agreement with HSRL for combined day and night ACAOD measurements ($R^2 = 0.27$, Figure 5.6) and ~68% of the CALIOP ACAOD values are outside the ±40% envelope of the CALIOP=HSRL line. The non-detection or underestimation of ACAOD (i.e. the total number of aerosol layers or specific tenuous aerosol layers above each cloud) is mostly due to tenuous aerosol layers with backscatter coefficients below the CALIOP detection threshold. A minority of the discrepancy seems to be due to a CALIOP-type misclassification or an error in the CALIOP modeled Sa. Compared to an erroneous Sa assumption, correcting for the CALIOP misdetection of aerosol vertical extent produces a larger root mean square (RMS) change from the initial CALIOP ACAOD, a smaller RMS difference between CALIOP and HSRL ACAOD, and a bigger reduction in RMS CALIOP–HSRL difference, measured as a percentage of the mean HSRL ACAOD of 0.047 (11%). The remaining source of error seems to be attributable to the spatial–temporal colocation of both instruments.

The CALIOP ACAOD values show discrepancies with the coincident HSRL values but no particular bias. It is important to emphasize that this study is mostly specific to the continental US and shows low ACAOD values. A similar analysis might lead to different conclusions over a region of strong occurrence of ACA with

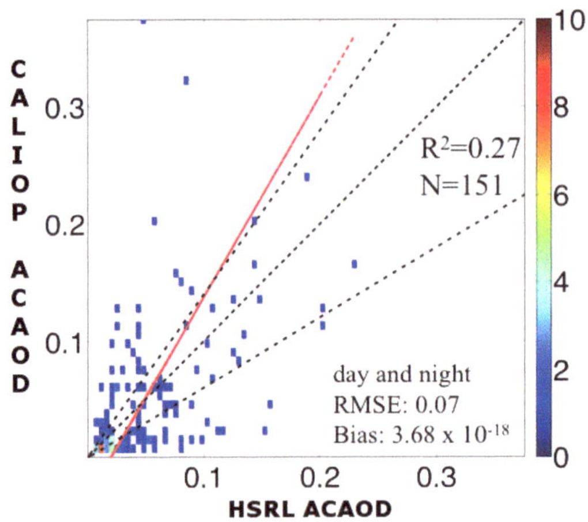

Fig. 5.6. CALIOP versus HSRL ACAOD at 532 nm with day and night measurements (red regression line: CALIOP ACAOD = 1.72 ± 0.23 HSRL ACAOD −0.03 ± 0.01, $R^2 = 0.27$; $N = 151$, RMSE = 0.07 and Bias = 3.68×10^{-18}). The cloud underlying the aerosol is defined by HSRL. The profiles are cloud-free and aerosol-free above the HSRL airplane according to CALIOP. The color bar shows the percentage of points in each cell compared to the total number of coincident CALIOP–HSRL ACA cases ($N = 151$). The dashed lines represent CALIOP ACAOD = HSRL ACAOD ± 40% HSRL ACAOD and the 1:1 line

higher AOD (e.g. offshore from the Namibian coast of Africa). This analysis underlines the need for additional suborbital field experiments in regions with high frequency of occurrence of ACA. The evaluation of the CALIOP standard or alternate detection and retrieval of ACA should be studied with a more extensive CALIOP–HSRL data set in different regions, once the appropriate HSRL data sets become available.

5.4.4 Spectrometers

When combined with sun photometer observations, aircraft measurements of upward and downward propagating irradiance can be used to determine the SSA and absorption aerosol optical depth (AAOD = $(1-\mathrm{SSA})*\mathrm{AOD}$) of specific aerosol layers. Schmidt et al. (2009) compared modeled spectral irradiance of shallow cumulus clouds to observations by the Solar Spectral Flux Radiometer (SSFR; Pilewskie et al., 2003) and found agreement only if aerosols were included in the simulations. An example of the use of the SSFR to observe ACA is described by Bergstrom et al. (2007), who use coupled SSFR and AATS-14 sun photometer observations (see section 5.4.2) above and below an aerosol layer to determine the spectrally resolved SSA and AAOD. This is done by optimizing a solar RT model, so that output for portions of the spectrum that are insensitive to gaseous absorption match the observations (Bergstrom et al., 2003). This method is appropriate for any aerosol layer that has been observed from above and below, and the first application to ACA was shown with MILAGRO/INTEX-B (Megacity Initiative-Local and Global Research Observations/Phase B of the Intercontinental Chemical Transport Experiment; Molina et al., 2010) data in the Gulf of Mexico and the Mexico City metropolitan area (Bergstrom et al., 2010). During MILAGRO/INTEX-B, the SSFR and AATS-14 were integrated on the same aircraft. Observations from this campaign are also described by Coddington et al. (2008); Livingston et al. (2009); Redemann et al. (2009); Knobelspiesse et al. (2011), and others referenced in Molina et al., 2010.

Figure 5.7 illustrates the spectrally resolved products of this technique: fractional aerosol absorption, SSA, and AAOD. Considering that some other ACA remote sensing techniques are minimally sensitive to SSA (or must assume SSA values in the retrieval algorithm), this approach has potential validation utility for that aspect of ACA. However, this technique is complicated by the operational difficulty of over and under flying an aerosol layer, and the need to merge different instrument observations.

5.4.5 Airborne polarimeters

Airborne polarimeters have the potential to retrieve ACA properties in a manner similar to POLDER (see section 5.3.4). For example, the Research Scanning Polarimeter (RSP) simultaneously retrieved ACA and cloud optical properties off the coast of Veracruz, Mexico during the MILAGRO/INTEX-B field campaign (Knobelspiesse et al., 2011). The RSP, which is an airborne prototype of the APS, overflew a optically thin and somewhat absorbing ACA layer (ACAOD(555 nm) = 0.14, SSA(532 nm) = 0.87) suspended over a marine stratocumulus cloud. Like the APS, the RSP has many viewing angles (roughly 150), polarization sensitivity, and nine

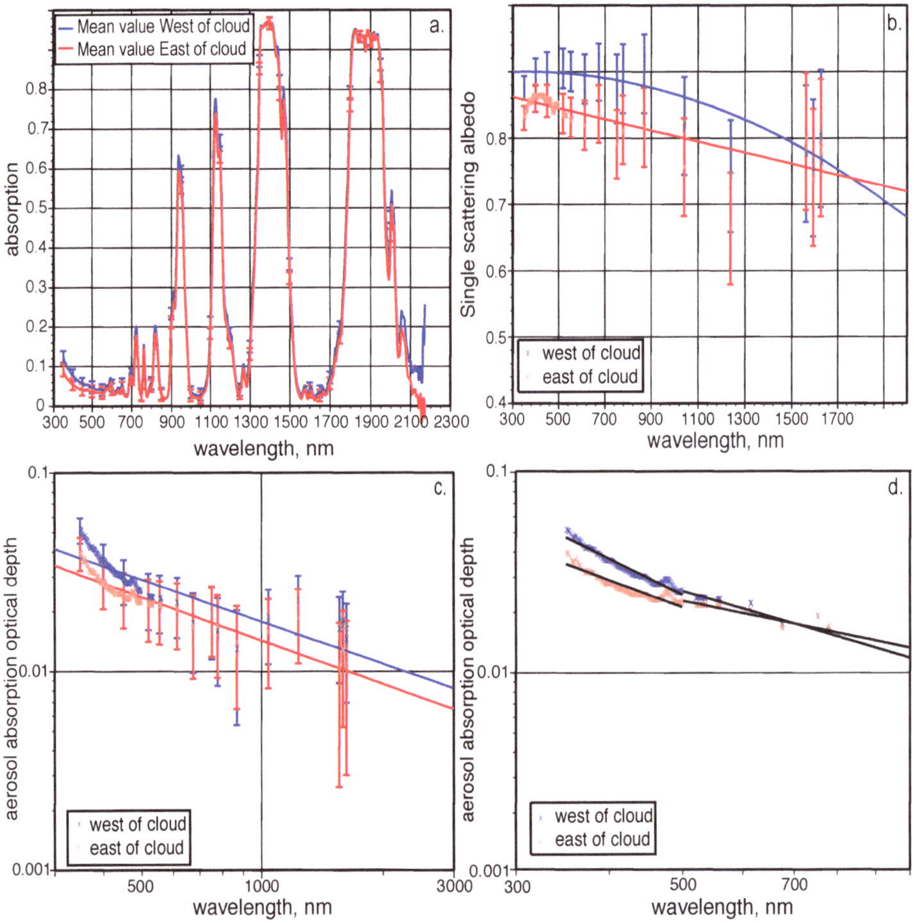

Fig. 5.7. Example of a retrieval of ACA optical properties from a scene in MILAGRO/ INTEX-B. Panel (a) is the fractional absorption of the ACA, (b) is the SSA, and (c) and (d) are the AAOD. Figure is adapted from Bergstrom et al. (2010).

spectral channels (from 410 to 2250 nm), seven of which are devoted to aerosol and cloud remote sensing.

Provided external information about cloud and ACA vertical distribution, the RSP could determine the spectral ACAOD, cloud droplet size distribution and ACA fine size mode size distribution accurately. Using two parameters to parameterize aerosol imaginary refractive index (and thus absorption), the spectral SSA was also determined, although with a large assessed uncertainty. On March 13, 2006, the RSP was a payload on the same aircraft as the SSFR and AATS-14 instruments described in sections 5.4.4 and 5.4.2, respectively. Comparisons of RSP retrieval results to observations by those instruments (Fig. 5.8) are within RSP assessed uncertainties. In the case of SSA, differences between SSFR and RSP retrievals are far smaller than assessed RSP uncertainty.

Knobelspiesse et al. (2011) also tests the relationship between observational uncertainty and measurement conditions. They found that the SSA retrieval error

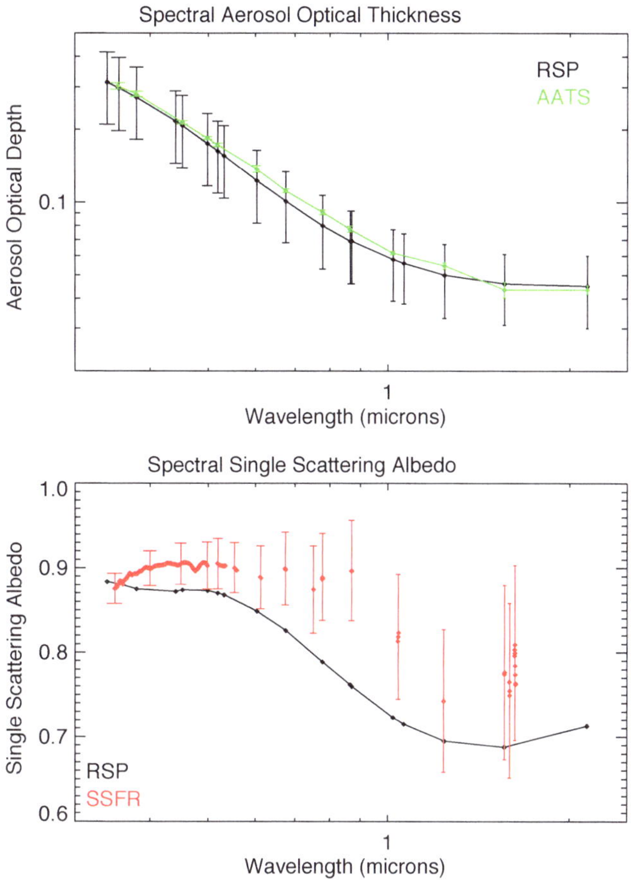

Fig. 5.8. (from Knobelspiesse et al., 2011) Comparison of RSP retrieval results to AATS-14 spectral optical depth (top panel) and SSFR spectral SSA (bottom panel). SSFR results are from Bergstrom et al. (2010), and the large RSP retrieval error bars for SSA are omitted for clarity.

(assessed at 0.45 for the MILAGRO case) is reduced as ACAOD increases. The assessed uncertainty becomes lower than 0.03 when the mid-VIS ACAOD is 0.8 or larger. Obviously, this is an uncommonly large value for AOD, whose clear-sky, over-ocean, global mean value is about 0.13 at 550 nm (Remer et al., 2008). This uncertainty could possibly be reduced if the total reflectance, in addition to polarized reflectance, were incorporated into the retrieval algorithm. To date, this capability has been tested theoretically, but not with observed data (Hasekamp, 2010).

Unlike spectrometers such as SSFR and sun photometers such as AATS-14, polarimeters such as RSP can determine ACA optical properties with a single fly-over (analogous to an orbital spacecraft). Their operational capability therefore lies between airborne instruments described earlier in this section or the orbital instruments described in the previous section. Despite the history of airborne polarimeter measurements, little ACA data exist. It is likely this is because they have been typically deployed in North America (RSP, AirMSPI) or Europe (the Micropolarimeter, MICROPOL; Waquet et al., 2005), where ACA are less common, and because they have not been incorporated into field campaigns specifically targeting ACA. Future campaigns will be successful if the polarimeter incorporates both total and polarized reflectance, enough viewing angles to determine the cloud-bow maximum scattering angle, and have access to external information about cloud and aerosol vertical distribution. For this reason, combinations of instruments such as the RSP and the HSRL (Hair et al., 2008) on the same aircraft would be a particularly powerful combination for the remote sensing of ACA.

5.4.6 RF assessment using observational data and regional climate models

Global RF due to aerosols stems both from the direct effect of scattering and absorption of aerosols and from semi-direct and indirect aerosol effects on clouds. While DRE for clear-sky conditions is relatively understood and yields an overall net negative forcing (e.g. Jacobson, 2001; Bellouin et al., 2005; Yu et al., 2006; Forster et al., 2007; Sena et al., 2013), the all-sky forcing sign and magnitude still remain uncertain (Yu et al., 2012). This is in part due to the lack of operational observations that contain both aerosols and clouds in the same column (Yu and Zhang, 2013), and the biases in retrievals of cloud and aerosol properties for ACA cases (e.g. Wilcox et al., 2009; Jethva et al., 2013). The uncertainty in all-sky RF is especially large when absorbing aerosols (i.e. smoke from biomass-burning events or dust) reside above cloud layers. Over the south-eastern Atlantic, for example, where biomass-burning smoke advection from African savannas forms a layer over the stratocumulus cloud deck, there is no agreement between models as to the sign or magnitude of direct aerosol forcing for all-sky conditions (e.g. Schultz et al., 2006). This discrepancy is attributed to both the lack of representative cloud fields and cloud cover below the aerosol layer and reliable SSA values for the various aerosol types. Observationally derived estimates, assuming SSA, indicate that aerosol TOA direct forcing over the south-eastern Atlantic switches sign from positive to negative when cloud cover exceeds 40% (Keil and Haywood, 2003; Chand et al., 2009).

The thermal interplay between aerosols and clouds that is the semi-direct effect is substantial (Johnson et al., 2004; Wilcox, 2012). The semi-direct aerosol effect is caused by a change in cloud cover or cloud liquid water path due to heating from aerosol absorption of solar radiation (Sakaeda et al., 2011). Johnson et al. (2004) found that the sign of direct and semi-direct aerosol effects depends on whether the absorbing aerosol is located above, in, or both in and above the boundary layer.

Satellite observations are needed to constrain global climate models to improve the closure and understanding of aerosol direct and semi-direct effects for all-sky conditions. To date, there are several investigations that followed this path. A recent modeling study by Sakaeda et al. (2011) that examined biomass-burning direct and semi-direct effects from South African fires used MODIS clear-sky AOD and aerosol layers above clouds from CALIPSO above land and ocean to constrain their simulations with the global Community Atmospheric Model (CAM). In their investigation, they found that, over the ocean, where the aerosol layers are primarily located above cloud decks, negative TOA semi-direct effects are dominant over the weak all-sky DRE. In contrast, they found that, over land, where aerosols are often below or within cloud layers, reduction in cloud liquid water path leads to a positive semi-direct effect that dominates over a near zero DRE.

Wilcox (2010, 2012) obtained similar conclusions when assessing the radiative heating effect of biomass-burning aerosol above a stratocumulus marine deck in the south-eastern Atlantic using liquid water path (LWP) and sea surface temperature (SST) products from the Advanced Microwave Scanning Radiometer-EOS (AMSR-E), Atmospheric Infrared Sounder (AIRS) temperature profiles, and OMIAI to show the temperature increase of the biomass-burning aerosol layer above the clouds and the increase in LWP for such clouds. RF calculations attributed net cooling due to the semi-direct effect of aerosol above clouds to be -5.9 ± 3.5 W/m^2, which counteracts more than 60% of the DRE in this region (which was 9.2 ± 6.6 W/m^2). After accounting for the probability of occurrences of overlaying aerosols above overcast scenes, they have estimated the average forcing to be 1.0 ± 0.7 W/m^2, illustrating the need for accurate characterization of these phenomena. For example, this work used the OMI AI, which does not uniquely express aerosol absorption.

In a regional modeling study over the Amazon, Ten Hoeve et al. (2012) studied the effect of biomass-burning (BB) aerosol loading on cloud properties and their radiative effects using a GCM. Results were compared with MODIS AOD and COD observations, together with validation of AOD and aerosol properties from the AERONET network. They found an increase in COD with increasing AOD for AOD below roughly 0.3, due to a combination of aerosol microphysical effects and destabilization of the aerosol layer from aerosol heating, as suggested above. In addition, they found that COD decreases with increasing AOD for AOD between 0.3 and 0.9 due to radiative effects, which evaporate clouds and stabilize the lower boundary layer. The authors concluded that the similarity between MODIS trends and the model results, together with similar behavior observed in other independent investigations (e.g. Koren et al., 2008; Jiang and Feingold, 2006), suggests that these AOD/COD correlations are physical and not retrieval artifacts such as brightening of aerosols near clouds, cloud contamination of aerosol, or artificial darkening of clouds below an absorbing aerosol layer.

In sum, modeling combined with observational analysis can shed light on the complex processes of aerosol–cloud interaction, specifically for aerosols above and within low-level clouds. Nevertheless, more work is needed to improve retrieval approaches to gain a better sense of the real aerosol–cloud interface properties. Also, additional modeling work on a global scale might allow the generalization of the phenomenon observed on specific regional domains, as detailed above.

5.5 The future for ACA retrievals

In order to preface the opportunities from planned and future orbital instruments and the fusion of different measurements, we will here briefly recap what signals from aerosols above clouds are currently available. The most obvious candidate for detecting and characterizing aerosols above clouds is Lidar (e.g. Winker et al.,

2009), which can readily separate the signals of clouds and aerosols with any significant vertical separation. In passive remote sensing, the signal of absorbing aerosols above clouds is apparent in their effect on the observed color of cloud scenes, from the ultraviolet to the mid-visible, when aerosols have spectrally varying absorption optical depths (e.g. Torres et al., 2012; Jethva et al., 2013). This is the case for both dust aerosols and brown carbon aerosols, as a result of spectral variations in the imaginary refractive index. For accumulation mode aerosols containing black carbon, spectral variability is also a result of their size relative to wavelength. With regard to polarization, accumulation mode aerosols generate substantial side scattering in the near-infrared that is readily differentiated from the signal of molecular scattering (Waquet et al., 2009; Knobelspiesse et al., 2011). Over Stratiform water clouds, the attenuation of the rainbow by larger particles such as dust or volcanic ash can also be used to estimate ACAOD (Waquet et al., 2013).

5.5.1 Upcoming orbital opportunities

Here we will discuss upcoming orbital opportunities including identification of those sensors already on orbit that may provide opportunities beyond those reviewed above. Table 5.3 is a summary of those sensors.

The Suomi NPP spacecraft was launched on October 28, 2011, and has both the Ozone Mapping Profiler Suite (OMPS) and VIIRS on board. These sensors are successors to TOMS/OMI and MODIS, respectively. The OMPS sensor is very similar to OMI and can be used to detect absorbing aerosols above clouds in the manner described in section 5.3.1, but has an advantage over OMI in that it is located on the same platform VIIRS, allowing simultaneous collocation of the observations from both sensors. The VIIRS sensor, although similar to MODIS, lacks many of the spectral bands that were used for cloud detection and characterization. Furthermore, there are concerns about its radiometric performance, specifically for the 412, 445, and 488 nm spectral bands for ocean color research. However, the fact that these bands have dual gains means that they do not saturate over clouds and can therefore be used to similar, or potentially better, effect than the method described in section 5.3.2.

The Greenhouse gases Observing SATellite (GOSAT) was launched on January 23, 2009, and its payload is the Cloud and Aerosol Imager (CAI) and the Thermal And Near-infrared Sensor for carbon Observation (TANSO), which includes a Fourier Transform Spectrometer (FTS) for mapping the column concentrations of carbon dioxide (Sano et al., 2010b; Kuze et al., 2009). The CAI has similar spatial resolution to MODIS and a spectral range extending into the UV. CAI has the potential to provide excellent observations of ACA, since it has both higher spatial resolution than OMI (reducing the probability of partially cloudy pixels) and spectral sensitivity at shorter wavelengths than MODIS (Sano et al., 2010b). In addition, the FTS has high (0.2 per cm) spectral resolution for observations of the Oxygen-A band (758–775 nm), which provides additional constraints on aerosol vertical extent and optical depth. The 10 km spatial resolution of the FTS measurements, however, means that sub-pixel clouds would need to be evaluated using the CAI.

The Sentinel-5 Precursor mission is scheduled to launch in summer 2015, with the Tropospheric Ozone Monitoring Instrument (TropOMI) sensor on board. The

Table 5.3. Summary of new or forthcoming sensors that can be used to detect and characterize ACA. Bold face type indicates spectral bands for which polarization is measured.

Sensor	Mission	Measurement type	Spectral range/sample	Spatial resolution
OMPS	Suomi NPP	Spectral absorption	300–380 nm	50 km
VIIRS	Suomi NPP	Spectral absorption	412, 445, 488, 555, 672, 865, 1240, 1378, 1610, 2250 nm	1 km
CAI (TANSO)	GOSAT	Spectral absorption	380, 674, 870, 1600 nm (Oxygen-A band spectrometer)	0.5 km (10 km)
CATS	ISS	Lidar	Backscatter, depolarization: 355, 532, 1064 nm. Extinction: 532 nm	350 m (60 m vertical)
TropOMI	Sentinel 5P	Spectral absorption	270–490, 710–790, 2300 nm	7 km
SGLI	GCOM-C1	Spectral absorption + polarization	380, 412, 443, 490, 530, 555, **673**, **868**, 1050, 1380, 1630, 2210 nm	250 m (1 km polarized)
ATLID	EarthCARE	Lidar	355 nm extinction, backscatter, and depolarization	280 m (100 m vertical)
3MI	MSG	Spectral absorption + polarization	**410**, **443**, **490**, **555**, **670**, 763, 754, **865**, 910, **1370**, **1650**, **2130** nm	4 km

primary function of TropOMI is to extend trace gas observations and provide continuity with the SCHIAMACHY and Global Ozone Monitoring Experiment (GOME) instruments. The spectral range of the primary spectrometer of TropOMI is similar to OMPS and OMI, but TropOMI has additional measurements in the Oxygen-A band at 760 nm. These help estimate the AOD and vertical distribution of ACA, thereby reducing the uncertainty in the UV absorption estimates (Boesch et al., 2008). Moreover, the 7-km spatial resolution of the TropOMI instrument will provide increased capability compared with OMI and OMPS to quantitatively interpret parameters such as the aerosols index.

The Second-Generation Global Land Imager (SGLI) is scheduled to launch in December 2015 on the Global Change Observation Mission–Climate (GCOM-C) satellite (Shimoda, 2010). This sensor is unique in that it provides relatively high spatial resolution (250 m) observations from the UV to the SWIR, allowing methods developed for retrieving aerosols above clouds for OMI and MODIS to be applied, but with the additional spectral capabilities of both CAI and VIIRS that were noted above. Moreover, the SGLI sensor suite includes a gimbaled polarimetric imager that can be tilted to look forward or aft at 45° from nadir to provide scattering angle geometries that are close to 90°, thereby providing sensitivity to both absorbing and non-absorbing accumulation mode aerosols (Tanaka et al., 2010). The combination of polarized NIR and UV/blue/visible total radiance observations will allow retrievals of ACA with fewer assumptions than has thus far been the case for passive sensors (Sano et al., 2010a).

The first EUMETSAT Second Generation satellite is expected to launch in 2020 with the 3MI sensor as its payload (Marbach et al., 2013). Although the development of 3MI is still in its early stages, the planned capabilities for ACA retrievals would be greater than any currently funded space-borne sensor. The VIS to SWIR spectral coverage, Oxygen-A band sensitivity, moderate spatial resolution of 4 km, and polarized measurements over a wide range of spectra and viewing angles could be used in a comprehensive ACA retrieval scheme.

The Cloud and Aerosol Transport System (CATS) is expected to launch in 2014 and will be mounted on the Japanese Experiment Module Exposed Facility of the International Space Station (ISS; Chuang et al., 2013). The spatial resolution of the downlinked data is expected to be 60 m vertically and 350 m horizontally, with between one and three beams, depending on operational mode. This Lidar system will allow continuity of the observing capability provided by CALIPSO, with the addition of depolarization at both 532 and 1064 nm in most operational modes, valuable for characterizing dust. In addition, CATS will have 355-nm backscatter and 532-nm HSRL capability, which allows the direct determination of the aerosol extinction. While these two features of the system are primarily technology demonstrations, if they perform as expected, they would provide a significant improvement in our ability to detect and characterize ACA compared to CALIPSO and the Ice Cloud and land Elevation Satellite (ICESat, see below), but with the caveat that global coverage will be limited to low and mid-latitudes due to the ISS orbit.

The ATmospheric LIDar (ATLID) will be launched on the Earth Clouds, Aerosols and Radiation Explorer (EarthCARE) observatory in 2015 (Hélière et al., 2007). It operates at a wavelength of 355 nm and has a HSRL receiver and depolarization channel. This will provide aerosol extinction above cloud observations, while the measurements of the Lidar ratio will provide a constraint on aerosol type and absorption (Sugimoto et al., 2011). Unfortunately, the shortest wavelength on the Multi-Spectral Imager on EarthCARE is 670 nm, which limits the value of combining Lidar and imager data for ACA retrievals and essentially precludes the possibility of actually retrieving aerosol absorption from EarthCARE observations alone.

Lastly, we note that the ICESat-2 satellite to be launched in 2016 will carry the Advanced Topography Laser Altimeter System (ATLAS) (Abdalati et al., 2010). This laser system is focused on measuring changes in ice thickness and vegetation canopies. While it will therefore necessarily detect aerosols, the system it is not optimized to characterize them.

5.5.2 Data fusion

The primary existing orbital opportunity for data fusion is provided by the A-Train set of platforms and sensors. The passive MODIS, POLDER, and OMI sensors and the active CALIPSO Lidar all provide sensitivity to ACA as described in previous sections. Two studies by Costantino and Bréon (2010, 2013) make use of

CALIPSO retrievals to identify when aerosols are above clouds. Interestingly, they do not actually use CALIPSO retrievals of ACAOD, or any other direct retrieval of ACAOD, instead searching for clear-sky aerosol retrievals from MODIS within 10 km (Costantino and Bréon, 2013) or 150 km (Costantino and Bréon 2010). Nonetheless, the results they obtain through this statistical analysis demonstrate that improved ACA retrievals would have significant scientific value. A contrasting study, which focuses on ACA retrievals, rather than interactions between aerosols and clouds, is presented by Yu et al. (2012). In this work, the feasibility of combining OMI and MODIS observations to derive ACAOD is examined. The ACAOD from CALIPSO is used to determine whether there is a robust relationship between the UV AI from OMI and the AOD at 532 nm. The authors conclude that the ratio of AOD at 532 nm to AI is independent of aerosol type (dust or biomass-burning) and depends primarily on COD, which suggests that developing an empirical relationship between AI from OMI and COD from MODIS may allow the ACAOD to be retrieved. This would allow ACAOD to be retrieved over a much larger area than is possible with statistical analysis of CALIPSO observations, facilitating the study of aerosol–cloud interaction studies without the need to aggregate observations over extended periods. However, the relationship between AI and AOD at 532 nm should in principle depend on aerosol composition and in particular the size distribution and amount of black carbon contained in the aerosols (Torres et al., 2012). It therefore remains to be seen how accurately the ACAOD can be retrieved by empirically combining MODIS and OMI observations.

As we noted at the beginning of this section, the effect of absorbing ACA on the spectral variation of the UV and blue reflected radiation is substantial, and primarily dependent on the strength and spectral variation of absorption. In contrast, polarization observations at side scattering angles for accumulation mode aerosols (Waquet et al., 2009; Knoblespiesse et al., 2012) and attenuation of the rainbow for coarse mode aerosols (Waquet et al., 2013) are more sensitive to the AOD than to aerosol absorption. We would therefore speculate, based on recent work (de Graaf et al., 2012; Jethva and Torres, 2011; Jethva et al., 2013) that the most robust retrievals of the properties of ACA from orbital observations would be provided by a combination of OMI, POLDER, and MODIS observations, with aerosol vertical profiles being constrained by CALIPSO.

5.5.3 Recommendations for future instruments

The sensitivities of various measurements to ACA outlined at the beginning of this section define the type of future instruments that can provide improved observations of ACA. Certainly, passive instruments that measure total radiance into the blue or UV, in combination with longer wavelength measurements for cloud characterization, are the most sensitive and direct means for determining the radiative effects of ACA at the top of the atmosphere (de Graaf et al., 2012; Jethva et al., 2013). However, such measurements alone do not eliminate the indeterminacy between the strength of absorption and the total aerosol load. Furthermore, other types of observation are required to determine where in the atmosphere the incident radiation is absorbed. Measurements of polarized radiance reduce this indeterminacy substantially, with the ACAOD uncertainty (Knobelspiesse et al.,

2011) and SSA thereby being reduced to levels that allow for the accurate partitioning of radiation between surface and atmospheric absorption (Hasekamp, 2010). The remaining uncertainty in ACA retrievals is associated with the vertical profile of the aerosols. A precise determination of cloud top height and an estimate of these properties requires observations from a Lidar (e.g. CALIPSO), or a high-spectral-resolution A-band spectrometer (e.g. GOSAT, OCO-2). Beyond reducing uncertainties in retrievals and partitioning the absorption of radiation between the surface and atmosphere properly, the capability to clearly differentiate between multiple aerosol types above cloud requires a Lidar system with greater capability than the type of elastic backscatter Lidar on CALIPSO. Multi-wavelength HSRLs that provide this capability (Müller et al., 2000) are currently being flown on aircraft (Rogers et al., 2009), and could also be flown in space.

5.6 Conclusion

Neglected until recently, the remote sensing of above cloud aerosols is currently undergoing a renaissance of instrument design and algorithm development. Existing instruments, be they passive imagers, spectrometers, multi-angle polarimeters, or Lidars, are now being used to determine the loading and optical properties of ACA. Furthermore, new instruments are being tested as airborne prototypes, and upcoming orbital missions promise to expand the information available to parameter retrieval algorithms. The diversity of observation configurations results in a variety of retrieval algorithms, with unique products and associated uncertainties. This rapidly developing and important topic is the subject of this review.

The scientific community is still in the early stages of understanding and validating the diverse observations of ACA. Clearly, different instruments have access to different pieces of information about a scene. Combinations of data from different instruments could potentially be utilized in a retrieval algorithm more capable than the sum of the individual instrument algorithms. However, creating such an algorithm is a complicated task, requiring an intimate understanding of the instruments involved and the physical nature of a scene. Future instruments could be designed to combine the ACA observational capability of contemporary instruments. However, there is still much to learn before this can be done with confidence, highlighting the need for field tests with airborne prototype instruments and the acquisition of *in situ* validation data. Certainly, this topic is on the frontier of Earth remote sensing, and will remain so for quite some time.

Acronyms and symbols

α	Ångström exponent
λ	Wavelength (μm)
ω	Single-scattering albedo
τ_a	Aerosol absorption optical depth
τ_o	Reference aerosol optical depth
f	Fraction of nonspherical aerosol particles
m_r	Real refractive index

m_i	Imaginary refractive index
r_{eff}	Effective radius (μm)
R_s	Surface albedo
S_a	Extinction-to-backscatter ratio (sr)
v_{eff}	Effective variance
3MI	Multi-directional, Multi-polarization and Multispectral Instrument
4STAR	Spectrometer for Sky-Scanning, Sun-Tracking Atmospheric Research
AAOD	absorption aerosol optical depth
AATS-14	Ames Airborne Tracking Sunphotometer, 14 channel
ACA	Above cloud aerosol
ACAOD	Above cloud aerosol optical depth
ACE	Aerosol, Clouds, Ecosystems
ADEOS	Advanced Earth Observing System
AERONET	AErosol RObotic NETwork
AI	Aerosol Index
AirMSPI	Airborne Multiangle SpectroPolarimetric Imager
AIRS	Atmospheric Infrared Sounder
AMSR-E	Advanced Microwave Scanning Radiometer-EOS
AOD	Aerosol optical depth
APS	Aerosol Polarimetry Sensor
AR4	4th Assessment Report (of the IPCC)
ARCTAS	Arctic Research of the Composition of the Troposphere from Aircraft and Satellites
ATLAS	Advanced Topography Laser Altimeter System
ATLID	Atmospheric LIDar
AVHRR	Advanced Very High Resolution Radiometer
BB	Biomass-burning
CALIOP	Cloud Aerosol Lidar with Orthogonal Polarization
CALIPSO	Cloud-Aerosol Lidar and Infrared Pathfinder Satellite Observations
CAI	Cloud Aerosol Imager
CAM	Community Atmospheric Model
CATS	Cloud and Aerosol Transport System
CCR	Cloud color ratio
CNES	Centre National D'Études Spatiales
COD	Cloud optical depth
CR	Color ratio
DoLP	Degree of Linear Polarization
DR	Depolarization ratio
DRE	Direct Radiative Effect
EarthCARE	Earth Clouds, Aerosols and Radiation Explorer
Envisat	Environmental Satellite
ESA	European Space Agency
EPS-SG	EUMETSAT Polar System–Second Generation
EUMETSAT	European Organisation for the Exploitation of Meteorological Satellites
FFSSP	Fast Forward Scattering Spectrometer Probe
FTS	Fourier Transform Spectrometer
GCM	General Circulation Model
GCOM-C	Global Change Observation Mission–Climate

GOME	Global Ozone Monitoring Experiment
GOSAT	Greenhouse gases Observing SATellite
HARP	HyperAngular Rainbow Polarimeter
HSRL	High Spectral Resolution Lidar
ICESat	Ice, Cloud and land Elevation Satellite
InVEST	In-space Validation of Earth Science Technologies
IPCC	Intergovernmental Panel on Climate Change
ISCCP	International Satellite Cloud Climatology Project
LOSU	Level of scientific understanding
LUT	Lookup Table
LWP	Liquid water path
MICROPOL	MICROPOLarimeter
MILAGRO/INTEX-B	Megacity Initiative-Local and Global Research Observations/Phase B of the Intercontinental Chemical Transport Experiment
MISR	Multi-angle Imaging SpectroRadiometer
MODIS	Moderate Resolution Imaging Spectometer
NASA	National Aeronautics and Space Administration
NIR	Near-infrared
NPP	National Polar-orbiting Partnership
OMI	Ozone Monitoring Instrument
OMPS	Ozone Mapping Profiler Suite
PACE	Pre-Aerosol, Clouds, and ocean Ecosystem
PARASOL	Polarization and Anisotropy of Reflectances for Atmospheric Sciences Coupled with Observations from a Lidar
PCASP	Passive Cavity Aerosol Spectrometer Probe
POLDER	POLarization and Directionality of the Earth's Reflectances
PSAP	Passive Soot Absorption Photometer
RF	Radiative forcing (W/m^2)
RMS	Root mean square
RSP	Research Scanning Polarimeter
RT	Radiative transfer
SAFARI	Southern African Fire–Atmosphere Research Initiative)
SCIAMACHY	Scanning Imaging Absorption Spectrometer for Atmospheric Chartography
SGLI	Second-generation Global Land Imager
SNR	Signal-to-noise ratio
SSA	Single-scattering albedo
SSFR	Solar Spectral Flux Radiometer
SST	Sea surface temperature
SWIR	ShortWave Infrared
TANSO	Thermal And Near-infrared Sensor for carbon Observati
TARFOX	Tropospheric Aerosol Radiative Forcing Observational eXperiment
TCAP	Two Column Aerosol Project
TOA	Top of Atmosphere
TOMS	Total Ozone Mapping Spectrometer
TropOMI	Tropospheric Ozone Monitoring Instrument
UV	Ultraviolet
VIIRS	Visible-Infrared Imager Radiometer Suite
VIS	Visible

References

Abdalati, W., Zwally, H., Bindschadler, R., Csatho, B., Farrell, S.L., Fricker, H.A., Harding, D., Kwok, R., Lefsky, M., Markus, T., et al. (2010) 'The ICESat-2 laser altimetry mission', *Proc. IEEE*, **98**, 735–51.

Ackerman, A.S., Toon, O.B., Stevens, D.E., Heymsfield, A.J., Ramanathan, V. and Welton, E. (2000) 'Reduction of tropical cloudiness by soot', *Science*, **288**, 1042.

Albrecht, B.A. (1989) 'Aerosols, cloud microphysics, and fractional cloudiness', *Science*, **245**, 1227.

Alexandrov, M.D., Cairns, B. and Mishchenko, M.I. (2012) 'Rainbow Fourier transform', *J. Quant. Spectrosc. Radiat. Transfer*, **113**, 2521–35.

Anderson, T.L., Charlson, R.J., Bellouin, N., Boucher, O., Chin, M., Christopher, S.A., Haywood, J., Kaufman, Y.J., Kinne, S., Ogren, J.A., Remer, L.A., Takemure, T., Tanré, D., Torres, O., Trepte, C.R., Wielicki, B.A., Winker, D.M. and Yu, H. (2005) 'An "A-Train" strategy for quantifying direct climate forcing by anthropogenic aerosols', *Bull. Amer. Meteor. Soc.*, **86**, 1795–809.

Atwater, M.A. (1970) 'Planetary albedo changes due to aerosols', *Science*, **170**, 64–6.

Bellouin, N., Boucher, O., Haywood, J. and Reddy, M.S. (2005) 'Global estimate of aerosol direct radiative forcing from satellite measurements', *Nature*, **438**, 1138–41.

Bergstrom, R.W. and Russell, P.B. (1999) 'Estimation of aerosol radiative effects over the mid-latitude North Atlantic region from satellite and *in situ* measurements', *Geophys. Res. Lett.*, **26**, 1731–4.

Bergstrom, R.W., Pilewskie, P., Russell, P.B., Redemann, J., Bond, T.C., Quinn, P.K. and Sierau, B. (2007) 'Spectral absorption properties of atmospheric aerosols', *Atmos. Chem. Phys.*, **7**, 5937–43.

Bergstrom, R.W., Pilewskie, P., Schmid, B. and Russell, P.B. (2003) 'Estimates of the spectral aerosol single scattering albedo and aerosol radiative effects during SAFARI 2000', *Journal of Geophysical Research: Atmospheres (1984–2012)*, **108**, D13.

Bergstrom, R.W., Schmidt, K.S., Coddington, O., Pilewskie, P., Guan, H., Livingston, J.M., Redemann, J. and Russell, P.B. (2010) 'Aerosol spectral absorption in the Mexico City area: Results from airborne measurements during MILAGRO/INTEX B', *Atmos. Chem. Phys.*, **10**, 6333–43.

Boesche, E., Stammes, P. and Bennartz, R. (2008) 'Aerosol influence on polarization and intensity in near-infrared O2 and CO2 absorption bands observed from space', *J. Quant. Spectrosc. Radiat. Transfer.*, **110**, 223–39, doi:10.1016/j.jqst.2008.09.019

Bovensmann, H., Burrows, J., Buchwitz, M., Frerick, J., Noël, S., Rozanov, V., Chance, K. and Goede, A. (1999) 'SCIAMACHY: Mission objectives and measurement modes', *J. Atmos. Sci.*, **56**, 127–50.

Bréon, F.M. and Deschamps, P.-Y. (1993) 'Optical and physical parameter retrieval from POLDER measurements over the ocean using an analytical model', *Remote Sens. Environ.*, **43**, 193–207.

Bréon, F.M. and Goloub, P. (1998) 'Cloud droplet effective radius from spaceborne polarization measurements', *Geophys. Res. Lett.*, **25**, 1879–82.

Bréon, F.-M., Boucher, O., Lifermann, A., Morcrette, J.-J., Peuch, V.-H., Phulpin, T., Riedi, J., Stamnes, P., Tanré, D. and Thepaut, J.-N. (2011) *The 3MI Mission for Operational Monitoring of Aerosols from EPS-SG*, Centre national d'études spatiales.

Burton, S.P., Ferrare, R.A., Hostetler, C.A., Hair, J.W., Rogers, R.R., Obland, M.D., Butler, C.F., Cook, A.L., Harper, D.B. and Froyd, K.D. (2012) 'Aerosol classification using airborne high spectral resolution Lidar measurements: Methodology and examples', *Atmos. Meas. Tech.*, **5**, 73–98.

Burton, S.P., Ferrare, R.A., Vaughan, M.A., Omar, A.H., Rogers, R.R., Hostetler, C.A. and Hair, J.W. (2013) 'Aerosol classification from airborne HSRL and comparisons with the CALIPSO vertical feature mask', *Atmos. Meas. Tech. Discus.*, **6**, 1815–58.

Cattrall, C., Reagan, J., Thome, K. and Dubovik, O. (2005) 'Variability of aerosol and spectral Lidar and backscatter and extinction ratios of key aerosol types derived from selected Aerosol Robotic Network locations', *J. Geophys. Res.-Atmospheres*, **110**, D10S11.

Chand, D., Anderson, T., Wood, R., Charlson, R., Hu, Y., Liu, Z. and Vaughan, M. (2008) 'Quantifying above-cloud aerosol using spaceborne Lidar for improved understanding of cloudy-sky direct climate forcing', *J. Geophys. Res.*, **113**, D13206.

Chand, D., Wood, R., Anderson, T., Satheesh, S. and Charlson, R. (2009) 'Satellite-derived direct radiative effect of aerosols dependent on cloud cover', *Nature Geoscience*, **2**, 181–4.

Charlson, R. and Pilat, M. (1969) 'Climate: The influence of aerosols', *J. Appl. Meteorol.*, **8**, 1001–2.

Chuang, T., Burns, P., Walters, E.B., Wysocki, T., Deely, T., Losse, A., Le, K., Drumheller, B., Schum, T., Hart, M., Puffenberger, K., Ziegler, W. and Hovis, F. (2013) 'Space-based, multi-wavelength solid-state lasers for NASA's Cloud Aerosol Transport System for International Space Station (CATS–ISS)', *Proc. SPIE*, 85990N.

Coakley, J.A., Jr., Cess, R.D. and Yurevich, F.B. (1983) 'The effect of tropospheric aerosols on the Earth's radiation budget: A parameterization for climate models', *J. Atmos. Sci.*, **40**, 116–38.

Coddington, O., Pilewskie, P., Redemann, J., Platnick, S., Russell, P., Schmidt, K., Gore, W., Livingston, J., Wind, G. and Vukicevic, T. (2010) 'Examining the impact of overlying aerosols on the retrieval of cloud optical properties from passive remote sensing', *J. Geophys. Res.*, **115**, D10211.

Coddington, O., Schmidt, K., Pilewskie, P., Gore, W., Bergstrom, R., Román, M., Redemann, J., Russell, P., Liu, J. and Schaaf, C. (2008) 'Aircraft measurements of spectral surface albedo and its consistency with ground-based and space-borne observations', *J. Geophys. Res.*, **113**, D17209.

Costantino, L. and Bréon, F.-M. (2010) 'Analysis of aerosol–cloud interaction from multisensory satellite observations', *Geophys. Res. Lett.*, **37**, L11801.

Costantino, L. and Bréon, F.-M. (2013) 'Aerosol indirect effect on warm clouds over south-east Atlantic, from co-located MODIS and CALIPSO observations', *Atmos. Chem. Physics*, **13**, 69–88.

de Graaf, M., Stammes, P. and Aben, E. (2007) 'Analysis of reflectance spectra of UV-absorbing aerosol scenes measured by SCIAMACHY', *J. Geophys. Res.*, **112**, D02206.

de Graaf, M., Tilstra, L., Wang, P. and Stammes, P. (2012) 'Retrieval of the aerosol direct radiative effect over clouds from spaceborne spectrometry', *J. Geophys. Res.*, **117**, D07207.

Del Castillo (Chair), C. (2012) 'Pre-Aerosol, Clouds, and ocean Ecosystems (PACE) Mission Science Definition Team Report', *National Aeronautics and Space Administration*, available online at *http://decadal.gsfc.nasa.gov/pace_documentation/PACE_SDT_Report_final.pdf*.

Deschamps, P.Y., Bréon, F.M., Leroy, M., Podaire, A., Bricaud, A., Buriez, J.C. and Seze, G. (1994) 'The POLDER mission: Instrument characteristics and scientific objectives', *Geoscience and Remote Sensing, IEEE Transactions on Geoscience and Remote Sensing*, **32**, 598–615.

Deuzé, J.L., Bréon, F.M., Devaux, C., Goloub, P., Herman, M., Lafrance, B., Maignan, F., Marchand, A., Nadal, F., Perry, G. and Tanre, D. (2001) 'Remote sensing of aerosols over land surfaces from POLDER-ADEOS-1 polarized measurements', *J. Geophys. Res.*, **106**, 4913.

Deuzé, J.L., Goloub, P., Herman, M., Marchand, A., Perry, G., Susana, S. and Tanré, D. (2000) 'Estimate of the aerosol properties over the ocean with POLDER', *J. Geophys. Res.*, **105**, 15329–46.

Devasthale, A. and Thomas, M. (2011) 'A global survey of aerosol–liquid water cloud overlap based on four years of CALIPSO–CALIOP data', *Atmos. Chem. Phys.*, **11**, 1143–54.

Diner, D.J., Xu, F., Garay, M.J., Martonchik, J.V., Rheingans, B.E., Geier, S., Davis, A., Hancock, B.R., Jovanovic, V.M., Bull, M.A., Capraro, K., Chipman, R.A. and McClain, S.C. (2013) 'The Airborne Multiangle SpectroPolarimetric Imager (AirMSPI): A new tool for aerosol and cloud remote sensing', *Atmos. Measurement Techniques Discussions*, **6**, 1717–69.

Dubovik, O., Holben, B.N., Eck, T.F., Smirnov, A., Kaufman, Y.J., King, M.D., Tanré, D. and Slutsker, I. (2002) 'Variability of absorption and optical properties of key aerosol types observed in worldwide locations', *J. Atmos. Sci.*, **59**, 590–608.

Dubovik, O., Smirnov, A., Holben, B., King, M., Kaufman, Y., Eck, T. and Slutsker, I. (2000) 'Accuracy assessments of aerosol optical properties retrieved from AERONET sun and sky-radiance measurements', *J. Geophys. Res*, **105**, 9791–806.

Dunagan, S.E., Johnson, R., Zavaleta, J., Russell, P.B., Schmid, B., Flynn, C., Redemann, J., Shinozuka, Y., Livingston, J. and Segal-Rosenhaimer, M. (2013) '4STAR spectrometer for sky-scanning sun-tracking atmospheric research: Instrument technology', *Remote Sensing (Special Issue: 'Optical Remote Sensing of the Atmosphere')*, **5**, 3872–95.

Forster, P., Ramaswamy, V., Artaxo, P., Berntsen, T., Betts, R., Fahey, D.W., Haywood, J., Lean, J., Lowe, D.C., Myhre, G., Nganga, J., Prinn, R., Raga, G., Schulz, M. and Dorland, R.V. (2007) 'Changes in atmospheric constituents and in radiative forcing', in Solomon, S., Qin, D., Manning, M., Chen, Z., Marquis, M., Averyt, K.B., Tignor, M. and Miller, H.L. (eds), *Climate Change 2007: The Physical Science Basis. Contribution of Working Group I to the Fourth Assessment Report of the Intergovernmental Panel on Climate Change*, Cambridge, United Kingdom, and New York, NY, USA, Cambridge University Press.

Fraser, R.S. and Kaufman, Y.J. (1985) 'The relative importance of aerosol scattering and absorption in remote sensing', *Geoscience and Remote Sensing, IEEE Transactions on Geoscience and Remote Sensing*, **5**, 625–33.

Goloub, P., Deuzé, J., Herman, M. and Fouquart, Y. (1994) 'Analysis of the POLDER polarization measurements performed overcloud covers', *IEEE Trans. Geosci. Remote Sens.*, **32**, 78–88.

Goloub, P., Herman, M., Chepfer, H., Riedi, J., Brogniez, G., Couvert, P. and Sèze, G. (2000) 'Cloud thermodynamical phase classification from the POLDER spaceborne instrument', *J. Geophys. Res.*, **105**, 14747–60.

Hair, J.W., Hostetler, C.A., Cook, A.L., Harper, D.B., Ferrare, R.A., Mack, T.L., Welch, W., Isquierdo, L.R. and Hovis, F.E. (2008) 'Airborne high spectral resolution Lidar for profiling aerosol optical properties', *Appl. Optics*, **47**, 6734–52.

Hansen, J., Sato, M. and Ruedy, R. (1997) 'Radiative forcing and climate response', *J. Geophys. Res.: Atmospheres*, **102**, 6831–64.

Hansen, J.E. and Travis, L.D. (1974) 'Light scattering in planetary atmospheres', *Space Science Reviews*, **16**, 527–610.

Harmel, T. and Chami, M. (2011) 'Influence of polarimetric satellite data measured in the visible region on aerosol detection and on the performance of atmospheric correction procedure over open ocean waters', *Optics Express*, **19**, 20960–83.

Hasekamp, O.P. (2010) 'Capability of multi-viewing-angle photo-polarimetric measurements for the simultaneous retrieval of aerosol and cloud properties', *Atmos. Meas. Tech.*, **3**, 839–51.

Haywood, J. and Boucher, O. (2000) 'Estimates of the direct and indirect radiative forcing due to tropospheric aerosols: A review', *Rev. Geophys.*, **38**, 513–43.

Haywood, J., Roberts, D., Slingo, A., Edwards, J. and Shine, K. (1997) 'General circulation model calculations of the direct radiative forcing by anthropogenic sulfate and fossil-fuel soot aerosol', *J. Climate*, **10**, 1562–77.

Haywood, J.M., Osborne, S.R. and Abel, S.J. (2004) 'The effect of overlying absorbing aerosol layers on remote sensing retrievals of cloud effective radius and cloud optical depth', *Quarterly J. Royal Meteorological Society*, **130**, 779–800.

Hélière, A., Lefebvre, A., Wehr, T., Bezy, J. and Durand, Y. (2007) 'The EarthCARE mission: Mission concept and Lidar instrument pre-development', *Proc. Geoscience Remote Sensing Symposium (IGARSS)*, 4975–8.

Herman, J., Bhartia, P., Torres, O., Hsu, C., Seftor, C. and Celarier, E. (1997a) 'Global distribution of UV-absorbing aerosols from Nimbus 7/TOMS data', *J. Geophys. Res.*, **102**, 16911–16.

Herman, M., Deuzé, J.L., Devaux, C., Goloub, P., Bréon, F.M. and Tanré, D. (1997b) 'Remote sensing of aerosols over land surfaces including polarization measurements and application to POLDER measurements', *J. Geophys. Res.*, **102**, 17039–50.

Herman, M., Deuzé, J.L., Marchand, A., Roger, B. and Lallart, P. (2005) 'Aerosol remote sensing from POLDER/ADEOS over the ocean: Improved retrieval using a nonspherical particle model', *J. Geophys. Res.*, **110**, D10S02.

Holben, B., Vermote, E., Kaufman, Y.J., Tanre, D. and Kalb, V. (1992) 'Aerosol retrieval over land from AVHRR data-application for atmospheric correction', *Geoscience and Remote Sensing, IEEE Transactions on Geoscience and Remote Sensing*, **30**, 212–22.

Hostetler, C.A., Liu, Z., Reagan, J., Vaughan, M., Winker, D., Osborn, M., Hunt, W.H., Powell, K.A. and Trepte, C. (2006) *CALIOP Algorithm Theoretical Basis Document, Calibration and Level 1 Data Products, PC-SCI-201*, Hampton, VA, USA, NASA Langley Research Center, available online at *www-calipso.larc.nasa.gov/resources/project/documentation.php*.

Hu, Y., Vaughan, M., Liu, Z., Powell, K. and Rodier, S. (2007) 'Retrieving optical depths and Lidar ratios for transparent layers above opaque water clouds from CALIPSO Lidar measurements', *Geoscience Remote Sensing Letters, IEEE*, **4**, 523–6.

Jacobson, M.Z. (2001) 'Global direct radiative forcing due to multicomponent anthropogenic and natural aerosols', *J. Geophys. Res.*, **106**, 1551–68.

Jethva, H. and Torres, O. (2011) 'Satellite-based evidence of wavelength-dependent aerosol absorption in biomass burning smoke inferred from Ozone Monitoring Instrument', *Atmos. Chem. Phys.*, **11**, 10541–51.

Jethva, H., Torres, O., Remer, L.A. and Bhartia, P.K. (2013) 'A color ratio method for simultaneous retrieval of aerosol and cloud optical thickness of above-cloud absorbing aerosols from passive sensors: Application to MODIS measurements', *IEEE Trans. Geosci. Remote Sens.*, **99**, 1–9.

Jiang, H. and Feingold, G. (2006) 'Effect of aerosol on warm convective clouds: Aerosol–cloud–surface flux feedbacks in a new coupled large eddy model', *J. Geophys. Res.: Atmospheres (1984–2012)*, **111**, D1.

Johnson, B., Shine, K. and Forster, P. (2004) 'The semi-direct aerosol effect: Impact of absorbing aerosols on marine stratocumulus', *Quarterly J. Royal Meteorological Society*, **130**, 1407–22.

Kacenelenbogen, M., Redemann, J., Vaughan, M.A., Omar, A.H., Russell, P.B., Burton, S., Rogers, R.R., Ferrare, R.A. and Hostetler, C.A. (2014) 'An evaluation of CALIOP/CALIPSO's aerosol-above-cloud (AAC) detection and retrieval capability', *J. Geophys. Res.*, **119**, 230–44.

Kacenelenbogen, M., Vaughan, M.A., Redemann, J., Hoff, R.M., Rogers, R.R., Ferrare, R.A., Russell, P.B., Hostetler, C.A., Hair, J.W. and Holben, B.N. (2011) 'An accuracy assessment of the CALIOP/CALIPSO version 2/version 3 daytime aerosol extinction product based on a detailed multi-sensor, multi-platform case study', *Atmos. Chem. Phys.*, **11**, 3981–4000, doi:10.5194/acp-11-3981-2011, 2011.

Kaufman, Y.J., Remer, L.A., Tanré, D., Li, R., Kleidman, R., Mattoo, S., Levy, R.C., Eck, T.F., Holben, B.N., Ichoku, C., et al. (2005) 'A critical examination of the residual cloud contamination and diurnal sampling effects on MODIS estimates of aerosol over ocean', *IEEE Trans. Geosci. Remote Sens.*, **43**, 2886–97.

Keil, A. and Haywood, J.M. (2003) 'Solar radiative forcing by biomass burning aerosol particles during SAFARI 2000: A case study based on measured aerosol and cloud properties', *J. Geophys. Res.*, **108**, 8467.

Kim, S.-W., Berthier, S., Raut, J.-C., Chazette, P., Dulac, F. and Yoon, S.-C. (2008) 'Validation of aerosol and cloud layer structures from the space-borne Lidar CALIOP using a ground-based Lidar in Seoul, Korea', *Atmos. Chem. Phys.*, **8**, 3705–20.

Kinne, S. (2009) 'Remote sensing data combinations-superior global maps for aerosol optical depth', in Kokhanovsky, A. and de Leeuw, G. (eds), *Satellite Aerosol Remote Sensing Over Land*, Berlin, Springer.

Knobelspiesse, K., Cairns, B., Mishchenko, M., Chowdhary, J., Tsigaridis, K., van Diedenhoven, B., Martin, W., Ottaviani, M. and Alexandrov, M. (2012) 'Analysis of fine-mode aerosol retrieval capabilities by different passive remote sensing instrument designs', *Opt. Express*, **20**, 21457–84.

Knobelspiesse, K., Cairns, B., Redemann, J., Bergstrom, R.W. and Stohl, A. (2011), 'Simultaneous retrieval of aerosol and cloud properties during the MILAGRO field campaign', *Atmos. Chem. Phys.*, **11**, 6245–63.

Koch, D. and Del Genio, A. (2010) 'Black carbon absorption effects on cloud cover, review and synthesis', *Atmos. Chem. Phys.*, **10**, 7685–96.

Kokhanovsky, A., Mayer, B., von Hoyningen-Huene, W., Schmidt, S. and Pilewskie, P. (2006) 'Retrieval of cloud spherical albedo from top-of-atmosphere reflectance measurements performed at a single observation angle', *Atmos. Chem. Phys. Discuss.*, **6**, 2175–88.

Koren, I., Martins, J.V., Remer, L.A. and Afargan, H. (2008) 'Smoke invigoration versus inhibition of clouds over the Amazon', *Science*, **321**, 946–9.

Kuze, A., Suto, H., Nakajima, M. and Hamazaki, T. (2009) 'Thermal and near infrared sensor for carbon observation Fourier-transform spectrometer on the Greenhouse Gases Observing Satellite for greenhouse gases monitoring', *Appl. Opt.*, **48**, 6716–33.

Lafrance, B. and Herman, M. (1998) 'Correction of the stratospheric aerosol radiative influence in the POLDER measurements', *IEEE Trans. Geosci. Remote Sens.*, **36**, 1599.

Leroy, M., Deuzé, J., Bréon, F., Hautecoeur, O., Herman, M., Buriez, J., Tanré, D., Bouffies, S., Chazette, P. and Roujean, J. (1997) 'Retrieval of atmospheric properties and surface bidirectional reflectances over land from POLDER/ADEOS', *J. Geophys. Res.*, **102**, 17023–37.

Liu Z., Kuehn, R., Vaughan, M., Winker, D., Omar, A., Powell, K., Trepte, C., Hu, Y. and Hostetler, C. (2010) 'The CALIPSO cloud and aerosol discrimination: Version 3, Algorithm and test results', *25th International Laser and Radar Conference.*

Livingston, J., Redemann, J., Russell, P., Torres, O., Veihelmann, B., Veefkind, P., Braak, R., Smirnov, A., Remer, L., Bergstrom, R., et al. (2009) 'Comparison of aerosol optical depths from the Ozone Monitoring Instrument (OMI) on aura with results from airborne sunphotometry, other space and ground measurements during MILAGRO/ INTEX-B', *Atmos. Chem. Phys.*, **9**, 6743–65.

Livingston, J.M., Redemann, J., Shinozuka, Y., Johnson, R., Russell, P.B., Zhang, Q., Mattoo, S., Remer, L., Levy, R., Munchak, L. and Ramachandran, S. (2014) 'Comparison of MODIS 3-km and 10-km resolution aerosol optical depth retrievals over land with airborne 2 Sunphotometer measurements during ARCTAS summer', *Atmos. Chem. Phys. Discuss.*, **14**, 2015–38.

Lohmann, U. and Feichter, J. (2005) 'Global indirect aerosol effects: A review', *Atmos. Chem. Phys.*, **5**, 715–37.

Marbach, T., Phillips, P. and Schlüssel, P. (2013) '3MI: The multi-viewing multi-channel multi-polarization imaging mission of the EUMETSAT Polar System-Second Generation (EPS-SG) dedicated to aerosol characterization', *American Institute of Physics Conf. Proc.*, **1531**, 344–7, doi:http://dx.doi.org/10.1063/1.4804777.

McCormick, R.A. and Ludwig, J.H. (1967) 'Climate modification by atmospheric aerosols', *Science*, **156**, 1358–9.

Mishchenko, M.I., Cairns, B., Kopp, G., Schueler, C., Fafaul, B., Hansen, J., Hooker, R., Itchkawich, T., Maring, H. and Travis, L. (2007) 'Accurate monitoring of terrestrial aerosols and total solar irradiance: Introducing the Glory mission', *Bull. Amer. Meteorol. Soc.*, **88**, 687–91.

Mitchell, J.M., Jr. (1971) 'The effect of atmospheric aerosols on climate with special reference to temperature near the Earth's surface', *J. Appl. Meteorol.*, **10**, 703–14.

Molina, L.T., Madronich, S., Gaffney, J.S., Apel, E., de Foy, B., Fast, J., Ferrare, R., Herndon, S., Jimenez, J.L., Lamb, B., Osornio-Vargas, A.R., Russell, P., Schauer, J.J., Stevens, P.S., Volkamer, R. and Zavala, M. (2010) 'An overview of the MILAGRO 2006 Campaign: Mexico City emissions and their transport and transformation', *Atmos. Chem. Phys.*, **10**, 8697–760.

Müller, D., Veselovskii, I., Kolgotin, A., Tesche, M., Ansmann, A. and Dubovik, O. (2000) 'Microphysical particle parameters from extinction and backscatter Lidar data by inversion with regularization: Experiment', *Applied Optics.*, **39**, 1879–92.

Nadal, F. and Bréon, F.-M. (1999) 'Parameterization of surface polarized reflectance derived from POLDER spaceborne measurements', *IEEE Trans. Geosci. Remote Sens.*, **37**, 1709.

Nagaraja Rao, C., Stowe, L. and McClain, E. (1989) 'Remote sensing of aerosols over the oceans using AVHRR data: Theory, practice and applications', *International J. of Remote Sensing*, **10**, 743–9.

National Research Council (US) (2007) *Earth Science and Applications from Space: National Imperatives for the Next Decade and Beyond*, Committee on Earth Science, Applications from Space: A Community Assessment, & Strategy for the Future, National Academy Press.

Omar, A., Winker, D., Kittaka, C., Vaughan, M., Liu, Z., Hu, Y., Trepte, C., Rogers, R., Ferrare, R., Kuehn, R. and Hostetler, C. (2009) 'The CALIPSO automated aerosol classification and Lidar ratio selection algorithm', *J. Atmos. Ocean. Tech.*, **26**, 1994–2014.

Pappalardo, G., Wandinger, U., Mona, L., Wandinger, U., Mona, L., Hiebsch, A., Mattis, I., Amodeo, A., Ansmann, A., Seifert, P., Linné, H., Apituley, A., Arboledas, L.A., Balis, D., Chaikovsky, A., D'Amico, G., De Tomasi, F., Freudenthaler, V., Giannakaki, E., Giunta, A., Grigorov, I., Iarlori, M., Madonna, F., Mamouri, R., Nasti, L., Papayannis, A., Pietruczuk, A., Pujadas, M., Rizi, V., Rocadenbosch, F., Russo, F.,

Schnell, F., Spinelli, N., Wang, X. and Wiegner, M. (2010) 'EARLINET correlative measurements for CALIPSO: First intercomparison results', *J. Geophys. Res.*, **115**, D00H19.

Parol, F., Buriez, J., Vanbauce, C., Riedi, J., C.-Labonnote, L., Doutriaux-Boucher, M., Vesperini, M., Sèze, G., Couvert, P., Viollier, M. and Bréon, F.M. (2004) 'Review of capabilities of multi-angle and polarization cloud measurements from POLDER', *Advances in Space Research*, **33**, 1080–8.

Parol, F., Buriez, J.-C., Vanbauce, C., Couvert, P., Seze, G., Goloub, P. and Cheinet, S. (1999) 'First results of the POLDER "Earth Radiation Budget and Clouds" operational algorithm', *Geoscience and Remote Sensing, IEEE Transactions on Geoscience and Remote Sensing*, **37**, 1597–612.

Peters, K., Quaas, J. and Bellouin, N. (2011) 'Effects of absorbing aerosols in cloudy skies: A satellite study over the Atlantic Ocean', *Atmos. Chem. Phys.*, **11**, 1393–404.

Pilewskie, P., Pommier, J., Bergstrom, R., Gore, W., Howard, S., Rabbette, M., Schmid, B., Hobbs, P. and Tsay, S. (2003) 'Solar spectral radiative forcing during the southern african regional science initiative', *J. Geophys. Res*, **108**, 8486.

Podgorny, I. and Ramanathan, V. (2001) 'A modeling study of the direct effect of aerosols over the tropical Indian Ocean', *J. Geophys. Res.*, **106**, 24097–105.

Powell, K.A., Hostetler, C.A., Liu, Z., Vaughan, M.A., Kuehn, R.E., Hunt, W.H., Lee, K., Trepte, C.R., Rogers, R.R., Young, S.A. and Winker, D.M. (2009) 'CALIPSO lidar calibration algorithms: Part I—Nighttime 532 nm parallel channel and 532nm perpendicular channel', *J. Atmos. Ocean. Tech.*, **26**, 2015–33.

Redemann, J., Zhang, Q., Livingston, J., Russell, P., Shinozuka, Y., Clarke, A., Johnson, R. and Levy, R. (2009) 'Testing aerosol properties in modis collection 4 and 5 using airborne sunphotometer observations in INTEX-B/MILAGRO', *Atmos. Chem. Phys.*, **9**, 8159–72.

Remer, L.A. (2009) 'Atmospheric science: Smoke above clouds', *Nature Geoscience*, **2**, 167–8.

Remer, L.A., Kleidman, R.G., Levy, R.C., Kaufman, Y.J., Tanré, D., Mattoo, S., Martins, J.V., Ichoku, C., Koren, I., Yu, H. and Holben, B.N. (2008) 'Global aerosol climatology from the MODIS satellite sensors', *J. Geophys. Res.*, **113**, D14S07.

Rogers, R.R., Hair, J.W., Hostetler, C.A., Ferrare, R.A., Obland, M.D., Cook, A.L., Harper, D. B., Burton, S.P., Shinozuka, Y., McNaughton, C.S., Clarke, A.D., Redemann, J., Russell, P.B., Livingston, J.M., Clayton, M.F., Fast, J.D. and Kleinman, L.I. (2009) 'NASA LaRC airborne high spectral resolution Lidar aerosol measurements during MILAGRO. Observations and intercomparisons', *Atmos. Chem. Phys.*, **9**, 4811–26.

Rogers, R.R., Hostetler, C.A., Hair, J.W., Ferrare, R.A., Liu, Z., Obland, M.D., Harper, D.B., Cook, A.L., Powell, K.A., Vaughan, M.A. and Winker, D.M. (2011) 'Assessment of the CALIPSO Lidar 532 nm attenuated backscatter calibration using the NASA LaRC airborne High Spectral Resolution Lidar', *Atmos. Chem. Phys.*, **11**, 1295–311.

Rosenfeld, D. and Lensky, I.M. (1998) 'Satellite-based insights into precipitation formation processes in continental and maritime convective clouds', *Bulletin of the American Meteorological Society*, **79**, 2457–76.

Rossow, W.B. and Schiffer, R.A. (1999) 'Advances in understanding clouds from ISCCP', *Bull. Am. Meteorol. Soc.*, **80**, 2261–88.

Russell, P., Livingston, J., Schmid, B., Eilers, J., Kolyer, R., Redemann, J., Ramirez, S., Yee, J.-H., Swartz, W., Shetter, R., Trepte, C., Risley, A., Jr., Wenny, B., Zawodny, J., Chu, W., Pitts, M., Lumpe, J., Fromm, M., Randall, C., Hoppel, K. and Bevilacqua, R. (2005) 'Aerosol optical depth measurements by airborne sun photometer in SOLVE

II: Comparisons to SAGE III, POAM III and airborne spectrometer measurements', *Atmos. Chem. Phys.*, **5**, 1311–39.

Russell, P., Redemann, J., Schmid, B., Bergstrom, R., Livingston, J., McIntosh, D., Ramirez, S., Hartley, S., Hobbs, P., Quinn, P., et al. (2002) 'Comparison of aerosol single scattering albedos derived by diverse techniques in two North Atlantic experiments', *J. Atmos. Sciences*, **59**, 609–19.

Russell, P.B., Bergstrom, R.W., Shinozuka, Y., Clarke, A.D., DeCarlo, P.F., Jimenez, J.L., Livingston, J.M., Redemann, J., Dubovik, O. and Strawa, A. (2010) 'Absorption Ångström Exponent in AERONET and related data as an indicator of aerosol composition', *Atmos. Chem. Phys.*, **10**, 1155–69.

Russell, P.B., Hobbs, P.V. and Stowe, L.L. (1999) 'Aerosol properties and radiative effects in the United States east coast haze plume: An overview of the Tropospheric Aerosol Radiative Forcing Observational Experiment (TARFOX)', *J. Geophys. Res.*, **104**, 2213–22.

Russell, P.B., Swissler, T.J. and McCormick, M.P. (1979) 'Methodology for error analysis and simulation of Lidar aerosol measurements', *Appl. Optics*, **18**, 3783–97.

Sakaeda, N., Wood, R. and Rasch, P.J. (2011) 'Direct and semidirect aerosol effects of southern African biomass burning aerosol', *J. Geophys. Res.*, **116**, D12205.

Sano, I., Nakata, M., Mukai, S. and Holben, B. (2010a) 'Algorithm for aerosol retrieval based on radiance and polarimetry for SGLI', *International Archives of the Photogrammetry, Remote Sensing and Spatial Information Science*, Vol. XXXVIII, Part 8, Kyoto, Japan, 380–5.

Sano, I., Tanabe, M., Takamasa, K., Nakata, M. and Mukai, S. (2010b) 'Carbonaceous aerosols over Siberia and Indonesia with GOSAT/CAI', *SPIE Proceedings*, 7859, doi:10.1117/12.869625.

Schmidt, K., Feingold, G., Pilewskie, P., Jiang, H., Coddington, O. and Wendisch, M. (2009) 'Irradiance in polluted cumulus fields: Measured and modeled cloud-aerosol effects', *Geophysical Research Letters*, **36**, L07804.

Schulz, M., Textor, C., Kinne, S., Balkanski, Y., Bauer, S., Berntsen, T., Berglen, T., Boucher, O., Dentener, F., Guibert, S., et al. (2006) 'Radiative forcing by aerosols as derived from the AeroCom present-day and pre-industrial simulations', *Atmos. Chem. Phys.*, **6**, 5246.

Schuster, G.L., Vaughan, M., MacDonnell, D., Su, W., Winker, D., Dubovik, O., Lapyonok, T. and Trepte, C. (2012) 'Comparison of CALIPSO aerosol optical depth retrievals to AERONET measurements, and a climatology for the Lidar ratio of dust', *Atmos. Chem. Phys.*, **12**, 10.5194.

Sena, E., Artaxo, P. and Correia, A. (2013) 'Spatial variability of the direct radiative forcing of biomass burning aerosols and the effects of land use change in Amazonia', *Atmos. Chem. Phys.*, **13**, 1261–75.

Shimoda, H. (2010) 'Overview of GCOM', *Proc. of the Geoscience and Remote Sensing Symposium (IGARSS)*, 1351–4.

Sugimoto, N., Nishizawa, T., Shimizu, A. and Okamoto, H. (2011) 'AEROSOL classification retrieval algorithms for EarthCARE/ATLID, CALIPSO/CALIOP, and ground-based Lidars', *Geoscience and Remote Sensing Symposium (IGARSS)*, doi:10.1109/IGARSS.2011.6050137.

Swap, R.J., Annegarn, H.J., Suttles, J.T., King, M.D., Platnick, S., Privette, J.L. and Scholes, R.J. (2003) 'Africa burning: A thematic analysis of the Southern African Regional Science Initiative (SAFARI 2000)', *J. Geophys. Res.: Atmospheres (1984–2012)*, **108**, D13.

Tanaka, K., Okamura, Y., Amano, T., Hiramatsu, M. and Shiratama, K. (2010) 'Operation concept of the second-generation global imager (SGLI)', *Proc. SPIE*, **7862**, 786209.

Tanré, D., Bréon, F.M., Deuzé, J.L., Dubovik, O., Ducos, F., Francois, P., Goloub, P., Herman, M., Lifermann, A. and Waquet, F. (2011) 'Remote sensing of aerosols by using polarized, directional and spectral measurements within the A-Train: The PARASOL mission', *Atmospheric Measurement Techniques*, **4**, 1383–95.

Ten Hoeve, J.E., Jacobson, M.Z. and Remer, L.A. (2012) 'Comparing results from a physical model with satellite and *in situ* observations to determine whether biomass burning aerosols over the Amazon brighten or burn off clouds', *J. Geophys. Res.: Atmospheres (1984–2012)*, **117**, D8.

Torres, O., Bhartia, P., Herman, J., Ahmad, Z. and Gleason, J. (1998) 'Derivation of aerosol properties from satellite measurements of backscattered ultraviolet radiation: Theoretical basis', *J. Geophys. Res.: Atmospheres (1984–2012)*, **103**, 17099–110.

Torres, O., Jethva, H. and Bhartia, P.K. (2012) 'Retrieval of aerosol optical depth above clouds from OMI observations: Sensitivity analysis and case studies', *J. Atmos. Sci.*, **69**, 1037–53.

Twomey, S. (1972) 'The effect of cloud scattering on the absorption of solar radiation by atmospheric dust', *J. Atmos. Sci.*, **29**, 1156–9.

Twomey, S. (1977) 'The influence of pollution on the shortwave albedo of clouds', *J. Atmos. Sci.*, **34**, 1149–52.

Vanbauce, C., Buriez, J.C., Parol, F., Bonnel, B., Seze, G. and Couvert, P. (1998) 'Apparent pressure derived from ADEOS-POLDER observations in the oxygen A-band over ocean', *Geophys. Res. Lett.*, **25**, 3159–62.

Várnai, T. and Marshak, A. (2009) 'MODIS observations of enhanced clear sky reflectance near clouds', *Geophys. Res. Lett.*, **36**, L06807.

Vaughan, M., Garnier, A., Liu, Z., Josset, D., Hu, Y., Lee, K., Hunt, W., Vernier, J.P. Rodier, S., Pelon, J. and Winker, D. (2012) 'Chaos, consternation and CALIPSO calibration: New strategies for calibrating the CALIOP 1064 nm channel', *ILRC Conference.*

Vaughan, M., Powell, K., Kuehn, R., Young, S., Winker, D., Hostetler, C., Hunt, W., Liu, Z., McGill, M. and Getzewich, B. (2009) 'Fully automated detection of cloud and aerosol layers in the CALIPSO Lidar measurements', *J. Atmos. Oceanic Technol.*, **26**, 2034–50.

Vernier, J.P., Pommereau, J.P., Garnier, A., Pelon, J., Larsen, N., Nielsen, J., Christensen, T., Cairo, F., Thomason, L.W., Leblanc, T. and McDermid, I.S. (2009) 'The tropical stratospheric aerosol layer from CALIPSO Lidar observations', *J. Geophys. Res.*, **114**, D00H10, doi:10.1029/2009JD011946.

Waquet, F., Cornet, C., Deuzé, J.-L., Dubovik, O., Ducos, F., Goloub, P., Herman, M., Lapionak, T., Labonnote, L., Riedi, J., Tanré, D., Thieuleux, F. and Vanbauce, C. (2013) 'Retrieval of aerosol microphysical and optical properties above liquid clouds from POLDER/PARASOL polarization measurements', *Atm. Meas. Tech.*, **6**, 991–1016.

Waquet, F., Léon, J.-F., Goloub, P., Pelon, J., Tanré, D. and Deuzé, J.-L. (2005) 'Maritime and dust aerosol retrieval from polarized and multispectral active and passive sensors', *J. Geophys. Res.*, **110**, D10S10.

Waquet, F., Riedi, J., Labonnote, L., Goloub, P., Cairns, B., Deuzé, J.L. and Tanré, D. (2009) 'Aerosol remote sensing over clouds using A-Train observations', *J. Atmos. Sci.*, **66**, 2468–80.

Wen, G., Marshak, A., Cahalan, R.F., Remer, L.A. and Kleidman, R.G. (2007) '3-D aerosol–cloud radiative interaction observed in collocated MODIS and ASTER images of cumulus cloud fields', *J. Geophys. Res.*, **112**, D13204.

Wilcox, E. (2010) 'Stratocumulus cloud thickening beneath layers of absorbing smoke aerosol', *Atmos. Chem. Phys.*, **10**, 11–769.

Wilcox, E. (2012) 'Direct and semi-direct radiative forcing of smoke aerosols over clouds', *Atmos. Chem. Phys.*, **12**, 139–49.

Wilcox, E.M., Harshvardhan and Platnick, S. (2009) 'Estimate of the impact of absorbing aerosol over cloud on the MODIS retrievals of cloud optical thickness and effective radius using two independent retrievals of liquid water path', *J. Geophys. Res.*, **114**, D5.

Winker, D.M., Pelon, J. and McCormick, M.P. (2003) 'The CALIPSO mission: Space borne Lidar for observation of aerosols and clouds', *SPIE Proceedings*, **4893**, 1–11.

Winker, D.M., Tackett, J.L., Getzewich, B.J., Liu, Z., Vaughan, M.A. and Rogers, R.R. (2013) 'The global 3-D distribution of tropospheric aerosols as characterized by CALIOP', *Atmos. Chem. Phys.*, **13**, 3345–61.

Winker, D.M., Vaughan, M.A., Omar, A., Hu, Y., Powell, K.A., Liu, Z., Hunt, W.H. and Young, S.A. (2009) 'Overview of the CALIPSO mission and CALIOP data processing algorithms', *J. Atmos. Oceanic Technology*, **26**, 2310–23.

Young, S.A. and Vaughan, M.A. (2009) 'The retrieval of profiles of particulate extinction from Cloud Aerosol Lidar Infrared Pathfinder Satellite Observations (CALIPSO) data: Algorithm description', *J. Atmos. Oceanic Technol.*, **26**, 1105–19.

Yu, H. and Zhang, Z. (2013) 'New directions: Emerging satellite observations of above-cloud aerosols and direct radiative forcing', *Atmos. Environ.*, **72**, 36–40.

Yu, H., Kaufman, Y., Chin, M., Feingold, G., Remer, L. anderson, T., Balkanski, Y., Bellouin, N., Boucher, O., Christopher, S., DeCola, P., Kahn, R., Koch, D., Loeb, N., Reddy, M. S., Schulz, M., Takemura, T. and Zhou, M. (2006) 'A review of measurement-based assessments of the aerosol direct radiative effect and forcing', *Atmos. Chem. Phys.*, **6**, 613–66.

Yu, H., Zhang, Y., Chin, M., Liu, Z., Omar, A., Remer, L.A., Yang, Y., Yuan, T. and Zhang, J. (2012) 'An integrated analysis of aerosol above clouds from A-Train multi-sensor measurements', *Remote Sensing of Environment*, **121**, 125–31.

Zhang, J., Reid, J.S. and Holben, B.N. (2005) 'An analysis of potential cloud artifacts in MODIS over ocean aerosol optical thickness products', *Geophys. Res. Lett.*, **32**, L15803.

Part III

Polarimetry

6 Principles of the Mueller matrix measurements

Sergey N. Savenkov

6.1 Introduction

The primary physical characteristics associated with a quasimonochromatic light are the intensity, wavelength, and polarization. All of them are the channels to provide the corresponding information about the object under consideration. In particular, polarization provides the researcher with information about surface features, shape, shading, roughness, anisotropy, etc. of the object. At that, polarization provides information that is largely unobtainable by spectral and intensity measurements, and thereby could enhance the information capability of optical metrology. Nevertheless, polarization channel and intensity channel are complementary in the information they provide about a particular scattering scene since, ultimately, a polarization channel consists of a series of intensity measurements.

It is important that polarization measurements can principally be carried out in any region of the electromagnetic spectrum (van Zyl and Ulaby, 1990; Boerner, 1992; Guissard, 1994; Mott, 2007). Because of that, in spite of substantial differences in the physics to build the polarimeters in different spectral regions, many important concepts of polarimetry remain the same.

Polarimetric measurements have to date a long-term history (Collett, 1993; Brosseau, 1998). The first apparatus for polarization measurement was constructed by Arago. It is interesting that G.G. Stokes himself could in principle measure his parameters. The measurement method that G.G. Stokes could use for that is the so-called null method (Azzam and Bashara, 1987). This means that being able to measure the Stokes vector, G.G. Stokes could measure the Mueller matrix elements as well.

The evidences of fruitfulness of polarimetric measurements are numerous (for only few examples, Refs.: Boerner, 1992; Benoit et al., 2001; Smith, 2001; DeBoo et al., 2005; Lopatin et al., 2004; Tuchin et al., 2006; Goldstein, 2008; Nunziata, 2008; Antonelli et al., 2010; Carrieri et al., 2010). However, in all of the discussion to follow, we are interested in the Mueller matrix measurements only. Therefore, the extensive results and many of the corresponding references concerning other aspects of polarimetric measurements, such as passive Stokes polarimetry, are beyond the scope of the chapter. At the same time, the passive Stokes polarimetry is of essential interest for us as an important constituent of the Mueller polarimetry. Indeed, the

Stokes polarimeter is the polarization state analyzer (PSA) in tim e-sequential Mueller matrix measurement strategy. We also limit the scope of this study to nonimaging polarimeters, though the discussed methods and results are relevant to imaging polarimeters as well.

The present chapter is intended to outline state-of-the-art and some modern trends on the field of the Mueller matrix measurements.

6.2 Mathematics of the Mueller matrix method

In the Mueller matrix calculus, the polarization state of light can be completely characterized by a Stokes vector, while the polarization transforming properties of a medium can be completely characterized by a Mueller matrix (Bohren and Huffman, 1983; Azzam and Bashara, 1987; Collett, 1993):

$$\mathbf{S}^{out} = \mathbf{M}\mathbf{S}^{inp}, \tag{6.1}$$

where the four-component Stokes column vector (with '*out*' and '*inp*' denoting the Stokes vectors of the output and input light, respectively) consists of the following parameters:

$$\mathbf{S} = \begin{pmatrix} I \\ Q \\ U \\ V \end{pmatrix} = \begin{pmatrix} s_1 \\ s_2 \\ s_3 \\ s_4 \end{pmatrix} = \begin{pmatrix} \langle | E_x |^2 + | E_y |^2 \rangle \\ \langle | E_x |^2 - | E_y |^2 \rangle \\ \langle E_x^* E_y + E_x E_y^* \rangle \\ \mathrm{i}\langle E_x^* E_y + E_x E_y^* \rangle \end{pmatrix}, \tag{6.2}$$

with $\mathrm{i} = (-1)^{1/2}$. Among the pioneering contributions to this field of research, we note those by Soleillet (1929), Perrin (1942), Mueller (1948), and Parke (1948, 1949).

The Stokes parameter I is proportional to the total energy flux of the light beam. The Stokes parameters Q and U represent the differences between two components of the flux in which the electric vectors oscillate in mutually orthogonal directions. The Stokes parameter V is the difference between two oppositely circularly polarized components of the flux. As indicated by the angular brackets, the Stokes parameters s_i are ensemble averages (or time averages in the case of ergodic, stationary processes). This implies that no coherence effects are considered.

The Stokes vectors and Mueller matrices represent operations on intensities and their differences, namely incoherent superpositions of light beams; they are not adequate to describe either interference or diffraction effects. However, they are well suited to describe partially polarized and unpolarized light. Extensive lists of various Mueller matrices have been presented by several authors (e.g. Shurcliff, 1962; Kliger et al., 1990; Gerrard and Burch, 1975).

To be the Stokes vector, the real parameters of a 4×1 vector should obey the inequality

$$s_1^2 \geq s_2^2 + s_3^2 + s_4^2. \tag{6.3}$$

This inequality is called the Stokes–Verdet criterion and is a consequence of the Schwartz (or Couchy–Buniakovski) theorem (Barakat, 1963, 1987; Brosseau, 1998).

The degree of polarization p is defined by (Simon, 1987; Chipman, 1999, 2005)

$$p = \sqrt{s_2^2 + s_3^2 + s_4^2}/s_1. \tag{6.4}$$

In Eq. (6.3), the equality holds for a completely polarized (pure) beam of light. In this case, $p = 1$. Another limiting case, $p = 0$, occurs when $s_2^2 + s_3^2 + s_4^2 = 0$, namely when the electric vector vibrates in all directions randomly and with no preferential orientation. An intermediate case, $0 < p < 1$, implies that light contains both polarized and depolarized components and is, therefore, called partially polarized (Bohren and Huffman, 1983).

The inequality in Eq. (6.3) plays an important role in polarimetry because it allows one to classify the character of the light–object interaction. Assume first that the input light is completely polarized. In this case, the equality in Eq. (6.3) for output light implies that the medium is non-depolarizing. Note that the terms 'non-depolarizing' and 'deterministic' or 'pure' are not, in general, identical. The term 'deterministic' means that the Mueller matrix describing such a medium can be derived from the corresponding Jones matrix (Simon, 1982; Gil and Bernabeu, 1985, 1986; Azzam and Bashara, 1987; Anderson and Barakat, 1994; Gopala Rao et al., 1998a, 1998b). This condition is more rigid than the condition of a non-depolarizing medium (Savenkov and Yushtin, 2000a). Hereinafter, we call this class of matrices pure Mueller matrices (Hovenier, 1994). If the output light results in an inequality in Eq. (6.3), then the scattering medium is not deterministic. If, in addition, the transformation matrix in Eq. (6.1) can be represented as a convex sum of deterministic Mueller matrices (Cloude, 1986, 2010; Cloude and Pottier, 1995; Gil, 2000, 2007), then the result is a depolarizing Mueller matrix (hereinafter Mueller matrix); otherwise, the result is a Stokes transformation matrix—that is, the transformation matrix ensures the fulfillment of the Stokes–Verdet criterion only. The properties of matrices transforming Stokes vectors into Stokes vectors, namely those satisfying the Stokes–Verdet criterion, have been studied by many authors (Xing, 1992; van der Mee, 1993; van der Mee and Hovenier, 1992; Sridhar and Simon, 1994; Nagirner, 1993; Givens and Kostinski, 1993; Gopala Rao et al., 1998a, 1998b).

Any pure Mueller matrix $\mathbf{M}$ can be transformed to the corresponding Jones matrix $\mathbf{T}$ using the following relation (Parke III, 1949; Kim et al., 1987; Azzam and Bashara, 1987; Dubois and Norikane, 1987):

$$\mathbf{M} = \mathbf{A}(\mathbf{T} \otimes \mathbf{T}^*)A^{-1}, \tag{6.5}$$

where the asterisk denotes the complex-conjugate value,

$$\mathbf{T} = \begin{pmatrix} t_1 & t_4 \\ t_3 & t_2 \end{pmatrix}, \tag{6.6}$$

$$\mathbf{A} = \begin{pmatrix} 1 & 0 & 0 & 1 \\ 1 & 0 & 0 & -1 \\ 0 & 1 & 1 & 0 \\ 0 & \mathrm{i} & -\mathrm{i} & 0 \end{pmatrix}, \tag{6.7}$$

the t_i are, in general complex, and $\otimes$ is the tensorial (Kronecker) product.

Since the element m_{11} is a gain for unpolarized incident light, it must evidently satisfy the following inequality:

$$m_{11} > 0. \tag{6.8}$$

Furthermore, the elements of the Mueller matrix must obey the following conditions (Xing, 1992):

$$m_{11} \geq |m_{ij}|, \tag{6.9}$$

$$\mathrm{Tr}(\mathbf{M}) \geq 0, \tag{6.10}$$

$$\mu \mathbf{T} \sim |\mu|^2 \mathbf{M}, \tag{6.11}$$

where 'Tr' denotes the trace operation and μ is an arbitrary real or complex constant.

The relations in Eqs (6.8)–(6.11) define the ability of the Mueller matrices to represent a 'physically realizable' medium (Lu and Chipman, 1994; Anderson and Barakat, 1994; Gil, 2007) as well and implies the physical restriction according to which the ratio g of the intensities of the emerging and incident light beams (the gain or intensity transmittance) must always be in the interval $0 \leq g \leq 1$. This condition is called the gain or transmittance condition and can be written in terms of the elements of the Mueller matrix as follows (Barakat, 1987):

$$\begin{aligned} m_{11} + (m_{12}^2 + m_{13}^2 + m_{14}^2)^{1/2} \leq 1, \\ m_{11} + (m_{21}^2 + m_{31}^2 + m_{41}^2)^{1/2} \leq 1. \end{aligned} \tag{6.12}$$

Note that, if the Mueller matrix fulfils the relations in Eqs (6.8)–(6.11), then it fulfills the Stokes–Verde criterion in Eq. (6.3) as well. However, the contrary is not always true (Gil, 2007). On the other hand, no method has been quoted to physically realize a Stokes matrix that cannot be represented as a convex sum of deterministic Mueller matrices.

While a Jones matrix has generally eight independent parameters, the absolute phase is lost in Eq. (6.5), yielding only seven independent elements for a pure Mueller matrix. Evidently, this results in the existence of interrelations for the elements of a general pure Mueller matrix. This fact was pointed out for the first time, although without a derivation of their explicit form, by van de Hulst (1957). Since then, this subject has been studied by many authors (e.g. Abhyankar and Fymat, 1969; Fry and Kattawar, 1981; Hovenier et al., 1986). In the most complete and refined form, these interrelations are presented in Hovenier (1994).

Alongside the problem of measuring the Mueller matrix, one more very important problem in polarimetry is the analysis of the Mueller matrix to characterize a particular scattering process.

There are two main concepts, which are currently used for object modeling. First, an object can be modeled as a discrete ensemble of scatterers (van de Hulst, 1957; Tuchin et al., 2006); and, second, the object can be modeled as a medium with a continuous distribution of optical parameters (Azzam and Bashara, 1987; Tuchin, 1994; Schmitt and Kumar, 1996; Brosseau, 1998). In turn, the choice of concept is completely determined by both the structural features of the object in question and the type of scattering characteristics that are to be obtained. In the first case, the Mueller matrix contains information related to the optical properties, size, shape, and composition of the constituent scatterers (van de Hulst, 1957; Bohren and

Huffman, 1983) while, in the second case, the Mueller matrix contains information related to the anisotropy of the medium, viz., linear and circular dichroism and birefringence (i.e. amplitude and phase anisotropies) (Landau et al., 1984; Berry and Dennis, 2003). In both cases, the Mueller matrix can contain information on depolarization (Chipman, 1995; Mishchenko and Hovenier, 1995; Mishchenko and Travis, 2000; Gil, 2007). The methods of interpretation of the Mueller matrices have been developed by many authors (Hurwitz and Jones, 1941; Whitney, 1971; Cloude, 1986; Gil and Bernabeu, 1987; Lu and Chipman, 1994, 1996; Savenkov et al., 2005, 2006, 2007b; Ossikovski, 2008, 2009). In spite of that, the spaces of Stokes vectors and Mueller matrices are not formal vector spaces because of the requirement that s_1 and m_{11} be positive (Bickel et al., 1976; Collett, 1993; Brosseau, 1998), and some properties of vector spaces, therefore, may not apply to Stokes vectors or Mueller matrices; nevertheless, the tools of matrix analysis have proven to be quite useful both in studying of polarimeters and Mueller matrix interpretation. More details about the last subject can be found in Appendix A.

6.3 Complete Mueller polarimetry

An active polarimeter is composed of a polarization state generator (PSG) and a PSA, as shown in Fig. 6.1. The PSG forms a particular polarization state with which to probe the object under consideration. In general, the Stokes vector that describes this state is arbitrary, namely unpolarized (Savenkov, 2002; Mishchenko et al., 2010), partially polarized (Mahler and Chipman, 2011), and completely polarized light. The last one is the most common case in practice (Hauge, 1980; Azzam, 1997; Chipman, 1995).

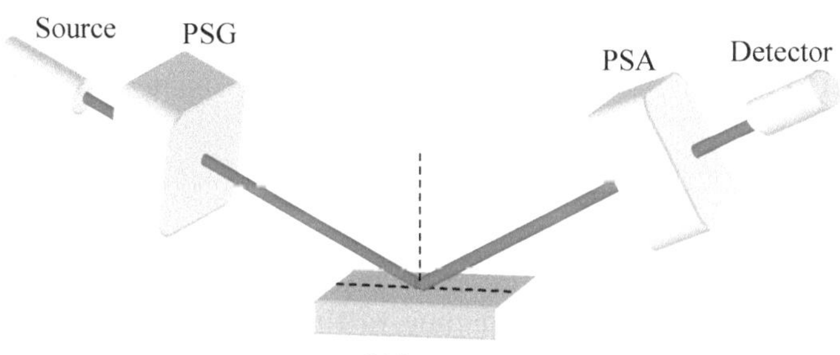

Fig. 6.1. General schematic of the Mueller matrix polarimeter.

The PSA is a combination of retarders and diattenuators that is capable of analyzing the polarization state of the scattered light. Continually, the PSA is build to measure the full Stokes vector of the scattered light. However, when only a subset of the Mueller matrix is of interest, the PSA can measure the corresponding incomplete Stokes vector.

For either of the Mueller matrix measurement strategies, namely time-sequential or dynamic, complete or incomplete polarimetry, etc., the PSA is nevertheless built as it is shown in Fig. 6.2:

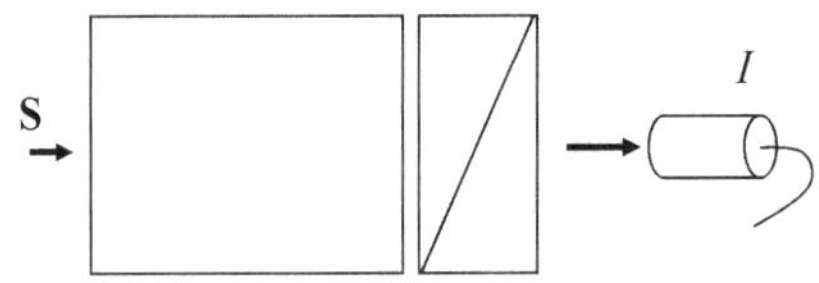

Fig. 6.2. Scheme of polarization state analyzer (PSA).

The rectangle disposed before the polarizer in Fig. 6.2 designates the controlled polarization element or group of polarization elements carrying out the polarization transformation (modulation) only. The most general case of polarization transformation without light amplitude altering is determined by the first Jones equivalent theorem (Hurwitz and Jones, 1941) and described by three degrees of freedom. However, three degrees of freedom are redundant here. The PSG and PSA for complete Mueller matrix measurements need only two degrees of freedom each. One of the great advantages of the configuration presented in Fig. 2 is that the polarization sensitivity of the detector is not important because the orientation of the final polarizer is fixed.

For any of measurement strategies the main principal of the Mueller matrix measurement can be presented in the following form:

$$I = f(m_{11}, \ldots, m_{44}). \tag{6.13}$$

This means that, to measure the Mueller matrix (Stokes vector), the information about matrix elements (Stokes parameters) should be presented in the first Stokes parameter of light impinging on the photodetector.

For any PSG and PSA, the total flux measured by the detector is

$$g = \mathbf{QML} = \sum_{i=1}^{4}\sum_{j=1}^{4} q_i m_{ij} l_j, \tag{6.14}$$

where $\mathbf{L}$ is the Stokes vector produced by PSG; $\mathbf{M}$ is the object Mueller matrix and $\mathbf{Q}$ is the Stokes vector corresponding to the first row of the Mueller matrix representing the PSA.

To measure the complete Mueller matrix, $N \geq 16$ flux measurements Eq. (6.14) are required. Flattening the Mueller matrix $\mathbf{M}$ into a 16×1 Mueller vector of the form $\vec{\mathbf{M}} = [m_{11}\ m_{12}\ m_{13}\ m_{14}\ \cdots\ m_{43}\ m_{44}]^T$, the polarimetric measurement equation can be represented as follows:

$$\mathbf{G} = \mathbf{W}\vec{\mathbf{M}} = \begin{pmatrix} q_1^1 l_1^1 & q_1^1 l_2^1 & q_1^1 l_3^1 & \cdot & q_4^1 l_4^1 \\ q_1^2 l_1^2 & q_1^2 l_2^2 & q_1^2 l_3^2 & \cdot & q_4^2 l_4^2 \\ q_1^3 l_1^3 & q_1^3 l_1^3 & q_1^3 l_3^3 & \cdot & q_4^3 l_4^3 \\ \cdot & \cdot & \cdot & \cdot & \cdot \\ q_1^N l_1^N & q_1^N l_2^N & q_1^N l_3^N & \cdot & q_4^N l_4^N \end{pmatrix} \begin{pmatrix} m_{11} \\ m_{12} \\ m_{13} \\ \cdot \\ m_{44} \end{pmatrix}, \tag{6.15}$$

where $\mathbf{G}$ is the $N \times 1$ vector, whose components are the fluxes measured by the detector and $\mathbf{W}$ is the $N \times 16$ general characteristic or data-reduction matrix with elements $w_{ij}^N = q_i^N l_j^N$.

Equation (6.15) is a system of generally N algebraic equations for Mueller matrix elements m_{ij}. The simplest case of the system Eq. (6.15) occurs when 16 independent measurements are performed. In this case $N = 16$, $\mathbf{W}$ is of rank 16, and the inverse matrix $\mathbf{W}^{-1}$ is unique. Then all 16 Mueller matrix elements are

$$\vec{\mathbf{M}} = \mathbf{W}^{-1}\mathbf{G}. \tag{6.16}$$

Most Mueller matrix polarimeters are configured so that $N > 16$. This makes $\vec{\mathbf{M}}$ overdetermined, and $\mathbf{W}^{-1}$ does not exist. The optimal (least-squares) estimation of $\vec{\mathbf{M}}$ can be obtained using the pseudoinverse matrix $\tilde{\mathbf{W}}$ of $\mathbf{W}$ (Lankaster and Tismenetsky, 1985):

$$\vec{\mathbf{M}} = \tilde{\mathbf{W}}\mathbf{G} = \left(\mathbf{W}^T\mathbf{W}\right)^{-1}\mathbf{W}^T\mathbf{G}. \tag{6.17}$$

In mathematics, there exist a variety of pseudoinverse matrices, such as one-sided inverse, Drazin inverse, group inverse, Bott-Duffin inverse, etc. (Moore, 1920; Bjerhammar, 1951; Penrose, 1955). Here we use the so-called Moore–Penrose pseudoinverse matrix (Horn and Johnson, 1986). Note, at the same time, that the characteristic matrix $\mathbf{W}$ in Eq. (6.15) is or can evidently be reduced to those of the full column rank.

This provides a very general point of view on the object under consideration. This approach assumes only linearity. A very important perspective is that the approach allows the inclusion of imperfections of all parts of polarimeter, namely source of radiation, polarization elements, photodetectors, etc., into the analysis, applying readily to the calibration of hardware and reconstruction of experimental data. Contrariwise, this is an interesting mathematical problem. Indeed, an examination of the rank, range, and null space of $\mathbf{W}$ allows the determination of whether the polarimeter is suitable for measuring the complete Mueller matrix $\mathbf{M}$ or one or another set of matrix elements. It is evident that the rank of $\mathbf{W}$ should be greater than or equal to the number of degrees of freedom of $\mathbf{M}$, such as 16 for a complete Mueller matrix and no more than 7 for a pure Mueller matrix (Hovenier, 1994), under consideration.

The approach in Eqs (6.15)–(6.17) is named the complete Mueller polarimetry. The theory of operation and calibration of Mueller matrix polarimetry were developed in Chipman (1995). This procedure can be repeated at different scattering angles in order to determine the angular profile of the Mueller matrix.

For a time-sequential measurement strategy, PSA is generally the complete Stokes polarimeter. Then, representing the elements of vector $\mathbf{G}$ in Eq. (6.15) as parameters of the Stokes vectors measured by PSA, the polarimetric measurement equation of the time-sequential measurement strategy in the case of $N = 16$ is given by (Savenkov, 2002, 2007):

$$\begin{pmatrix} l_1^1 & l_2^1 & l_3^1 & l_4^1 & 0 & 0 & 0 & 0 & 0 & 0 & 0 & 0 & 0 & 0 & 0 & 0 \\ l_1^2 & l_2^2 & l_3^2 & l_4^2 & 0 & 0 & 0 & 0 & 0 & 0 & 0 & 0 & 0 & 0 & 0 & 0 \\ l_1^3 & l_2^3 & l_3^3 & l_4^4 & 0 & 0 & 0 & 0 & 0 & 0 & 0 & 0 & 0 & 0 & 0 & 0 \\ l_1^4 & l_2^4 & l_3^4 & l_4^4 & 0 & 0 & 0 & 0 & 0 & 0 & 0 & 0 & 0 & 0 & 0 & 0 \\ 0 & 0 & 0 & 0 & l_1^1 & l_2^1 & l_3^1 & l_4^1 & 0 & 0 & 0 & 0 & 0 & 0 & 0 & 0 \\ 0 & 0 & 0 & 0 & l_1^2 & l_2^2 & l_3^2 & l_4^2 & 0 & 0 & 0 & 0 & 0 & 0 & 0 & 0 \\ 0 & 0 & 0 & 0 & l_1^3 & l_2^3 & l_3^3 & l_4^3 & 0 & 0 & 0 & 0 & 0 & 0 & 0 & 0 \\ 0 & 0 & 0 & 0 & l_1^4 & l_2^4 & l_3^4 & l_4^4 & 0 & 0 & 0 & 0 & 0 & 0 & 0 & 0 \\ 0 & 0 & 0 & 0 & 0 & 0 & 0 & 0 & l_1^1 & l_2^1 & l_3^1 & l_4^1 & 0 & 0 & 0 & 0 \\ 0 & 0 & 0 & 0 & 0 & 0 & 0 & 0 & l_1^2 & l_2^2 & l_3^2 & l_4^2 & 0 & 0 & 0 & 0 \\ 0 & 0 & 0 & 0 & 0 & 0 & 0 & 0 & l_1^3 & l_2^3 & l_3^3 & l_4^3 & 0 & 0 & 0 & 0 \\ 0 & 0 & 0 & 0 & 0 & 0 & 0 & 0 & l_1^4 & l_2^4 & l_3^4 & l_4^4 & 0 & 0 & 0 & 0 \\ 0 & 0 & 0 & 0 & 0 & 0 & 0 & 0 & 0 & 0 & 0 & 0 & l_1^1 & l_2^1 & l_3^1 & l_4^1 \\ 0 & 0 & 0 & 0 & 0 & 0 & 0 & 0 & 0 & 0 & 0 & 0 & l_1^2 & l_2^2 & l_3^2 & l_4^2 \\ 0 & 0 & 0 & 0 & 0 & 0 & 0 & 0 & 0 & 0 & 0 & 0 & l_1^3 & l_2^3 & l_3^3 & l_4^3 \\ 0 & 0 & 0 & 0 & 0 & 0 & 0 & 0 & 0 & 0 & 0 & 0 & l_1^4 & l_2^4 & l_3^4 & l_4^4 \end{pmatrix} \begin{pmatrix} m_{11} \\ m_{12} \\ m_{13} \\ m_{14} \\ m_{21} \\ m_{22} \\ m_{23} \\ m_{24} \\ m_{31} \\ m_{32} \\ m_{33} \\ m_{34} \\ m_{41} \\ m_{42} \\ m_{43} \\ m_{44} \end{pmatrix} = \begin{pmatrix} s_1^1 \\ s_1^2 \\ s_1^3 \\ s_1^4 \\ s_2^1 \\ s_2^2 \\ s_2^3 \\ s_2^4 \\ s_3^1 \\ s_3^2 \\ s_3^3 \\ s_3^4 \\ s_4^1 \\ s_4^2 \\ s_4^3 \\ s_4^4 \end{pmatrix}, \quad (6.18)$$

where l_i^k is the i-th parameter of k-th Stokes vector, $k = \overline{1,4}$, generated by PSG; s_i^k is the i-th parameter of k-th Stokes vector measured by PSA, respectively.

Let us consider the polarimetric measurement equation for the dynamic measurement strategy (Savenkov and Yushtin, 2001). Taking into account the explicit form of the detector intensity

$$I(t) = \sum_{k=0}^{2} \sum_{n=-2}^{2} \left(A_{k,n} \cos\left(2t\,(k\,\omega_1 + n\,\omega_2)\right) + B_{k,n} \sin\left(2t\,(k\,\omega_1 + n\,\omega_2)\right)\right), \quad (6.19)$$

where ω_1 and ω_2 are angular frequencies of wave plate rotation in PSG and PSA correspondingly and the most widespread measurement configuration

$$\mathbf{M}_1^{LA}\left(0, \frac{\theta}{2}\right) \mathbf{M}_1^{LP}\left(\frac{\alpha_2}{2}, \delta_2\right) \mathbf{M}^{Sample}\, \mathbf{M}_2^{LP}\left(\frac{\alpha_1}{2}, \delta_1\right) \mathbf{M}_2^{LA}(0, 0) \quad (6.20)$$

in the dynamic measurement strategy the polarimetric measurement Eq. (6.15) takes the form (Chipman, 1995) shown in Table 6.1.

Table 6.1. General polarimetric measurement equation for the dynamic measurement strategy

$\mathbf{A}_{0,2}$	$\frac{1}{4}(1-\cos(\delta_2))((2m_{21}+m_{22}(1+\cos(\delta_1)))\cos(\theta)- -(2m_{31}+m_{32}(1+\cos(\delta_1)))\sin(\theta))$
$\mathbf{B}_{0,2}$	$\frac{1}{4}(1-\cos(\delta_2))((2m_{31}+m_{32}(1+\cos(\delta_1)))\cos(\theta)- -(2m_{21}+m_{22}(1+\cos(\delta_1)))\sin(\theta))$
$\mathbf{A}_{0,1}$	$\left(m_{41}+\frac{1}{2}m_{42}(1+\cos(\delta_1))\right)\sin(\delta_2)\sin(\theta)$
$\mathbf{B}_{0,1}$	$-\left(m_{41}+\frac{1}{2}m_{42}(1+\cos(\delta_1))\right)\sin(\delta_2)\cos(\theta)$
$\mathbf{A}_{0,0}$	$m_{11}+\frac{1}{2}m_{12}(1+\cos(\delta_1))+\frac{1}{2}(2m_{21}+m_{22}(1+\cos(\delta_1)))\times \times\cos(\theta)\cos^2\left(\frac{\delta_2}{2}\right)+\frac{1}{4}(2m_{31}+m_{32}(1+\cos(\delta_1)))\sin(\theta)\cos^2\left(\frac{\delta_2}{2}\right)$
$\mathbf{A}_{1,2}$	$-\frac{1}{2}\sin(\delta_1)\sin^2\left(\frac{\delta_2}{2}\right)(m_{34}\cos(\theta)+m_{24}\sin(\theta))$
$\mathbf{B}_{1,2}$	$\frac{1}{2}\sin(\delta_1)\sin^2\left(\frac{\delta_2}{2}\right)(m_{24}\cos(\theta)-m_{34}\sin(\theta))$
$\mathbf{A}_{1,1}$	$\frac{1}{2}m_{44}\cos(\theta)\sin(\delta_1)\sin(\delta_2)$
$\mathbf{B}_{1,1}$	$\frac{1}{2}m_{44}\sin(\theta)\sin(\delta_1)\sin(\delta_2)$
$\mathbf{B}_{0,2}$	$\frac{1}{4}((2m_{21}+m_{22})\sin(\theta)-(2m_{31}+m_{32})\cos(\theta))$
$\mathbf{B}_{1,0}$	$\sin(\delta_1)\left(m_{14}+\frac{1}{2}m_{24}(1+\cos(\delta_2))\cos(\theta)+\frac{1}{2}m_{34}(1+\cos(\delta_2))\sin(\theta)\right)$
$\mathbf{A}_{1,-1}$	$-\frac{1}{2}m_{44}\cos(\theta)\sin(\delta_1)\sin(\delta_2)$
$\mathbf{B}_{1,-1}$	$\frac{1}{2}m_{44}\sin(\theta)\sin(\delta_1)\sin(\delta_2)$
$\mathbf{A}_{1,-2}$	$\frac{1}{2}\sin(\delta_1)\sin^2\left(\frac{\delta_2}{2}\right)(m_{34}\cos(\theta)+m_{24}\sin(\theta))$
$\mathbf{B}_{1,-2}$	$\frac{1}{2}\sin(\delta_1)\sin^2\left(\frac{\delta_2}{2}\right)(m_{24}\cos(\theta)-m_{34}\sin(\theta))$
$\mathbf{A}_{2,2}$	$\frac{1}{2}\sin^2\left(\frac{\delta_1}{2}\right)\sin^2\left(\frac{\delta_2}{2}\right)((m_{22}-m_{33})\cos(\theta)-(m_{23}+m_{32})\sin(\theta))$
$\mathbf{B}_{2,2}$	$\frac{1}{2}\sin^2\left(\frac{\delta_1}{2}\right)\sin^2\left(\frac{\delta_2}{2}\right)((m_{23}+m_{32})\cos(\theta)+(m_{22}-m_{33})\sin(\theta))$
$\mathbf{A}_{2,1}$	$\frac{1}{2}\sin^2\left(\frac{\delta_1}{2}\right)\sin(\delta_2)(m_{43}\cos(\theta)+m_{42}\sin(\theta))$

Table 6.1. Continued

$\mathbf{B}_{2,1}$	$-\frac{1}{2}\sin^2\left(\frac{\delta_1}{2}\right)\sin(\delta_2)(m_{42}\cos(\theta)-m_{43}\sin(\theta))$
$\mathbf{A}_{2,-1}$	$-\frac{1}{2}\sin^2\left(\frac{\delta_1}{2}\right)\sin(\delta_2)(m_{43}\cos(\theta)-m_{42}\sin(\theta))$
$\mathbf{B}_{2,-1}$	$\frac{1}{2}\sin^2\left(\frac{\delta_1}{2}\right)\sin(\delta_2)(m_{42}\cos(\theta)+m_{43}\sin(\theta))$
$\mathbf{A}_{2,-2}$	$\frac{1}{2}\sin^2\left(\frac{\delta_1}{2}\right)\sin^2\left(\frac{\delta_2}{2}\right)((m_{22}+m_{33})\cos(\theta)+(m_{23}-m_{32})\sin(\theta))$
$\mathbf{B}_{2,-2}$	$\frac{1}{2}\sin^2\left(\frac{\delta_1}{2}\right)\sin^2\left(\frac{\delta_2}{2}\right)((m_{23}-m_{32})\cos(\theta)-(m_{22}+m_{33})\sin(\theta))$

6.4 Physical realizability of the experimental Mueller matrix

One more important problem in polarimetry results from the fact that, due to measurement errors, the experimental Mueller matrix may misfit the class of inverse problem under consideration or even be physically unrealizable provided that the measurement procedure is fully correct.

Indeed, let us consider the measurements of pure Mueller matrix. The elements of pure Mueller matrix satisfy the following relation:

$$\sum_{i=1}^{4}\sum_{j=1}^{4} m_{ij}^2 = 4m_{11}^2. \tag{6.21}$$

This equality was obtained for the first time by Fry and Kattawar (1981). However, the question of whether this is a sufficient condition for $\mathbf{M}$ to be a pure Mueller matrix has been the subject of extensive discussions (see, e.g. Simon, 1982, 1987; Hovenier, 1994; Kim et al., 1987; Kostinski, 1992; Kostinski et al., 1993; Gil and Bernabeu, 1985; Anderson and Barakat, 1994; Brosseau, 1990; Brosseau et al., 1993; Gil, 2000). Under the premise that the Mueller matrix in question can be represented as a convex sum of pure Mueller matrices (Cloude and Pottier, 1995), Eq. (6.21) is both a necessary and a sufficient condition for $\mathbf{M}$ to be a pure Mueller matrix (Gil, 2007).

All the Mueller matrices satisfying the relation in Eq. (6.21) form the sphere in the Cartesian space spanned by all of the matrix elements m_{ij} normalized by m_{11}.

All points inside the sphere, for which the relation in Eq. (6.13) takes the form

$$\sum_{i=1}^{4}\sum_{j=1}^{4} m_{ij}^2 < 4m_{11}^2, \tag{6.22}$$

correspond to the depolarizing Mueller matrices.

The origin of coordinates corresponds to the Mueller matrix of ideal depolarizer (Shindo, 1995; Chipman, 1999, 2005; Ossikovski, 2010a, 2010b)

$$\mathbf{M}^{ID} = \begin{pmatrix} m_{11} & 0 & 0 & 0 \\ 0 & 0 & 0 & 0 \\ 0 & 0 & 0 & 0 \\ 0 & 0 & 0 & 0 \end{pmatrix}. \tag{6.23}$$

The points outside the sphere, namely when

$$\sum_{i=1}^{4}\sum_{j=1}^{4} m_{ij}^2 > 4m_{11}^2, \tag{6.24}$$

correspond to the cases when

$$p^{out} > 1 \tag{6.25}$$

at least for one input polarization.

Thus, the cross-section of the region formed by all results of the single Mueller matrix measurements (the dimension of the region is evidently determined by measurement error ΔM) by the sphere Eq. (6.21) is a range formed by all Mueller matrix measurements satisfying Eq. (6.21). In general, this is not necessarily an equatorial cross-section. This can schematically be represented as shown in Fig. 6.3:

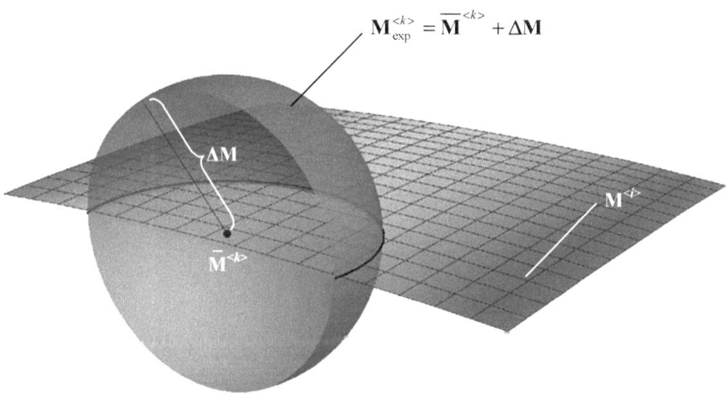

Fig. 6.3. Distribution of single measurements of the pure Mueller matrix.

Thus, among the results of single Mueller matrix measurements, there are the results which are physically realizable at least in principle—that is, pure and depolarizing Mueller matrices, and non-realizable ones. Furthermore, the averaged Mueller matrix can satisfy the relations in Eqs. (6.24) and (6.25) as well (Puentes et al., 2005). This means that, in the presence of measurement errors, the problem in Eq. (6.15) is ill-conditioned and an even more severely ill-posed problem. In other words, the experimental Mueller matrix can be irrelevant to the class of inverse problem under consideration—that is, as we have seen above, the experimental Mueller matrix of a homogeneous anisotropic medium can be a depolarizing or even physically unrealizable one.

Numerous attempts have been made to develop algebraic criteria that can determine a priori whether a given 4×4 matrix is a physically realizable Mueller matrix (Gil and Bernabeu, 1985, 1986; Barakat, 1987; Brosseau, 1990; Cloude, 1990; Brosseau et al., 1993; Givens and Kostinski, 1993; Kostinski et al., 1993; Gil, 2000, 2007). So many attempts have also been made to develop the methods to find the physically realizable Mueller matrix which is 'nearest' to the given physically unrealizable Mueller matrix (Cloude, 1990, 1997; Anderson and Barakat, 1994; Savenkov and Yushtin, 2000b; Ahmad and Takakura, 2008). Since we are interested in Mueller matrix measurement, then, evidently, the last aspect is determinative.

One of the examples of the physically unrealizable Mueller matrix that is measured using the correct procedure is a well known matrix (Howell, 1979) of the form

$$\begin{pmatrix} 0.7599 & -0.0623 & 0.0295 & 0.1185 \\ -0.0573 & 0.4687 & -0.1811 & 0.1863 \\ 0.0384 & -0.1714 & 0.5394 & 0.0282 \\ 0.1240 & -0.2168 & -0.0120 & 0.6608 \end{pmatrix}. \tag{6.26}$$

The physical unrealizability of this matrix is analyzed in numerous papers (see, e.g. Xing, 1992; Anderson and Barakat, 1994).

Indeed, to characterize the depolarization properties of the matrix in Eq. (6.26), let us take advantage of the following value (Barakat, 1987):

$$R = s_1^2 - s_2^2 - s_3^2 - s_4^2. \tag{6.27}$$

This value is a square of the intensity of the depolarized component of light. Thus, it means that a physically realizable Stokes vector has to give a non-negative value of R:

$$R = s_1^2 \left(1 - p^2\right). \tag{6.28}$$

The matrix in Eq. (6.26) has the next minimal value of R for output light:

$$R_{\text{min}} = -0.14. \tag{6.29}$$

It takes place for the next input Stokes vector:

$$\mathbf{S}^{inp} = \begin{pmatrix} 1 & -0.61 & 0.38 & -0.69 \end{pmatrix}^T. \tag{6.30}$$

At the same time, in the (Howell, 1979) paper, the information about the value of the measurement error is absent. Because of that, it will be of interest to find the maximum value of the measurement error which still secures the physical realisability of agiven Mueller matrix.

Suppose that the m_{ij} elements of the Mueller matrix in Eq. (6.26) are affected by uncorrelated error Δm, then the value R is consequently affected by errors $\Delta R\left(\theta, \varepsilon\right)$, given by Landi Degl'Innocenti and del Toro Iniesta (1998):

$$\Delta R\left(\theta, \varepsilon\right) = -\sum_{i=1}^{4} \sum_{j=1}^{4} \left| \frac{\partial R\left(\theta, \varepsilon\right)}{\partial m_{ij}} \right| \Delta m \approx 0.12. \tag{6.31}$$

This gives the following maximum estimation of the measurement error of Howell's Mueller matrix: $\Delta m = 0.012$.

The criterion for a given 4×4 real matrix $\mathbf{M}$ to be a Mueller matrix is non-negativity of all four eigenvalues λ_i of the coherency matrix associated with $\mathbf{M}$ (Cloude, 1986, 1990 1997, 2010; see Appendix A for more details). If the experimental matrix overpolarizes the output light at least for one input polarization in Eq. (6.25), this results in negativity of one of the eigenvalue of coherency matrix in Eq. (6.24A). The best approximation of $\mathbf{M}$ in this case will be the Mueller matrix in Eq. (6.26A) without a term corresponding to the negative eigenvalue. The pure Mueller matrix which is a best approximation of $\mathbf{M}$ associates with the pure Mueller matrix in Eq. (6.26A) corresponding to the maximal eigenvalue of the coherency matrix. This approximation makes sense in the case of $H < 0.5$ only.

The eigenvalues of Cloude's coherency matrix in Eq. (6.24A) for the Mueller matrix in Eq. (6.26) are as follows:

$$\lambda_i = (0.6478, \quad 0.1051, \quad 0.0801, \quad -0.0731)\,, \tag{6.32}$$

and entropy Eq. (6.25A) $H = 0.6207$.

Thus, the best approximation for the Mueller matrix in Eq. (6.26) will be the Mueller matrix obtained in Eq. (6.26A) excluding the eigenvalue $\lambda_4 = -0.0731$:

$$\begin{pmatrix} 0.8330 & -0.0631 & 0.0353 & 0.1161 \\ -0.0518 & 0.4888 & -0.1111 & 0.1811 \\ 0.0359 & -0.1021 & 0.5741 & 0.0389 \\ 0.1256 & -0.2079 & -0.0197 & 0.5889 \end{pmatrix}. \tag{6.33}$$

The degree of polarization for the matrix in Eq. (6.33) behaves as shown in Fig. 6.4.

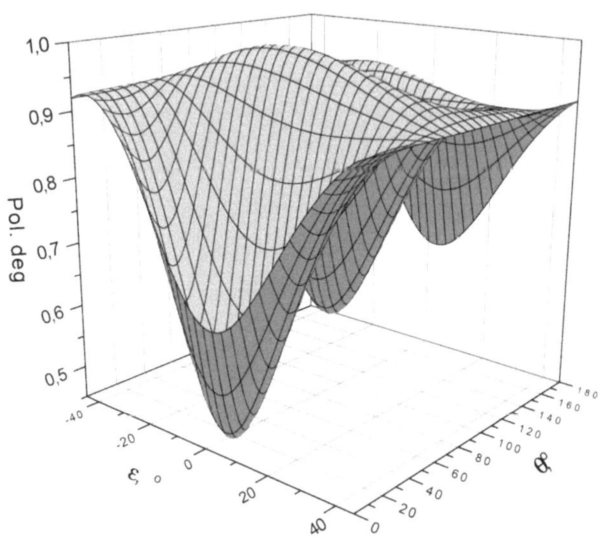

Fig. 6.4. Degree of polarization of output light for the Mueller matrix in Eq.(6.33) as a function of ellipticity ε and azimuth θ of polarization input light.

In the case of $0 < E < 1$, the finding of the best approximation of experimental matrix $\mathbf{M}$ needs detailed information about the classes of depolarizing object characterized by two and three non-zero eigenvalues of Cloude's coherency matrix. These questions have been analyzed in Aiello and Woerdman (2005), Puentes et al. (2005), Aiello et al. (2006), and Savenkov et al. (2009).

At the same time, the physically realizable Mueller matrix $\mathbf{N}(\mathbf{X})$ that best approximates the experimental Mueller matrix $\mathbf{M}$ can be defined in the sense of minimizing (Anderson and Barakat, 1994; Savenkov and Yushtin, 2000b)

$$\|\mathbf{M} - \mathbf{N}(\mathbf{X})\| \to \min . \tag{6.34}$$

As a matrix, $\mathbf{N}(\mathbf{X})$ can in principle be used in any multiplicative and additive Mueller matrix model of an object under consideration (Cloude, 1986; Gil and Bernabeu, 1987; Lu and Chipman, 1996; Ossikovski, 2009). Then $\mathbf{X}$ is a vector formed by physical parameters of the model.

To find the best approximation matrix in the class of pure Mueller matrices for $\mathbf{N}(\mathbf{X})$, Eq. (6.5) can be used matrix (Anderson and Barakat, 1994; Savenkov and Yushtin, 2000b).

Taking into account that, in this case, $\mathbf{X}$ can be considered a 8×1 vector containing eight parameters, real and imagine, parts of the Jones matrix elements

$$\mathbf{T} = \sqrt{2} \begin{pmatrix} t_1 + i\,t_2 & t_3 + i\,t_4 \\ t_5 + i\,t_6 & t_7 + i\,t_8 \end{pmatrix}, \tag{6.35}$$

and representing $N(X)$ as

$$\begin{aligned}
n_{11} &= 1/2\left(t_1^2 + t_2^2 + t_3^2 + t_4^2 + t_5^2 + t_6^2 + t_7^2 + t_8^2\right), \\
\mathrm{n}_{12} &= 1/2\left(t_1^2 + t_2^2 - t_3^2 - t_4^2 + t_5^2 + t_6^2 - t_7^2 - t_8^2\right), \\
\mathrm{n}_{13} &= t_1t_3 + t_2t_4 + t_5t_7 + t_6t_8, \\
\mathrm{n}_{14} &= -t_1t_4 + t_2t_3 - t_5t_8 + t_6t_7, \\
\mathrm{n}_{21} &= 1/2\left(t_1^2 + t_2^2 + t_3^2 + t_4^2 - t_5^2 - t_6^2 - t_7^2 - t_8^2\right), \\
\mathrm{n}_{22} &= 1/2\left(t_1^2 + t_2^2 - t_3^2 - t_4^2 - t_5^2 - t_6^2 + t_7^2 + t_8^2\right), \\
\mathrm{n}_{23} &= t_1t_3 + t_2t_4 - t_5t_7 - t_6t_8, \\
\mathrm{n}_{24} &= -t_1t_4 + t_2t_3 + t_5t_8 - t_6t_7, \\
\mathrm{n}_{31} &= t_1t_5 + t_2t_6 + t_3t_7 + t_4t_8, \\
\mathrm{n}_{32} &= t_1t_5 + t_2t_6 - t_3t_7 - t_4t_8, \\
\mathrm{n}_{33} &= t_1t_7 + t_2t_8 + t_3t_5 + t_4t_6, \\
\mathrm{n}_{34} &= -t_1t_8 + t_2t_7 + t_3t_6 - t_4t_5, \\
\mathrm{n}_{41} &= t_1t_6 - t_2t_5 + t_3t_8 - t_4t_7, \\
\mathrm{n}_{42} &= t_1t_6 - t_2t_5 - t_3t_8 + t_4t_7, \\
\mathrm{n}_{43} &= t_1t_8 - t_2t_7 + t_3t_6 - t_4t_5, \\
\mathrm{n}_{44} &= t_1t_7 + t_2t_8 - t_3t_5 - t_4t_6,
\end{aligned} \tag{6.36}$$

in (Savenkov and Yushtin, 2000b), it is shown that Eq. (6.34) takes the form

$$
\begin{aligned}
&(m_{11}+m_{12}+m_{21}+m_{22})\,t_1+(m_{13}+m_{23})\,t_3-(m_{14}+m_{24})\,t_4+(m_{31}+m_{32})\,t_5+\\
&\quad+(m_{41}+m_{42})\,t_6+(m_{33}+m_{44})\,t_7+(m_{43}-m_{34})\,t_8=4\,\|\mathbf{t}\|^2\,t_1,\\
&(m_{11}+m_{12}+m_{21}+m_{22})\,t_2+(m_{14}+m_{24})\,t_3+(m_{13}+m_{23})\,t_4-(m_{41}+m_{42})\,t_5+\\
&\quad+(m_{31}+m_{32})\,t_6-(m_{43}-m_{34})\,t_7+(m_{33}+m_{44})\,t_8=4\,\|\mathbf{t}\|^2\,t_2,\\
&(m_{13}+m_{23})\,t_1+(m_{14}+m_{24})\,t_2+(m_{11}-m_{12}+m_{21}-m_{22})\,t_3+(m_{33}-m_{44})\,t_5+\\
&\quad+(m_{34}+m_{43})\,t_6+(m_{31}-m_{32})\,t_7+(m_{41}+m_{42})\,t_8=4\,\|\mathbf{t}\|^2\,t_3,\\
&-(m_{14}+m_{24})\,t_1+(m_{13}+m_{23})\,t_2+(m_{11}-m_{12}+m_{21}-m_{22})\,t_4-(m_{34}+m_{43})\,t_5+\\
&\quad+(m_{33}-m_{44})\,t_6-(m_{41}+m_{42})\,t_7+(m_{31}-m_{32})\,t_8=4\,\|\mathbf{t}\|^2\,t_4,\\
&(m_{31}+m_{32})\,t_1-(m_{41}+m_{42})\,t_2+(m_{33}-m_{44})\,t_3-(m_{34}+m_{43})\,t_4+\\
&\quad+(m_{11}+m_{12}-m_{21}-m_{22})\,t_5+(m_{13}-m_{23})\,t_7+(m_{24}-m_{14})\,t_8=4\,\|\mathbf{t}\|^2\,t_5,\\
&(m_{41}+m_{42})\,t_1+(m_{31}+m_{32})\,t_2+(m_{34}+m_{43})\,t_3+(m_{33}-m_{44})\,t_4+\\
&\quad+(m_{11}+m_{12}-m_{21}-m_{22})\,t_6+(m_{14}-m_{24})\,t_7+(m_{13}-m_{23})\,t_8=4\,\|\mathbf{t}\|^2\,t_6,\\
&(m_{33}+m_{44})\,t_1+(m_{34}-m_{43})\,t_2+(m_{31}-m_{32})\,t_3+(m_{42}-m_{41})\,t_4+\\
&\quad+(m_{13}-m_{23})\,t_5+(m_{14}-m_{24})\,t_6+(m_{11}-m_{12}-m_{21}+m_{22})\,t_7=4\,\|\mathbf{t}\|\,t_7,\\
&(m_{43}-m_{34})\,t_1+(m_{33}+m_{44})\,t_2+(m_{41}-m_{42})\,t_3+(m_{31}-m_{32})\,t_4+\\
&\quad+(m_{24}-m_{14})\,t_5+(m_{13}-m_{23})\,t_6+(m_{11}-m_{12}-m_{21}+m_{22})\,t_8=4\,\|\mathbf{t}\|^2\,t_8,
\end{aligned}
\tag{6.37}
$$

where

$$
\|\mathbf{t}\|=\sqrt{t_1^2+t_2^2+t_3^2+t_4^2+t_5^2+t_6^2+t_7^2+t_8^2}. \tag{6.38}
$$

Or, in matrix form, we finally have

$$
\mathbf{D}\mathbf{t}=\|\mathbf{t}\|\,\mathbf{t}. \tag{6.39}
$$

From a mathematical point of view, Eq. (6.39) is a spectral problem (Horn and Johnson, 1986) for matrix **D**. Thus, the best approximation $\mathbf{N}(\mathbf{X})$ corresponds to the Jones matrix **T**, the real and imaginary part t_i of which elements form a vector t with minimal length.

6.5 Partial Mueller polarimetry

All of the analyses of the Mueller polarimeters presented the in preceding sections are carried out for measurement of the complete Mueller matrix. However, in many applications, there are physical circumstances that reduce the number of degrees of freedom of the Mueller matrix from 16 to something less than 16 and, hence, measurement of the complete Mueller matrix is not necessary (Savenkov, 2002, 2007; Tyo et al., 2010; Oberemok and Savenkov, 2003; Hoover and Tyo, 2007). This corresponds to the third case in Eq. (6.15) occurring when $N < 16$ and **W** is of rank less than 16. The optimal estimation of $\vec{\mathbf{M}}$ is again obtained using the pseudoinverse matrix $\tilde{\mathbf{W}}$. However, only 15 or fewer Mueller matrix elements can be determined from the system in Eq. (6.15)—that is, polarimetry is 'incomplete' or 'partial'.

First of all, some subset of matrix elements might completely describe the scattering which is of interest (Hovenier, 1969, 1970, 1994; Hovenier and Mackowski, 1998; Hovenier and van der Mee, 2000) and, hence, these subsets can be considered as initial information for the solution of corresponding classes of inverse problems.

Another reason making the measurement of the complete Mueller matrix unnecessary is that the symmetries might dictate linear relationships among Mueller matrix elements. An illustrative example is the pure Mueller matrix with symmetry determined by the first Jones equivalence theorem (Hurwitz and Jones, 1941; Hu et al., 1987; Hovenier, 1994):

$$\begin{pmatrix} m_{11} & 0 & 0 & 0 \\ 0 & m_{22} & m_{23} & m_{24} \\ 0 & m_{32} & m_{33} & m_{34} \\ 0 & m_{42} & m_{43} & m_{44} \end{pmatrix}. \tag{6.40}$$

This matrix is extremely widespread in the literature (de Boer et al., 1997; Tang and Kwok, 2001; Swami et al., 2006; Herreros-Cedres et al., 2006; Wood et al., 2007; Lin, 2008, Ghosh et al. 2008, 2009a) because it describes linear crystal optics without absorption. This approach is termed incomplete or partial Mueller polarimetry (Savenkov, 2007; Tyo et al., 2010).

One more prominent example in this sense is the block-diagonal Mueller matrix of the form (van de Hulst, 1957)

$$\begin{pmatrix} m_{11} & m_{12} & 0 & 0 \\ m_{21} & m_{22} & 0 & 0 \\ 0 & 0 & m_{33} & m_{34} \\ 0 & 0 & m_{43} & m_{44} \end{pmatrix}. \tag{6.41}$$

This matrix has extensive bibliography and plays a key role in many light-scattering problems. The structure in Eq. (6.41) can be caused by symmetry of a single particle and a collection of particles in single and multiple scattering (van de Hulst, 1957; Mishchenko and Travis, 2000) and by illumination-observation geometry for backward (Zubko et al., 2004) and forward (Voss and Fry, 1984; Savenkov et al., 2009a) scattering.

The model of a medium described by the Mueller matrix of Eq. (6.41) has been used in studies of optical characteristics of oceanic water (Voss and Fry, 1984; Kokhanovsky, 2001); ensembles of identical, but randomly oriented, fractal particles (Kokhanovsky, 2003); dense spherical particle suspensions in the multiple-scattering regime (Kaplan et al., 2001); ice clouds consisting of nonspherical ice crystals in the multiple-scattering regime (Lawless et al., 2006); polydisperse, randomly oriented ice crystals modeled by finite circular cylinders with different size distributions (Xu et al., 2002); cylindrically shaped radially inhomogeneous particles (Manickavasagam and Menguc, 1998); and small spherical particles (ranging in diameter from 0.2 to 1.5 μm) sparsely seeded on the surface of a crystalline silicon c-Si wafer (Kaplan and Drevillon, 2002). Other applications included measurements of the complex refractive index of isotropic materials as matrices of isotropic and ideal metal mirror reflections (Deibler and Smith, 2001); the development of a symmetric three-term product decomposition of a Mueller–Jones matrix

(Ossikovski, 2008); and the description of very general and practically important cases of (i) randomly oriented particles with a plane of symmetry (Hovenier and van der Mee, 2000) and/or (ii) equal numbers of particles and their mirror particles (Mishchenko et al., 2002). This list of applications can be extended significantly.

An example of the situation in which the Mueller matrix has the structure of Eq. (6.41) and contains information on the strong dependence of depolarization on the polarization state of the input light is the exact forward-scattering of polarized light by a slab of inhomogeneous linear birefringent medium (Savenkov et al., 2007a).

A comparison of the matrix in Eq. (6.41) and the structures of incomplete Mueller matrices measured in framework of time-sequential (Eq. (6.18)), and dynamic measurement strategies (Table 6.1) shows that the most appropriate strategy for measurement of block-diagonal scattering matrix is the time-sequential strategy (Chipman, 1995).

It can be seen that for block-diagonal matrix, Eq. (6.41), the polarimetric measurement Eq. (6.18) is reduced to the form

$$\begin{pmatrix} m_{11}l_1^1 + m_{12}l_2^1 \\ m_{11}l_1^2 + m_{12}l_2^2 \\ m_{11}l_1^3 + m_{12}l_2^3 \\ \vdots \\ m_{43}l_3^4 + m_{44}l_4^4 \end{pmatrix} = \begin{pmatrix} s_1^1 \\ s_1^2 \\ s_1^3 \\ s_1^4 \\ \vdots \\ s_4^4 \end{pmatrix}. \tag{6.42}$$

Evidently, the set of equations in Eq. (6.42) is overdetermined and non-zero elements of the matrix Eq. (6.41) can be measured using only two elliptical input polarizations. Indeed, we have

$$\begin{cases} m_{11}l_1^1 + m_{12}l_2^1 = s_1^1 \\ m_{11}l_1^2 + m_{12}l_2^2 = s_1^2 \end{cases}, \quad \begin{cases} m_{21}l_1^1 + m_{22}l_2^1 = s_2^1 \\ m_{21}l_1^2 + m_{22}l_2^2 = s_2^2 \end{cases}, \tag{6.43}$$

$$\begin{cases} m_{33}l_3^1 + m_{34}l_4^1 = s_3^1 \\ m_{33}l_3^2 + m_{34}l_4^2 = s_3^2 \end{cases}, \quad \begin{cases} m_{43}l_3^1 + m_{44}l_4^1 = s_4^1 \\ m_{43}l_3^2 + m_{44}l_4^2 = s_4^2 \end{cases}, \tag{6.44}$$

with following characteristic matrices

$$\mathbf{V}_1 = \begin{pmatrix} l_1^1 & l_2^1 \\ l_1^2 & l_2^2 \end{pmatrix}, \tag{6.45}$$

$$\mathbf{V}_2 = \begin{pmatrix} l_3^1 & l_4^1 \\ l_3^2 & l_4^2 \end{pmatrix}. \tag{6.46}$$

It can be seen that rows of characteristics matrices, $\mathbf{V}_1$ and $\mathbf{V}_2$ are formed by Stokes parameters of two input elliptical polarizations.

The exact sets of matrix elements, namely structures of incomplete Mueller matrices, which can be measured in the framework of any of measurement strategies (time-sequential, dynamic, etc.), are also determined by the structure of the data-reduction matrix of polarimetric measurement Eq. (6.15).

In the framework of the time-sequential measurement strategy in Eq. (6.18), one can independently measure the following incomplete Mueller matrices (Oberemok and Savenkov, 2003):

(i) using three input polarizations:

$$\left(r_1^k,\ r_2^k,\ r_3^k,\ 0\right)^T, \quad \text{or} \quad \left(r_1^k,\ r_2^k,\ 0,\ r_4^k\right)^T, \quad \text{or} \quad \left(r_1^k,\ 0,\ r_3^k,\ r_4^k\right)^T$$

three-column structures, consisting of twelve matrix elements:

$$\underset{(a)}{\begin{pmatrix} m_{11} & m_{12} & m_{13} & \bullet \\ m_{21} & m_{22} & m_{23} & \bullet \\ m_{31} & m_{32} & m_{33} & \bullet \\ m_{41} & m_{42} & m_{43} & \bullet \end{pmatrix}} \underset{(b)}{\begin{pmatrix} m_{11} & m_{12} & \bullet & m_{14} \\ m_{21} & m_{22} & \bullet & m_{24} \\ m_{31} & m_{32} & \bullet & m_{34} \\ m_{41} & m_{42} & \bullet & m_{44} \end{pmatrix}} \underset{(c)}{\begin{pmatrix} m_{11} & \bullet & m_{13} & m_{14} \\ m_{21} & \bullet & m_{23} & m_{24} \\ m_{31} & \bullet & m_{33} & m_{34} \\ m_{41} & \bullet & m_{43} & m_{44} \end{pmatrix}} \tag{6.47}$$

(ii) using two input polarizations:

$$\left(r_1^k,\ r_2^k,\ 0,\ 0\right)^T, \quad \text{or} \quad \left(r_1^k,\ 0,\ 0,\ r_4^k\right)^T, \quad \text{or} \quad \left(r_1^k,\ 0,\ r_3^k,\ 0\right)^T,$$

two-column structures consisting of eight elements:

$$\underset{(a)}{\begin{pmatrix} m_{11} & m_{12} & \bullet & \bullet \\ m_{21} & m_{22} & \bullet & \bullet \\ m_{31} & m_{32} & \bullet & \bullet \\ m_{41} & m_{42} & \bullet & \bullet \end{pmatrix}} \underset{(b)}{\begin{pmatrix} m_{11} & \bullet & \bullet & m_{14} \\ m_{21} & \bullet & \bullet & m_{24} \\ m_{31} & \bullet & \bullet & m_{34} \\ m_{41} & \bullet & \bullet & m_{44} \end{pmatrix}} \underset{(c)}{\begin{pmatrix} m_{11} & \bullet & m_{13} & \bullet \\ m_{21} & \bullet & m_{23} & \bullet \\ m_{31} & \bullet & m_{33} & \bullet \\ m_{41} & \bullet & m_{43} & \bullet \end{pmatrix}}. \tag{6.48}$$

In Eqs (6.47) and (6.48) m_{ij} are the elements of the object Mueller matrix; symbol $\bullet$ defines the matrix elements which are not measured in corresponding modes of the time-sequential measurement strategy.

(iii) using one input polarization:

$$\begin{pmatrix} r_1^1 & 0 & 0 & 0 \\ 0 & r_1^1 & 0 & 0 \\ 0 & 0 & r_1^1 & 0 \\ 0 & 0 & 0 & r_1^1 \end{pmatrix} \begin{pmatrix} m_{11} \\ m_{21} \\ m_{31} \\ m_{41} \end{pmatrix} = \begin{pmatrix} s_1^1 \\ s_2^1 \\ s_3^1 \\ s_4^1 \end{pmatrix}. \tag{6.49}$$

Case **(iii)** is commonly applied in astronomical polarimetry for non-polarized (natural) input light (Mishchenko et al., 2010).

In the framework of dynamic measurement strategy (Chipman, 1995) and, in particular, in the most commonly used dual rotated retarders mode in Eq. (6.20) (Collins and Koh, 1999; Smith, 2002; Zallat et al., 2006; Takakura and Ahmad,

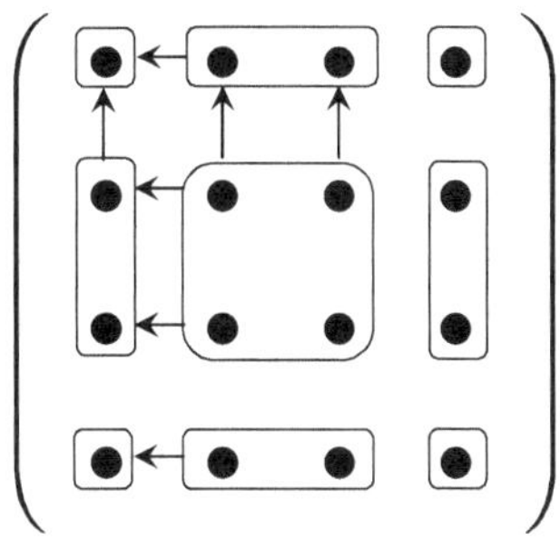

Fig. 6.5. The sets of Mueller matrix elements which can be independently measured in the scope of dynamic measurement strategy.

2007; Vaughn and Hoover, 2008), the sets of matrix elements shown in Fig. 6.5 can be independently measured.

The relations between the matrix elements occurred in measurement (indirect measurements) are marked with corresponding arrows (Savenkov and Yushtin, 2001).

6.6 Mueller polarimeter optimization

The general polarimetric measurement equation, see section 6.3, has been applied by many authors to the optimization of Mueller matrix polarimeters in the presence of noise and measurement error (Savenkov, 2002; Smith, 2002; De Martino et al., 2003; Twietmeyer and Chipman, 2008).

The optimal choice of four configurations of PSA for Stokes polarimeters is first discussed in Azzam et al. (1988) with demonstration that the optimal choice of Stokes vectors corresponds to a regular tetrahedron inscribed into the Poincaré sphere. Tyo contributed considerably to the optimization of a whole number of Stokes polarimeters, including rotating retarder, variable retarder, multichannel linear, and hyperspectral dual variable retarder polarimeters (Tyo, 1998, 2000). DeMartino et al. and Garcia-Caurel et al. optimize the spectroscopic polarimeters using photoelastic modulators and liquid crystal retarders (De Martino et al., 2003, 2004; Garcia-Caurel et al., 2004). Zallat et al. discussed random and systematic errors in Mueller matrix polarimeters (Zallat et al., 2006). R. Kleim et al. (1994) showed for a dual rotating retarder Mueller matrix polarimeter thet all azimuthal-angle errors can be eliminated by a two-zone measurement and that the natural optical rotation of the rotating quartz compensator has no effect on the measurements. The effect of systematic error in a dual rotating retarder Mueller matrix polarimeter and the conclusion that errors in the retarder have a stronger effect than errors in the analyzer or polarizer are considered in Piller et al. (2008). Vaughn and Hoover (2008) analyzed noise reduction for non-symmetric dual rotating retarder polarimeters. Broch and Johann minimized the effect of random noise in an incomplete Mueller matrix polarimeter by optimization of the orienations of the polarizer and analyzer (Broch and Johann, 2008).

Further, we discuss some cases of optimization in more detail. In the scope of the time-sequential measurement strategy, the question thereof is what exact polarizations of input light or what exact alignment should be used for measurement of the Mueller matrix?

If the values of Stokes parameters l_i^k, s_i^k in Eq. (6.18) are known accurately without measurement errors, then the answer to this question is any polarizations giving non-zero determinant $\det(\mathbf{W}) \neq 0$. However, in the presence of measurement errors, not every different Stokes vector providing non-zero determinants of characteristic matrix $\mathbf{W}$ can be used as a set of input polarizations. In other words, when solving Eq. (6.18), the matrix $\mathbf{W}^{-1}$ can exist, but the resulting measurement errors are unacceptable.

Optimal choice of input polarizations is connected with conditioning of the characteristic matrix $\mathbf{W}$. In previous optimizations, different figures of merit to estimate conditioning of the characteristic matrices have been used (Sabatke et al., 2000). Here, as a figure of merit, we use the value of condition number $cond\,(\mathbf{V}_i)$ of the matrix $\mathbf{V}_i$ based on the Frobhenius norm (Horn and Johnson, 1986):

$$cond\,(\mathbf{V}_i) = \|\mathbf{V}_i\| \left\|\mathbf{V}_i^{-1}\right\|, \tag{6.50}$$

where $\|\mathbf{V}_i\|$ and $\left\|\mathbf{V}_i^{-1}\right\|$ are the norms of the direct and inverse matrix $\mathbf{V}_i$, respectively, which is defined for the Frobhenius norm as

$$\|\mathbf{V}\| = \sqrt{\sum_{i,j} |\mathbf{V}_{ij}|^2}. \tag{6.51}$$

The properties of the Frobhenius norm that make us to choose it for further analysis are the fact that it has very transparent geometrical interpretation.

Then, the upper bound of the value of the Mueller matrix element error $\Delta\mathbf{M}$ can be estimated as (Horn and Johnson, 1986)

$$\Delta\mathbf{M} = \frac{2\,cond\,(\mathbf{V}_i)\,\Delta\mathbf{S}}{1 - cond\,(\mathbf{V}_i)\,\Delta\mathbf{S}}, \tag{6.52}$$

where

$$\Delta\mathbf{S} = \frac{\|\mathbf{S}_{exact} - \mathbf{S}_{\exp}\|}{\|\mathbf{S}_{exact}\|} \quad \text{and} \quad \Delta\mathbf{M} = \frac{\|\mathbf{M}_{exact} - \mathbf{M}_{\exp}\|}{\|\mathbf{M}_{exact}\|}. \tag{6.53}$$

The quantities $\mathbf{S}_{\exp}$ and $\mathbf{S}_{exact}$ are the measured and exact Stokes vectors, respectively. $\mathbf{M}_{\exp}$ and $\mathbf{M}_{exact}$ are the measured and exact Mueller matrices of the studied object. Experimentally, estimation of the value $\Delta\mathbf{S}$ ($\Delta\mathbf{M}$) can be obtained by measuring the Stokes vector (Mueller matrix) of known polarization states (reference objects, namely empty space, prism polarizers, etc.).

For the value $cond\,(\mathbf{V}_{4\times 4})$

$$\begin{aligned} cond\,(\mathbf{V}_{4\times 4}) = D\ \big(\, & r_2^3 r_3^2 r_4^1 - r_2^4 r_3^2 r_4^1 - r_2^2 r_3^3 r_4^1 + r_2^4 r_3^3 r_4^1 + r_2^2 r_3^4 r_4^1 - \\ & - r_2^3 r_3^4 r_4^1 - r_2^3 r_3^1 r_4^2 + r_2^4 r_3^1 r_4^2 + r_2^1 r_3^3 r_4^2 - r_2^4 r_3^3 r_4^2 - \\ & - r_2^1 r_3^4 r_4^2 + r_2^3 r_3^4 r_4^2 + r_2^2 r_3^1 r_4^3 - r_2^4 r_3^1 r_4^3 - r_2^1 r_3^2 r_4^3 + \\ & + r_2^4 r_3^2 r_4^3 + r_2^1 r_3^4 r_4^3 - r_2^2 r_3^4 r_4^3 - r_2^2 r_3^1 r_4^4 + r_2^3 r_3^1 r_4^4 + \\ & + r_2^1 r_3^2 r_4^4 - r_2^3 r_3^2 r_4^4 - r_2^1 r_3^3 r_4^4 + r_2^2 r_3^3 r_4^4 \big)^{-1}, \end{aligned} \tag{6.54}$$

for *cond* $(\mathbf{V}_{3\times 3})$

$$cond\,(\mathbf{V}_{3\times 3}) = \frac{\sqrt{6}}{r_2^2\,r_3^3 - r_3^2\,r_2^3 - r_2^1\,r_3^3 + r_3^1\,r_2^3 + r_2^1\,r_3^2 - r_3^1\,r_2^2} \times$$
$$\times \left(\sqrt{\left(r_2^2\,r_3^3 - r_3^2\,r_2^3\right)^2 + \left(r_2^1\,r_3^3 - r_3^1\,r_2^3\right)^2 + \left(r_2^1\,r_3^2 - r_3^1\,r_2^2\right)^2 + \left(r_3^2 - r_3^3\right)^2 + \right.$$
$$\overline{\left. + \left(r_3^3 - r_3^1\right)^2 + \left(r_3^2 - r_3^1\right)^2 + \left(r_2^2 - r_2^3\right)^2 + \left(r_2^1 - r_2^3\right)^2 + \left(r_2^1 - r_2^2\right)^2} \right) \,, \tag{6.55}$$

and finally for *cond* $(\mathbf{V}_{2\times 2})$

$$cond\,(\mathbf{V}_{2\times 2}) = \frac{2 + \left(r_2^1\right)^2 + \left(r_2^2\right)^2}{\left(r_2^2 - r_2^1\right)}. \tag{6.56}$$

Thus, Eqs (6.54)–(6.56) show that minimization of the condition number in all these cases is equivalent to maximization of the volume, area, and length of the corresponding geometrical figures inscribed into the Poincaré sphere (see Fig. 6.6).

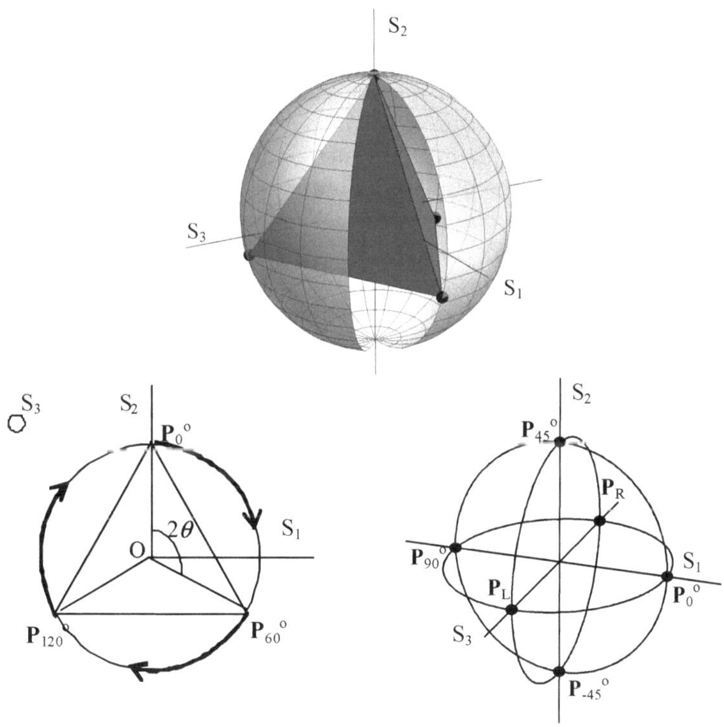

(b) (c)

Fig. 6.6. Geometrical interpretation of the problem of condition number minimization in the scope of the time-sequential strategy: (a) regular tetrahedron in four input polarizations mode; (b) regular triangle in three input polarizations mode; (c) line segment as the diameter of principal cross-section of the Poincaré sphere in the two input polarizations mode.

The minimal values of the condition numbers in the three measurement modes of the time-sequential measurement strategy are

$$\begin{aligned} cond\left(\mathbf{V}_{4\times 4}\right) &= 4{,}472, \\ cond\left(\mathbf{V}_{3\times 3}\right) &= 3{,}162, \\ cond\left(\mathbf{V}_{2\times 2}\right) &= 2. \end{aligned} \tag{6.57}$$

The effect of the choice of input polarizations on the Mueller matrix measurement errors in the four input polarizations mode is presented in Fig. 6.7. In this experiment, three input polarizations are taken as optimal and the ellipticity and azimuth of fourth input polarization is altered so that the block submatrix in Eq. (6.18) has the form

$$\mathbf{V}_{4\times 4}(\gamma,\varepsilon) = \begin{pmatrix} 1 & \cos(2\gamma)\cos(2\varepsilon) & \sin(2\gamma)\cos(2\varepsilon) & \sin(2\varepsilon) \\ 1 & -0{,}333 & -0{,}817 & 0{,}471 \\ 1 & -0{,}333 & 0 & -0{,}943 \\ 1 & -0{,}333 & 0{,}817 & 0{,}471 \end{pmatrix}. \tag{6.58}$$

Thus, Fig. 6.7 shows that the choice of input polarization affects considerably the Mueller matrix measurement errors. Furthermore, as can be seen, input polarizations may be different but measurement errors corresponding to these polarizations are inadmissible.

The results of analog analysis for a number of incomplete Mueller matrices resulting from Fig. 5 and measured in the scope of the dynamic measurement strategy are presented in Table 6.2 (Savenkov and Yushtin, 2001). These results are notable compared with the minimum obtainable value of the condition number for the measurement of the complete Mueller matrix in the scope of the dynamic measurement strategy $cond(\mathbf{W}) \approx 22.5$.

In section 6.5, it was shown that block-diagonal Mueller matrix Eq. (6.41) can be measured using two input elliptical polarizations only. Let us determine two optimal input elliptical polarizations for this case.

The connection between the characteristic matrices of the sub-systems in Eqs (6.43) and (6.44) dictates that the choice of input polarizations l_i^k should minimize the condition numbers of both characteristics matrices $\mathbf{V}_1$ and $\mathbf{V}_2$ in Eqs (6.45) and (6.46) simultaneously. Mathematically, this can be accomplished by minimization of the condition number for the product of matrices $\mathbf{V}_1\mathbf{V}_2 = \mathbf{V}$. The condition number is then a function of four variables: two azimuths and two ellipticities of two input polarizations:

$$\begin{aligned} &cond\left(\mathbf{V}\right) = \\ &\Bigg[\Big((s\varepsilon_1)^2 + c\theta_1\, S\varepsilon_1\, s\varepsilon_2 + c\theta_2\, s\varepsilon_1\, S\varepsilon_2 + (c\theta_2\, S\varepsilon_2)^2 + (c\varepsilon_1)^2\big(2(s\theta_1)^2 + c\varepsilon_2\, S\theta_1\, s\theta_2\big) + \\ &+\big((s\varepsilon_2)^2 + (c\varepsilon_2)^2\,(s\theta_2)^2\big)(c\varepsilon_1\, c\theta_1)^2 + c\varepsilon_1\,(c\varepsilon_2)^2\, s\theta_1\, S\theta_2 + (c\varepsilon_2)^4\,(S\theta_2)^2\Big) \times \\ &\times \frac{\Big((s\varepsilon_1 + c\varepsilon_1\, c\theta_1\, s\varepsilon_2)^2 + \big(s\varepsilon_1 + \frac{c\theta_2\, S\varepsilon_2}{2}\big)^2 + (c\varepsilon_1)^2(s\theta_1 + c\varepsilon_2\, c\theta_1\, s\theta_2)^2 + \big(c\varepsilon_1\, s\theta_1 + \frac{(c\varepsilon_2)^2\, S\theta_2}{2}\big)^2\Big)}{(c\varepsilon_1\, c\theta_1 - c\varepsilon_2\, c\theta_2)^2\,(c\varepsilon_1\, s\varepsilon_2\, s\theta_1 - c\varepsilon_2\, s\varepsilon_1\, s\theta_2)^2}\Bigg]^{\frac{1}{2}}, \end{aligned} \tag{6.59}$$

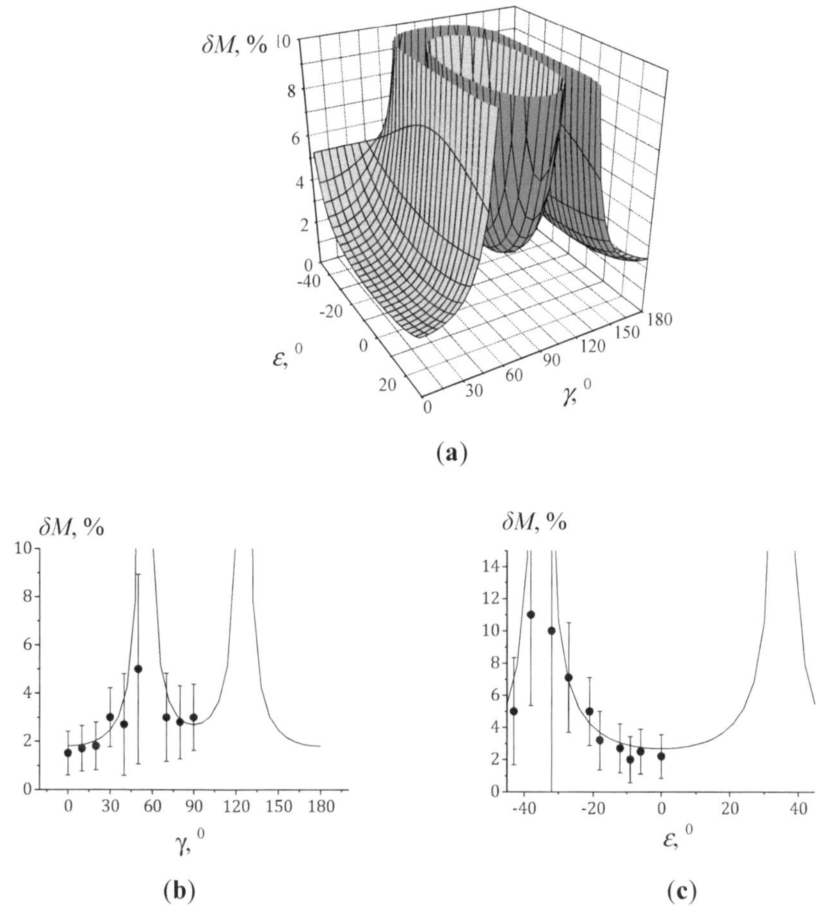

Fig. 6.7. (a) Dependence of the measurement errors on the option of input polarizations for the time-sequential measurement strategy; (b) and (c): the cross-sections of the dependence (a) by planes $\varepsilon = 0^0$ and $\gamma = 0^0$, correspondingly. The points are the experimental results.

where

$$
\begin{aligned}
s\varepsilon_i &= \sin(2\varepsilon_i); \quad c\varepsilon_i = \cos(2\varepsilon_i); \\
s\theta_i &= \sin(2\theta_i); \quad c\theta_i = \cos(2\theta_i); \\
S\varepsilon_i &= \sin(4\varepsilon_i); \quad S\theta_i = \sin(4\theta_i);
\end{aligned}
\tag{6.60}
$$

where $\varepsilon_{1,2}$ and $\theta_{1,2}$ are the ellipticities and azimuths of first and second input polarizations, respectively.

The minimal value of the function in Eq. (6.59) is $cond(\mathbf{V}) = 2$ with corresponding partial values $cond(\mathbf{V}_1)|_{\min} = cond(\mathbf{V}_2)|_{\min} = 3.18$. Note that the pair of values for ellipticity and azimuth providing this solution are not unique. Figure 6.8 illustrates this result.

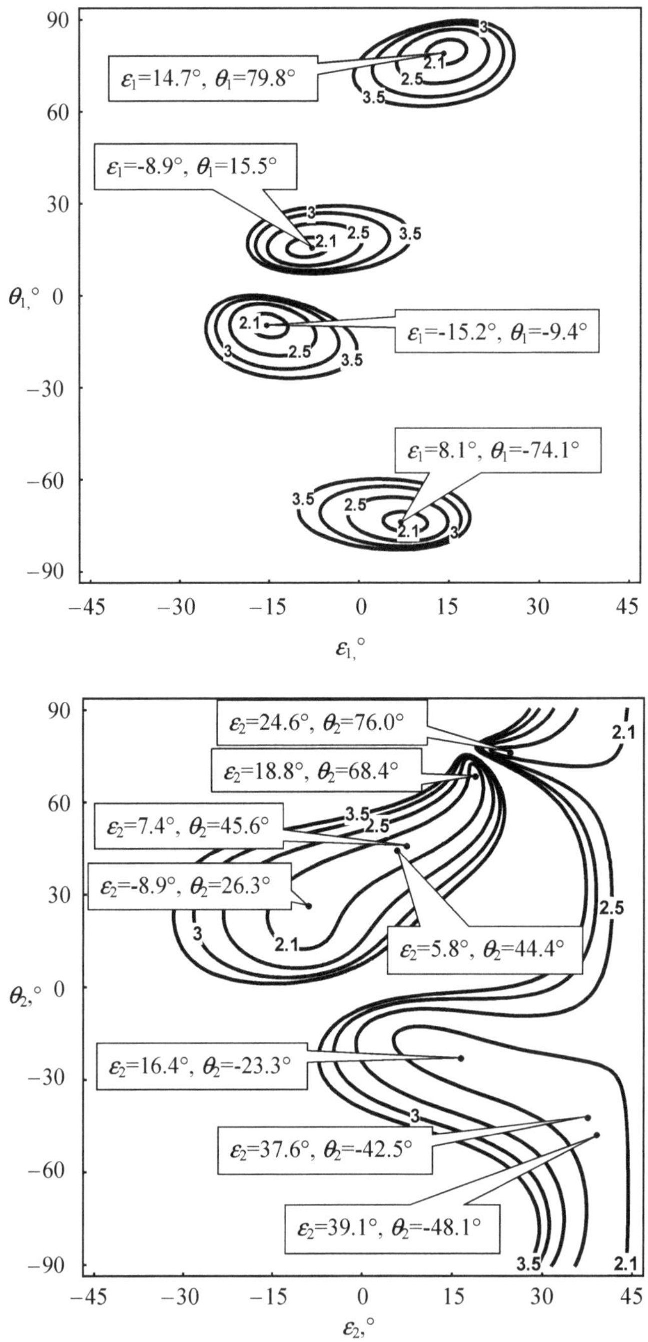

Fig. 6.8. Dependences of condition number $cond(\mathbf{V})$ on values of ellipticities $\varepsilon_{1,2}$ and azimuths $\theta_{1,2}$ for (a) $\varepsilon_2 = 37.6°$ and $\theta_2 = -42.5°$, and (b) $\varepsilon_1 = 14.7°$ and $\theta_1 = 79.8°$ of input polarizations.

Table 6.2. Condition numbers for some structures of incomplete Mueller matrices for the dynamic measurement strategy.

	Incomplete Mueller matrices	$cond(\mathbf{W})$	$\delta_1,^0$	$\delta_2,^0$	$\Delta\theta,^0$
1	$\begin{pmatrix} \bullet & \bullet & \bullet & \bullet \\ \bullet & m_{22} & m_{23} & m_{24} \\ \bullet & m_{32} & m_{33} & m_{34} \\ \bullet & m_{42} & m_{43} & m_{44} \end{pmatrix}$	9,0	116,5	116,5	90
2	$\begin{pmatrix} m_{11} & m_{12} & m_{13} & \bullet \\ m_{21} & m_{22} & m_{23} & \bullet \\ m_{31} & m_{32} & m_{33} & \bullet \\ \bullet & \bullet & \bullet & \bullet \end{pmatrix}$	9,54	180	180	90
3	$\begin{pmatrix} m_{11} & m_{12} & m_{13} & \bullet \\ m_{21} & m_{22} & m_{23} & \bullet \\ m_{31} & m_{32} & m_{33} & \bullet \\ m_{41} & m_{42} & m_{43} & \bullet \end{pmatrix}$	14,75	180	123,2	–
4	$\begin{pmatrix} \bullet & m_{12} & m_{13} & m_{14} \\ \bullet & m_{22} & m_{23} & m_{24} \\ \bullet & m_{32} & m_{33} & m_{34} \\ \bullet & m_{42} & m_{43} & m_{44} \end{pmatrix}$	15,52	125,2	119,9	90

Dependences presented in Fig. 6.8 correspond to the following cases: Fig. 6.8(a) $\varepsilon_2 = 37.6^\circ$ and $\theta_2 = -42.5^\circ$; Fig. 6.8(b) $\varepsilon_1 = 14.7^\circ$ and $\theta_1 = 79.8^\circ$. Explicit forms of the characteristic matrices in Eqs (6.45) and (6.46) for this pair of input polarizations are

$$\mathbf{V}_1 = \begin{pmatrix} 1 & -0.8164 \\ 1 & 0.0218 \end{pmatrix}, \tag{6.61}$$

$$\mathbf{V}_2 = \begin{pmatrix} 0.3043 & 0.4907 \\ -0.2539 & 0.9670 \end{pmatrix}. \tag{6.62}$$

Under some circumstances, the elements of a block-diagonal matrix satisfy the following conditions:

$$m_{12} = m_{21} \qquad m_{34} = -m_{43} \qquad m_{33} = m_{44}. \tag{6.63}$$

Such symmetry of the scattering matrix in Eq. (6.41) occurs either in the case of a spherically symmetrical scatterer or when nonspherical particles are randomly oriented and form a macroscopically isotropic and mirror-symmetric scattering medium (Mishchenko et al., 2006), for Rayleigh scattering by optically inactive particles with or without depolarization effects (van de Hulst, 1957) and some others. Symmetry in Eq. (6.63) implies that, for the elements m_{33} and m_{34}, the set of equations with characteristic matrix $\mathbf{V}_2$ in Eq. (6.46) is overdetermined. In this case, for measurement of the block-diagonal matrix, two linear input polarizations can be used. The optimal pair of input linear polarizations can be obtained by minimization of the condition number of the characteristic matrix $\mathbf{V}_1$ in Eq. (6.45) with the additive requirements that parameters l_3^1 and l_3^2 are equal and maximum to be simultaneously satisfied. This yields a pair of linear polarizations with azimuths 22.5^0 and 67.5^0.

If conditions in Eq. (6.63) become more rigid, namely

$$m_{12} = m_{21} = 0 \quad \text{and} \quad m_{34} = m_{43} = 0, \tag{6.64}$$

then, for measurement of the non-zero matrix elements, one input polarization is sufficient. Indeed, in this case, Eq. (6.18) takes the form

$$\begin{pmatrix} m_{11}r_1 \\ m_{22}r_2 \\ m_{33}r_3 \\ m_{44}r_4 \end{pmatrix} = \begin{pmatrix} s_1 \\ s_2 \\ s_3 \\ s_4 \end{pmatrix}. \tag{6.65}$$

Equation (6.65) describes the direct measurement of diagonal matrix elements m_{ii} using only one polarization of input light. Evidently, parameters of the Stokes vector of optimal input polarization have to satisfy

$$|r_2| \approx |r_3| \approx |r_4|. \tag{6.66}$$

Example matrices of this type are forward light scattering by a collection of randomly oriented identical particles, each of which has a plane of symmetry, or by a mixture of such collections (Hovenier and Mackowski, 1998), scattering by virus-infected and healthy wheat under normal and microgravity conditions (Savenkov et al., 2004), multiple scattering of light by spherical point-like diffusers (Rayleigh regime) (Bicout et al., 1994), Mueller matrix model of anisotropic depolarizer (Ossikovski, 2010b), etc.

Note thet all of the above optimal pairs of input polarizations (see Fig. 6.8), are equivalent. This results from the fact that we used integral estimations of errors of Stokes vector and Mueller matrix measurements of the form of Eq. (6.53). It is not the case when individual measurement errors for each matrix elements are taken into account.

Indeed, from the polarimetric measurement Eqs (6.15) and (6.18), it can be seen that the elements of the characteristic matrix W when measuring the block-diagonal scattering matrix are functions of the following quantities: ε_k, θ_k, ellipticities and azimuths of input polarizations. If these quantities are affected by uncorrelated errors $\Delta\varepsilon$, $\Delta\theta$, and s_i^k, measured parameters of the Stokes vectors, by ΔS, respectively (principally, in this consideration variance of the intensity of input radiation ΔI can also be included), the matrix elements are consequently affected by errors Δm_{ij}, given by

$$\Delta m_{ij} = \sqrt{\sum_{k=1}^{2}\sum_{i=1}^{4}\sum_{j=1}^{4}\left(\left(\frac{\partial m_{ij}}{\partial \varepsilon_k}\Delta\varepsilon\right)^2 + \left(\frac{\partial m_{ij}}{\partial \theta_k}\Delta\theta\right)^2 + \left(\frac{\partial m_{ij}}{\partial I_k}\Delta I\right)^2 + \sum_{r=1}^{4}\left(\frac{\partial m_{ij}}{\partial s_r^n}\Delta S\right)^2\right)}. \tag{6.67}$$

Taking as an example the scattering block-diagonal matrix for water droplets measured at 520 nm (scattering angle 150°; $r_g = 0.8$ μm; sigma = 1.5) (Munoz et al., 2010) and, for the sake of clarity, for the following values of variances $\Delta\varepsilon = 0.2°$, $\Delta\theta = 0.2°$, and $\Delta S = 0.002$, for four pairs of optimal input polarizations presented in Fig. 6.8(a) from Eq. (6.67), the average non-zero element errors $\overline{\Delta m_{ij}}$ are: (a) for $\varepsilon_1 = 14.7°$ and $\theta_1 = 79.8°$: $\overline{\Delta m_{ij}} = 0.0117$; (b) for $\varepsilon_1 = -8.9°$ and $\theta_1 = 15.5°$: $\overline{\Delta m_{ij}} = 0.0109$; (c) for $\varepsilon_1 = -15.2°$ and $\theta_1 = -9.4°$: $\overline{\Delta m_{ij}} = 0.0104$; (d) for $\varepsilon_1 = 8.1°$ and $\theta_1 = -74.1°$: $\overline{\Delta m_{ij}} = 0.0128$. So, the difference between the best and the worst cases of $\overline{\Delta m_{ij}}$ is about 25 %. It is substantial that, for other realizations of the block-diagonal matrix, the best pairs among optimal input polarizations can be different.

Note that, unlike the case of the two input polarizations mode of the time-sequential strategy measuring the two-column structures of incomplete Mueller matrices in Eq. (6.46), this is evidently another mode. The former assumes no a priori information on the Mueller matrix, whilst the latter assumes the block-diagonal structure of the Mueller matrix to be measured accurately.

Analysis of individual matrix element errors when measuring complete or incomplete Mueller matrices is an interesting and important subject for future research. Generally speaking, the polarimetric measurement in Eq. (6.15) shows that errors of Mueller matrix measurements depend also on anisotropy properties of the studied object (Savenkov and Yushtin, 2003). For example, the dependencies of Mueller matrix measurement errors on values of anisotropy parameters for the time-sequential measurement strategy providing that PSA is a complete Stokes polarimeter have the forms shown in Fig. 6.9.

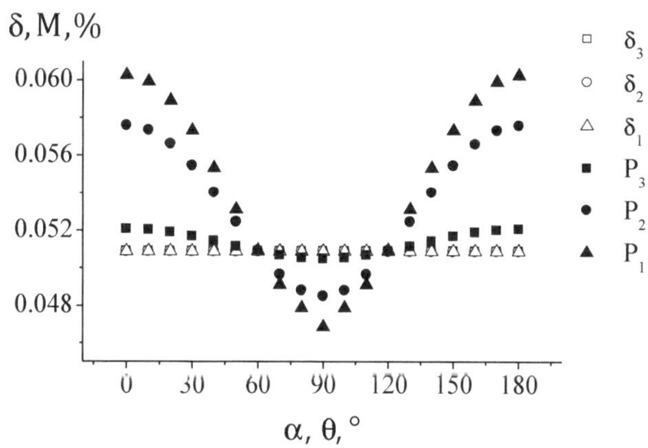

Fig. 6.9. Dependencies of Mueller matrix measurement errors on values of linear phase and amplitude anisotropy for the time-sequential measurement strategy.

In this experiment, the Mueller matrices of the following objects are measured: dichroic with values of linear amplitude anisotropy $P_1 = 0$; $P_2 = 0.3$; $P_3 = 0.7$; and birefringent with values of linear phase anisotropy $\delta_1 = 20°$; $\delta_2 = 50°$; $\delta_3 = 90°$ (see Appendix A). Azimuths of anisotropy are altering in the range from 0° to 180°.

In particular, these results make clear the fact that is observed in practice: the matter is that calibration of this type of polarimeters by measuring the Mueller matrices of, say, 'empty space' only, which is the most widespread calibration option (Tyo et al., 2010), does not ensure the acceptable accuracy when measuring the dichroic objects.

6.7 Conclusions

In this chapter, we have discussed a number of important developments in Mueller matrix polarimetry. We have intended to survey the literature that covers these subjects, but our reference list should by no means be considered exhaustive. This is only an attempt to refer the interested reader to the initial guiding lines in further reading.

How can the above results be exploited for arrangement and optimization of the polarimeter?

Undoubtedly, one can arrange a polarimeter capable of measuring one or another incomplete Mueller matrix. Such an approach, assuming measurement of only a particular type of anisotropy or some group of matrix elements, is quite common in remote sensing and polarimetry applications. The reason for that is, in this case, the design of the polarimeter becomes potentially simpler. However, in doing so, one has to be convinced that the measured incomplete Mueller matrix meets the inverse problem under consideration, that it is indeed a pure Mueller matrix or of block-diagonal structure, for example.

However, since, in practice, one encounters the situation when the Mueller matrices of the object are unknown, it would be reasonable to have the possibility to accomplish firstly the general calibration measurements of the complete Mueller matrix and only afterwards to choose the measurement mode in the framework of any of the measurement strategies which correspond best with the scattering scene under consideration. Thus, realizing this approach, polarimeter adapts ('teaches' itself) to the given experimental scene and, thereby, gains in intellectuality. Because of that, this approach could be named the adaptive Mueller polarimetry.

Appendix A: Some multiplicative and additive Mueller matrix models

In this section, we consider the problem of Mueller matrix interpretation in the framework of the approach wherein the medium studied is modeled as a medium with a continuous (and possibly random) distribution of optical parameters. The polarization of light changes if the amplitudes and phases of the components of the electric vector **E** change separately or simultaneously (Shurcliff, 1962; Azzam and Bashara, 1977; Brosseau, 1998). It is, therefore, customary to distinguish between the corresponding classes of anisotropic media: dichroic (or possessing amplitude anisotropy), influencing only the amplitudes; birefringent (or possessing phase anisotropy), influencing only the phases; and 'all other' (possessing both amplitude and phase anisotropy) affecting both the amplitudes and the phases of the components of the electric field vector. Among these classes, four types of anisotropic mechanisms are recognized as basic or, after Jones, elementary (Jones, 1941, 1942, 1947, 1956; Hurwitz and Jones, 1941): linear and circular phase and linear and circular amplitude anisotropies.

Linear birefringence is described by the following pure Mueller matrix:

$$\mathbf{M}^{LP} = \begin{pmatrix} 1 & 0 & 0 & 0 \\ 0 & \cos^2 2\alpha + \sin^2 2\alpha \cos\delta & \cos 2\alpha \sin 2\alpha\,(1-\cos\delta) & -\sin 2\alpha \sin\delta \\ 0 & \cos 2\alpha \sin 2\alpha\,(1-\cos\delta) & \sin^2 2\alpha + \cos^2 2\alpha \cos\delta & \cos 2\alpha \sin\delta \\ 0 & \sin 2\alpha \sin\delta & -\cos 2\alpha \sin\delta & \cos\delta \end{pmatrix} \tag{6.A1}$$

where δ is the phase shift between two orthogonal linear components of the electric field vector and α is the azimuth of the anisotropy.

The Mueller matrix describing linear dichroism is

$$\mathbf{M}^{LA} = \begin{pmatrix} 1+P & (1-P)\cos 2\theta & (1-P)\sin 2\theta & 0 \\ (1-P)\cos 2\theta & \cos^2 2\theta\,(1+P) + 2\sin^2 2\theta\sqrt{P} & \cos 2\theta \sin 2\theta\,(1-\sqrt{P})^2 & 0 \\ (1-P)\sin 2\theta & \cos 2\theta \sin 2\theta\,(1-\sqrt{P})^2 & \sin^2 2\theta\,(1+P) + 2\cos^2 2\theta\sqrt{P} & 0 \\ 0 & 0 & 0 & 2\sqrt{P} \end{pmatrix}, \tag{6.A2}$$

where P is the relative absorption of two linear orthogonal components of the electric vector and θ is the azimuth of the anisotropy.

The Mueller matrix describing circular birefringence is

$$\mathbf{M}^{CP} = \begin{pmatrix} 1 & 0 & 0 & 0 \\ 0 & \cos 2\varphi & \sin 2\varphi & 0 \\ 0 & -\sin 2\varphi & \cos 2\varphi & 0 \\ 0 & 0 & 0 & 1 \end{pmatrix}, \tag{6.A3}$$

where φ is the induced phase shift between two orthogonal circular components of the electric vector.

Finally, in terms of the Mueller matrix calculus, circular amplitude anisotropy is described by the following matrix:

$$\mathbf{M}^{CA} = \begin{pmatrix} 1+R^2 & 0 & 0 & 2R \\ 0 & 1-R^2 & 0 & 0 \\ 0 & 0 & 1-R^2 & 0 \\ 2R & 0 & 0 & 1+R^2 \end{pmatrix}, \tag{6.A4}$$

where R is the magnitude of anisotropy, namely the relative absorption of two orthogonal circular components of the electric vector. The six quantities α, Δ, P, γ, φ, and R are called anisotropy parameters.

It can be seen that the matrices describing linear and circular birefringence belong to the class of unitary matrices (in the case of matrices with real-valued elements—orthogonal matrices). The matrices of linear (Eq. (6.A2)) and circular (Eq. (6.A4)) dichroism belong to the class of Hermitian matrices (in the case of matrices with real-valued elements—symmetric matrices).

The Mueller matrices of Eqs (6.A1)–(6.A4) represent media exhibiting individual types of anisotropy. Experimental measurements of these matrices or of the corresponding informative matrix elements allow one to interpret and characterize anisotropy properties of media. However, more often, two or more types of anisotropy are exhibited by a medium simultaneously. Evidently, such cases require the development of more sophisticated polarimetric matrix models (Hurwitz and Jones, 1941; Cloude, 1986; Gil and Bernabeu, 1987; Lu and Chipman, 1996; Savenkov et al., 2006; Ossikovski, 2008, 2009).

The matrix model that is studied (Morio and Goudail, 2004; Ossikovski et al., 2007; Ghosh et al., 2010) and used most extensively in optical polarimetry for decoupling constituent polarization properties of optical medium (Chung et al., 2007; Ghosh et al., 2008; Wood et al., 2009; Li and Yao, 2009) is the polar decomposition proposed by Lu and Chipman (1996). This model is based on the so-called polar decomposition theorem (Lancaster and Tismenetsky, 1985), according to which an arbitrary matrix $\mathbf{M}$ can be represented by a product

$$\mathbf{M} = \mathbf{M}_P \mathbf{M}_R \quad \text{or} \quad \mathbf{M} = \mathbf{M}_R \mathbf{M}'_P, \tag{6.A5}$$

where $\mathbf{M}_P$ and $\mathbf{M}'_P$ are Hermitian matrices and $\mathbf{M}_R$ is a unitary one. The Hermitian matrix is associated with amplitude anisotropy, while the unitary matrix describes phase anisotropy (Whitney, 1971). The matrices $\mathbf{M}_P$ and $\mathbf{M}_R$ are called the dichroic and the phase polar forms (Whitney, 1971; Gil and Bernabeu, 1987; Lu and Chipman, 1996).

The polar decomposition was first employed by Whitney (1971) without finding explicit expressions for $\mathbf{M}_P$ and $\mathbf{M}_R$. They were proposed later, independently by Gil and Bernabeu (1987) and Lu and Chipman (1996). Alternatively, the dichroic and phase polar forms can be derived using spectral methods of linear algebra (Azzam and Bashara, 1977).

The phase polar form $\mathbf{M}_R$ (using notation from Lu and Chipman (1996)) is given by

$$\begin{gathered}\mathbf{M}_R = \begin{pmatrix} 1 & \vec{0}^T \\ \vec{0} & \mathbf{m}_R \end{pmatrix}, \\ (m_R)_{ij} = \delta_{ij} \cos R + a_i a_j (1 - \cos R) + \textstyle\sum_{k=1}^{3} \varepsilon_{ijk} a_k \sin R,\end{gathered} \tag{6.A6}$$

where $\vec{0}$ is the 3×1 zero vector; $[\,1 \;\; a_1 \;\; a_2 \;\; a_3\,]^T = [\,1 \;\; \widehat{R}^T\,]^T$ is the normalized Stokes vector for the fast axis of $\mathbf{M}_R$; δ_{ij} is the Kronecker delta; ε_{ijk} is the Levi–Civita permutation symbol; m_R is the 3×3 submatrix of $\mathbf{M}_R$ obtained by striking out the first row and the first column of $\mathbf{M}_R$; and R is the birefringence given by

$$R = \arccos\left(\frac{1}{2} \operatorname{Tr} \mathbf{M}_R - 1\right). \tag{6.A7}$$

The dichroic polar form $\mathbf{M}_P$ is as follows:

$$\begin{gathered}\mathbf{M}_P = T_u \begin{pmatrix} 1 & \vec{\mathbf{D}}^T \\ \vec{\mathbf{D}} & \mathbf{m}_P \end{pmatrix}, \\ \mathbf{m}_P = \sqrt{1 - D^2}\, \mathbf{I} + \left(1 - \sqrt{1 - D^2}\right) \widehat{\mathbf{D}} \widehat{\mathbf{D}},\end{gathered} \tag{6.A8}$$

where $\mathbf{I}$ is the 3×3 identity matrix; $\widehat{\mathbf{D}} = \vec{\mathbf{D}}/|\vec{\mathbf{D}}|$ is the unit vector in the direction of the diattenuation vector $\vec{\mathbf{D}}$; T_u is the transmittance for unpolarized light; and the value of diattenuation can be obtained as

$$D = \{1 - 4|\det(\mathbf{T})|^2/[\mathrm{Tr}(\mathbf{T}^*\mathbf{T})]^2\}^{1/2}. \tag{6.A9}$$

The models of anisotropic media based on the polar decomposition contain six independent parameters: three for the phase polar form $\mathbf{M}_R$ and three for the dichroic polar form $\mathbf{M}_P$. It can be seen that the phase polar form is a unitary (orthogonal) matrix and the dichroic polar form is a Hermitian (symmetric) matrix. Note that unitarity (orthogonality) of the phase polar form (Eq. (6.A6)), is in complete agreement with the first Jones equivalence theorem (Hurwitz and Jones, 1941), and is a general model of elliptically birefringent media. The situation with the dichroic polar form is more complex (Savenkov et al., 2005, 2007b). Mathematically, the complexity originates from the fact that, in contrast to unitary matrices, the product of Hermitian matrices is generally not a Hermitian matrix (Lancaster and Tismenetsky, 1985).

The characterization of depolarization is of considerable importance owing to the fact that depolarization phenomena are encountered in many theoretical and experimental applications of polarimetry to discrete random media and media with bulk and surface inhomogeneities. Note that the light–medium interaction with depolarization is heretofore studied in considerably less detail than the problem described by pure Mueller matrices discussed above.

Depolarization properties of the object under consideration can be characterrized by depolarization metrics providing a single scalar number that varies from zero, thereby corresponding to a totally depolarized output light, to a certain positive number corresponding to a totally polarized output light. All intermediate values are associated with partial polarization.

The first depolarization metric, the depolarization index, was apparently suggested by Gil and Bernabeu (1985, 1986):

$$DI(\mathbf{M}) = \sqrt{\sum_{i,j=1}^{4} m_{ij}^2 - m_{11}^2} \Big/ (\sqrt{3}m_{11}). \tag{6.A10}$$

The depolarization index is bounded according to $0 \leq DI(\mathbf{M}) \leq 1$. The extreme values of $DI(M)$ correspond to the case of unpolarized and totally polarized output light, respectively.

The average degree of polarization was defined by Chipman (2005) as

$$Average\ DoP(\mathbf{M}) = \frac{1}{4\pi} \int_0^{\pi} \int_{-\pi/2}^{\pi/2} p\left[\mathbf{MS}(\varepsilon, \zeta)\right] \cos\varepsilon \, d\varepsilon \, d\zeta. \tag{6.A11}$$

The term $\cos\varepsilon\, d\varepsilon\, d\zeta$ scans the incident polarization state over the Poincaré sphere, with the latitude ε and longitude ζ. The Stokes vector $S(\varepsilon, \zeta)$ is a function of ellipticity and orientation azimuth of the polarization ellipse of light:

$$\mathbf{S}(\varepsilon, \zeta) = \begin{pmatrix} 1 & \cos 2\varepsilon \cos 2\zeta & \cos 2\varepsilon \sin 2\zeta & \sin 2\varepsilon \end{pmatrix}^T. \tag{6.A12}$$

The so-called $Q(\mathbf{M})$ metrics is defined as (Espinosa-Luna and Bernabeu, 2007)

$$Q(\mathbf{M}) = \sum_{i=2}^{4}\sum_{j=1}^{4} m_{ij}^2 \Big/ \sum_{j=1}^{4} m_{1j}^2$$
$$= \left(3\,[DI(\mathbf{M})]^2 - [D(\mathbf{M})]^2\right) \Big/ \left(1 + [D(\mathbf{M})]^2\right), \tag{6.A13}$$

where $D(\mathbf{M}) = (m_{12}^2 + m_{13}^2 + m_{14}^2)^{1\ 2}$ is the diattenuation parameter and $0 \leq D(\mathbf{M}) \leq 1$. The metric $Q(\mathbf{M})$ is bounded according to $0 \leq Q(M) \leq 3$. Specifically, $Q(M) = 0$ corresponds to a totally depolarizing medium; $0 < Q(M) < 1$ describes a partially depolarizing medium; $1 \leq Q(M) < 3$ represents a partially depolarizing medium if, in addition, $0 < DI(M) < 1$; otherwise, it represents a non-depolarizing diattenuating medium; finally, $Q(M) = 3$ for a non-depolarizing non-diattenuating medium.

Thus, the depolarization metrics provide a summary of the depolarizing property of a medium via a single number. The depolarization index $DI(\mathbf{M})$ and the $Q(\mathbf{M})$ metrics are directly related to the Mueller matrix elements only and, in contrast to the average degree of polarization *Average DoP*, require no scan of the whole Poincaré sphere of the input polarizations. Furthermore, $Q(\mathbf{M})$ provides more detailed information about depolarization properties of a medium.

Some media depolarize all polarization states equally. Other depolarizing media partially depolarize most polarization states but may not depolarize one or some incident states. Depolarization depends significantly on the polarization state of the input light in the multiple-scattering regime (Bicout et al., 1994; Rojas-Ochoa et al., 2004; Kim et al., 2006, and references therein). In particular, Bicout et al. (1994) studied numerically and experimentally how depolarization evolves for linear and circular input polarizations as the size of the particles increases from very small (Rayleigh regime) to large (Mie regime) in the case of a forward-scattering geometry.

Below in this section, in the scope of the additive matrix model of the arbitrary object based on Cloude's coherency matrix the depolarization metric called entropy is presented (see Eq. (6.A24)).

One more group of depolarization metrics are introduced in Ossikovski (2010a). This is the first and second Lorentz depolarization index

$$L1 = \sqrt{\frac{tr\,(\mathbf{N}) - \rho_{\max}\,(\mathbf{N})}{3\rho_{\max}\,(\mathbf{N})}}, \tag{6.A14}$$

$$L2 = \sqrt{\frac{4tr\,(\mathbf{N}^2) - tr^2\,(\mathbf{N})}{3tr^2\,(\mathbf{N})}}, \tag{6.A15}$$

respectively, and Lorentz entropy

$$HL = -\sum_{i=1}^{4} \rho_i \log_4 \rho_i, \tag{6.A16}$$

where $tr\,(\mathbf{N})$, ρ_i and $\rho_{\max}\,(\mathbf{N})$ are trace, eigenvalues, and maximal eigenvalue of the matrix $\mathbf{N} = \mathbf{G}\mathbf{M}^T\mathbf{G}\mathbf{M}$ associated with studied Mueller matrix $\mathbf{M}$ and $\mathbf{G} = diag\,(1,\ -1,\ -1,\ -1)$ is the Minkowski metric.

The first Lorentz depolarization index $L1$ is equal to 1 for a pure $\mathbf{M}$ and to less than 1 otherwise; it will be equal to zero for the ideal depolarizer $\mathbf{M}^{ID}$ in Eq. (6.23). The second Lorentz depolarization index $L2$ will equal zero for a pure $\mathbf{M}$, 1 for $\mathbf{M}^{ID}$, and take intermediate values otherwise. The entropy HL is equal to 1 for a pure $\mathbf{M}$ and less than 1 for depolarizing $\mathbf{M}$.

Note that each single-number depolarization metric providing a summary of depolarization by a medium cannot generally give detailed information about all features of depolarization. Because of that, it would be reasonable to test the capability of each existing depolarization metric for a given Mueller matrix (Savenkov et al., 2012). Furthermore, in a specific range of cases, some of the depolarization criteria can even incorrectly interpret experimental polarimetric data (Ossikovski, 2010a).

Detailed information about all features of depolarization information can only be obtained from Mueller matrix models of depolarization. The case when, for all polarizations of the input light, the degree of polarization p of the output light is the same is called isotropic depolarization. When the degree of polarization of the output light is a function of parameters of the input polarization, one speaks of anisotropic depolarization.

The most general expression for the Mueller matrix describing depolarization was suggested by Lu and Chipman (1996):

$$\begin{pmatrix} 1 & \vec{\mathbf{O}}^T \\ \vec{\mathbf{P}}_\Delta & \mathbf{m}_\Delta \end{pmatrix}, \quad \mathbf{m}_\Delta^T = \mathbf{m}_\Delta, \tag{6.A17}$$

where $\vec{\mathbf{P}}_\Delta$ denotes the so-called polarizance vector. The polarizance vector describes the state of polarization generated by this Mueller matrix from unpolarized incident light. The Mueller matrix of Eq. (6.A17) has nine degrees of freedom, and this is of interest because this matrix along with a generalized deterministic Mueller matrix are jointly characterized by 16 degrees of freedom. This means that, in this way, one obtains the generalized Mueller matrix of an arbitrary medium that has 16 degrees of freedom and linearly interacts with polarized light.

The product of Mueller matrices of the polar forms in Eqs (6.A6) and (6.A8) and the depolarizing Mueller matrix in Eq. (6.A17)

$$\mathbf{M} = \mathbf{M}_\Delta \mathbf{M}_R \mathbf{M}_P \tag{6.A18}$$

is the generalized polar decomposition and a multiplicative matrix model of an arbitrary Mueller matrix (Lu and Chipman, 1996; Gil, 2000, 2007).

The product of the phase polar form and the depolarizing matrices can then be obtained as

$$\mathbf{M}_\Delta \mathbf{M}_R = \mathbf{M}' = \mathbf{M}\mathbf{M}_P^{-1}. \tag{6.A19}$$

Then

$$\vec{\mathbf{P}}_\Delta = (\vec{\mathbf{P}} - \mathbf{m}\vec{\mathbf{D}})/(1 - D^2), \tag{6.A20}$$

where $\vec{\mathbf{P}} = (1/m_{11})\,[\, m_{21} \;\; m_{31} \;\; m_{41} \,]^T$ and $\mathbf{m}$ is the submatrix of the initial matrix $\mathbf{M}$. The $\mathbf{m}'$ is the submatrix of $\mathbf{M}'$ and can be written as

$$\mathbf{m}' = \mathbf{m}_\Delta \mathbf{m}_R. \tag{6.A21}$$

The submatrix $\mathbf{m}_\Delta$ can be calculated as

$$\mathbf{m}_\Delta = \pm\left[\mathbf{m}'(\mathbf{m}')^T + (\sqrt{\lambda_1\lambda_2} + \sqrt{\lambda_2\lambda_3} + \sqrt{\lambda_1\lambda_3})\,\mathbf{I}\right]^{-1} \times\left[(\sqrt{\lambda_1} + \sqrt{\lambda_2} + \sqrt{\lambda_3})\,\mathbf{m}'(\mathbf{m}')^T + \sqrt{\lambda_1\lambda_2\lambda_3}\,\mathbf{I}\right], \tag{6.A22}$$

where λ_i are the eigenvalues of $\mathbf{m}'(\mathbf{m}')^T$. The sign '+' or '−' is determined by the sign of the determinant of $\mathbf{m}'$. The net depolarization coefficient Δ can be calculated according to

$$\Delta = 1 - \frac{1}{3}\left|\,\mathrm{Tr}\,(\mathbf{M}_\Delta) - 1\,\right|. \tag{6.A23}$$

In conclusion of this section, we consider the additive Mueller matrix model of the depolarizing object suggested by Cloude (1986) and extensively employed in optical and radar polarimetry (see, e.g. Savenkov et al., 2003, 2004; Munoz et al., 2001, 2002, 2004, 2007; Volten et al., 2001; Cloude and Pottier, 1995, 1996, 1997). The Cloude coherence matrix J is derived from the corresponding arbitrary Mueller matrix as

$$\begin{array}{ll}
J_{11}=(1/4)(m_{11}+m_{22}+m_{33}+m_{44}) & J_{12}=(1/4)(m_{12}+m_{21}-im_{34}+im_{43})\\
J_{13}=(1/4)(m_{13}+m_{31}+im_{24}-im_{42}) & J_{14}=(1/4)(m_{14}-im_{23}+im_{32}+m_{41})\\
J_{21}=(1/4)(m_{12}+m_{21}+im_{34}-im_{43}) & J_{22}=(1/4)(m_{11}+m_{22}-m_{33}-m_{44})\\
J_{23}=(1/4)(im_{14}+m_{23}+m_{32}-im_{41}) & J_{24}=(1/4)(-im_{13}+im_{31}+m_{24}+m_{42})\\
J_{31}=(1/4)(m_{13}+m_{31}-im_{24}+im_{42}) & J_{32}=(1/4)(-im_{14}+m_{23}+m_{32}+im_{41})\\
J_{33}=(1/4)(m_{11}-m_{22}+m_{33}-m_{44}) & J_{34}=(1/4)(im_{12}-im_{21}+m_{34}+m_{43})\\
J_{41}=(1/4)(m_{14}+im_{23}-im_{32}+m_{41}) & J_{42}=(1/4)(im_{13}-im_{31}+m_{24}+m_{42})\\
J_{43}=(1/4)(-im_{12}+im_{21}+m_{34}+m_{43}) & J_{44}=(1/4)(m_{11}-m_{22}-m_{33}+m_{44})
\end{array} \tag{6.A24}$$

It can be seen that coherence matrix J is positive semidefinite Hermitian and, hence, always has four real eigenvalues. The eigenvalues of the coherence matrix λ_i can be combined to form a quantity that is a measure of the depolarization, depolarization metric, of the studied medium. This quantity is called entropy and is defined as

$$H = -\sum_{i=1}^{N}\left(\frac{\lambda_i}{\sum_j \lambda_j}\right)\log_N\left(\frac{\lambda_i}{\sum_j \lambda_j}\right) \tag{6.A25}$$

Given eigenvalues λ_i of coherence matrix J, we have for the initial Mueller matrix

$$\mathbf{M} = \sum_{k=1}^{4}\lambda_k \mathbf{M}_D^k; \qquad \mathbf{M}_D^k \Leftrightarrow \mathbf{T}^k, \tag{6.A26}$$

where $\mathbf{M}_D^k$ are the pure Mueller matrices obtained from the Jones matrices by Eq. (6.5).

The Jones matrix $\mathbf{T}$, in turn, is obtained in the following manner:

$$\begin{array}{ll}
t_{11}^{(k)} = \Psi_1^{(k)} + \Psi_2^{(k)}, & t_{12}^{(k)} = \Psi_3^{(k)} - i\Psi_4^{(k)} \\
t_{21}^{(k)} = \Psi_3^{(k)} + i\Psi_4^{(k)}, & t_{22}^{(k)} = \Psi_1^{(k)} - \Psi_2^{(k)} \qquad k = \overline{1,4}
\end{array}, \tag{6.A27}$$

where $\mathbf{\Psi}^{(k)} = \begin{pmatrix}\Psi_1 & \Psi_2 & \Psi_3 & \Psi_4\end{pmatrix}_k^T$ is k-th eigenvector of coherence matrix $\mathbf{J}$.

Thus, the substance of the Cloude's coherency matrix concept, which, in essence, is an additive matrix model of depolarizing Mueller matrix in Eq. (6.A26), is the representation of the initial depolarizing Mueller matrix as a weighted convex sum of four pure Mueller matrices.

If three of the eigenvalues of $\mathbf{J}$ vanish, then the entropy $H = 0$ and initial matrix $\mathbf{M}$ is a deterministic Mueller–Jones matrix. If all four eigenvalues of J are not equal to zero and $H \leq 0.5$, then the pure Mueller matrix, which corresponds to the maximal eigenvalue, is the dominant type of deterministic polarization transformation of the studied object. So, this model gives the possibility to study the anisotropy properties of depolarizing objects on the one hand and, on the other hand, is a necessary and sufficient criterion for a given 4×4 real matrix to be a Mueller matrix (the case when all four eigenvalues of J are non-negative) and a pure Mueller matrix (the case when three of the eigenvalues vanish) (Munoz et al., 2001, 2002, 2004; Volten et al., 2001).

References

Abhyankar, K.D. and Fymat, A.L. (1969) 'Relations between the elements of the phase matrix for scattering', *J. Math. Phys.*, **10**, 1935–8.

Ahmad, J.E. and Takakura, Y. (2008) 'Estimation of physically realizable Mueller matrices from experiments using global constrained optimization', *Opt. Exp.*, **16**, 14274–87.

Aiello, A. and Woerdman, J.P. (2005) 'Physical bounds to the entropy–depolarization relation in random light scattering', *Phys. Rev. Lett.*, **94**, 090406.

Aiello, A., Puentes, G., Voigt, D. and Woerdman, J.P. (2006) 'Maximum-likelihood estimation of Mueller matrices', *Optics Letters*, **31**, 817–19.

Anderson, D.G.M. and Barakat, R. (1994) 'Necessary and sufficient conditions for a Mueller matrix to be derivable from a Jones matrix', *J. Opt. Soc. Am. A.*, **11**, 2305–19.

Antonelli, M.-R., Pierangelo, A., Novikova, T., et al. (2010) 'Mueller matrix imaging of human colon tissue for cancer diagnostics: How Monte Carlo modeling can help in the interpretation of experimental data', *Opt. Express*, **18**, 10200–8.

Azzam, R.M. (1997) 'Mueller-matrix ellipsometry: A review', *Proc. SPIE*, **3121**, 396–405.

Azzam, R.M. and Bashara, N.M. (1977) *Ellipsometry and Polarized Light*, New York, North-Holland.

Azzam, R.M. and Bashara, N.M. (1987) *Ellipsometry and Polarized Light*, New York, North-Holland.

Azzam, R.M., Elminyawi, I.M. and El-Saba, A.M. (1988) 'General analysis and optimization of the four-detector photopolarimeter', *J. Opt. Soc. Am. A*, **5**, 681–9.

Barakat, R. (1963) 'Theory of the coherency matrix for light of arbitrary spectral bandwidth', *J. Opt. Soc. Am.*, **53**, 317–22.

Barakat, R. (1987) 'Conditions for the physical realizability of polarization matrices characterizing passive systems', *J. Mod. Opt.*, **34**, 1535–44.

Benoit, A.-M., Naoun, O.K., Louis-Dorr, V., et al. (2001) 'Linear dichroism of the retinal nerve fiber layer expressed with Mueller matrices', *Appl. Opt.*, **40**, 565–9.

Berry, M.V. and Dennis, M.R. (2003) 'The optical singularities of birefringent dichroic chiral crystals', *Proc. R. Soc. London, Ser. A.*, **459**, 1261–92.

Bickel, W.S., Davidson, J.F., Huffman, D.R. and Kilkson, R. (1976) 'Application of polarization effects in light scattering: A new biophysical tool', *Proc. Natl. Acad. Sci. USA*, **73**, 486–90.

Bicout, D., Brosseau, C., Martinez, A.S. and Schmitt, J.M. (1994) 'Depolarization of multiply scattered waves by spherical diffusers: Influence of the size parameter', *Phys. Rev. E.*, **49**, 1767–70.

Bjerhammar, A. (1951) 'Application of calculus of matrices to method of least squares, with special references to geodetic calculations', *Trans. Inst. Technol. (Stockholm)*, **49**, 1–86.

Boerner, W.-M. (1992) *Direct and Inverse Methods in Radar Polarimetry*, Dordrecht, Kluwer.

Bohren, C.F. and Huffman, E.R. (1983) *Absorption and Scattering of Light by Small Particles*, New York, Wiley.

Broch, L. and Johann, L. (2008) 'Optimizing precision of rotating compensator ellipsometry', *Phys. Status Solidi*, **C5**, 1036–40.

Brosseau, C. (1990) 'Analysis of experimental data for Mueller polarization matrices', *Optik*, **85**, 83–6.

Brosseau, C. (1998) *Fundamentals of Polarized Light: A Statistical Optics Approach*, New York, Wiley.

Brosseau, C., Givens, C.R. and Kostinski, A.B. (1993) 'Generalized trace condition on the Mueller–Jones polarization matrix', *J. Opt. Soc. Am. A.*, **10**, 2248–51.

Bueno, J.M. (2001) 'Indices of linear polarization for an optical system', *J. Opt. A: Pure Appl. Opt.*, **3**, 470–6.

Carrieri, A.H., Copper, J., Owens, D.J., Roese, E.S., Bottiger, J.R., Everly, R.D. and Hung, K.C. (2010) 'Infrared differential-absorption Mueller matrix spectroscopy and neural network-based data fusion for biological aerosol standoff detection', *Appl. Opt.*, **49**, 382–93.

Chipman, R.A. (1995) 'Polarimetry', in Chipman, R.A. (ed.), *Handbook of Optics*, Vol. II, New York, McGraw Hill.

Chipman, R.A. (1999) 'Depolarization', *Proc. SPIE*, **3754**, 14–20.

Chipman, R.A. (2005) 'Depolarization index and the average degree of polarization', *Appl. Opt.*, **44**, 2490–5.

Chung, J., Jung, W., Hammer-Wilson, M.J., et al. (2007) 'Use of polar decomposition for the diagnosis of oral precancer', *Appl. Opt.*, **46**, 3038–45.

Cloude, S.R. (1986) 'Group theory and polarization algebra', *Optik*, **7**, 26–36.

Cloude, S.R. (1990) 'Conditions for the physical realisability of matrix operators in polarimetry', *Proc. SPIE*, **1166**, 177–87.

Cloude, S.R. (1997) 'Physical interpretation of eigenvalue problems in optical scattering polarimetry', *Proc. SPIE*, **3121**, 88–99.

Cloude, S.R. (2010) *Polarization: Applications in Remote Sensing*, New York, Oxford University Press.

Cloude, S.R. and Pottier, E. (1995) 'Concept of polarization entropy in optical scattering', *Opt. Eng.*, **34**, 1599–610.

Cloude, S.R. and Pottier, E. (1996) 'A review of target decomposition theorems in radar polarimetry', *IEEE Trans. Geosci. Remote Sensing*, **34**, 498–518.

Cloude, S.R. and Pottier, E. (1997) 'An entropy based classification scheme for land applications of polarimetric SAR', *IEEE Transactions on Geoscience and Remote Sensing*, **35**, 68–78.

Collett, E. (1993) *Polarized Light: Fundamentals and Applications*, New York, Marcel Decker.

Collins, R.W. and Koh, J. (1999) 'Dual rotating-compensator multichannel ellipsometer: Instrument design for real-time Mueller matrix spectroscopy of surfaces and films', *J. Opt. Soc. Am. A*, **16**, 1997–2006.

de Boer, J.F., Milner, T.E., van Gemert, M.J.V. and Nelson, J.S. (1997) 'Two dimensional birefringence imaging in biological tissue by polarization-sensitive optical coherence tomography', *Opt. Lett.*, **22**, 934–6.

De Martino, A., Garcia-Caurel, E., Laude, B. and Drevillon, B. (2003) 'Optimized Mueller polarimeter with liquid crystals', *Opt. Lett.*, **28**, 616–18.

De Martino, A., Garcia-Caurel, E., Laude, B. and Drevillon, B. (2004) 'General methods for optimized design and calibration of Mueller polarimeters', *Thin Solid Films*, **455–456**, 112–19.

DeBoo, B.J., Sasian, J.M. and Chipman, R.A. (2005) 'Depolarization of diffusely reflecting man-made objects', *Appl. Opt.*, **44**, 5434–45.

Deibler, L.L. and Smith, M.H. (2001) 'Measurement of the complex refractive index of isotropic materials with Mueller matrix polarimetry', *Appl. Opt.*, **40**, 3659–67.

Dubois, P.C. and Norikane, I. (1987) 'Data volume reduction for imaging radar polarimetry', *Proc. IGARSS*, 691–6.

Espinosa-Luna, R. and Bernabeu, E. (2007) 'On the Q(M) depolarization metric', *Opt. Commun.*, **277**, 256–8.

Fry, E.S. and Kattawar, G.W. (1981) 'Relationships between elements of the Stokes matrix', *Appl. Opt.*, **20**, 2811–14.

Garcia-Caurel, E., De Martina, A. and Drevillon, B. (2004) 'Spectroscopic Mueller polarimeter based on liquid crystal devices', *Thin Solid Films*, **455–456**, 120–3.

Gerrard, A. and Burch, J.M. (1975) *Introduction to Matrix Methods in Optics*, New York, Wiley.

Ghosh, N., Wood, M.F.G. and Vitkin, I.A. (2008) 'Mueller matrix decomposition for extraction of individual polarization parameters from complex turbid media exhibiting multiple scattering, optical activity, and linear birefringence', *J. Biomed. Opt.*, **13**, 044036.

Ghosh, N., Wood, M.F.G. and Vitkin, I.A. (2009a) 'Polarimetry in turbid, birefringent, optically active media: A Monte Carlo study of Mueller matrix decomposition in the backscattering geometry', *J. Appl. Phys.*, **105**, 102023.

Ghosh, N., Wood, M.F.G. and Vitkin, I.A. (2010) 'Influence of the order of the constituent basis matrices on the Mueller matrix decomposition-derived polarization parameters in complex turbid media such as biological tissues', *Opt. Commun.*, **283**, 1200–8.

Ghosh, N., Wood, M.F.G., Li, S.-h., et al. (2009b) 'Mueller matrix decomposition for polarized light assessment of biological tissues', *J. Biophoton.*, **2**, 145–56.

Gil, J.J. (2000) 'Characteristic properties of Mueller matrices', *J. Opt. Soc. Am. A.*, **17**, 328–34.

Gil, J.J. (2007) 'Polarimetric characterization of light and media', *Eur. Phys. J. Appl. Phys.*, **40**, 1–47.

Gil, J.J. and Bernabeu, E. (1985) 'A depolarization criterion in Mueller matrices', *Opt. Acta*, **32**, 259–61.

Gil, J.J. and Bernabeu, E. (1986) 'Depolarization and polarization indexes of an optical system', *Opt. Acta*, **33**, 185–9.

Gil, J.J. and Bernabeu, E. (1987) 'Obtainment of the polarizing and retardation parameters of a non-depolarizing optical system from the polar decomposition of its Mueller matrix', *Optik*, **76**, 67–71.

Givens, C.R. and Kostinski, A.B. (1993) 'A simple necessary and sufficient condition on physically realizable Mueller matrices', *J. Mod. Opt.*, **40**, 471–81.

Goldstein, D.H. (2008) 'Polarization measurements of automobile paints', *Proc. of SPIE*, **6972**, 69720V-1.

Gopala Rao, A.V., Mallesh, K.S. and Sudha (1998a) 'On the algebraic characterization of a Mueller matrix in polarization optics. I. Identifying a Mueller matrix from its N matrix', *J. Mod. Opt.*, **45**, 955–87.

Gopala Rao, A.V., Mallesh, K.S. and Sudha (1998b) 'On the algebraic characterization of a Mueller matrix in polarization optics. II. Necessary and sufficient conditions for Jones-derived Mueller matrices', *J. Mod. Opt.*, **45**, 989–99.

Guissard, A. (1994) 'Mueller and Kennaugh matrices in radar polarimetry', *IEEE Transactionson Geoscience and Remote Sensing*, **32**, 590–7.

Hauge, P.S. (1980) 'Recent developments in instrumentation in ellipsometry', *Surface Sci.*, **96**, 108–40.

Herreros-Cedres, J., Hernandez-Rodrguez, C. and Guerrero-Lemus, R. (2006) 'Influence of the imperfect parallelism of crystal faces on high-accuracy universal polarimeter measurements', *J. Opt. A: Pure Appl. Opt.*, **8**, 44–8.

Hoover, B.G. and Tyo, J.S. (2007) 'Polarization Components Analysis for angle-invariant laser discrimination', *Appl. Opt.*, **46**, 8364–73.

Horn, R. and Johnson, C.R. (1986) *Matrix Analysis*, London, Cambridge University Press.

Hovenier, J.W. (1969) 'Symmetry relations for scattering of polarized light in a slab of randomly oriented particles', *J. Atmos. Sci.*, **26**, 488–99.

Hovenier, J.W. (1970) 'Principles of symmetry for polarization studies of planets', *Astron. Astrophys.*, **7**, 86–90.

Hovenier, J.W. (1994) 'Structure of a general pure Mueller matrix', *Appl. Opt.*, **33**, 8318–24.

Hovenier, J.W. and Mackowski, D.W. (1998) 'Symmetry relations for forward and backward scattering by randomly oriented particles', *J. Quant. Spectrosc. Radiat. Transfer*, **60**, 483–92.

Hovenier, J.W. and van der Mee, C.V.M. (2000) 'Basic relationships for matrices describing scattering by small particles', in Mishchenko, M.I., Hovenier, J.W. and Travis, L.D. (eds), *Light Scattering by Nonspherical Particles: Theory, Measurements, and Applications*, San Diego, Academic Press, pp. 61–85.

Hovenier, J.W., van de Hulst, H.C. and van der Mee, C.V.M. (1986) 'Conditions for the elements of the scattering matrix', *Astron. Astrophys.*, **157**, 301–10.

Howell, B.J. (1979) 'Measurement of the polarization effects of an instrument using partially polarized light', *Applied Optics*, **18**, 809–12.

Hu, C.-R., Kattawar, G.W., Parkin, M.E. and Herb, P. (1987) 'Symmetry theorems on the forward and backward scattering Mueller matrices for light scattering from a nonspherical dielectric scatterer', *Appl. Opt.*, **26**, 4159–73.

Hurwitz, H. and Jones, C.R. (1941) 'A new calculus for the treatment of optical systems. II. Proof of three general equivalence theorems', *J. Opt. Soc. Am.*, **31**, 493–9.

Jones, R.C. (1941) 'A new calculus for the treatment of optical systems. I. Description and discussion of the calculus', *J. Opt. Soc. Am.*, **31**, 488–93.

Jones, R.C. (1942) 'A new calculus for the treatment of optical systems. IV', *J. Opt. Soc. Am.*, **32**, 486–93.

Jones, R.C. (1947) 'A new calculus for the treatment of optical systems. V. A more general formulation and description of another calculus', *J. Opt. Soc. Am.*, **37**, 107–10.

Jones, R.C. (1956) 'A new calculus for the treatment of optical systems. VIII. Electromagnetic theory', *J. Opt. Soc. Am.*, **46**, 126–31.

Kaplan, B. and Drevillon, B. (2002) 'Mueller matrix measurements of small spherical particles deposited on a c-Si wafer', *Appl. Opt.*, **41**, 3911–18.

Kaplan, B., Ledanois, G. and Drevillon, B. (2001) 'Mueller matrix of dense polystyrene latex sphere suspensions: Measurements and Monte Carlo simulation', *Appl. Opt.*, **40**, 2769–77.

Kim, K., Mandel, L. and Wolf, E. (1987) 'Relationship between Jones and Mueller matrices for random media', *J. Opt. Soc. Am. A.*, **4**, 433–7.

Kim, Y.L., Pradhan, P., Kim, M.H. and Backman, V. (2006) 'Circular polarization memory effect in low-coherence enhanced backscattering of light', *Opt. Lett.*, **31**, 2744–6.

Kleim, R.., Kuntzler, L. and El Ghemmaz, A. (1994) 'Systematic errors in rotating-compensator ellipsometry', *J. Opt. Soc. Am. A*, **11**, 2550–9.

Kliger, D.S., Lewis, J.W. and Randall, C.E. (1990) *Polarized Light in Optics and Spectroscopy*, New York, Academic Press.

Kokhanovsky, A.A. (2001) *Light Scattering Media Optics: Problems and Solutions*, Chichester, UK, Praxis.

Kokhanovsky, A.A. (2003) *Polarization Optics of Random Media*, Chichester, UK, Praxis.

Kostinski, A.B. (1992) 'Depolarization criterion for incoherent scattering', *Appl. Optics*, **31**, 3506–8.

Kostinski, A.B., Givens, C.R. and Kwiatkowski, J.M. (1993) 'Constraints on Mueller matrices of polarization optics', *Appl. Opt.*, **32**, 1646–51.

Landau, L.D., Lifshitz, E.M. and Pitaevskii, L.P. (1984) *Electrodynamics of Continuous Media*, Oxford, Pergamon Press.

Landi Degl'Innocenti, E. and del Toro Iniesta, J.C. (1998) 'Physical significance of experimental Mueller matrices', *J. Opt. Soc. Am. A*, **15**, 533–7.

Lankaster, P. and Tismenetsky, M. (1985) *The Theory of Matrices*, San Diego, Academic Press.

Lawless, R., Xie, Y., Yang, P., et al. (2006) 'Polarization and effective Mueller matrix for multiple scattering of light by nonspherical ice crystals', *Opt. Express*, **14**, 6381–93.

Li, X. and Yao, G. (2009) 'Mueller matrix decomposition of diffuse reflectance imaging in skeletal muscle', *Appl. Opt.*, **48**, 2625–31.

Lin, J.-F. (2008) 'Concurrent measurement of linear and circular birefringence using rotating-wave-plate Stokes polarimeter', *Applied Optics*, **47**, 4529–39.

Lopatin, V.N., Priezzhev, A.V., Apanasenko, A.D., et al. (2004) *Methods of Light Scattering in the Analysis of Dispersed Biological Media*, Moscow, Fizmatlit (in Russian).

Lu, S.-Y. and Chipman, R.A. (1994) 'Homogeneous and inhomogeneous Jones matrices', *J. Opt. Soc. Am. A.*, **11**, 766–73.

Lu, S.-Y. and Chipman, R.A. (1996) 'Interpretation of Mueller matrices based on polar decomposition', *J. Opt. Soc. Am. A.*, **13**, 1106–13.

Mahler, A.-B. and Chipman, R.A. (2011) 'Polarization state generator: A polarimeter calibration standard', *Applied Optics*, **50**, 1726–34.

Manickavasagam, S. and Menguc, M.P. (1998) 'Scattering-matrix elements of coated infinite-length cylinders', *Appl. Opt.*, **37**, 2473–82.

Mishchenko, M.I. and Hovenier, J.W. (1995) 'Depolarization of light backscattered by randomly oriented nonspherical particles', *Opt. Lett.*, **20**, 1356–8.

Mishchenko, M.I. and Travis, L.D. (2000) 'Polarization and depolarization of light', in Moreno, F. and González, F. (eds), *Light Scattering from Microstructures*, Berlin, Springer, pp. 159–75.

Mishchenko, M.I., Rosenbush, V.K., Kiselev, N.N., Lupishko, D.F., Tishkovets, V.P., Kaydash, V.G., Belskaya, I.N., Efimov, Y.S. and Shakhovskoy, N.M. (2010) *Polarimetric Remote Sensing of Solar System Objects*, Kyiv, Akademperiodyka.

Mishchenko, M.I., Travis, L.D. and Lacis, A.A. (2002) *Scattering, Absorption, and Emission of Light by Small Particles*, Cambridge, UK, Cambridge University Press.

Mishchenko, M.I., Travis, L.D. and Lacis, A.A. (2006) *Multiple Scattering of Light by Particles. Radiative Transfer and Coherent Backscattering*, Cambridge, Cambridge University Press.

Moore, E.H. (1920) 'On the reciprocal of the general algebraic matrix', *Bull. Am. Math. Soc.*, **26**, 394–5.

Morio, J. and Goudail, F. (2004) 'Influence of the order of diattenuator, retarder, and polarizer in polar decomposition of Mueller matrices', *Opt. Lett.*, **29**, 2234–6.

Mott, H. (2007) *Remote Sensing with Polarimetric Radar*, New York, Wiley.

Mueller, H. (1948) 'The foundation of optics', *J. Opt. Soc. Am.*, **38**, 661.

Munoz, O., Moreno, F., Guirado, D., Ramos, J.L., Lopez, A., Girela, F., Jeronimo, J.M., Costillo, L.P. and Bustamante, I. (2010) 'Experimental determination of scattering matrices of dust particles at visible wavelengths: The IAA light scattering apparatus', *J. Quant. Spectrosc. Radiat. Transfer*, **111**, 187–96.

Munoz, O., Volten, H., de Haan, J.F., Vassen, W. and Hovenier, J.W. (2001) 'Experimental determination of scattering matrices of fly ash and clay particles at 442 and 633 nm', *J. Geophys. Res.*, **106**, 22833–44.

Munoz, O., Volten, H., de Haan, J.F., Vassen, W. and Hovenier, J.W. (2002) 'Experimental determination of the phase function and degree of linear polarization of El Chichon and Pinatubo volcanic ashes', *J. Geophys. Res.*, **107(D13)**, 4174.

Munoz, O., Volten, H., Hovenier, J.W., Nousiainen, T., Muinonen, K., Guirado, D., Moreno, F. and Waters, L.B.F.M. (2007) 'Scattering matrix of large Saharan dust particles: Experiments and computations', *J. Geophys. Res.*, **112**, D 13215.

Munoz, O., Volten, H., Hovenier, J.W., Veihelmann, B., van der Zande, W.J., Waters, L.B.F.M. and Rose, W.I. (2004) 'Scattering matrices of volcanic ash particles of Mount St. Helens, Redoubt, and Mount Spurr Volcanoes', *J. Geophys. Res.*, **109**, D 16201.

Nagirner, D.I. (1993) 'Constrains on matrices transforming Stokes vectors', *Astron. Astrophys.*, **275**, 318–24.

Nunziata, F., Gambardella, A. and Migliaccio, M. (2008) 'On the Mueller Scattering Matrix for SAR Sea Oil Slick Observation', *IEEE Geosci. Remote Sens. Let.*, **5**, 691–5.

Oberemok, E.A. and Savenkov, S.N. (2003) 'Structure of deterministic Mueller matrices and their reconstruction in the method of three input polarizations', *J. Appl. Spectrosc.*, **70**, 224–9.

Ossikovski, R. (2008) 'Interpretation of nondepolarizing Mueller matrices based on singular-value decomposition', *J. Opt. Soc. Am. A.*, **25**, 473–82.

Ossikovski, R. (2009) 'Analysis of depolarizing Mueller matrices through a symmetric decomposition', *J. Opt. Soc. Am. A.*, **26**, 1109–18.

Ossikovski, R. (2010a) 'Alternative depolarization criteria for Mueller matrices', *J. Opt. Soc. Am. A*, **27**, 808–14.

Ossikovski, R. (2010b) 'Canonical forms of depolarizing Mueller matrices', *J. Opt. Soc. Am. A*, **27**, 123–30.

Ossikovski, R., De Martino, A. and Guyot, S. (2007) 'Forward and reverse product decompositions of depolarizing Mueller matrices', *Opt. Lett.*, **32**, 689–91.

Parke III, N.G. (1948) 'Matrix optics', Ph.D. thesis, Massachusetts Institute of Technology, Cambridge, MA, USA.

Parke III, N.G. (1949) 'Optical algebra', *J. Math. Phys.*, **28**, 131–9.

Penrose, R. (1955) 'A generalized inverse for matrices', *Proc. Cambridge Philos. Soc.*, **151**, 406–13.

Perrin, F. (1942) 'Polarization of light scattering by isotropic opalescent media', *J. Chem. Phys.*, **10**, 415–27.

Piller, G., Broch, L. and Johann, L. (2008) 'Experimental study of the systematic errors for a Mueller matrix double rotating compensator ellipsometer', *Phys. Status Solidi*, **C5**, 1027–30.

Puentes, G., Voigt, D., Aiello, A. and Woerdman, J.P. (2005) 'Experimental observation of universality in depolarized light scattering', *Optics Letters*, **30**, 3216–18.

Rojas-Ochoa, L.F., Lacoste, D., Lenke, R., et al. (2004) 'Depolarization of backscattered linearly polarized light', J. *Opt. Soc. Am. A.*, **21**, 1799–804.

Sabatke, D.S., Locke, A.M., Descour, M.R., Sweatt, W.C., Garcia, J.P., Dereniak, E., Kemme, S.A. and Phipps, G.S. (2000) 'Figures of merit for complete Stokes polarimeters', *Proc. SPIE*, **4133**, 75–81.

Savenkov, S.N. (2002) 'Optimization and structuring of the instrument matrix for polarimetric measurements', *Opt. Eng.*, **41**, 965–72.

Savenkov, S.N. (2007) 'Analysis of generalized polarimetric measurement equation', *Proc. SPIE*, **6682**, 668214.

Savenkov, S.N. and Yushtin, K.E. (2000a) 'New classes of objects in polarimetry including the isotropic depolarization', *Radiotechnika*, **116**, 3–11 (in Russian).

Savenkov, S.N. and Yushtin, K.E. (2000b) 'Some peculiarities of the correspondence between experimental deterministic Mueller and Jones matrices', *Ukr. J. Phys.*, **45**, 159–63.

Savenkov, S.N. and Yushtin, K.E. (2001) 'Optimization of dynamic polarimeter for incomplete Mueller matrices measurement', *Radiotekhnika: All-Ukr. Sci. Interdep. Mag.*, **124**, 111–18.

Savenkov, S.N. and Yushtin, K.E. (2003) 'Mueller matrix elements error distribution for polarimetric measurements', *Proc. SPIE*, **5158**, 251–9.

Savenkov, S.N., Grygoruk, V.I., Muttiah, R.S., Oberemok, Y.A., Yushtin, K.E. and Yakubchak, V.V. (2009a) 'Effective dichroism in forward scattering by inhomogeneous birefringent medium', *J. Quant. Spectrosc. Radiat. Transfer*, **110**, 1441–7.

Savenkov, S.N., Marienko, V.V., Oberemok, E.A. and Sydoruk, O.I. (2006) 'Generalized matrix equivalence theorem for polarization theory', *Phys. Rev. E.*, **74**, 056607.

Savenkov, S.N., Mishchenko, L.T., Muttiah, R.S., Oberemok, Y.A. and Mishchenko, I.A. (2004) 'Mueller polarimetry of virus-infected and healthy wheat under field and microgravity conditions', *J. Quant. Spectrosc. Radiat. Transfer*, **88**, 327–43.

Savenkov, S.N., Muttiah, R.S. and Oberemok, Y.A. (2003) 'Transmitted and reflected scattering matrices from an English oak leaf', *Appl. Opt.*, **42**, 4955–62.

Savenkov, S.N., Muttiah, R.S., Yushtin, K.E. and Volchkov, S.A. (2007a) 'Mueller-matrix model of an inhomogeneous, linear, birefringent medium: Single scattering case', *J. Quant. Spectrosc. Radiat. Transfer*, **106**, 475–86.

Savenkov, S.N., Oberemok, Y.A., Yakubchak, V.V., Aulin, Y.V. and Barchuk, O.I. (2009b) 'Light depolarization by inhomogeneous linear birefringent media', *Semiconductor Physics, Quantum Electronics and Optoelectronics*, **12**, 86–90.

Savenkov, S.N., Priezzhev, A.V., Oberemok Ye, A., Silfsten, P., et al. (2012) 'Characterization of porous media by means of the depolarization metrics', *J. Quant. Spectrosc. Radiat. Transfer*, **113**, 2503–11.

Savenkov, S.N., Sydoruk, O.I. and Muttiah, R.S. (2005) 'The conditions for polarization elements to be dichroic and birefringent', *J. Opt. Soc. Am. A.*, **22**, 1447–52.

Savenkov, S.N., Sydoruk, O.I. and Muttiah, R.S. (2007b) 'Eigenanalysis of dichroic, birefringent, and degenerate polarization elements: A Jones-calculus study', *Appl. Opt.*, **46**, 6700–9.

Schmitt, J.M. and Kumar, G. (1996) 'Turbulent nature of refractive-index variations in biological tissue', *Opt. Lett.*, **21**, 1310–12.

Shindo, Y. (1995) 'Applications of polarized modulator technique in polymer science', *Opt. Eng.*, **34**, 3369–84.

Shurcliff, W.A. (1962) *Polarized Light: Production and Use*, Harvard, MA, Harvard University Press.

Simon, R. (1982) 'The connection between Mueller and Jones matrices of polarization optics', *Opt. Commun.*, **42**, 293–7.

Simon, R. (1987) 'Mueller matrices and depolarization criteria', *J. Mod. Opt.*, **34**, 569–75.

Smith, M.H. (2001) 'Interpreting Mueller matrix images of tissues', *Proc. SPIE.*, **4257**, 82–9.

Smith, M.H. (2002) 'Optimization of a dual-rotating-retarder Mueller matrix polarimeter', *Appl. Opt.*, **41**, 2488–95.

Solleillet, P. (1929) 'Sur les paramètres caractérisant la polarisation partielle de la lumière dans les phénomènes de fluorescence', *Ann. Phys. Biol. Med.*, **12**, 23–97.

Sridhar, R. and Simon, R. (1994) 'Normal form for Mueller matrices in polarization optics', *J. Mod. Opt.*, **41**, 1903–15.

Swami, M.K., Manhas, S., Buddhiwant, P., Ghosh, N., Uppal, A. and Gupta, P.K. (2006) 'Polar decomposition of 3x3 Mueller matrix: A tool for quantitative tissue polarimetry', *Opt. Express*, **14**, 9324–37.

Takakura, Y. and Ahmad, J.E. (2007) 'Noise distribution of Mueller matrices retrieved with active rotating polarimeters', *Appl. Opt.*, **46**, 7354–64.

Tang, S.T. and Kwok, H.S. (2001) '3x3 matrix for unitary optical systems', *J. Opt. Soc. Am. A*, **18**, 2138–45.

Tuchin, V.V. (ed.) 1994) *Selected Papers on Tissue Optics Applications in Medical Diagnostics and Therapy*, Bellingham, WA, SPIE Press.

Tuchin, V.V., Wang, L.V. and Zimnyakov, D.A. (2006) *Optical Polarization in Biomedical Applications*, Berlin, Springer.

Twietmeyer, K. and Chipman, R.A. (2008) 'Optimization of Mueller polarimeters in the presence of error sources', *Opt. Express*, **16**, 11589–603.

Tyo, J.S. (1998) 'Optimum linear combination strategy for an N-channel polarization-sensitive vision or imaging system', *J. Opt. Soc. Am. A*, **15**, 359–66.

Tyo, J.S. (2000) 'Noise equalization in Stokes parameter images obtained by use of variable retardance polarimeters', *Opt. Lett.*, **25**, 1198–2000.

Tyo, J.S., Wang, Z., Johnson, S.J. and Hoover, B.G. (2010) 'Design and optimization of partial Mueller matrix polarimeters', *Appl. Opt.*, **49**, 2326–33.

van de Hulst, H.C. (1957) *Light Scattering by Small Particles*, New York, Wiley.

van der Mee, C.V.M. (1993) 'An eigenvalue criterion for matrices transforming Stokes parameters', *J. Math. Phys.*, **34**, 5072–88.

van der Mee, C.V.M. and Hovenier, J.W. (1992) 'Structure of matrices transforming Stokes parameters', *J. Math. Phys.*, **33**, 3574–84.

van Zyl, J.J. and Ulaby, F.T. (1990) 'Scattering matrix representation for simple targets', in *Radar Polarimetry for Geosciences Applications*, Norwood, MA, Artech House.

Vaughn, I.J. and Hoover, B.G. (2008) 'Noise reduction in a laser polarimeter based on discrete waveplate rotations', *Opt. Exp.*, **16**, 2091–108.

Volten, H., Munoz, O., Rol, E., de Haan, J.F., Vassen, W., Hovenier, J.W., Muinonen, K. and Nousianen, T. (2001) 'Scattering matrices of mineral aerosol particles at 441.6nm and 632.8nm', *J. Geophys. Res.*, **106**, 17375–401.

Voss, K.J. and Fry, E.S. (1984) 'Measurement of the Mueller matrix for ocean water', *Appl. Opt.*, **23**, 4427–36.

Whitney, C. (1971) 'Pauli-algebraic operators in polarization optics', *J. Opt. Soc. Am.*, **61**, 1207–13.

Wood, M.F.G., Ghosh, N., Moriyama, E.H., et al. (2009) 'Proof-of-principle demonstration of a Mueller matrix decomposition method for polarized light tissue characterization *in vivo*', *J. Biomed. Opt.*, **14**, 014029.

Wood, M.F.G., Guo, X. and Vitkin, I.A. (2007) 'Polarized light propagation in multiply scattering media exhibiting both linear birefringence and optical activity: Monte Carlo model and experimental methodology', *J. Biomed. Opt.*, **12**, 014029.

Xing, Z.-F. (1992) 'On the deterministic and nondeterministic Mueller matrix', *J. Mod. Opt.*, **39**, 461–84.

Xu, L., Ding, J. and Cheng, A.Y.S. (2002) 'Scattering matrix of infrared radiation by ice finite circular cylinders', *Appl. Opt.*, **41**, 2333–48.

Zallat, J., Anouz, S. and Stoll, M.P. (2006) 'Optimal configurations for imaging polarimeters: Impact of image noise and systematic errors', *J. Opt. A*, **8**, 807–14.

Zubko, E., Shkuratov, Y.G. and Videen, G. (2004) 'Coherent backscattering effect for non-zero elements of Mueller matrix of discrete media at different illumination-observation geometries', *J. Quant. Spectrosc. Radiat. Transfer*, **89**, 443–52.

7 Reflectance and polarization characteristics of various vegetation types

Jouni I. Peltoniemi, Maria Gritsevich, and Eetu Puttonen

7.1 Introduction

An important factor for reducing the amount of data acquired by Earth Observation projects and for collecting more accurate knowledge on the Earth, universe, and environment is to provide a set of reliable references by measuring various known terrestrial and planetary targets. This requires strong technological bases and careful gathering of information about land surfaces and vegetation via remote sensing technologies supplemented by Earth surveying techniques. Many satellites and aircraft are observing Earth with a variety of sophisticated instruments (e.g. Short, 2011). While the quality and quantity of the remote sensing information are constantly growing, the need and demand for support studies, such as the collection of ground references, are growing even faster. At present, it is already possible to distinguish vegetation from non-vegetated surfaces. Ongoing research concentrates on getting more quantitative measurements of different forest types, tree species, understory types, biomass, diseases, phases of growth, and crop estimates. As an application, these data support land-use or climate change studies, monitoring and responding to natural disasters, including fires, floods, earthquakes, and storm damage. In addition to improved performance and resolution, new types of sensors are being introduced. These include lasers, spectral imagers and, multi-directional cameras. Polarization is already used as a valuable tool for many purposes.

New light-weight unmanned flying systems (UAV = unmanned aerial vehicle, UAS = unoccupied aerial system, etc.) have revolutionized low-altitude data collection. They can carry small sensors and fly fully automatically over selected targets with flexibility and proximity that have not been achievable before. This era will thus see many new applications.

It is now of utmost value to design instruments and systems optimally for their purposes, to observe the most indicative signals, and to yield the best procedures for extracting the desired pieces of information from huge data sets.

Fundamentally, remote sensing is the analysis of reflected/emitted radiation (light) from the target with account for external disturbances affecting signal record. The information source is of prime importance in order to locate targets, shapes, and borderlines from the images (photogrammetry and image analysis), which actually already provides much more useful information than all other remote sensing techniques together, from the past or in the future. However, significantly more additional information can be obtained using all measurable properties of the radiation: spectrum, polarization, anisotropy. Using these additional signals optimally requires understanding of the physics of the reflection and the full remote sensing process as a whole.

This review presents the state-of-the-art research concerning reflectance of various vegetation, especially on the lesser-known spectral, directional, and polarization signals. The emphasis is put on what kind of information can be extracted from the actual field measurement data. A brief review of polarization modeling is included. For unpolarized reflectance models, we refer to the RAdiation transfer Model Intercomparison (RAMI) experiment (Widlowski et al., 2006) and many other reviews.

7.2 Definitions

7.2.1 BRF, BRDF

The bidirectional reflectance factor (BRF for short, or R in equations) is defined as the ratio of the reflected light intensity of a given target to an ideal Lambertian reflector with a spherical albedo of 1.0 under the same incident irradiation (Nicodemus et al., 1977; Hapke, 1993; Liang, 2004). The BRF can be presented as

$$R(\mu, \mu_0, \phi, \phi_0) = \frac{\pi I(\mu, \phi)}{\mu_0 F_0(\mu_0, \phi_0)}, \tag{7.1}$$

where F_0 is the incident collimated flux and I the reflected radiance; ι and ϕ_0 are the zenith and azimuth angles of incidence, ϵ and ϕ are the zenith and azimuth angles of emergence, α is the scattering phase angle that is defined as a complement of the scattering angle ($\cos\alpha = \cos\iota\cos\epsilon + \sin\iota\sin\epsilon\cos(\phi - \phi_0)$), and $\mu = \cos\epsilon$, $\mu_0 = \cos\iota$ (Fig. 7.1). A related quantity is the bidirectional reflectance distribution function (BRDF), here denoted as $\mathcal{R} = R/\pi$. Experimentalists and practical users prefer using BRF for its more intuitive magnitude which is normalized to a perfect Lambertian reflector $R = 1$. Modelers prefer BRDF for its more natural mathematical interpretation and simpler equations. For example, the reflected radiation can be written with $\mathcal{R}$ as

$$\begin{aligned} I(\mu, \phi) &= \int \mathrm{d}\mu'\phi'\mathcal{R}(\mu, \phi, \mu', \phi')\mu' I(\mu', \phi') \\ &= \int \mathrm{d}\mu'\phi'\mathcal{R}(\mu, \phi, \mu', \phi')\mu' I_{\mathrm{Diff}}(\mu', \phi') \\ &\quad + \mathcal{R}(\mu, \phi, \mu_0, \phi_0)\mu_0 F_0(\mu_0, \phi_0), \end{aligned} \tag{7.2}$$

where I_{Diff} represents diffuse skylight.

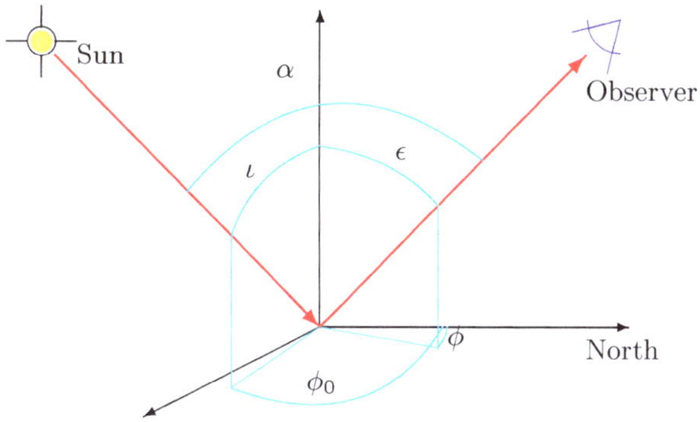

Fig. 7.1. Definition of the angles used in surface reflectance work: ϵ and ι are the zenith angles of the emergent (Observer) and incident (solar) radiation respectively (shorthand $\mu = \cos\epsilon$ and $\mu_0 = \cos\iota$ are also used). ϕ and ϕ_0 are the corresponding azimuths. The phase angle α is the angle between the observer and the Sun. The principal plane is fixed by the solar direction and the surface normal, while the cross plane is a vertical plane perpendicular to the principal plane.

7.2.2 Polarization

The polarization of electromagnetic radiation waves, such as light, is defined by the direction of the electric field. Natural light is a mixture of a very large number of various wave packets. Light that contains an equal amount of all polarization states, such as sunlight, is said to be unpolarized. In general, any interaction of light with matter polarizes light, and all interactions depend on the incident polarization, with only a few exceptions, such as transmission through symmetric medium or perfect reflection.

A convenient way to describe natural light with an arbitrary state of polarization is to use the Stokes parameters $[I, Q, U, V]$. These are related to electromagnetic fields as

$$I = I_{\parallel} + I_{\perp} \propto |E_{\parallel}|^2 + |E_{\perp}|^2, \tag{7.3}$$

$$Q = I_{\parallel} - I_{\perp} \propto |E_{\parallel}|^2 - |E_{\perp}|^2, \tag{7.4}$$

$$U = I_{\backslash} - I_{/} \propto E_{\parallel}E_{\perp}^* + E_{\parallel}^*E_{\perp}, \tag{7.5}$$

$$V = I_{+} - I_{-} \propto E_{\parallel}E_{\perp}^* - E_{\parallel}^*E_{\perp}, \tag{7.6}$$

where $E_{\parallel}$ and $E_{\perp}$ are electromagnetic field amplitudes parallel and perpendicular to a selected scattering plane, and $I_{\parallel}, I_{\perp}, I_{\backslash}, I_{/}$ the intensities measured at linear polarizer angles of 0°, 90°, 45°, and 135°, respectively, and I_+ and I_- with left and right circular filter. It is important to note that the Stokes parameters depend on this selected polarization plane, and this plane must be defined in order to uniquely use the Stokes parameters. The strength of the Stokes vectors is that the radiation equations are easily modified for polarization by replacing the scalar intensity (I) with a Stokes vector $\mathbf{I} = [I, Q, U, V]$, and the scalar reflectance $R \leftarrow$ with 4×4 Muller matrix $\mathbf{R}$. Thus

$$\mathbf{I}_r = \frac{\mu_0}{\pi}\mathbf{R}\mathbf{F}_0 \tag{7.7}$$

for incident flux $\mathbf{F}_0$. This applies to any other linear combination of Stokes vectors, and some papers still use $\mathbf{I} = [I_1, I_2, U, V]$, for example. For pure fields, apply $I^2 = Q^2+U^2+V^2$ but, for natural light, that is a mixture of fields $I^2 < Q^2+U^2+V^2$. Likewise, $I = I_{\|} + I_{\perp} = I_{\backslash} + I_{/} = I_1 + I_2 = I_{\min} + I_{\max}$.

The $R_p = \sqrt{R_{12}^2 + R_{13}^2}$ is also called polarized reflectance, and unpolarized reflectance is then $R_{\mathrm{unpol}} = R - R_p$. Note that the unpolarized reflectance is not the same as the scalar reflectance.

One can further define the degrees of linear polarization as

$$P_Q = -\frac{Q}{I}, \tag{7.8}$$

$$P_U = -\frac{U}{I}, \tag{7.9}$$

$$P_L = \frac{\sqrt{Q^2 + U^2}}{I}, \tag{7.10}$$

and degree of circular polarization

$$P_V = -\frac{V}{I}, \tag{7.11}$$

and degree of (full) polarization

$$P_F = \frac{\sqrt{Q^2 + U^2 + V^2}}{I}. \tag{7.12}$$

It should always be taken into account which of these quantities are actually used, and how the sign is defined, because there are different definitions and many applications. Polarization degrees are convenient since they are independent of the calibration of the sensor, and can thus be measured in conditions where reflectance factor cannot be normalized, as is usually the case in astronomical observations. The inconvenient factor is that they cannot be processed very much any more, so, when one wants to integrate them over wavelengths, one must integrate the original ratios separately for the nominator and denominator, as

$$P = -\frac{\int d\lambda f(\lambda) Q(\lambda)}{\int d\lambda f(\lambda) I(\lambda)}, \tag{7.13}$$

where f is the spectral sensitivity of the simulated channel of the camera or sensor.

7.3 Theory and modeling

The unpolarized reflectance modeling of vegetation and radiative transfer has already been addressed by many, such as Apresyan and Kravtsov (1983), Myneni and Ross (1991), Liang (2004), von Schönermark et al. (2004), Jin (2005), and Furfaro (2009). Here we concentrate on polarization.

The classical electromagnetic field ($\mathbf{E}$) is fully covered by the Maxwell equations and constitutive relations. The scattering problem can be further formulated using the wave equation

$$\nabla^2 \mathbf{E}(\mathbf{r}) - m^2(\mathbf{r})\mathbf{E}(\mathbf{r}) = 0, \tag{7.14}$$

with appropriate boundary conditions, or all compactly in the integral form

$$\mathbf{E}(\mathbf{r}) = \mathbf{E}_0(\mathbf{r}) + \int_V \mathrm{d}^3\mathbf{r}' \mathsf{G}(\mathbf{r}-\mathbf{r}') \cdot (m(\mathbf{r}')^2 - 1)\mathbf{E}(\mathbf{r}'), \tag{7.15}$$

where m is the refractive index and G is the Green tensor function, which can be written using dyads:

$$\mathsf{G}(\mathbf{R}) = [1 + \nabla\nabla]g(R) \tag{7.16}$$

$$= \left\{1\left[1 + \frac{\mathrm{i}}{R} - \frac{1}{R^2}\right] - \hat{\mathbf{R}}\hat{\mathbf{R}}\left[1 + \frac{3\mathrm{i}}{R} - \frac{3}{R^2}\right]\right\}g(R), \tag{7.17}$$

where $g(R) = \frac{\exp(\mathrm{i}R)}{4\pi R}$ and $R = |\mathbf{R}|$, $\hat{\mathbf{R}} = \frac{\mathbf{R}}{R}$, and 1 is a unit tensor.

The electromagnetic permittivity $\varepsilon = m^2$ is usually a scalar, but in a more general case a tensor. There are a few cases where the chirality may be of importance, such as chlorophyll.

The wave equations can be solved analytically only in extremely simple cases. Thus, real targets such as vegetation require heavy modeling, simplification, and approximations. At first, the problem is divided into two scales: single-scattering by selected elements, and multiple scattering between these elements. The division can be continued by assuming a certain volume of multiple-scattering to be a new single-scatterer for larger volumes (or solving single-scattering from an element with similar sub-division). With vegetation, the natural division line is that the leaves are considered single-scatterers, and trees and forests multiple-scatterers. Shoots or branches can be used as intermediate steps, and currently many details still require better characterization for different vegetation types.

The multiple-scattering problem can be solved using radiative transfer techniques. This assumes that the scatterers can be located to small points, that the scatterers are completely non-interfering electromagnetically, and that the propagation and scattering can be described statistically. The radiative transfer equation for polarized radiation vector $\mathbf{I}$ at point $\mathbf{r}$ and direction $\mathbf{k}$ is written as

$$\mathbf{k} \cdot \nabla\mathbf{I}(\mathbf{r}, \mathbf{k}) = -\beta(\mathbf{r})\mathbf{I}(\mathbf{r}, \mathbf{k}) + \int \mathrm{d}^2\mathbf{k}' \frac{\beta(\mathbf{r})\varpi_0 \mathbf{P}(\mathbf{k}, \mathbf{k}')}{4\pi}\mathbf{I}(\mathbf{r}, \mathbf{k}'), \tag{7.18}$$

where β is the extinction coefficient, ϖ_0 is the single-scattering albedo, and $\mathbf{P}$ the single-scattering phase matrix. This equation can be solved more easily if one makes a more radical assumption of horizontal homogeneity, but the errors of this assumption are already so big that most remote sensing applications cannot accept it. Solving the 3D radiative transfer equation is unfortunately much more demanding. Also, the turbidity assumption (small independent scatterers) fails at least near backscattering where the finite size of the scatterers causes visible effects—the light-scattering exactly backward can always return by the same path it came but, in other directions, the probability for obstruction is much greater. Although there are some extensions to radiative transfer models correcting for the backscattering effect (Marshak, 1989; Knyazikhin and Marshak, 1991; Peltoniemi, 1993), it is more popular to model using ray-tracing techniques. Ray-tracing allows more free positioning of scatterers to fixed or random locations, and rays can be traced between and up to the scatterers, to simulate the scatterings.

Also, at least near the backscattering direction, the assumption of electromagnetic incoherence breaks down. This is because, for every multiply scattered ray, there is a counter ray propagating the same path in the reverse direction, and interfering constructively (Hapke et al., 1996; Liang and Mishchenko, 1997; Muinonen, 2004; Mishchenko, 2008; Muinonen et al., 2012). This interference causes a strong and narrow brightness spike in the backscattering, and a branch of negative linear polarization a few degrees around.

Otherwise, multiple scattering by default scrambles the polarization directions which macroscopically depolarizes the light. Thus, most models assume that all the observed polarization comes from the singly scattered light, and thus avoid the complicated polarized radiative transfer problem. While this is certainly not exactly true, it can give a reasonable first-order approximation.

Solving the scattering from individual leaves is another not yet fully resolved problem. There is a reflection from the surface, and there is transmission through the leaf, and a lot of internal scattering inside the leaf from many structures. Since these structures cover all scales from nm to cm, it is not possible to make very complete models at all, but only basic-level approximations. Thus, the main line of polarization models simplify the picture even more, by assuming that the polarization comes only from the Fresnelian reflection of the wax-covered leaves, and all other scattering is unpolarized (Rondeaux and Herman, 1991; Bréon et al., 1995; Nadal and Bréon, 1999; Maignan et al., 2009; Dong-hui et al., 2010; Litvinov et al., 2010, 2011). The Fresnel reflection is no doubt the most dominant factor, and explains the overall shape well (see Fig. 7.2 and all polarization plots later and in the mentioned references). Of course, there are discrepancies in details, requiring at least the coherent backscattering, but also something from the now ignored other scattering components.

Zhang et al. (2012) are already doing numerical scene simulations with polarization.

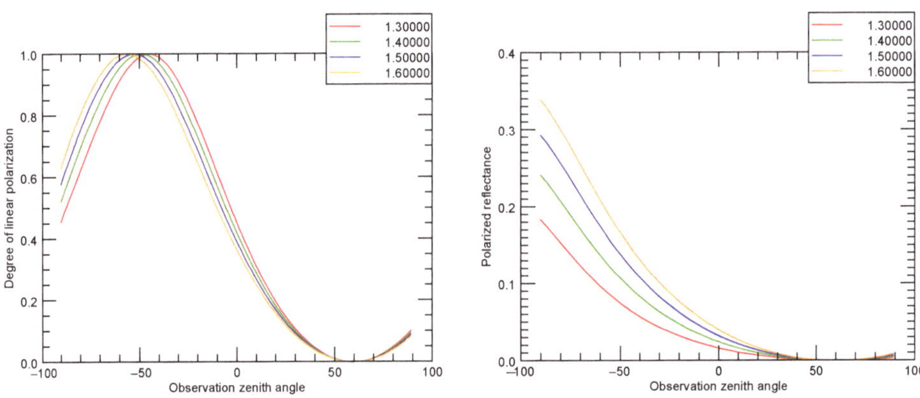

Fig. 7.2. The degree of linear polarization(P_Q) and polarized reflectance for pure Fresnel reflection as a function of the observation zenith angle, for refractive indices 1.3, 1.4, 1.5, 1.6, and solar zenith angle of 60°.

7.4 Field and laboratory measurements

For a good understanding of the polarization and proper validation of the models, very thorough laboratory and field measurements are needed. The measurements of vegetation are still not very numerous, but the number is growing. Since polarized sunglasses and camera filters have existed for a long time, we may assume that many observations have been made, and no big surprises or unknown features are to be expected, even though not everything is published. Non-vegetated objects have been studied for a very long time (see, e.g. Coulson et al. 1965). Thus, we can say it is well known that many surfaces polarize light in the forward direction or, more exactly, near the Brewster angle. And that otherwise vegetation and land surfaces are quite featureless, contrary to atmosphere with strong blue sky polarization and many features from clouds and rainbows. But more accurate measurements do reveal some signals even from the boring vegetation.

We should first remind ourselves that, to get a full understanding of the reflectance and all remote sensing signals, we must make as complete measurements as possible, taking the directional, spectral, polarization, spatial, and temporal dimensions into account. Polarization is an especially strong function of the scattering angle and, without the directional dimension, polarization tells very little. It is possible to measure the degree of polarization alone, as that does not need reference standards or calibration or control of incident radiation, but it is much better to measure it together with calibrated reflectance.

The simplest reflection polarimeters measure only linear polarization of the reflected light. This requires only a turning polarizer in front of the detector. When using monochromatic or narrow-band illumination, basic plate polarizers are sufficient but, for wide-band spectral measurements, more advanced prisms are used. Polarimetric imaging of soil and vegetation has been demonstrated by Curran (1981, 1982), Duggin et al. (1997). With laser source, the incident polarization can also be fully controlled. Under sunlight, we have only the very unpolarized Sun and occasionally polarized sky. Halogen lamps are also naturally unpolarized. Controlling the polarization from these is more complicated, because of the larger size of the beam and possible heat problems, especially, if one wants to keep the wide spectral range.

Instruments capable of measuring angular properties are generally called goniometers, and more specifically gonioradiometers, goniospectrometers, goniopolarimeters, or some combination of these or other properties. Some goniometers used in remote sensing or vegetation applications are EGO, Field Goniometer System (FIGOS), ASG, and FIGIFIGO (Hosgood et al., 2000; Sandmeier and Itten, 1999; Turner, 1998; Painter et al., 2003; Suomalainen et al., 2009a; Peltoniemi et al., 2005). Further measurements have been performed by Kriebel (1978), Guyot et al. (1980), Deering and Leone (1986), and Kuusk (1991). There are at least two field goniometers that can measure linear polarization: the FIGIFIGO in Finland (Suomalainen et al., 2009b; Peltoniemi et al., 2014) Figs. 7.3 and 7.4, and one in China (Sun and Zhao, 2011; Dong-hui et al., 2010). Both of these instruments use ASD FieldSpec Pro as the sensor and a rotating calcite wedge polarizer, and can thus measure detailed spectral polarization.

Fig. 7.3. FIGIFIGO measuring heather in Sodankylä, 2007. The optics is in the top of the arm and the light is guided to the spectrometer inside the body using a light cable. Under the arm is the Spectralon reference panel used to calibrate the measurements and, left side, one can see the pyranometer monitoring skylight changes. The sun is now close to the arm direction and there is some diffuse light coming from the clouds in the sky. (Actually, the photo is taken during a break in the measurements, while waiting for the sky to clear.)

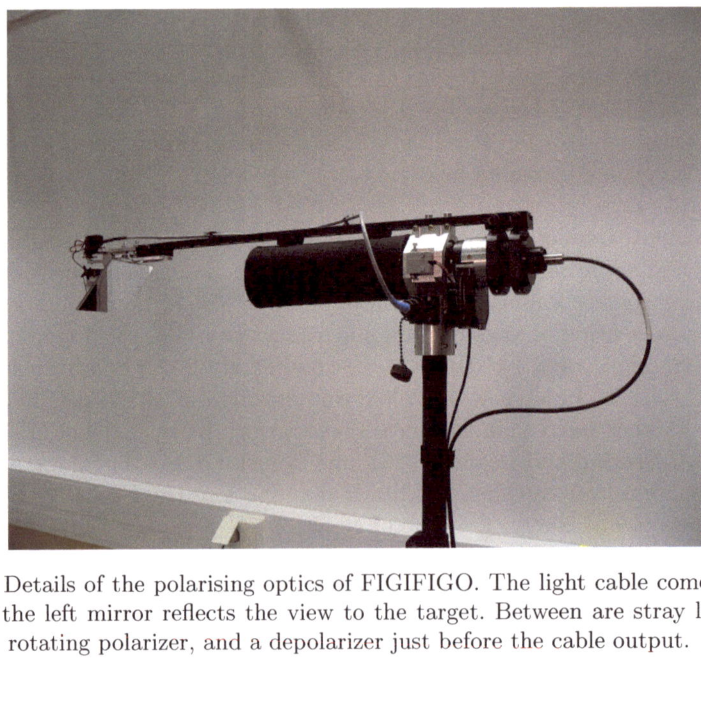

Fig. 7.4. Details of the polarising optics of FIGIFIGO. The light cable comes from the right and the left mirror reflects the view to the target. Between are stray light shield, front lens, rotating polarizer, and a depolarizer just before the cable output.

An alternative for larger homogeneous areas is to use a turning polarimeter on a platform. Rondeaux and Guyot (1990), Rondeaux and Herman (1991) measured polarization on a crane at a height of 20 m. Shibayama et al. (2011) mounted the imaging polarimeter on top of a small truck. Ghosh et al. (1993) used a tripod and tilted the spectrometers at three angles.

Individual leaves can be measured using smaller laboratory goniometers or ad hoc set-ups. Pospergelis (1969) measured the polarization of rocks and leaves. Many researchers (Vanderbilt et al., 1985, 1991; Vanderbilt and Grant, 1986; Grant et al., 1993; Raven et al., 2002) have performed extensive measurements on various leaves. Kharuk and Yegorov (1990) also used a Brewster angle radiometer with turning polarization filters (Fig. 7.5). Shibayama et al. (Shibayama and Akita, 2002; Shibayama, 2004; Shibayama and Watanabe, 2006, 2007, 2008; Shibayama et al., 2011) measured wheat canopies and leaf polarization to estimate the leaf inclination and other crop properties. Brakke (1994) measured polarized reflectance

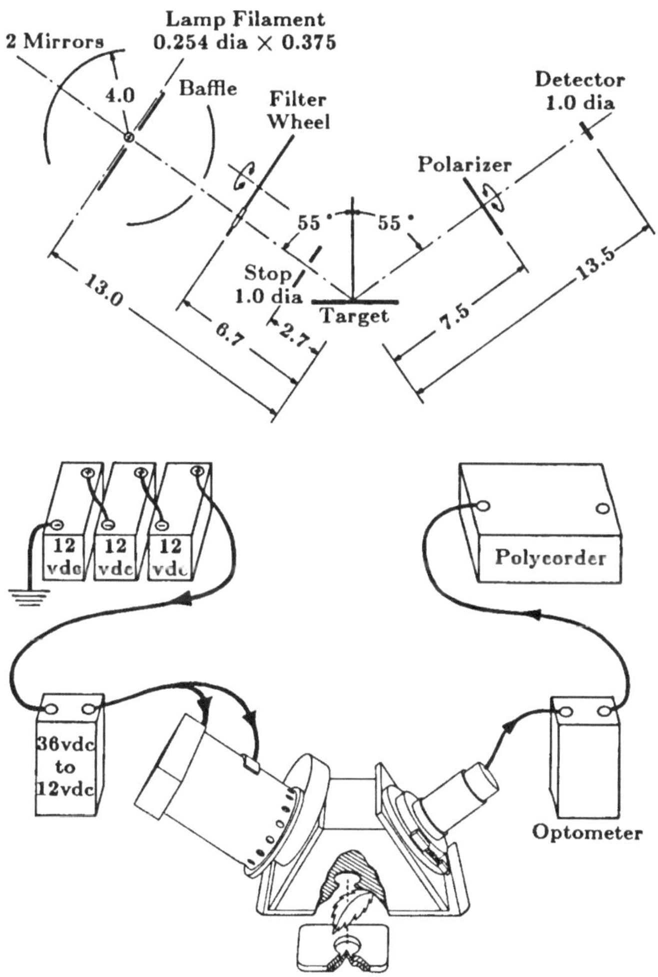

Fig. 7.5. A typical set-up to measure polarization of leaves in the Brewster angle (Grant et al., 1993).

and transmittance of oak and maple leaves. Typically, these are using laser light and can measure a few selected wavelengths. Georgiev et al. (2010) used NASA's scatterometer with ASD spectrometer to measure the BRF of several leaves, but this time polarising the incident beam. Some reviews of polarization measurements are given by Egan (1985) and Tyo et al. (2006).

Polarization is also measured by satellites, such as POLDER. However, all the polarization sensors have a pixel size of several kilometers, and are thus not suitable for land surface remote sensing. They still need an understanding of land surface polarization to compensate for the errors in atmospheric sensing, and thus several studies on land surface polarization using POLDER data have been done (Nadal and Bréon, 1999; Anjum and Ghosh, 2000; Sridhar et al., 2005).

Much better resolution is obtained by aerial sensors (Elias et al., 2004; Waquet et al., 2009; Gu et al., 2011). Airborne BOLDER has been used by Deuze et al. (1993), Bicheron et al. (1997), and Bréon et al. (1997).

Measurements of circular polarization of vegetation are very rare, and mostly motivated by astronomy (Kolokolova et al., 2011). Pospergelis (1969) had a system to measure full Stokes parameters of rocks and leaves spectrally. More recently, Sparks et al. (2009a, 2009b) measured circular polarization using diffused incident light. This measurement arrangement also neutralizes polarization, but is justified by their application (Fig. 7.6). Later, Martin et al. (2010) measured in several wavelengths, but only one fixed geometry, which, through experience of linear polarization is not necessarily the most significant (Fig. 7.7).

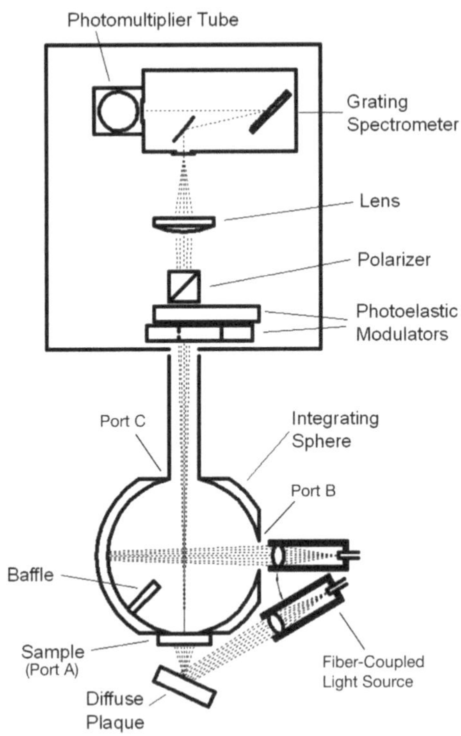

Fig. 7.6. The measurement set-up of Sparks et al. (2009a, 2009b).

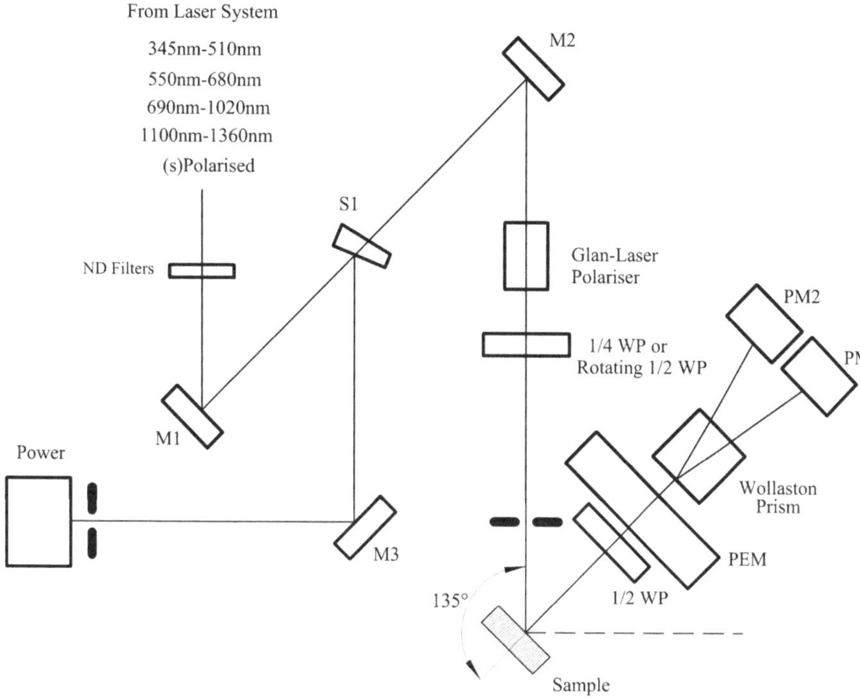

Fig. 7.7. The measurement set-up of Martin et al. (2010).

Table 7.1. List of polarization measurements from vegetation: first part.

Author	Target name	Instrument	Description
Dollfus (1961)	Grass	Photopolarimeter	
Pospergelis (1969)	Green leaf	Full Stokes spectrometer	
Vanderbilt et al. (1985)	Wheat		
Kharuk and Yegorov (1990)	Polluted leaves	Micron radiometer, polarising filters	Brewster angle
Rondeaux and Herman (1991)	Corn canopy		
Rondeaux and Herman (1991)	Soy bean		
Ghosh et al. (1993)	Wheat canopy	Manual filter radiometer	Time series
Grant et al. (1993)	18 Leaves	Polarising photometer	Only Brewster angle
Brakke (1994)	Oak and maple leaves		
Bréon et al. (1995)	Millet		
Bréon et al. (1995)	Grass		
Sridhar et al. (2005)	Wheat	POLDER	
Shibayama and Watanabe (2006)	Wheat		
Sparks et al. (2009a, 2009b)	Leaves	Integrating sphere spectrometer	Circular
Martin et al. (2010)	Leaves	Laser prism system	Circular
Gu et al. (2011)	Forest, shrub, soil	DPC	Flight at 4 km, resolution 4 m
Shibayama et al. (2011)	wheat	MBPI	

A set of the most referred polarization measurements is listed in Table 7.1. Table 7.2 adds also all respective FIGIFIGO measurements.

Table 7.2. List of polarization measurements from vegetation: second part—FIGIFIGO measurements.

Target name	Date	Location	Solar angles
Lingonberry	2007-07-03	Sodankylä	47.4°
Crowberry	2007-07-03	Sodankylä	54.2°
	2007-07-04	Sodankylä	44.5°
	2007-07-05	Sodankylä	45.3°
Heather	2007-07-03	Sodankylä	50.4°
	2007-07-05	Sodankylä	45.4°
	2007-07-05	Sodankylä	44.6°
	2007-07-05	Sodankylä	47.5°
	2007-07-05	Sodankylä	44.7°
Lichen	2009-08-12	Masala	38°
	2009-08-12	Masala	48°
	2009-07-13	Masala	62°
Lawn grass	2009-09-24	Masala	63.5 °
	2009-09-24	Masala	63.3°, 70.0°

7.5 Analysis

Firstly, we compare the vegetation against non-vegetation, to see the most evident features. In this way, four samples are studied in further detail: lawn grass and moose lichen, as measured in the FGI laboratory in Masala using FIGIFIGO, snow from Greenland, and sand from a beach (Fig. 7.8). We look at every scattering feature available in these sets.

Figure 7.9 presents a 3D plot of the BRF of the four samples in red color (670 nm). Clearly, the vegetation is very strong backscatterer. Lichen, in particular, is the strongest backscatterer of all samples in the FGI's Reflectance Library database (Peltoniemi et al., 2013). A strong bowl shape and forward-scattering in grass can also be detected. Snow is much more of a forward-scatter, and sand has a more symmetrical bowl shape. A more detailed view in the principal plane is given in Fig. 7.10. Here, one can see the differences of several wavelengths. The typical features remain in all channels, but the forward-scattering is even more enhanced in brighter wavelengths. However, based on these and many other published results, the BRF shape alone appears to be quite a limited indicator of green vegetation.

The spectrum (Fig. 7.11) marks all vegetation clearly: all leaves are green, except lichen is gray, but the typical NIR brightening is still present. Snow, sand, soil, and all artificial materials are easily ruled out with a few spectral tests.

What new perspective will the polarization give? In Fig. 7.12, the degree of linear polarization ($P_Q = -Q/I$) is shown in the principal plane in six wavelengths.

a) b) c) d)

Fig. 7.8. Four sample targets compared here: (a) grass, (b) lichen, (c) snow, and (d) sand. The red laser spots, about 10 cm apart, show the field of view (they are now turned off for the measurements, but some older data may show a small spike at 640 nm).

The typical feature for all samples is the small negative polarization in the backward direction, and a varying positive polarization in the forward direction. The position of the maximum polarization should be connected to the refractive index. From the dark SWIR band of snow, we can say it agrees well with the refractive index of 1.35. Sand is too bright to make any comment; multiple and internal scattering is much more dominant than single reflection in all wavelengths (Figs. 7.13 and 7.14). Grass would agree with refractive indices between 1.3 and 1.5. No conclusion can be made about lichen. The hemispherical plot of $P_L = \sqrt{Q^2 + V^2}/I$ is shown in Fig. 7.15. It is clear to see that the strongest signatures are in the principal plane.

Spectrally, polarization is quite smooth, as seen in the spectral plots of the degree of polarization in Fig. 7.16 and polarized reflectance in Fig. 7.17. Here, the backscattering spectrum is always taken 5° from the solar direction, and the forward spectrum is taken near the Brewster angle of 110° phase angle. The polarization follows inversely the brightness spectrum: the degree of polarization is strongest at the darkest wavebands—red and violet. The green vegetation can be clearly identified. Otherwise, the polarization spectrum brings little new information.

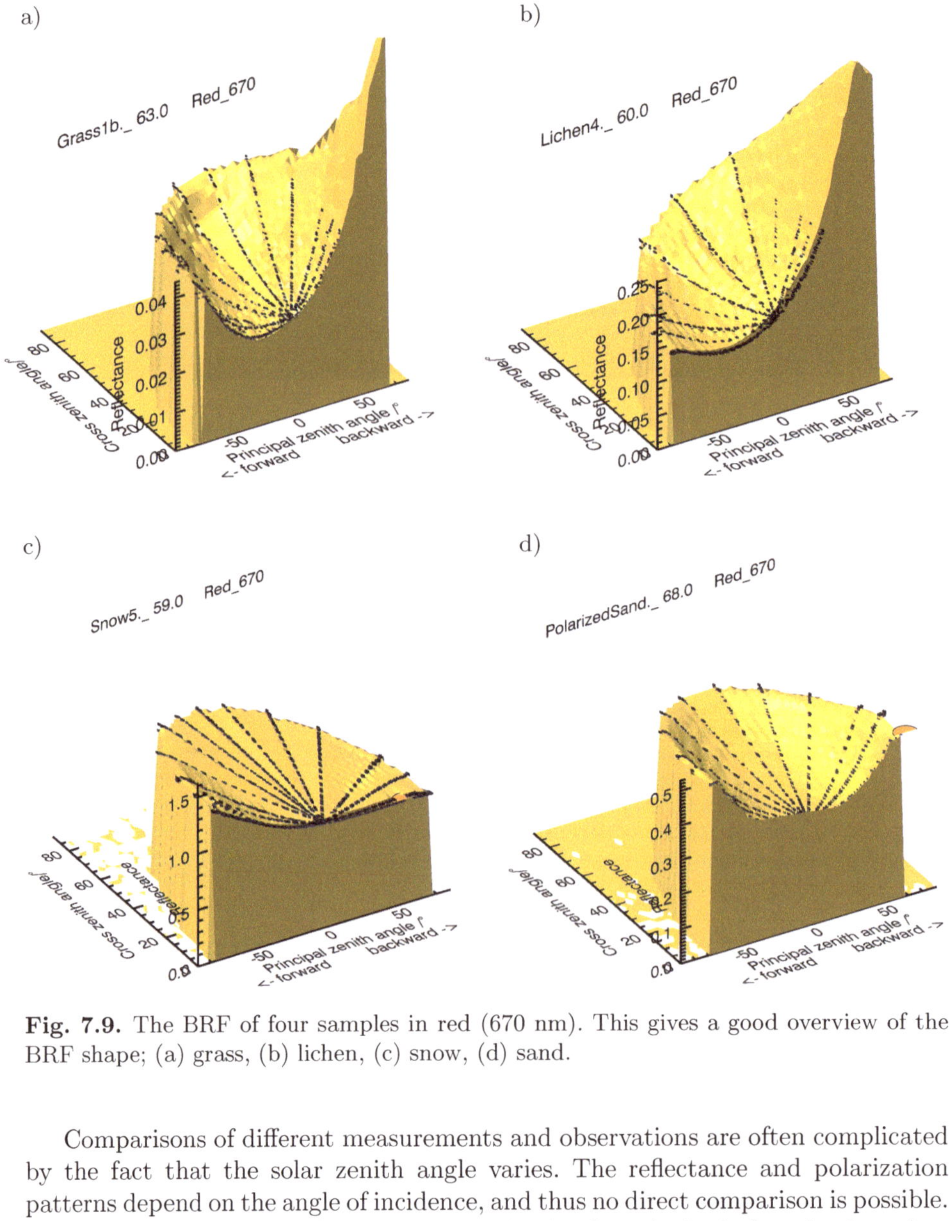

Fig. 7.9. The BRF of four samples in red (670 nm). This gives a good overview of the BRF shape; (a) grass, (b) lichen, (c) snow, (d) sand.

Comparisons of different measurements and observations are often complicated by the fact that the solar zenith angle varies. The reflectance and polarization patterns depend on the angle of incidence, and thus no direct comparison is possible. In Fig. 7.18, the degree of linear polarization in the principal plane is plotted at three solar zenith angles. The polarization signature looks more pronounced at the larger solar zenith angles of 60° to 70° than at 45°.

Of the circular polarization, we are unable to show comparable data—no compatible measurements yet. As an example, we present three plots of various leaves from Sparks et al. (2009a, 2009b) and Martin et al. (2010) (Figs. 7.19–7.21). Their measurement set-ups were described earlier. This demonstrates that, in these cases, the circular polarization is perhaps one-third of the linear polarization, or smaller. The geometry is not ideal for a maximum polarization signature, however.

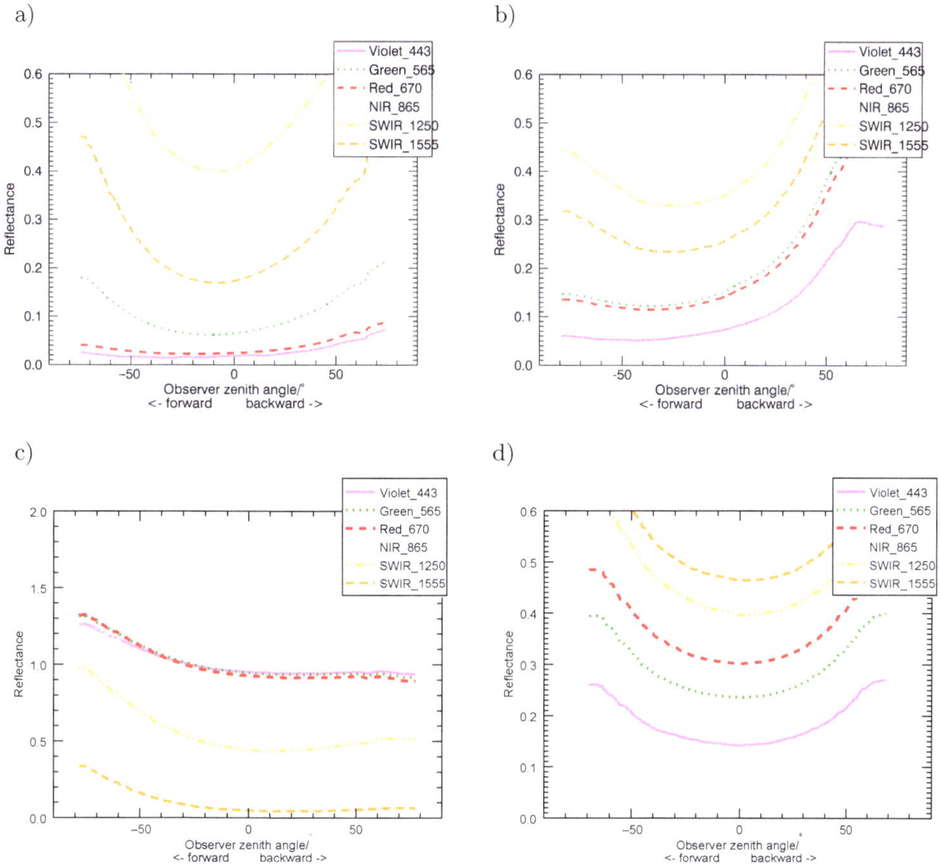

Fig. 7.10. The BRF of four samples in the principal plane in six wavelengths. Here, one can see more quantitative the differences between targets and channels; (a) grass, (b) lichen, (c) snow, (d) sand.

7.5.1 Special features by species

After the general overview, we examine more closely the differences between targets. However, before going to too far into conclusions regarding differences between different species, it is a good idea to look at how in principle the same target type varies. Heather was measured in seven different locations in Sodankylä during the 2007 research project. The conditions were quite similar in all cases. Four of these measurements are shown in Figs 7.22 and 7.23 with the reflectance and the degree of linear polarization in the principal plane. One can clearly see many small differences in the details. Some of these may be measurement errors, but certainly for the most part it is simply that the samples are that much different.

Next, we will discuss in further detail the specific features of different vegetation. In Fig. 7.24 two more samples—lingonberry and crowberry—are shown, also measured in Sodankylä in 2007 using FIGIFIGO. We also refer to many other measurements, of which two examples are shown in Fig. 7.25.

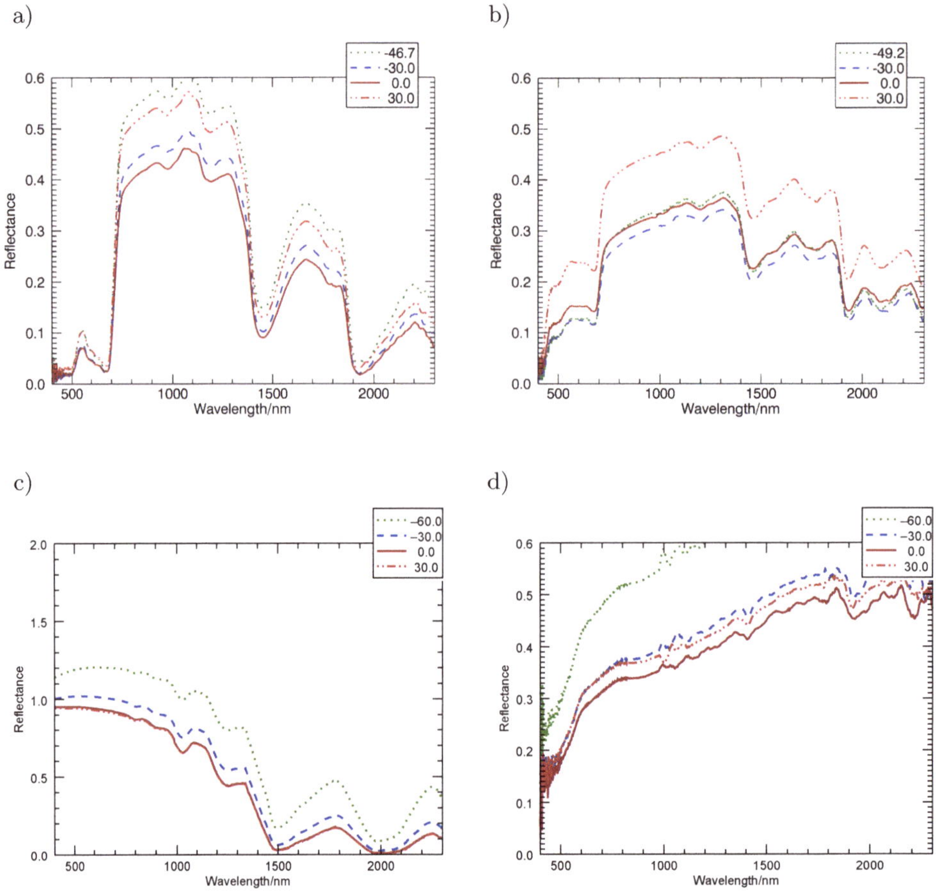

Fig. 7.11. The detailed spectrum of four samples in the principal plane in four directions; (a) grass, (b) lichen, (c) snow, (d) sand.

Lichen (*Cladina arbuscula* and *C. rangiferina*) of course looks much brighter in visual bands or is grayer than the green vegetation, and is easily identifiable. All lichen measured was a strong backscatterer, to a greater or lesser extent—actually among the strongest we have measured. The polarization is more neutral than with green vegetation, consistently with the brightness and multiple scattering. Lichen abundance should thus be rather invertible compared to spectral-directional data. From the available data, we are not yet in a position to promise signals for more detailed classifying of lichen subspecies.

Grass (Bréon et al., 1995; Dollfus, 1961) (including natural, garden, and football pitches) shows the 'purest' green vegetation spectrum of all the measured samples, namely sharpest green spike at 550 nm and brightest at near infrared (NIR) 800–1300 nm. Grass may have the strongest bowl shape of all measured samples. This may be due to its erectophile nature. Clover flowers make the spec-

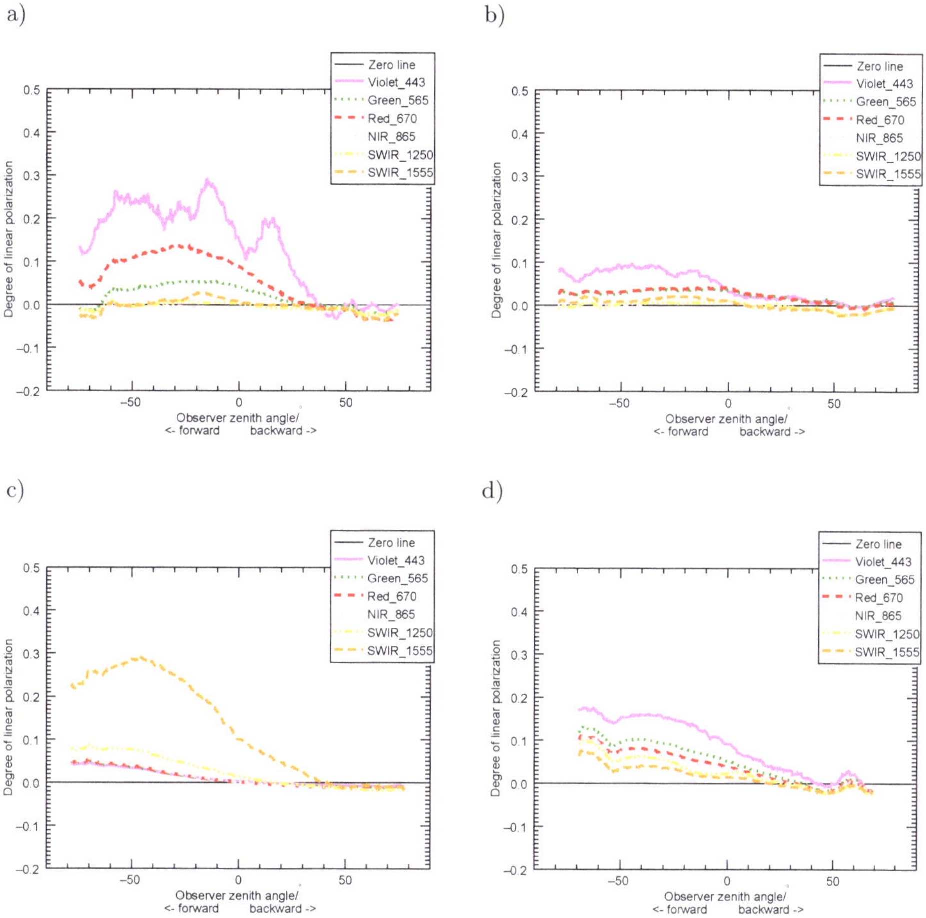

Fig. 7.12. The degree of linear polarization ($P_Q = -Q/I$) in the principal plane in six wavelengths; (a) grass, (b) lichen, (c) snow, (d) sand

trum a tiny bit whiter, and reduce the bowl shape. Polarization is quite average for green vegetation, without any evident grass-specific features.

Dwarf shrubs have many common features: all have some backward enhancement and bowl shape. **Bilberry** (*Vaccinium myrtillus*) and **lingonberry** (*Vaccinium vitis-idaea*) with their wax leaves have additionally clear forward brightening, which means they can be easily distinguished from **heather** (*Calluna vulgaris*), which is the strongest backscatterer among the shrubs measured. **Crowberry** (*Empetrum nigrum*) is also more of a backscatterer. Measurements for **dwarf birch** (*Betula nana*) are intermediate between lingonberry and crowberry. Some bilberry samples show a tiny enhancement around specular direction. The strong forward-scattering is connected to strong polarization, up to 50% with lingonberry. Crowberry is a slightly weaker polarizer, and heather already even more neutral than grass. Forward polarization measurement is certainly a good tool for identifying

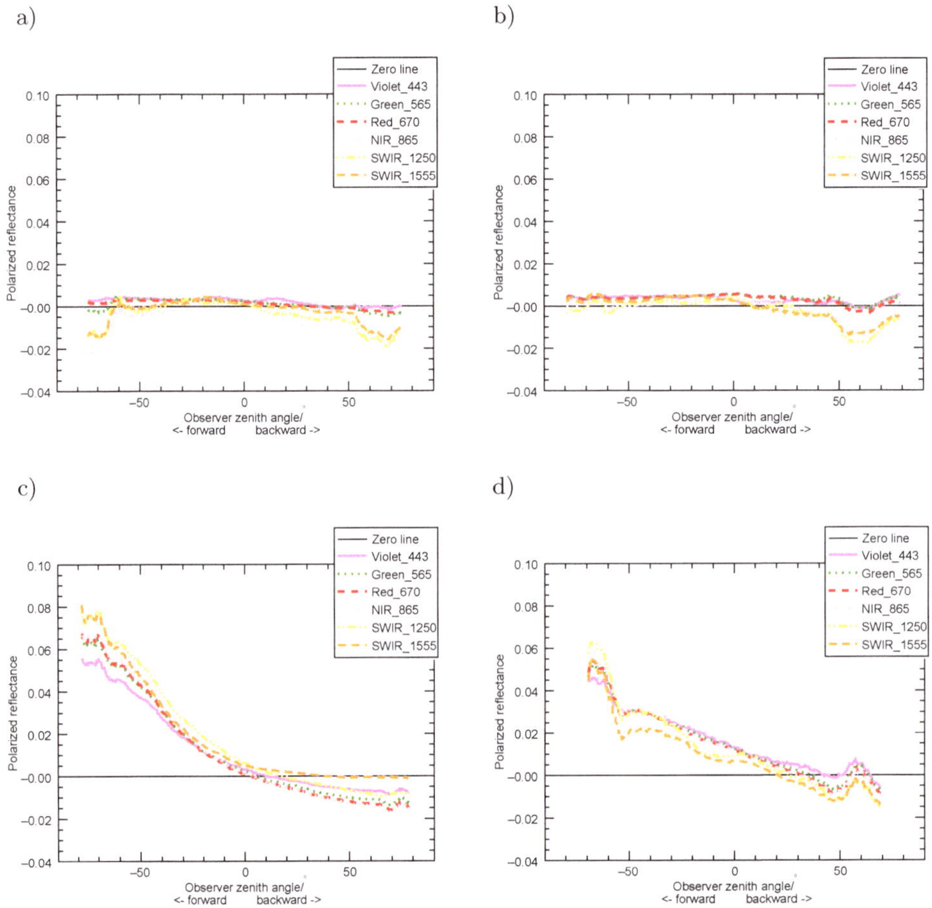

Fig. 7.13. The polarized reflectance in the principal plane in six wavelengths; (a) grass, (b) lichen, (c) snow, (d) sand.

forest understorey species, if only such forward directions can be observed inside forests, or at least for mountain and swamp vegetation.

Soybean (*Glycine max*) (Rondeaux and Herman, 1991) shows very strong forward polarization—up to 35%.

Corn canopy (Rondeaux and Herman, 1991) is also almost as high a polarizer as soybean. The polarization depends on the growing phase—the tassels slightly weaken the polarization.

Wheat canopy (*Triticum* genus) (Vanderbilt et al., 1985; Ghosh et al., 1993; Shibayama and Watanabe, 2006) is a forward polarizer, at least up to 10% to 20%. The polarization varies significantly during the growing phase. The headed canopy polarizes much less than preheaded. Recently, Shibayama et al. (2011) noted dependence on several other variables, but with only a less than 1% degree of

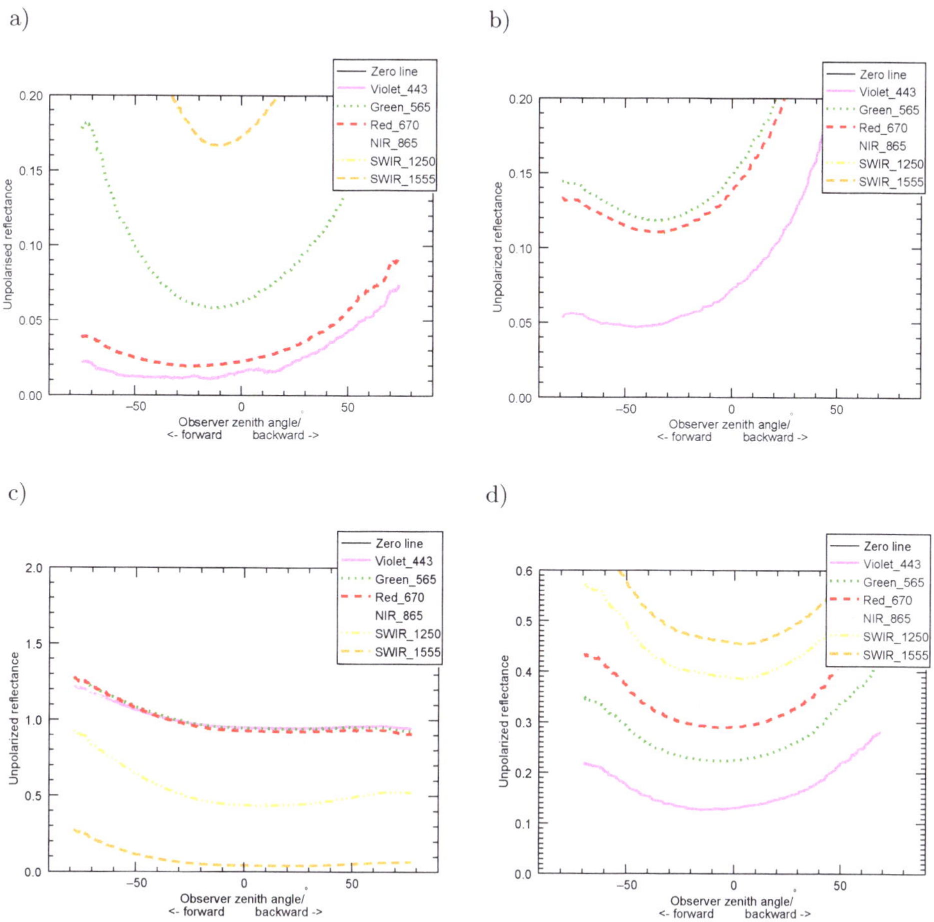

Fig. 7.14. The unpolarized reflectance in the principal plane in six wavelengths; (a) grass, (b) lichen, (c) snow, (d) sand.

polarization at off-maximum directions. Thus, polarization is a promising remote sensing tool and should not be ignored.

Headed millet (*Pennisetum glaucum*) (Bréon et al., 1995) polarizes more weakly than grass, but here we cannot compare them more quantitatively, as the measurements differ too greatly.

Individual leaves vary in a full range between weakly to almost fully polarising (Grant et al. 1993, Fig. 7.26), depending on many surface properties, and they are not yet fully analysed. Moisture, dirt, and damage affect polarization significantly (Kharuk and Yegorov, 1990; Figs 7.27 and 7.28).

Other samples tend to follow similar patterns. Thus, wax leaves cause the strongest polarization, smooth leaves are weaker, and tassels, heads, or flowers weaker still (Raven et al., 2002; Savenkov et al., 2003). Different measurements

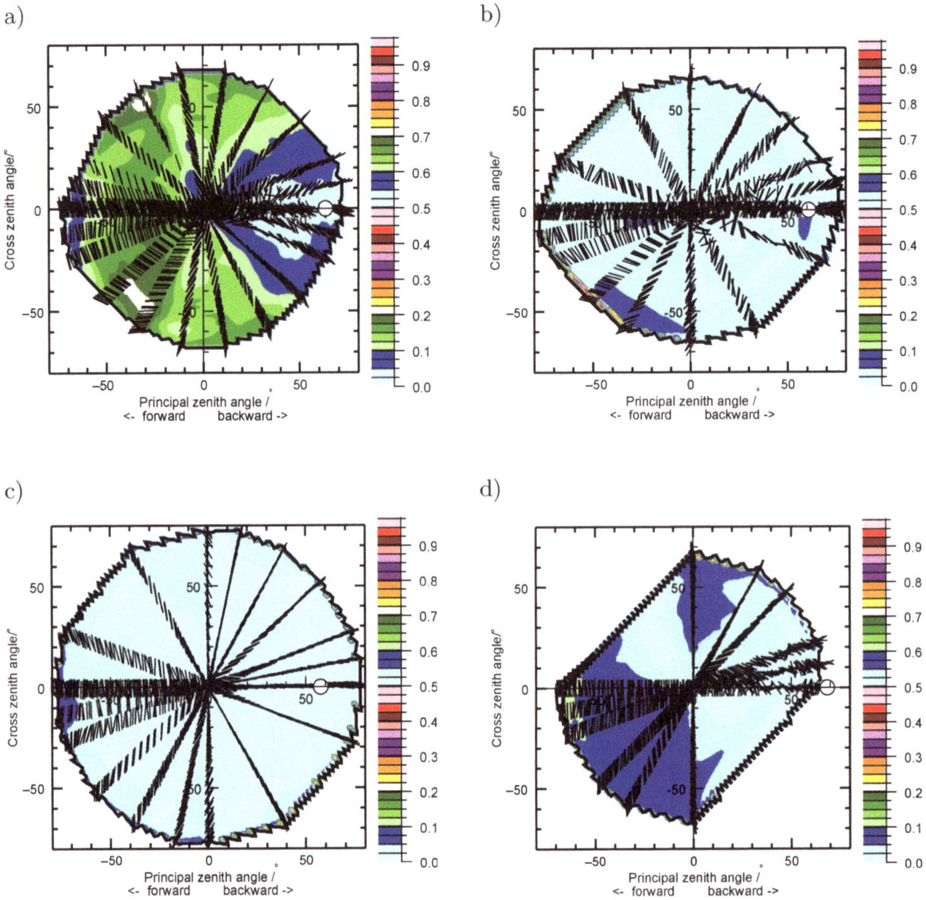

Fig. 7.15. The degree of linear polarization $P_L = \sqrt{Q^2 + V^2}/I$ in 670 nm. Colors indicate the strength of the polarization, and the small lines denote the direction. The small white disk marks the position of the Sun; (a) grass, (b) lichen, (c) snow, (d) sand.

may be incompatible with each other, and most space and airborne measurements have too large a footprint for heterogeneous land surfaces.

7.6 Discussion on specific remote sensing signatures

7.6.1 Heterogeneity, or spatial variations

All land surfaces are heterogeneous in many scales and are defined by many variables: topography, land cover, tree types, understory types, wetness, snow cover, etc. This structure must be first revealed with certain accuracy by classical photogrammetric tools, before one can apply more advanced remote sensing techniques. Otherwise, too many signals are mixed and do not allow for the interpretation of

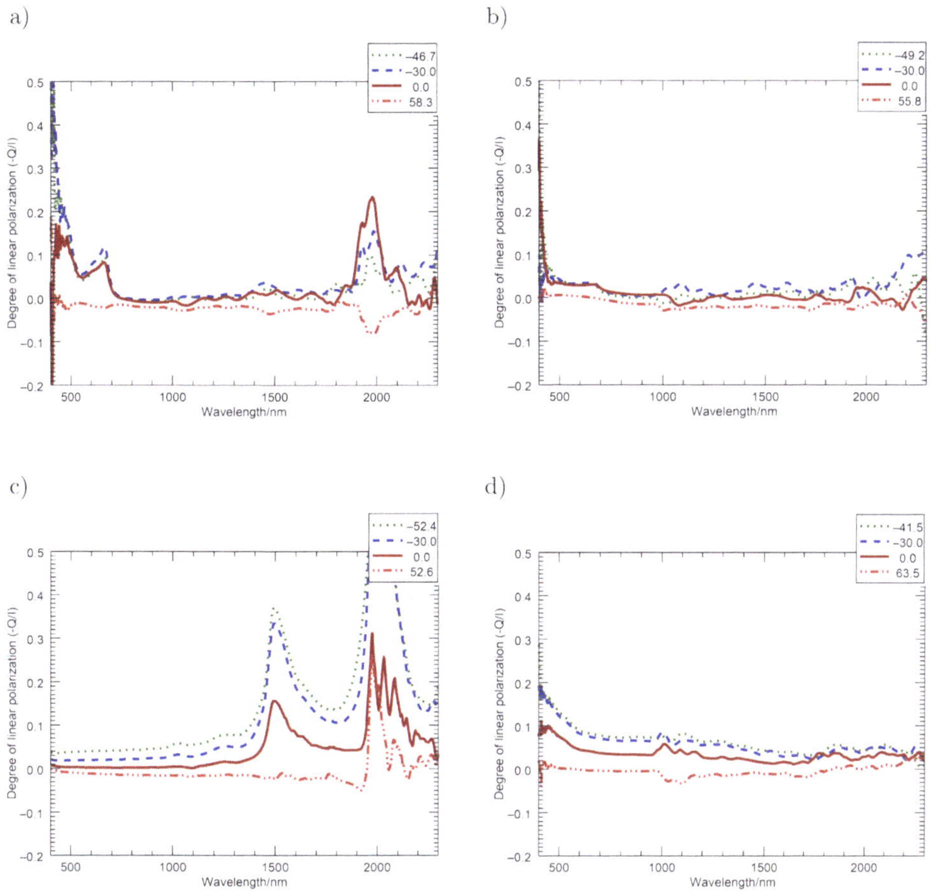

Fig. 7.16. The polarization spectrum in the principal plane, at a phase angle of 5° up from the Sun, nadir, 30° forward, and at 110° phase angle (near the Brewster angle). The legend gives the zenith angles; (a) grass, (b) lichen, (c) snow, (d) sand. The degree of polarization follows inversely the brightness.

measurement data in a unique way. Often, this primary information is already available in maps and existing databases. The spectrometric and polarimetric techniques are most efficient when one can locate fixed targets or closely homogeneous target areas. For the identification of trees or other species, knowledge of shapes and sizes of the crowns, and/or individual leaves, provides the most valuable input, though such data are not always available.

7.6.2 Anisotropy

The typical features of the anisotropy of green vegetation are:

1. Very strong backscattering at all wavelengths. We have not yet measured any exceptions to that, although it is possible that some rare flat planophile plants could be an exception to this rule.

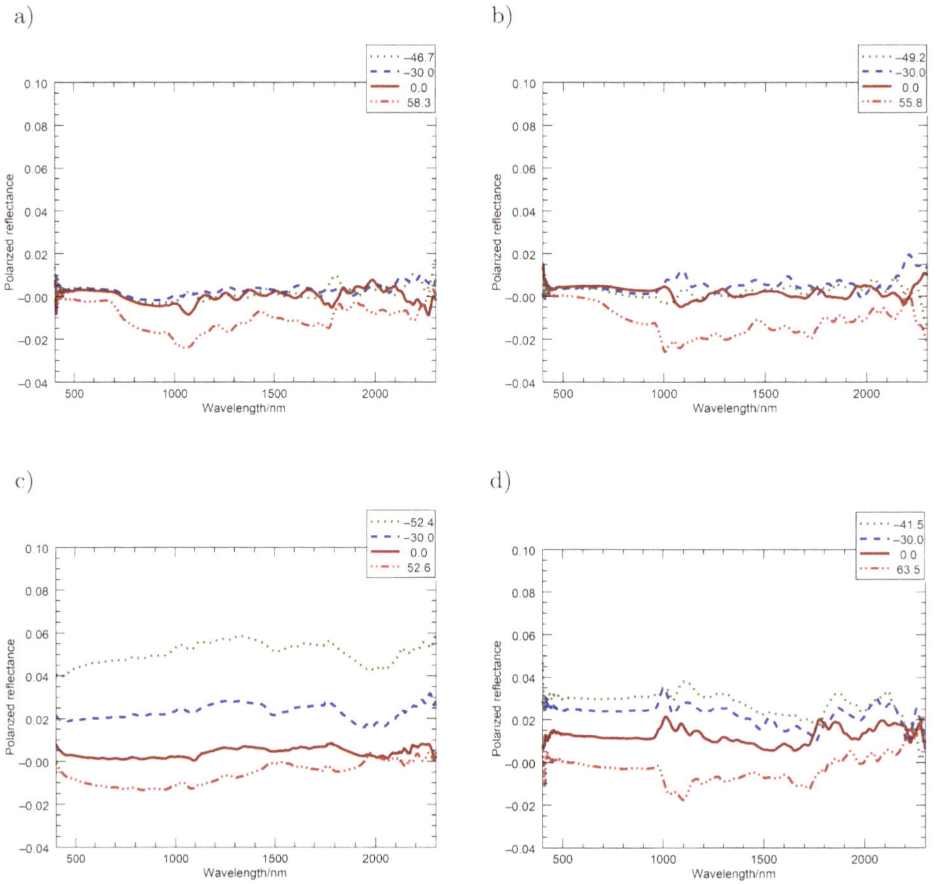

Fig. 7.17. The polarized reflectance spectrum, as in Fig. 7.16; (a) grass, (b) lichen, (c) snow, (d) sand. Looks quite featureless, only backscattering has some trends.

2. Forward brightening, as a function of the phase angle. This varies widely from non-existent to significant. By and large, this is assumed to be the result of the specular reflection from leaf surfaces.
3. The specular effect is rarely observable, presumably requiring strong planophile leaf orientation.
4. A bowl shape—that is, brightening from nadir to larger zenith angles, sometimes weaker, sometimes stronger. The reverse (bell shape) has not been seen in FGI's measurements of vegetation to date.
5. The darkest spot is usually either near nadir, or far forward.

Despite the fairly large differences in the BRF shape (anisotropy), it is quite difficult to suggest clear anisotropy signals to be used for remote sensing purposes—firstly because most instruments observe too small a directional range, and are missing the most indicative forward and backward directions, so that anisotropy signals are too mixed with heterogeneity signals; and secondly because the anisotropy de-

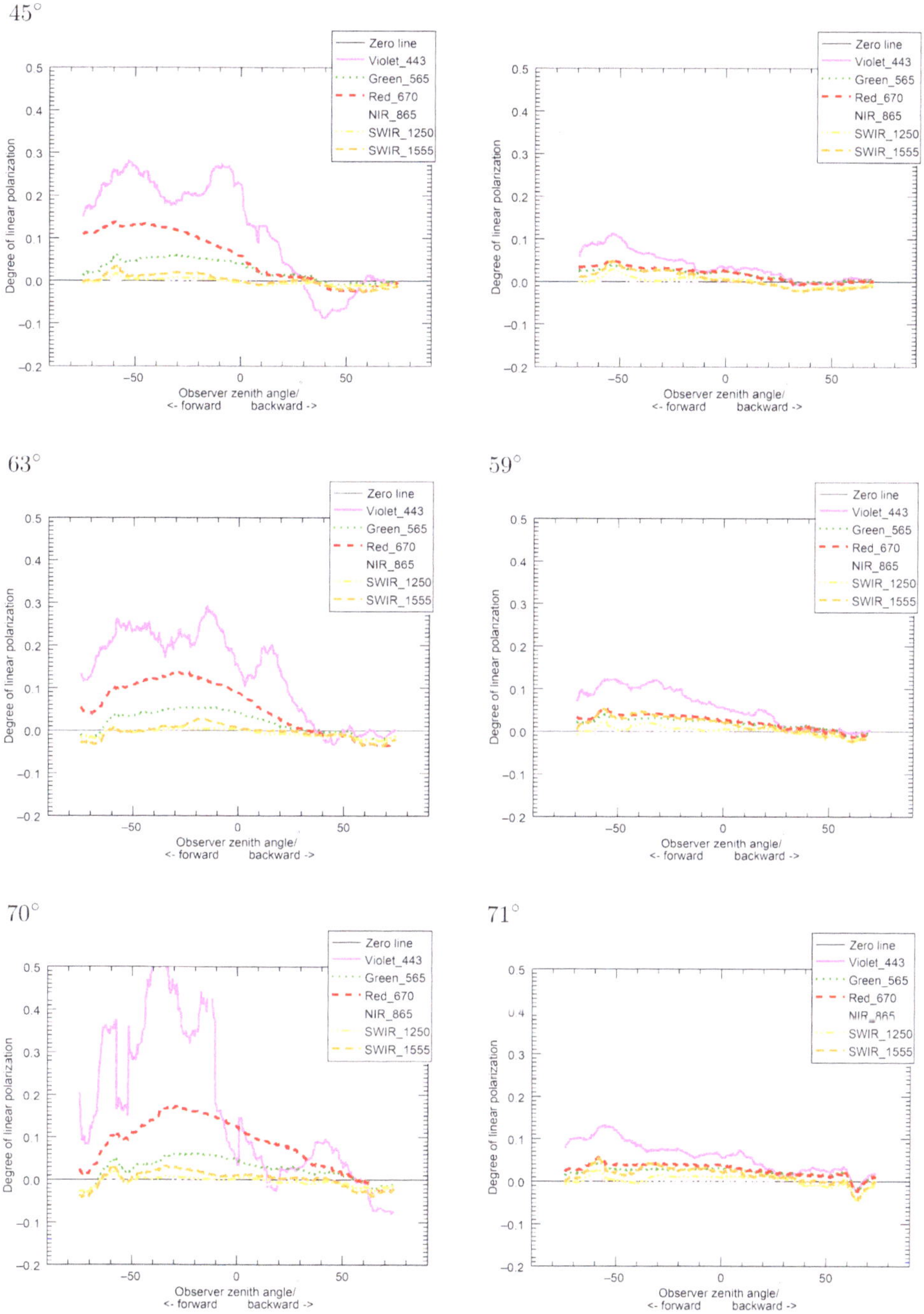

Fig. 7.18. Effect of the angle of incidence for the degree of linear polarization. On the left Grass at 45°, 63°and 70° angles of incidence, and right is Lichen at 45°, 59°, and 71°. The polarization maximum is shifted, but otherwise systematic effects are hidden behind fluctuations. At 60°–70° incidence angles, the polarization signals are stronger than at 45°, and more useful for remote sensing.

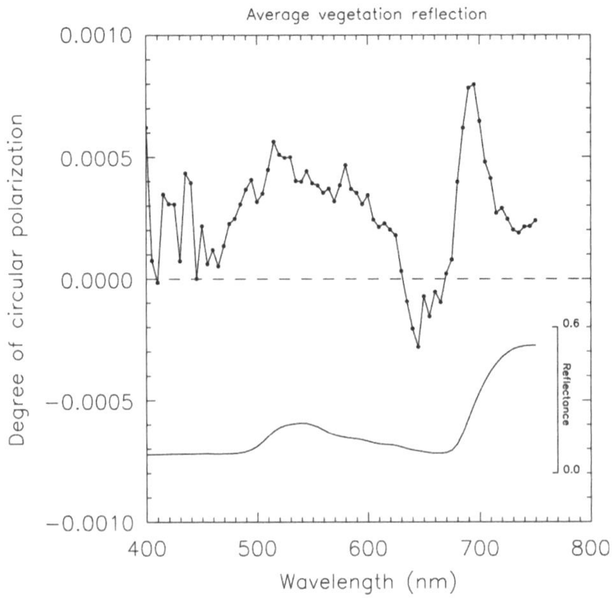

Fig. 7.19. Circular polarization and reflectance (low) of a leaf (measurements from Sparks et al. (2009b)).

pends on many parameters which cannot be determined by means of anisotropy measurements alone.

However, when a large enough angular range can be observed, the strengths of the forward and backward scattering can differentiate between some species. Clearly, strong forward-scatterers are moss (all measured species), grass, lingonberry, crowberry, and bilberry, which return the signal in approximately the same order. Heather, peat, cotton grass, and lichen have smaller forward brightening—some samples do not show brightening at all. Lichen types are all strong backscatterers.

7.6.3 Spectral signature

The typical spectral shape of all vegetation is very similar: very dark in red and blue wavelengths, a small bump in green, and very bright in NIR. Furthermore, in SWIR, there are signatures of water absorption, with a minimum amount at around 960 nm, 1200 nm, 1450 nm, and 1900 nm. Flowers, trunks, soil, and rocks all give their contribution to the spectra in various ways.

It is obvious that lichen is much whiter than green vegetation, and differences extend to all wavelengths, although the typical NIR brightening feature is present. Moss has some yellowness in the visible part and clear darkening in SWIR. Recently mowed grass may be the brightest in NIR, whereas crowberry and heather are the darkest. Many of the differences may be more related to dryness and ageing of the leaves than to real differences between species (Rautiainen et al., 2011). Unfortunately, we still do not have enough systematic data to separate all these effects.

a

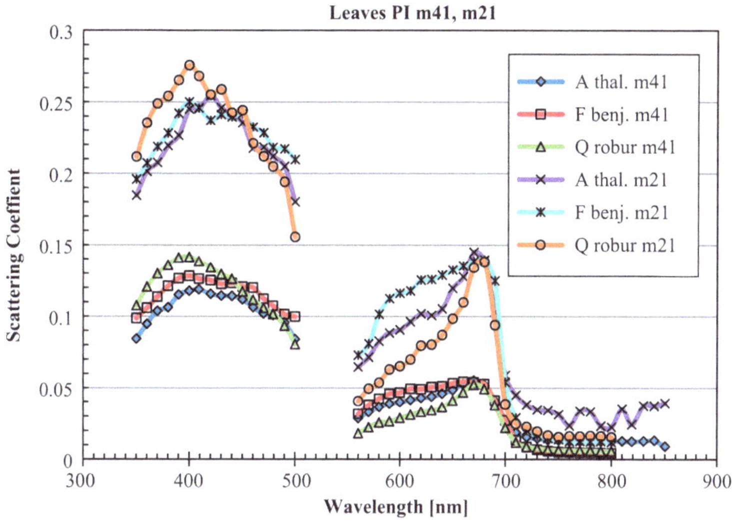

b

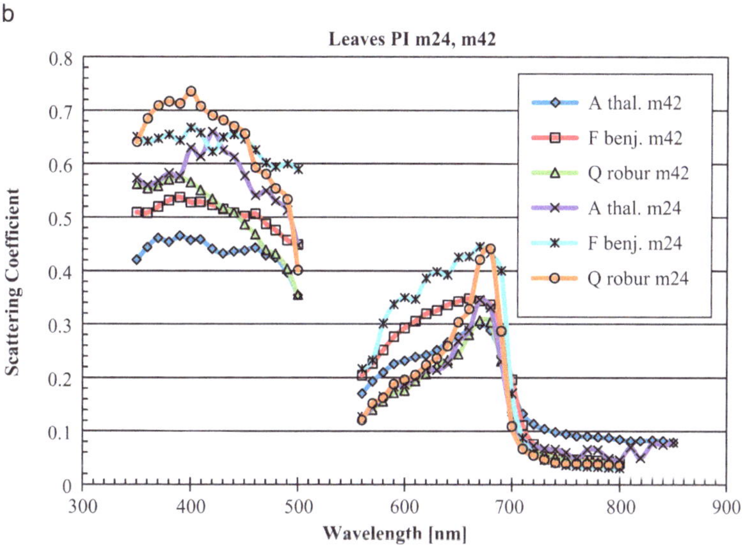

Fig. 7.20. Several elements of the scattering Muller matrix from different leaves, measured by Martin et al. (2010), at a fixed geometry. Elements m41 refer to circular polarization.

7.6.4 Polarization—any new signals?

Polarization measurements of vegetation are still rare, but some general trends can be noted. Spectrally, polarization contains fewer features than the scalar reflectance, and many of these signals are already better seen in the reflectance spectrum. For practical sensing, we see no reason for high resolution polarization spectrometry, but recommend that only one to six channels are selected as the most interesting bands.

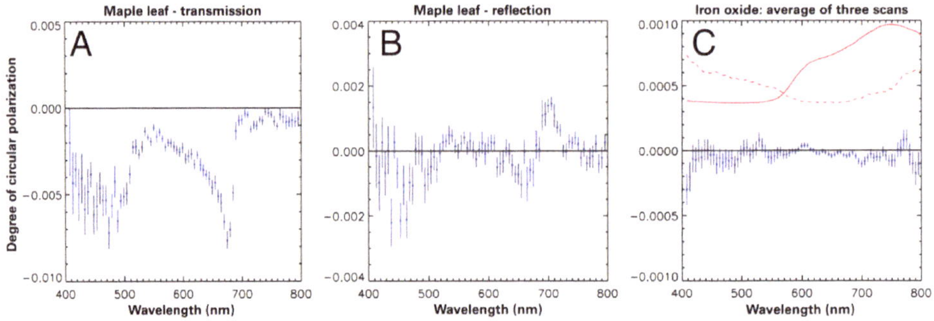

Fig. 7.21. Degree of circular polarization of a maple leaf, transmitted and reflected, compared to minerals (from Sparks et al. (2009a)).

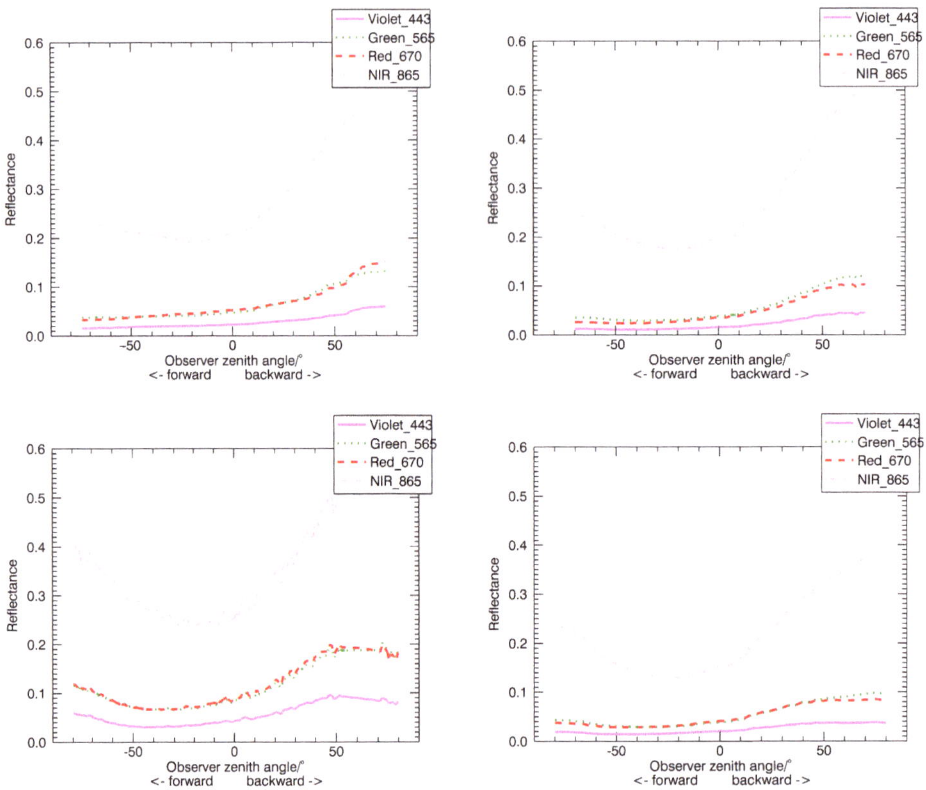

Fig. 7.22. To demonstrate the variability, here four different samples of heather taken at Sodankylä are shown. Firstly, the BRF is in the principal plane.

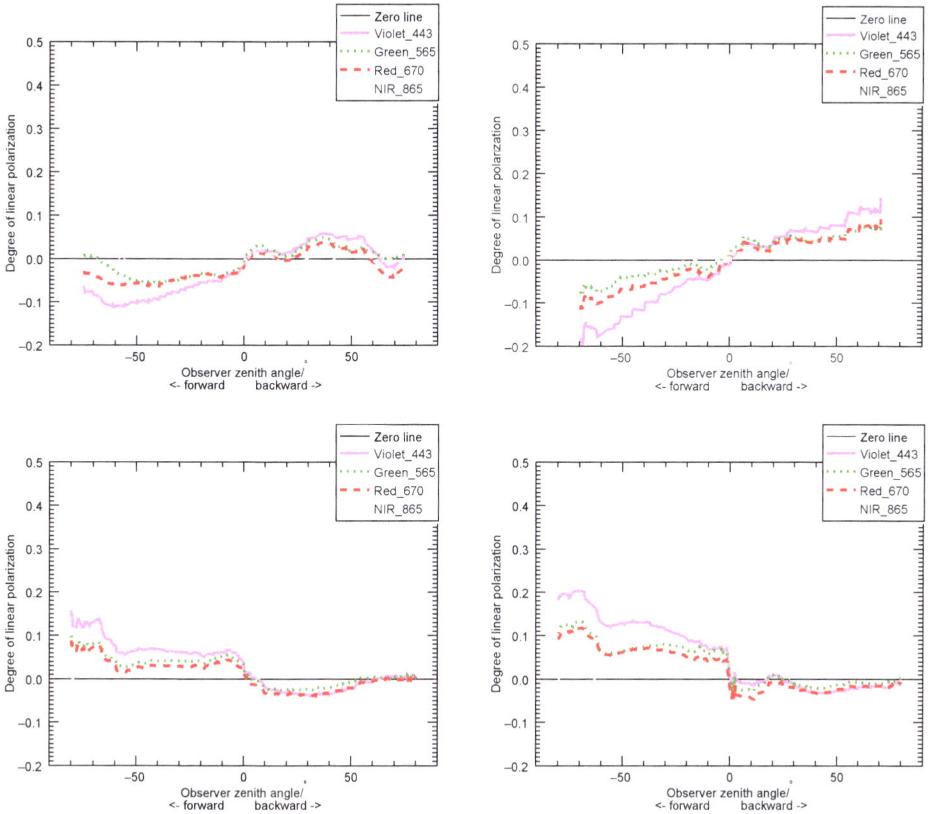

Fig. 7.23. Continuing the demonstration of variability from Fig. 7.22. Here, the degree of linear polarization in the principal plane is shown. The step at nadir is an artifact due to a non-optimal measurement sequence.

The most indicative number is the strength of the forward polarization. This can already separate various small vegetation types from each other well, and indicate the growing phase of many crops (Vanderbilt et al., 1985; Rondeaux and Herman, 1991; Ghosh et al., 1993; Shibayama and Watanabe, 2006). Theoretically, the accurate location of the maximum polarization would also connect the observation to chemical analysis but, to date, this is too uncertain with real varying targets.

The negative polarization around the backscattering interests astronomers in particular. However, applying near backscattering information to any operational work has some challenges. The signal is rather weak, requiring the instruments to have quite a high level of accuracy even to see the small differences between targets, if any. And the developed theoretical models are still not quite ready to say what the differences really mean. Also, the backscattering dome is too narrow to be systematically and economically observable in regular cases, but remains only a speciality for specific purposes. If active laser remote sensing can develop to multi-angular receiving, is another possibility we cannot yet predict. Other uses of Lidar backscattering polarization are not discussed here, due to a lack of data available.

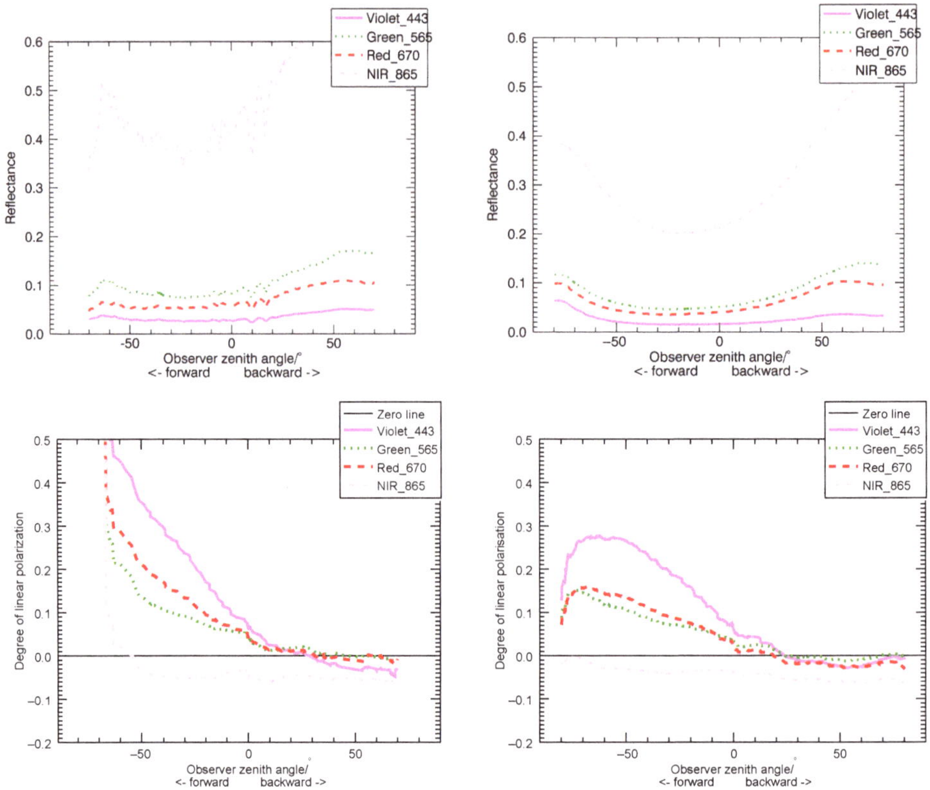

Fig. 7.24. To complement the picture, here lingonberry (left) and crowberry (right) BRF (top) and polarization (bottom) in the principal plane are also shown.

7.7 Discussion on measurement principles

Polarization measurement techniques are already frequently used in numerous physical measurements and practical analyses. The technology is mature, and most components are available off-the-self, so there are no principal difficulties in collecting the desired measurement, if known. Of course, including polarization further complicates measurements, and a compromise may need to be made. The non-polarizing alternative is usually also sought as a result. Here, we skip the most technical level, and discuss more idealistic levels of measuring.

Each polarization measurement system referred to here differs significantly from others. It is not possible to select any of these as a general reference system, as they all have their purposes that they fulfill. The drawback is that all data are also different and a direct comparison requires some effort, and is often not at all possible. There are reasons for every design and, below, we discuss some aspects that might be useful in future constructions.

Which polarization should be measured? Only two Stokes parameters $[I, Q]$, three $[I, Q, U]$, or also the circular polarization (V)? Most features can be already seen from just the two components in the principal plane but, to date, other com-

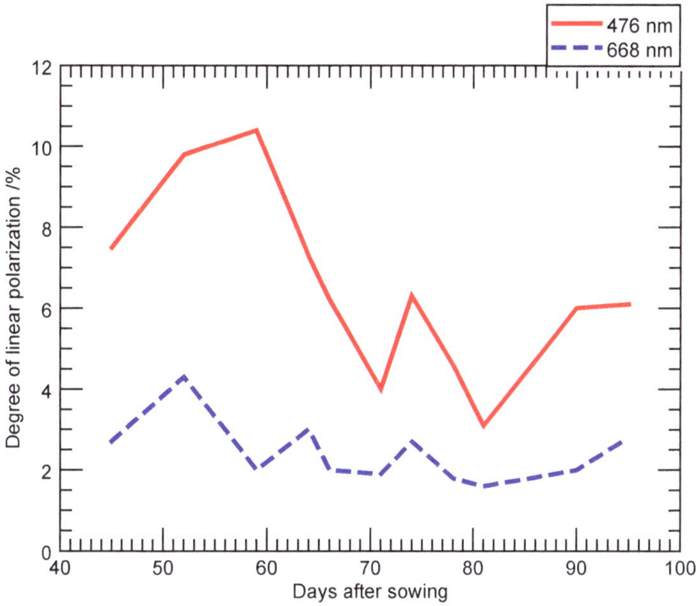

Fig. 7.25. Degree of linear polarization of a wheat canopy at two wavelengths, as a function of the days after sowing. (adapted from Ghosh et al. (1993)).

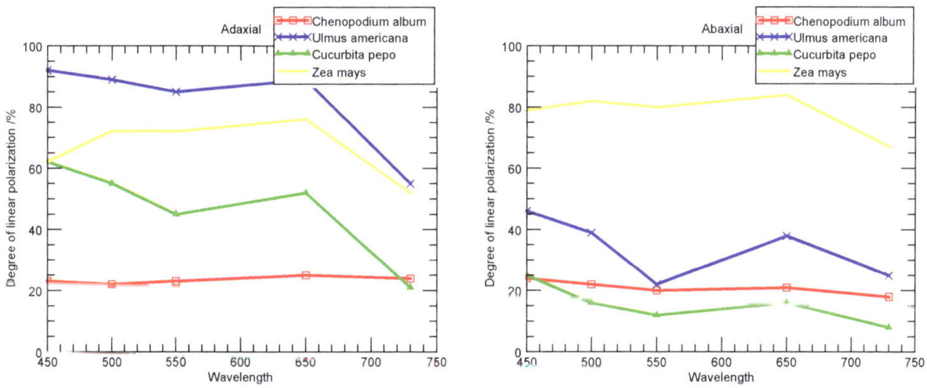

Fig. 7.26. Degree of linear polarization from four leaves at the Brewster angle, at adaxial and abaxial sides, adapted from Grant et al. (1993). The other 16 samples measured lie between these. The differences between different leaves and even different sides of the same leaves are remarkable, and the polarization has a very strong signature.

ponents are so under-researched that there is good reason to measure all. It would be even better if the full 4×4 Muller matrix could be measured. At least for model development and validation, such data are helpful, although, in practical remote sensing, full Muller matrix applications do not yet exist. This brings further challenges to the measurements set-up, doubling the complexity, or more, if the strong and wide incident beam must be polarized. On the other hand, in some cases, even

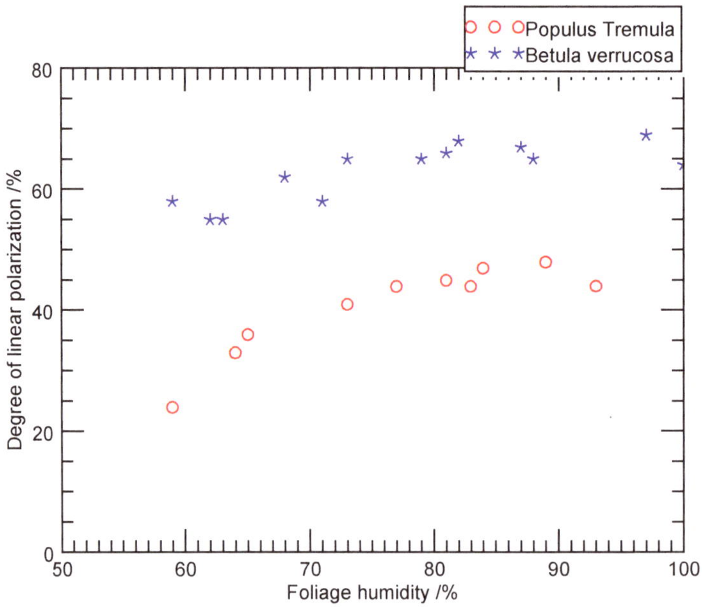

Fig. 7.27. The degree of linear polarization in two samples (*Betula* and *Populus*) as a function of foliage humidity (adapted from Kharuk and Yegorov (1990)).

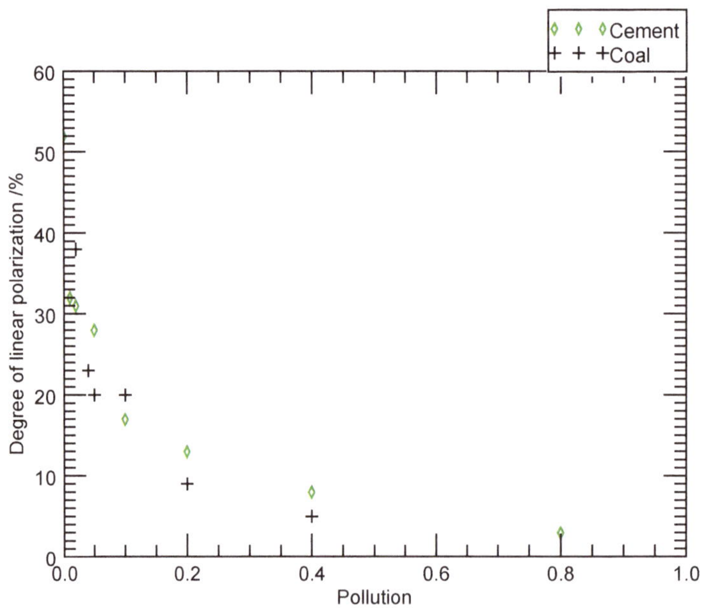

Fig. 7.28. The degree of linear polarization from a shoot of *Padus rademosa* as a function of contamination by cement and coal (adapted from Kharuk and Yegorov (1990)). The contamination makes a clear difference on polarization.

single polarization state $I_{\min}$ may be enough, to separate the unwanted externally reflected signal from the interesting internally scattered measurement.

Which directions to measure? The most promising signals are no doubt in the forward direction, near the Brewster angle, but we still recommend the widest angular range possible for measurement, varying also the incident direction, or the orientation of the scatterers. The polarization depends on the scattering angle, and this dependence is also good to know. Additionally, in practical remote sensing, it is not possible to fix observation angles too tightly, but the best possible solution must be sought. For model validation and development, the most extreme forward and backward directions are of specific interest, as are all locations of minima and maxima. The very backward direction requires specific instrumentation, since the sensor is normally shadowing the target. Typically, this is settled using beam splitters.

Which wavelengths should be measured? Based on this experience, we would recommend low to medium resolution multi-spectral measurements in a wide range from ultraviolet (UV) to NIR as long as convenient. High-resolution spectrometry is also useful for research purposes, especially if the spectrometer is available, as it is likely that not all features have yet been seen. Most measurements to date have only been in the visual domain, and the NIR part is shown as an almost white area on the map. The wider wavelength range brings new demands for the instruments. The basic polarizers are typically designed for a narrow wavelength range. To measure over a wide band, more specific systems must be used, which normally means using prisms instead of filters. Also, the sensors are usually less effective and optimal in NIR range, since they require more intense light or longer integration time. NIR sensors are not so extensively tested for polarimetric applications, and one must first check the inherent polarization sensitivity of the instrument itself, and decide what actions are needed, such as the use of a depolarizer.

Which target area should be measured? The turning polarimeters see a different part of the surface for all angles, but even center-looking goniometers have the problem that the field of view elongates with the changing zenith angle. If the sample has a 3D structure, more significant problems arise. The size of interesting targets varies from micrometers to kilometers. A single instrument is usually optimal for limited target sizes only.

Temporal aspect—living vegetation is changing in seasonal and diurnal cycles. It also reacts to ambient light, meteorological parameters and conditions (such as, temperature, water amount, cyclones) and natural hazards. Following these changes forms a separate set of research problems, and often one would need to have taken a snapshot of a constant state. If one outside measurement cycle takes hours, the end experimental conditions can already differ significantly from the start conditions. As a rule of thumb, a reasonable measurement should be able to be performed in about 10 minutes.

It is obvious that one cannot solve all these problems using a single instrument and protocol. Thus, a suitable combination of desired properties must be selected. Additional inputs can be supplemented with possible cooperative measurements or by coupling data with those obtained with several instruments and/or laboratories to get a more complete data set. Careful planning is needed to design a reasonably optimal measurement sequence.

One more important issue is calibration and the validation of each instrument. During field measurements, one cannot put too much trust in the factory calibration, but must use specific field standards. For spectrum and reflectance, there are already several commercial standards available, but the literature on polarization measurements did not refer to the use of any common standard. We leave this question open for further discussion.

Other technical aspects affecting the measurements—we noticed the following factors caused some disturbance to measurements or complicated them. The instrument itself often causes shadowing of the incident light, or its surfaces cause unwanted reflections. In outdoor measurements, diffuse light comes from everywhere, which is also strongly polarized, and must be taken into account. UV is particularly strong in the skylight and, in NIR, all vegetation and trees around are as bright as white walls. The temporal variability of sunlight, or actually any unstabilized light source, must be taken into account and compensated accurately. Successive polarization measurements are particularly easily corrupted by even small uncertainties in parameters describing incident light.

7.8 Conclusions

The use of polarization in optical remote sensing of land surfaces is still in its early stages. As the polarization signals are mostly relatively weak, they often vanish behind the atmospheric effects and noise. Astronomical and atmospheric applications are developing more rapidly, because the measured polarization signals are easier to interpret than those obtained from heterogeneous land surfaces. There are many satellites with polarization sensors that measure the atmosphere, but land surface measurements need a much higher resolution, on a meter scale. Nevertheless, the present sensors help to provide remarkable data for measuring vegetation and snow. More high-resolution airborne data are likely to be available soon.

Polarization measurements represent a complementary source of information with good perspective. Spatial, spectral, and directional signals are usually clearly stronger, and polarization measurements should be integrated with these other techniques to ensure the most productive output. Additionally, the instrumentation and observation protocol must be designed well. For operational polarimetric remote sensing, we provide the following suggestions: (i) concentrate on forward angles near the maximum, namely phase angles between 90° and 130°. Measure at least the darkest bands, red or violet, optionally a few more; (ii) fly on clear days with little diffuse light; best time is when the Sun is rather low, at 60° to 70° zenith angles. As always, the calibration of the system and proper atmospheric corrections are of great importance.

If polarization is the unwanted signal, then the most suitable scenario is to measure around nadir or in slightly backward directions. One may also want to use a single orientation polarization filter to remove the specular reflections to penetrate inside leaves or below the water surface. In sufficient light, such a polarization filters also smooth the anisotropy effects (BRDF) or reduce the forward spike, which may in some conditions be helpful to avoid overexposures and ease mosaicking of images.

The hot spot region, and the negative polarization around it, is of primary academic interest. Therefore, it must be carefully addressed in fundamental studies prior to further extension to wider practical applications and rough assumptions. The use of circular polarization remains also unresolved. The published data are far too limited to make general conclusions. Obviously, the circular polarization signal is weaker than linear polarization, but there are indications that it could provide some new information on chlorophyll. The current measurement set-ups require more advanced design to be able to reveal the most promising directions and wavelengths.

At present, polarization models offer explanations on some observed features. The Fresnel reflection can explain a major part of the high forward-scattering, and multiple-scattering explains the low polarization in NIR. However, the difficulties in electromagnetic modeling of scattering from structures that have a critical size of 1–100 wavelengths complicate further modeling, which would require the following improvements:

- full polarized 3D radiative transfer and/or ray-tracing modeling;
- extending electromagnetic scattering techniques to larger objects;
- finding good 'perturbed Fresnel' equations for semidiffuse reflection;
- developing coherent backscattering models to a more practical level;
- extending the models to more realistic cases (e.g. to account for proper shapes, structures, multi-scale phenomena, rich mixtures of everything, dependent statistics);
- making models more physical: all parameters should be uniquely and practically measurable from real targets;
- reliable extensive testing of developed models against experimental reference data.

All measurement-based papers share the conclusion that the polarization of vegetation is barely explored, and that more measurements are needed, from many different viewpoints. We recommend that new measurements are performed with a focus on:

- more complete polarization, including circular components (full Stokes vector), when possible measurements should be supported with polarized incident radiation making full Muller matrices available;
- more complete spectral and angular range, from very backscattering to very forward-scattering, from UV to TIR;
- systematic and comparable measurements of hundreds of different samples to quantify differences and to find specific features;
- more complete chain over different size scales, starting from individual leaves and their internal structures, and extending to shoots, branches, shrubs, individual trees, and to forests;
- systematic studies of the effects of different parameters, such as moisture, surface roughness, chemical composition, internal structures, size, shape, and orientation distributions, as well as multiple scattering;
- well-controlled artificial test targets to give elementary test data to validate the models.

Based on reviewed reflectance measurements and databases, and discussions raised, we conclude that polarization is a valuable aspect of electromagnetic radiation and may reveal differences between various vegetation types, as well as separate their different growing phases. It is a more difficult physical phenomenon in interpretation than many other features, and requires more careful measurements and more complete modeling to gain good understanding and optimal exploitation. We also advocate full understanding of the entire scattering and remote sensing process, not limiting to polarization or any other selected feature, but trying to find the most productive modeling and measurement concepts.

Acknowledgements

This research was supported by the Academy of Finland Projects Nos. 260027 (SIRONTA), 120949, and 265949, the Nordkalotten Satellite Evaluation co-operation Network (NorSEN), the ERC Advanced Grant No. 320773 (SAEMPL), the COST Action MP1104 'Polarization as a tool to study the Solar System and beyond', and the Finnish Ministry of Agriculture and Forestry.

We thank Teemu Hakala, Juha Suomalainen, and the many other colleagues and partners that supported the measurements and provided data and insight.

References

Anjum, V. and Ghosh, R. (2000) 'Angular and polarization response of vegetation, bare soil and water from ADEOS POLDER data over India', *International Journal of Remote Sensing*, **21**, 805–10.

Apresyan, L.A. and Kravtsov, Y.A. (1983) *Radiation Transfer Theory*, Moscow, Nauka (in Russian).

Bicheron, P., Leroy, M., Hautecoeur, O. and Bréon, F.M. (1997) 'Enhanced discrimination of boreal forest covers with directional reflectances from the airborne polarization and directionality of earth reflectances (POLDER) instrument', *J. Geophys. Res.*, **102**, 29517–28.

Brakke, T. (1994) 'Specular and diffuse components of radiation scattered by leaves', *Agricultural And Forest Meteorology*, **71**, 283–95.

Bréon, F.-M., Tanré, D., Lecomte, P. and Herman, M. (1995) 'Polarized reflectance of bare soils and vegetation: Measurements and models', *IEEE Trans. Geosci. Remote Sensing*, **33**, 478–99.

Bréon, M., Vanderbilt, V., Leroy, M., Bicheron, P., Walthall, C.L. and Kalshoven, E. (1997) 'Evidence of hot spot directional signature from airborne POLDER measurements', *IEEE Trans. Geosci. Remote Sensing*, **35**, 479–84.

Coulson, L., Bouricius, G.M.B. and Gray, E.L. (1965) 'Optical reflection properties of natural surfaces', *J. Geophys. Res.*, **70**, 4601–11.

Curran, P.J. (1981) 'Remote sensing: The use of polarized visible light (PVL) to estimate surface soil moisture', *Applied Geography*, **1**, 41–53.

Curran, P.J. (1982) 'Polarized light as an aid to vegetation classification', *Remote Sens. Environment*, **12**, 491–9.

Deering, D. and Leone, P. (1986) 'A sphere-scanning radiometer for rapid directional measurements of sky and ground radiance', *Remote Sens. Environment*, **19**, 1–24.

Deuze, J.L., Bréon, F.M., Roujean, J.L., Deschamps, P.Y., Devaux, C., Herman, M. and Podaire, A. (1993) 'Analysis of POLDER airborne instrument observations over land surfaces', *Remote Sens. Environment*, **45**, 137–54.

Dollfus, A. (1961) *Polarization Studies of Planets*, Chicago, University of Chicago Press.

Dong-hui, X., Pei-juan, W., Qi-jiang, Z. and Hong-min, Z. (2010) 'Modeling polarimetric BRDF of leaves surfaces', *Spectroscopy and Spectral Analysis*, **30**, 3324–8.

Duggin, M.J., King, G.J. and Schrader, M. (1997) 'Enhancement of vegetation mapping using Stokes parameter images', in Goldstein, D.H. and Chipman, R.A. (eds), *Polarisation: Measurement, Analysis, and Remote Sensing*, Proc. of SPIE, **3121**, 307–13.

Egan, G. (1985) *Photometry and Polarization in Remote Sensing*, New York, Elsevier.

Elias, T.G., Cairns, B. and Chowdhary, J. (2004) 'Surface optical properties measured by the airborne research scanning polarimeter during the CLAMS experiment', in Schaefer, K.P., Comeron, A., Carleer, M.R. and Picard, R.H. (eds), *Proc. of SPIE*, **5235**, 595–606.

Furfaro, R. (2009) *Radiative Transfer in Vegetation Canopies: Forward and Inverse Problems for Airborne Remote Sensing of Crop Fields*, Saarbrücken, VDM Publishing.

Georgiev, G., Thome, K., Ranson, K., King, M. and Butler, J. (2010) 'The effect of incident light polarization on vegetation bidirectional reflectance factor', *Proceedings of the 2010 IEEE International Geoscience and Remote Sensing Symposium (IGARSS)*, 25–30 July, Honolulu, HI, USA.

Ghosh, R., Sridhar, V., Venkatesh, H., Mehta, A. and Patel, K. (1993) 'Linear-polarization measurements of a wheat canopy', *International Journal of Remote Sensing*, **14**, 2501–8.

Grant, L., Daughtry, C.S.T. and Vanderbilt, V. (1993) 'Polarized and specular reflectance variation with leaf surface-features', *Physiologia Plantarum*, **88**, 1–9.

Gu, X., Cheng, T., Xie, D., Li, Z., Yu, T. and Chen, H. (2011) 'Analysis of surface and aerosol polarized reflectance for aerosol retrievals from polarized remote sensing in PRD urban region', *Atmospheric Environment*, **45**, 6607–12.

Guyot, G., Malet, P. and Baret, F. (1980) 'Analyse des indicatrices de réflexion de l'orge et du blé, possibilité de la stéréoradiométrie', *Proc. Int. Symp. I.S.P. Hamburg*, International Archives of Photogrammmetry, XXII-B7, Commission VII, pp. 372–81.

Hapke, B. (1993) *Theory of Reflectance and Emittance Spectroscopy*, Cambridge, Cambridge University Press.

Hapke, B., DiMucci, D., Nelson, R. and Smythe, W. (1996) 'The cause of the hot spot in vegetation canopies and soils, shadow-hiding versus coherent backscatter', *Remote Sens. Environment*, **58**, 63–8.

Hosgood, B., Sandmeier, S., Piironen, J., Andreoli, G. and Koechler, C. (2000) 'Goniometers', in Webster, J. (eds), *Encyclopedia of Electronics Engineering*, Vol. 8, New York, John Wiley, pp. 424–33.

Jin, Y.-Q. (2005) *Theory and Approach of Information Retrievals from Electromagnetic Scattering and Remote Sensing*, Springer.

Kharuk, V.I. and Yegorov, V.V. (1990) 'Polarimetric indication of plant stress', *Remote Sens. Environment*, **33**, 35–40.

Knyazikhin, Y. and Marshak, A. (1991) *Fundamental Equations of Radiative Transfer in Leaf Canopies and Iterative Methods for Their Solution*, Berlin, Springer–Verlag, pp. 9–44.

Kolokolova, L., Sparks, W. and Mackowski, D. (2011) 'Astrobiological remote sensing with circular polarization', in Mischenko, M.I., Yatskiv, Y.S., Rosenbush, V.K., and Videen, G. (eds), *Polarimetric Detection, Characterization and Remote Sensing (NATO Science for Peace and Security Series C: Environmental Security)*, Springer, pp. 277–94.

Kriebel, K.T. (1978) 'Measured spectral bidirectional reflection properties of four vegetated surfaces', *Appl. Opt.*, **17**, 253–9.

Kuusk, A. (1991) 'Determination of vegetation canopy parameters from optical measurements', *Remote Sens. Environment*, **37**, 207–18.

Liang, S. (2004) *Quantitative Remote Sensing of Land Surfaces*, Wiley Series in Remote Sensing, New Jersey, Wiley.

Liang, S. and Mishchenko, M. (1997) 'Calculations of the soil hot-spot effect using the coherent backscattering theory', *Remote Sens. Environment*, **60**, 163–73.

Litvinov, P., Hasekamp, O. and Cairns, B. (2011) 'Models for surface reflection of radiance and polarized radiance: Comparison with airborne multi-angle photopolarimetric measurements and implications for modeling top-of-atmosphere measurements', *Remote Sens. Environment*, **115**, 781–92.

Litvinov, P., Hasekamp, O., Cairns, B. and Mishchenko, M. (2010) 'Reflection models for soil and vegetation surfaces from multiple-viewing angle photopolarimetric measurements', *J. Quant. Spectrosc. Radiat. Transfer*, **111**, 529–39.

Maignan, F., Bréon, F.-M., Fedele, E. and Bouvier, M. (2009) 'Polarized reflectances of natural surfaces: Spaceborne measurements and analytical modeling', *Remote Sens. Environment*, **113**, 2642–50.

Marshak, A. (1989) 'The effect of the hot spot on the transport equation in plant canopies', *J. Quant. Spectrosc. Radiat. Transfer*, **42**, 615–30.

Martin, W.E., Hesse, E., Hough, J.H., Sparks, W.B., Cockell, C.S., Ulanowski, Z., Germer, T.A. and Kaye, P.H. (2010) 'Polarized optical scattering signatures from biological materials', *J. Quant. Spectrosc. Radiat. Transfer*, **111**, 2444–59.

Mishchenko, M.I. (2008) 'Multiple scattering, radiative transfer, and weak localization in discrete random media: Unified microphysical approach', *Reviews of Geophysics*, **46**, RG2003.

Muinonen, K., Mishchenko, M.I., Dlugach, J.M., Zubko, E., Penttilä, A. and Videen, G. (2012) 'Coherent backscattering verified numerically for a finite volume of spherical particles', *The Astrophysical Journal*, **760**, 11.

Muinonen, K.O. (2004) 'Coherent backscattering of light by complex random media of spherical scatterers: Numerical solution', *Waves in Random Media*, **14**, 365–88.

Myneni, R.B. and Ross, J. (eds) (1991) *Photon–Vegetation Interactions*, Berlin, Springer–Verlag.

Nadal, F. and Bréon, F.-M. (1999) 'Parametrization of surface polarized reflectance derived from POLDER spaceborne measurements', *IEEE Trans. Geosci. Remote Sensing*, **37**, 1709–18.

Nicodemus, F.E., Richmond, J.C., Hsia, J.J., Ginsberg, I.W. and Limperis, T. (1977) *Geometrical Considerations and Nomenclature for Reflectance: Technical Report*, Washington, DC, Institute for Basic Standards, National Bureau of Standards.

Painter, T.H., Paden, B. and Dozier, J. (2003) 'Automated spectro-goniometer: A spherical robot for the field measurement of the directional reflectance of snow', *Review of Scientific Instruments*, **74**, 5179–88.

Peltoniemi, J.I., Hakala, T., Suomalainen, J., Honkavaara, E., Markelin, L., Gritsevich, M., Eskelinen, J., Jaanson, P. and Ikonen, E. (2014) 'Technical notes: A detailed study for the provision of measurement uncertainty and traceability for goniospectrometers', *J. Quant. Spectrosc. Radiat. Transfer*, **146**, 376–90.

Peltoniemi, J., Kaasalainen, S., Näränen, J., Matikainen, L. and Piironen, J. (2005) 'Measurement of directional and spectral signatures of light reflectance by snow', *IEEE Trans. Geosci. Remote Sensing*, **43**, 2294–304.

Peltoniemi, J., Gritsevich, M., and Hakala, T. (2013) *FGI's Reflectance Library and Future Perspectives*, Int. conference 'Spectro-polarimetric experiments for remote sensing', Abstracts, 19–21 August 2013, Masala, Finland.

Peltoniemi, J.I. (1993) 'Radiative transfer in stochastically inhomogeneous media', *J. Quant. Spectrosc. Radiat. Transfer*, **50**, 655–71.

Pospergelis, M.M. (1969) 'Spectroscopic measurements of the four Stokes parameters for light scattered by natural objects', *Soviet Astronomy*, **12**, 973–7.

Rautiainen, M., Mõttus, M., Heiskanen, J., Akujarvi, A., Majasalmi, T. and Stenberg, P. (2011) 'Seasonal reflectance dynamics of common understory types in a northern European boreal forest', *Remote Sens. Environment*, **115**, 3020–8.

Raven, P., Jordan, D. and Smith, C. (2002) 'Polarized directional reflectance from laurel and mullein leaves', *Optical Engineering*, **41**, 1002–12.

Rondeaux, G. and Guyot, G. (1990) 'Polarisation de la lumiere par les couverts végétaux: Possibilités d'applications agronomiques', *Canadian Journal of Remote Sensing*, **16**, 32–41.

Rondeaux, G. and Herman, M. (1991) 'Polarization of light reflected by crop canopies', *Remote Sens. Environment*, **38**, 63–75.

Sandmeier, S.R. and Itten, K.I. (1999) 'A field goniometer system (FIGOS) for acquisition of hyperspectral BRDF data', *IEEE Trans. Geosci. Remote Sensing*, **37**, 648–58.

Savenkov, S., Muttiah, R. and Oberemok, Y. (2003) 'Transmitted and reflected scattering matrices from an English oak leaf', *Appl. Opt.*, **42**, 4955–62.

Shibayama, M. (2004) 'Seasonal profiles of polarized reflectance and leaf inclination distribution of wheat canopies', *Plant Production Science*, **7**, 397–405.

Shibayama, M. and Akita, S. (2002) 'A portable spectropolarimeter for field crop canopies: Distinguishing species and cultivars of fully developed canopies by polarized light', *Plant Production Science*, **5**, 311–19.

Shibayama, M. and Watanabe, Y. (2006) 'Remote assessment of wheat canopies under various cultivation conditions using polarized reflectance', *Plant Production Science*, **9**, 312–22.

Shibayama, M. and Watanabe, Y. (2007) 'Estimating the leaf inclination angle of wheat canopies using reflected polarized light', *Plant Production Science*, **10**, 329–42.

Shibayama, M. and Watanabe, Y. (2008) 'Testing polarisation measurements with adjusted view zenith angles in varying illumination conditions for detecting leaf orientation in wheat canopy', *Plant Production Science*, **11**, 498–506.

Shibayama, M., Sakamoto, T. and Kimura, A. (2011) 'A multiband polarimetric imager for field crop survey-instrumentation and preliminary observations of heading-stage wheat canopies', *Plant Production Science*, **14**, 64–74.

Short, M. (2011) *The LANDSAT Tutorial Workbook: Basics of Satellite Remote Sensing*, 20 September, NASA Reference Publication 1078, NASA.

Sparks, W.B., Hough, J., Germer, T.A., Chen, F., DasSarma, S., DasSarma, P., Robb, F.T., Manset, N., Kolokolova, L., Reid, N., Macchetto, F.D. and Martin, W. (2009a) 'Detection of circular polarization in light scattered from photosynthetic microbes', *PNAS*, **106**, 7816–21.

Sparks, W.B., Hough, J.H., Kolokolova, L., Germer, T.A., Chen, F., DasSarma, S., DasSarma, R., Robb, F.T., Manset, N., Reid, I.N., Macchetto, F.D. and Martin, W. (2009b) 'Circular polarization in scattered light as a possible biomarker', 11th Conference on Electromagnetic and Light Scattering, University of Hertfordshire, Hatfield, England, 7–12 September 2008, *J. Quant. Spectrosc. Radiat. Transfer*, **110**, 1771–9.

Sridhar, V.N., Singh, R.P., Mahtab, A. and Ghosh, R. (2005) 'Crop polarisation behavior derived from multi-temporal ADEOS–POLDER data', *Photonirvachak—Journal of the Indian Society of Remote Sensing*, **33**, 339–44.

Sun, Z. and Zhao, Y. (2011) 'The effects of grain size on bidirectional polarized reflectance factor measurements of snow', *J. Quant. Spectrosc. Radiat. Transfer*, **112**, 2372–83.

Suomalainen, J., Hakala, T., Peltoniemi, J. and Puttonen, E. (2009a) 'Polarised multiangular reflectance measurements using Finnish Geodetic Institute field goniospectrometer', *Sensors*, **9**, 3891–907.

Suomalainen, J., Hakala, T., Puttonen, E. and Peltoniemi, J. (2009b) 'Polarised bidirectional reflectance factor measurements from vegetated land surfaces', *J. Quant. Spectrosc. Radiat. Transfer*, **110**, 1044–56.
Turner, M. (1998) 'The Sandmeier field goniometer: A measurement tool for bi-directional reflectance', in *NASA Commercial Remote Sensing Verification and Validation Symposium*, pp. 1–6.
Tyo, J.S., Goldstein, D.H., Chenault, D.B. and Shaw, J.A. (2006) 'Review of passive imaging polarimetry for remote sensing applications', *Appl. Opt.*, **45**, 5453–69.
Vanderbilt, V.C. and Grant, L. (1986) 'Polarization photometer to measure bidirectional reflectance factor R(55°, 0°; 55°, 180°) of leaves', *Optical Engineering*, **25**, 566–71.
Vanderbilt, V.C., Grant, L. and Ustin, S.L. (1991) 'Polarization of light by vegetation', in Myneni, R.B. and Ross, J. (eds), *Photon–Vegetation Interactions*, Berlin, Springer–Verlag, Chapter 7, pp. 191–228.
Vanderbilt, V.C., Grant, L., Biehl, L.L. and Robinson, B.F. (1985) 'Specular, diffuse, and polarised light scattered by two wheat canopies', *Appl. Opt.*, **24**, 2408–18.
von Schönermark, M., Geiger, B. and Röser, H.-P. (eds) (2004) *Reflection Properties of Vegetation and Soil with a BRDF-Database*, Berlin, Wissenschaft & Technik.
Waquet, F., Leon, J.F., Cairns, B., Goloub, P., Deuze, J.L. and Auriol, F. (2009) 'Analysis of the spectral and angular response of the vegetated surface polarization for the purpose of aerosol remote sensing over land', *Appl. Opt.*, **48**, 1228–36.
Widlowski, J.-L., Taberner, M., Pinty, B., Bruniquel-Pinel, V., Disney, M., Fernandes, R., Gastellu-Etchegorry, J.-P., Gobron, N., Kuusk, A., Lavergne, T., Leblanc, S., Lewis, P., Martin, E., Mõttus, M., North, P.J.R., Qin, W., Robustelli, M., Rochdi, N., Ruiloba, R., Soler, C., Thompson, R., Verhoef, W., Verstraete, M.M. and Xie, D. (2006) 'The third RAdiation transfer Model Intercomparison (RAMI) exercise: Documenting progress in canopy reflectance modelling', *J. Geophys. Res.*, **112**, D09111.
Zhang, J., Chen, J., Zou, B. and Zhang, Y. (2012) 'Modeling and simulation of polarimetric hyperspectral imaging process', *IEEE Trans. Geosci. Remote Sensing*, **50**, 2238–53.

Part IV

Radiative Forcing

8 Diurnally averaged direct aerosol-induced radiative forcing from cloud-free sky field measurements performed during seven regional experiments

Claudio Tomasi, Christian Lanconelli, Angelo Lupi, and Mauro Mazzola

8.1 Introduction

Aerosol particles suspended in the atmosphere may originate from either natural or anthropic sources, or through mixed processes involving their variable combinations. Among the primary natural emissions, the most important are those leading to the formation of (i) mineral dust through wind erosion of natural soil and (ii) sea-salt particles from the ocean surface forced by winds. In addition, significant emission processes include biological particles released by plants and animals, combustion particles forming in forest fires and biomass-burning smokes, and volcanic debris ejections. Primary emissions of anthropogenic aerosol arise mainly from the erosion of soil debris in semi-arid regions exploited for sheep-breeding activities, and in ploughed soils and terrains used for agriculture, where mineral dust particles are subsequently mobilized by winds. Anthropogenic aerosols are also generated by industrial activities, fossil fuel smokes often containing soots, combustion particles from forest fires lit for deforestation purposes, and waste-burning smokes, which often contain black carbon and soluble or insoluble organic compounds. Additionally, secondary particles are formed in the troposphere from a variety of natural atmospheric gases, such as sulfur dioxide, nitrogen compounds, and biogenic volatile organic compounds (VOCs), while in the stratosphere they mainly form through the oxidation of sulfur dioxide injected at such high altitudes by particularly strong volcanic eruptions. Secondary products are also due to industrial activities and their gaseous emissions, including sulfur dioxide and other sulfur compounds, nitrogen compounds, such as NO_x and NH_3, hydrocarbons and VOCs, as pointed out by Seinfeld and Pandis (1998, 2006).

Estimates of the annual global emissions of natural aerosols from various sources were made by Andreae and Rosenfeld (2008), who found an overall value of between 4.2×10^2 and 2.3×10^4 Tg/yr. Including also the average estimate of annual volcanic debris emission, varying between 25 and 150 Tg/yr, the average global value of all natural emissions can be reliably estimated at around 1.2×10^4 Tg/yr. It is, however, important to point out that the overall particle emission flux can vary greatly from on year to another, depending on the intensity of the most variable sources, like volcanic eruptions, wind fields over oceans, and wind mobilization of soil debris. The annual global emissions of anthropogenic aerosols from various man-made sources (including industrial dust and secondary sulfates, nitrates, and organic substances) have been estimated to vary on average between 150 and 370 Tg/yr (Seinfeld and Pandis, 1998; Liao et al., 2003, 2004; Tsigaridis et al., 2006; Andreae and Rosenfeld, 2008). Therefore, anthropogenic aerosol emissions provide a very limited contribution compared to the annual production of natural aerosols.

Because of their varied origins, atmospheric aerosol particles present multimodal size-distributions over a very large size range, from less than 2×10^{-3} to more than 10^2 μm, with various modes consisting of ultrafine, fine, accumulation, and coarse particles. Among such size classes, accumulation particles (with equivalent diameters ranging from 0.1 to about 2.5 μm) and coarse particles (with sizes varying from 2.5 to more than 20 μm) mainly contribute to produce scattering and absorption of the incoming solar radiation, presenting spectral features that are closely related to the optical parameters of the various particle components. In fact, the optical parameters of aerosol particles differ appreciably with the origin of particulate matter, exhibiting marked differences on passing from sea salt to mineral dust or from industrial dust to biomass-burning smoke. Since aerosol particle sizes are comparable with the wavelengths of incoming solar radiation spectrum (89% of which lies over the 0.4–2.5-μm range), the incoming short-wave radiation interacts strongly with airborne aerosol particles, as shown by the Mie (1908) electromagnetic theory, which correctly evaluates the extent to which the intensity of particle scattering and absorption depends on the size spectrum of aerosol spherical particles and their optical characteristics.

Furthermore, aerosol particles directly emitted or formed in the atmosphere have residence times varying usually between less than one day and about two weeks, depending on their origin and the effectiveness of the removal processes occurring during the growth of aerosol particles by nucleation, coagulation, and condensation, and subsequently during the dry deposition phase regulated by gravitational settling or through wet scavenging episodes in clouds and removal by rains and solid precipitations (snow and hail). The intensity of such removal processes is subject to marked variations from one region to another, according to the prevailing directions of aerosol transport by winds, and is therefore closely associated with the general circulation features and precipitation regimes. Consequently, the atmospheric aerosol residence time can vary considerably from one region to another, also depending very closely on the aerosol hygroscopic properties. Thus, the mass concentration of aerosol particles is expected to vary with the aerosol chemical composition and differ in the various regions of the planet, presenting average global aerosol mass burdens in the atmosphere equal to 18 ± 5 Tg for mineral dust, 15 Tg for sea salt, 2.8 Tg for sulfates, 0.5 Tg for nitrates, and 1.1 Tg for industrial dust (Pósfai and Buseck, 2010).

Bearing in mind that the optical characteristics of aerosol particles can vary greatly as a function of their origin, it is evident that the interactions between aerosol particles and solar radiation can exhibit different features from one region to another. In fact, they usually induce (i) particularly strong direct aerosol radiative forcing (hereinafter referred to as DARF) effects in the mid-latitude industrial areas and in the more densely populated regions, because of the very high aerosol mass concentrations due to both natural processes and human activities (Chylek and Wong, 1995; Schwartz, 1996; Chung and Seinfeld, 2005); and (ii) relatively weak DARF effects in the remote oceanic and polar regions, where low particle mass concentrations are generally monitored in the atmosphere, apart from cases related to intense transport of dust, forest fire smoke, and anthropogenic aerosol coming from the most industrially developed and densely populated mid-latitude areas (Bates et al., 1998a, 1998b; Takemura et al., 2005).

The climate effects directly induced by columnar aerosol particles are caused by scattering and absorption of incoming solar radiation, which in general produce a marked decrease in the flux density of global solar radiation reaching the surface, a certain short-wave radiation absorption in the atmosphere, and a relative increase in the solar radiation fraction reflected back to space (Patadia et al., 2008). Thus, in cases of significant mass contents of soots and other organic substances within the particles, a not negligible fraction of incoming solar radiation is absorbed, leading to an appreciable warming of the atmospheric layer containing such strongly absorbing particles (Kaufman et al., 2002; Dubovik et al., 2004, 2008; Torres et al., 2005). The DARF effects on the radiation budget of the surface–atmosphere system can vary considerably as the atmospheric particulate mass content (and, hence, the aerosol optical thickness) assumes increasing values, while the variations in the particulate matter optical characteristics lead to a considerable change in the ratio between volume scattering and extinction coefficients of columnar aerosol. Moreover, the variations in DARF at the Top-of-Atmosphere (ToA) and the Bottom-of-Atmosphere (BoA) levels can also vary considerably as a function of the spectral and directional characteristics of surface reflectance (Charlson et al., 1992; Chylek and Coakley, 1974; Coakley and Chylek, 1975). In fact, surface albedo can vary markedly from one region to another, assuming rather low values over oceans, intermediate values over terrains covered by vegetation and forests, rather high values over desert and semi-arid land regions, and very high values in the polar regions covered by snow fields and glaciers.

As a result of the aforementioned radiative transfer processes, DARF effects can vary greatly from one region to another, as a function of the atmospheric content of aerosol particles, their size-distribution shape-parameters, optical characteristics of columnar particulate matter, and surface reflectance properties. In view of the great spatial and temporal variability in aerosol concentration and composition observed over the global scale (Takemura et al., 2002), a variety of regional experiments have been conducted over the past two decades, with the principal aim of investigating the impact of columnar aerosol loads on the radiation budget of the surface–atmosphere system, and evaluating the intensities of the regional radiative forcing effects directly induced by both natural and anthropogenic aerosols as a function of their physico-chemical and optical features.

The studies have been carried out employing different measurement techniques over land and ocean regions:

(1) Ground-based measurements of aerosol optical thickness $\tau_a(\lambda)$ at various visible and near-infrared wavelengths, taken throughout the day for different values of solar zenith angle θ_o, together with simultaneous measurements of the most significant radiative parameters of columnar aerosol (i.e. complex refractive index $n(\lambda) - ik(\lambda)$, phase function $P(\Theta)$ depending on scattering angle Θ, asymmetry factor $g(\lambda)$, and single-scattering albedo $\omega(\lambda)$) at some visible and near-infrared wavelengths. These measurements are usually carried out

employing passive remote sensing instruments, such as sun-photometers, sun-/sky-radiometers, and pyranometers), and completed with additional ground-based measurements taken with active remote sensing instruments, such as Lidars, nephelometers, aethalometers, and absorption photometers. Numerous evaluations of these parameters are available in the literature, as obtained from:

(a) field experiments over land, including: (i) SCAR-A, over the eastern mid-Atlantic coast in the US (Kinne and Pueschel, 2001); (ii) ACE-Asia, in southern Korea (Bush and Valero, 2003); (iii) SCAR-B, in Brazil (Christopher et al., 2000; Kinne and Pueschel, 2001); (iv) the Skyradiometer Network (SKYNET), in China and Southeast Asia (Kim et al., 2005a); and (v) the Aerosol Robotic Network (AERONET), in South America (Yu et al., 2006); and

(b) field experiments conducted over ocean, among which the most important were: (i) TARFOX, over the north Atlantic Ocean and along the US eastern coast (Hignett et al., 1999; Russell et al., 1999a, 1999b); (ii) INDOEX, in the tropical Indian Ocean, Bay of Bengala and Arabian Sea (Satheesh and Ramanathan, 2000; Conant, 2000; Satheesh, 2002; Satheesh and Srinivasan, 2002); (iii) ACE-2, in the subtropical north-eastern Atlantic Ocean (Kinne and Pueschel, 2001); (iv) ACE-Asia, in the north-west Pacific Ocean (Seinfeld et al., 2004a, 2004b); and (v) PRIDE, in the Caribbean region (Christopher et al., 2003).

(2) Combined satellite-based observations and ground-based radiometric and Lidar measurements performed over land and ocean regions, including:

(a) those conducted over land, based on (i) Geostationary Operational Environmental Satellite GOES-8, Clouds and Earth's Radiant Energy System (CERES) and Discrete Ordinate Radiative Transfer observations combined with AERONET data, in South America (Christopher and Zhang, 2002); (ii) Moderate Resolution Imaging Spectroradiometer (MODIS)/Global Ozone Chemistry Aerosol Radiation and Transport (GOCART) observations combined with AERONET data, on the global scale (Liu et al., 2005); (iii) CERES and MODIS observations validated through ground-based measurements, in the Atlantic Ocean adjacent to the African coast (Li et al., 2004); and (iv) MODIS/Terra data, combined with ground-based radiometric and Lidar data collected in the Himalayan region (Ramana et al., 2004); and

(b) experiments conducted over ocean, among which it is worth mentioning the experiments based on the use of (i) NOAA-14-AVHRR (Advanced Very High Resolution Radiometer) data and INDOEX ground-based data, in the northern Indian Ocean and over the Bay of Bengal and Arabian Sea (Tahnk and Coakley, 2002); (ii) CERES and MODIS observations over the global ocean (Loeb and Manalo-Smith, 2005); and (iii) MODIS/Terra observations combined with the measurements performed on-board the R/V Sagar Kanya vessel, in the Bay of Bengal (Sumanth et al., 2004).

To investigate the dependence of DARF effects on aerosol optical parameters and surface reflectance characteristics, we have calculated here (a) the DARF effects at the ToA- and BoA-levels and within the atmosphere (Atm) produced by columnar aerosol throughout the day, and (b) the corresponding efficiencies of these radiative effects for sets of aerosol extinction measurements performed in different areas of the planet, during several field campaigns conducted by our research group at the ISAC-CNR Institute (Bologna, Italy) for maritime, continental, anthropogenic, Saharan dust, and volcanic aerosol loads, and by other research groups for oceanic, desert dust, continental, urban, and forest fire smoke particles. Before presenting the DARF calculations, the definition of diurnally averaged DARF terms and the corresponding efficiencies are provided in section 8.2. The evaluations of the various DARF parameters derived from field measurements are presented in section 8.3, as obtained during the following seven field experiments: (i) the CLEARCOLUMN campaign, conducted at Sagres (Portugal) in June and July 1997, as part of the ACE-2 project (Vitale et al., 2000a, 2000b; Tomasi et al., 2003); (ii) the PRIN-2004 project campaign, performed by us in cooperation with the Department of Physics of Salento University (Lecce, Italy) in southern Puglia from March 2003 to January 2004 (Tafuro et al., 2007; Lanconelli, 2007); (iii) the AEROCLOUDS field campaign conducted by our group from early May 2007 to March 2008 at San Pietro Capofiume (SPC) in the Po Valley (northern Italy) (Mazzola et al., 2010); (iv) the pre- and post-Pinatubo eruption campaigns conducted by us in the summer months of 1991 and 1992 at the high-altitude CNR Pyramid Laboratory (Himalaya), as part of the Ev-K2-CNR project (Tomasi et al., 1997); (v) the polar aerosol data recorded over the 2000–2012 period at various Arctic and Antarctic sites, as a part of the POLAR-AOD project (Tomasi et al., 2007, 2012); (vi) the Aerosols99 cruise measurements performed in various sectors of the Atlantic Ocean, from the US eastern coast to Cape Town in South Africa, during which maritime, dust, and biomass-burning smoke particles were sampled and monitored with optical methods (Bates et al., 2001); and (vii) the Department of Energy (DOE)/ARM/Aerosol Intensive Operating Period (AIOP) project campaign conducted in north-central Oklahoma (US) in May 2003, in the presence of prevailing contents of continental polluted aerosol particles mixed with smaller fractions of industrial dust or forest fire smoke particles (Ferrare et al., 2006).

8.2 Definitions of diurnally averaged DARF at the ToA- and BoA-levels and within the atmosphere

The effects of airborne aerosols on the radiation budget in the surface–atmosphere system are mostly determined by neglecting the particulate scattering and absorption processes at infrared wavelengths, and taking into account only those induced at visible and near-infrared wavelengths. This is because airborne particles have considerably smaller sizes than the terrestrial radiation wavelengths and, therefore, their interactions with long-wave radiation are less intense than those affecting short-wave radiation, as clearly indicated by the Mie (1908) theory. For instance, the estimates of short-wave DARF effects at the ToA- and BoA-levels derived from the field sun-photometer measurements of $\tau_a(\lambda)$ and the other aerosol optical parameters performed at Lecce (Italy) were compared by Tafuro et al. (2007) with the infrared (4–80 μm) DARF terms obtained at both levels through radiative transfer model simulations. The comparison showed that (i) the ratio between the monthly mean values of long-wave and short-wave DARF at the ToA-level varied between about 10^{-3} and 2×10^{-2} throughout the year and (ii) the analogous ratio between DARF terms at the surface-level varied on average from 0.10 to no more than 0.25.

Conversely, as a result of the intense scattering and absorption effects on solar radiation due to aerosol particles, the net (downwelling minus upwelling) flux of short-wave radiation at the ToA-level (or at another level close to the tropopause) is subject to vary considerably, thus causing a significant change in the energy available for 'governing' the climate system. Since the net flux variation constitutes the radiative forcing induced by columnar particles at the ToA-level, it implies a cooling (or warming) effect, depending on the negative (or positive) sign of the net flux change (Chylek and Coakley, 1974; Coakley and Chylek, 1975). The perturbation in the radiation budget of the surface–atmosphere system is evaluated by assuming that it is due to a change in the overall aerosol columnar content and/or in the optical characteristics of particulate matter, while all the other atmospheric constituents are kept unperturbed. Therefore, the DARF at the ToA-level can be calculated as the change in the net (downwelling minus upwelling) radiation flux of the climate system at the ToA-level, which causes an energy deficit or a surplus (Hansen et al., 1997). At any given time, the change depends on (i) the optical parameters of columnar aerosol, (ii) the intensity of short-wave radiation, varying as a function of solar zenith angle throughout the day, and (iii) the surface reflectance characteristics.

Similar concepts are valid for the DARF effect determined at the BoA-level, by taking into account the perturbation in the short-wave radiation field at the surface, due to the scattering and absorption of solar radiation passing through the atmosphere. In other words, the DARF effect evaluated at the BoA-level gives a measure of the change caused by columnar aerosol in the net flux of solar radiation at the surface. Correspondingly, the net flux change in the atmosphere can be calculated as the difference between the DARF at the ToA-level and that simultaneously produced at the BoA-level (Ramanathan et al., 2001a, 2001b).

8.2.1 The instantaneous DARF effects at ToA- and BoA-levels and in the atmosphere

On the basis of the above remarks, the instantaneous DARF term $\Delta F(t)$ is commonly evaluated at a certain time t of the day by considering only the short-wave (solar) radiation flux change. It is generally determined as the instantaneous forcing induced by a certain columnar aerosol for a given value of the solar zenith angle θ_o and well-defined characteristics of surface reflectance (Bush and Valero, 2003). The instantaneous DARF term $\Delta F_{ToA}(t)$ can be calculated simply at the ToA-level as the difference between the net radiative flux for the turbid atmosphere containing a certain columnar load of aerosol particles, and the net radiative flux calculated at the same time in a pristine atmosphere without aerosols, where (i) the net radiative flux for the turbid atmosphere is given by the difference between the short-wave downwelling flux and the short-wave upwelling flux, both determined at the ToA-level for an atmosphere including all its constituents, and (ii) the net radiative flux in a pristine atmosphere is given by the difference between the corresponding short-wave downwelling and upwelling fluxes, both calculated in the aforesaid pristine atmosphere without aerosols. Thus, the instantaneous $\Delta F_{ToA}(t)$

at the ToA-level is given by the difference between the upwelling solar radiation flux emerging from the pristine atmosphere without aerosols and the upward solar radiation flux emerging from the atmosphere in the presence of aerosols. According to the WMO (1986) definition of radiative forcing in the atmosphere, negative values of $\Delta F_{ToA}(t)$ imply that the upwelling flux of solar radiation has increased, being associated with an increase in the overall albedo of the surface–atmosphere system and, hence, causing direct cooling effects on the climate system. Conversely, positive values of $\Delta F_{ToA}(t)$ indicate that the upwelling solar radiation flux has decreased due to the columnar aerosol load, thus causing a decrease in the overall albedo of the surface–atmosphere system, associated with direct warming effects (Christopher et al., 2006; Zhao et al., 2008).

Similarly, the instantaneous DARF term $\Delta F_{BoA}(t)$ at the BoA-level is the perturbation in the net (downwelling minus upwelling) flux at the surface, which is induced by the airborne aerosol load at a certain time t of the day. Therefore, $\Delta F_{BoA}(t)$ can be defined as the instantaneous difference between the net flux at surface-level in the atmosphere with aerosols and the net flux at surface-level in the atmosphere without aerosols (Satheesh and Ramanathan, 2000; Bush and Valero, 2002, 2003), where the two net fluxes are both given by the differences between downwelling flux and upwelling flux of short-wave radiation.

Besides the instantaneous direct aerosol-induced radiative forcing terms $\Delta F_{ToA}(t)$ and $\Delta F_{BoA}(t)$ at the ToA- and BoA-levels, respectively, it is obvious that a corresponding instantaneous radiative forcing $\Delta F_{Atm}(t)$ takes place within the atmosphere, induced by the columnar aerosol load through thermodynamic processes. It can be determined as the difference between the instantaneous DARF term $\Delta F_{ToA}(t)$ and the instantaneous DARF term $\Delta F_{BoA}(t)$, and is not radiative in nature, as pointed out by Ramanathan et al. (2001a, 2001b). In practice, it represents the change in the amount of latent heat released in the atmosphere by aerosol particles at a certain time, causing atmospheric heating effects that impact on local circulation and atmospheric heat balance, especially within the aerosol layer (Kinne and Pueschel, 2001). Therefore, such thermodynamic forcing does not modify the net energy budget of the surface–atmosphere system, but rather redistributes internally the latent heat released by aerosols, in this way affecting the temperature gradients within the lower part of the troposphere and influencing the local dynamics of the atmosphere.

It was shown by Tomasi et al. (2013) that the instantaneous direct radiative forcing induced by columnar aerosol is subject to vary throughout the day as a function of (i) time variations in $\tau_a(\lambda)$ at all visible and near-infrared wavelengths, giving a measure of the overall extinction effects produced by aerosol particles along the vertical path of the atmosphere; (ii) time variations in the aerosol optical characteristics of columnar aerosol, represented by means of various parameters (i.e. real and imaginary parts of particulate refractive index, phase function, asymmetry factor, and single-scattering albedo), which can vary appreciably as a function of altitude, depending on the vertical profiles of the above optical parameters of particulate matter; (iii) the solar zenith angle θ_o, which in practice defines the geometric configuration of the atmospheric path described by the incoming solar radiation; and (iv) the variations in the spectral and spatial features of the underlying surface reflectance, which was represented by Tomasi et al. (2013) by means of two sets of surface reflectance models, the first consisting of 16 Bidirectional Reflectance Distribution Function (BRDF) non-Lambertian models relative to ocean, vegetation-covered, bare soil, and polar snow-covered surfaces, and the latter of 16 corresponding Lambertian (isotropic) surface reflectance models defined for the same surface coverage features.

8.2.2 Diurnally averaged DARF and aerosol fractional forcing

The so-called 'diurnally averaged aerosol forcing' (hereinafter referred to as ΔDF) is given by the integral of instantaneous radiative forcing $\Delta F(t)$ calculated at a significant atmospheric level over the sunlit period (from sunrise to sunset) and divided by the 24-hour period (Bush and Valero, 2003; Xu et al., 2003; Kim et al., 2005b):

$$\Delta DF_{ToA} = \int_{sunrise}^{sunset} \Delta F_{ToA}(t)\, dt/24hr\,, \tag{8.1}$$

and

$$\Delta DF_{BoA} = \int_{sunrise}^{sunset} \Delta F_{BoA}(t)\, dt/24hr\,, \tag{8.2}$$

for the DARF at the ToA- and BoA-levels, respectively. Clearly, the sunlit period varies as a function of latitude and season. Therefore, evaluations of ΔDF_{ToA} and ΔDF_{BoA} obtained from field measurements performed at a site very far from the equator may be subject to significant variations throughout the year. However, the concept underlying the definition of these two physical quantities in terms of Eqs (8.1) and (8.2) is correct, since the resulting evaluations of ΔDF_{ToA} and ΔDF_{BoA} provide a realistic measure of the forcing produced by solar radiation on the surface–atmosphere system on each day of the year.

The above evaluations of ΔDF_{ToA} and ΔDF_{BoA}, obtained in terms of Eqs (8.1) and (8.2), respectively, can be also used to calculate the so-called 'diurnally averaged aerosol fractional forcing at the ToA-level' (hereinafter referred to as AFF_{ToA}), which is defined as the ratio between the diurnal average radiative forcing ΔDF_{ToA} and the incoming flux $I_S\downarrow$ of solar radiation at the ToA-level, averaged over the sunlit period (Bush and Valero, 2003). In practice, AFF_{ToA} yields an evaluation of the response of the climate system to the incident solar radiation, as induced by the aerosol particles suspended within the vertical column of the atmosphere.

8.2.3 DARF efficiency

The absolute magnitude of the DARF terms at the ToA-level closely depends on the amount of radiation entering the atmosphere, but also on the columnar particle amount and the optical characteristics of airborne aerosols perturbing the environment. An appropriate parameter for defining the columnar particle amount is $\tau_a(\lambda)$, which provides a measure of the monochromatic extinction of solar radiation along the vertical atmospheric path. Thus, in cases where $\tau_a(\lambda)$ increases, the radiative forcing is expected to increase over a certain surface of known reflectance through features depending on the optical characteristics of the atmosphere. Therefore, the absolute magnitude of the ToA-level DARF effect is given by the amount of upwelling solar radiation reflected back by the surface–atmosphere system towards space, as a result of aerosol scattering processes. This quantity depends closely on the number of aerosol particles that attenuate the incoming solar radiation, which is approximately proportional to $\tau_a(\lambda)$. Based on this concept, the rate at which the surface–atmosphere system is forced at the ToA-level per unit $\tau_a(\lambda)$ is known as the 'aerosol forcing efficiency' E_{ToA} (according to the definitions given by Bush and Valero, 2003; Xu et al., 2003; Kim et al., 2005b). This parameter can be evaluated over the whole 24-hour period as the ratio between the aerosol radiative forcing ΔDF_{ToA} defined in Eq. (8.1) and the daily mean value of $\tau_a(\lambda)$ measured at wavelength $\lambda = 0.55$ μm (or at another visible wavelength, when measurements of $\tau_a(0.55\ \mu\text{m})$ are not directly available from sun-photometer data). Similarly, the aerosol forcing efficiency E_{BoA} at the BoA-level can be calculated on each measurement day as the ratio between the diurnal average value of ΔDF_{BoA} determined in terms of Eq. (8.2) from field measurements or through radiative transfer model simulations, and the daily mean value of $\tau_a(0.55\mu\text{m})$, calculated over the sunlit period (Anderson et al., 2005).

Once the values of E_{ToA} and E_{BoA} are known, the DARF efficiency E_{Atm} for the atmosphere, substantially due to the atmospheric heating produced by aerosol particles, can be simply calculated as the difference

$$E_{Atm} = E_{ToA} - E_{BoA}\,. \tag{8.3}$$

As previously mentioned, the value of $\tau_a(\lambda)$ used in these calculations is commonly determined at a visible central wavelength (conventionally, chosen as equal to 0.55 μm) or alternatively averaged over the 0.30–0.70-μm broadband spectral range, as done during the TARFOX experiment (Hignett et al., 1999; Russell et al., 1999a, 1999b).

The calculations of the DARF efficiency parameters are in general very useful for characterizing the DARF effects within the surface–atmosphere system, since the absolute magnitude of the DARF terms at the ToA- and BoA-levels and within the atmosphere depend not only on the amount of energy entering the atmosphere, but also on the columnar amount and microphysical and radiative characteristics of aerosols causing the perturbation in the energy budget of the surface–atmosphere system (Bush and Valero, 2003). New calculations of such DARF parameters are presented in section 8.3, as obtained from both field measurements of $\tau_a(\lambda)$ and aerosol optical charactersitics, and radiation transfer simulations. Particular attention is paid to defining (i) the aerosol radiative parameters measured within the vertical atmospheric column by means of ground-based remote sensing techniques and *in situ* sampling measurements and (ii) the spectral and directional characteristics of surface reflectance, which are particularly important for achieving realistic estimates of the DARF effects (Chylek and Coakley, 1974). For the latter purpose, the BRDF non-Lambertian models determined by Tomasi et al. (2013) were used to represent the surface reflectance characteristics in the Atlantic Ocean and Mediterranean Sea, in the European and North American continental areas, and in the remote Himalayan, Arctic, and Antarctic regions.

8.3 Field measurements and calculations of the diurnally averaged DARF at the ToA- and BoA-levels and in the atmosphere, with corresponding efficiency estimates

As pointed out by Chylek and Coakley (1974), the DARF forcing induced at the ToA-level by a certain load of airborne aerosol particles can be evaluated using a two-stream approximation formula, where the aerosol radiative effects are considered separately from the Rayleigh scattering effects due to air molecules and the absorption effects caused by minor atmospheric gases (mainly water vapor, ozone, nitrogen dioxide, and carbon dioxide). Numerous articles are available in the literature, which illustrate this approach (Charlson et al., 1991; Nemesure et al., 1995; Haywood and Shine, 1995; Chylek and Wong, 1995), showing that the daily mean DARF effect due to an aerosol layer of known optical characteristics can generally be estimated using simplified formulas such as

$$\Delta DF_{ToA} = -\frac{1}{2} S_o T_{atm}^2 (1 - A_c) \tau_a \left[\omega B (1 - R_s)^2 - 2(1 - \omega) R_s \right] , \qquad (8.4)$$

proposed by Russell et al. (2002), where:

(1) factor 1/2 is used to take into account that 12 hours of the day present average solar insolation conditions during the year, while the remaining 12 hours are assumed to be without incoming solar radiation;
(2) S_o is the solar constant for the annual mean Earth–Sun distance, estimated to have slowly decreased over the past two decades from 1365.4 ± 1.3 W/m^2 in the 1990s to 1360.8 ± 0.5 W/m^2 in 2008 (Kopp and Lean, 2011);
(3) T_{atm} is the partial clear-sky atmospheric transmittance averaged over the whole solar spectrum, relative to Rayleigh scattering and gaseous absorption only;
(4) A_c is the cloud cover fraction;
(5) τ_a is the mean value of $\tau_a(\lambda)$ determined over the visible and near-infrared spectral range, most frequently evaluated in terms of its monochromatic value $\tau_a(0.55\ \mu\text{m})$;
(6) ω is the single-scattering albedo (hereinafter referred to as SSA) of columnar aerosol calculated over the same wavelength range;
(7) B is the hemispheric upward-scattered fraction of solar radiation, determined over the 2π solid angle; and
(8) R_s is the average surface reflectance in the visible and near-infrared.

Used for average values of the above parameters calculated over long time-periods, Eq. (8.4) can only provide reliable estimates of ΔF_{ToA} over long periods of the year. It is also important to note that only the three parameters τ_a, ω, and B used in the calculations of Eq. (8.4) depend on the scattering and absorption features of the columnar aerosol particles. In fact, the first two parameters can be determined as averages over long periods, while the last one takes into account the backscattering characteristics of atmospheric aerosols for clear-sky conditions. Conversely, R_s depends closely on the structural characteristics of the surface, and A_c can vary appreciably from one day to another, according to the cloud coverage parameters, cloud height, and optical characteristics of the cloud droplets and aerosols, being null only on totally cloudless days. The term given in Eq. (8.4) within the square bracket defines the sign of the aerosol-induced change flux, which is crucial for distinguishing cooling effects from warming ones: (i) in cases with negative sign, is $\omega B(1 - R_s)^2 > 2R_s(1 - \omega)$, and cooling effects are expected to occur in the surface–atmosphere system; and (ii) in cases with positive sign, giving $\omega B/(1-\omega) > 2R_s/(1-R_s)^2$, warming effects take place in the atmosphere. This implies that, in the critical condition defined by equation $\omega B/(1-\omega) = 2R_s/(1-R_s)^2$, the atmospheric aerosol load does not induce radiative flux changes at the ToA-level.

An example of the variability of ΔF_{ToA} as a function of R_s and ω is presented in Fig. 8.1, as obtained in terms of Eq. (8.4). It can be noted that a variation in ω equal to 0.07 (like that from 0.80 to 0.87, as indicated by the black vertical arrow) causes an absolute change in ΔF_{ToA} from +2.1 to +0.2 W/m^2 for $R_s = 0.30$. As shown in Fig. 8.1, such a sensitivity is expected to be relatively weak in cases with $R_s < 0.3$, rather strong for $R_s > 0.30$, and very strong for the more marked features of snow- and ice-covered surfaces (Bergstrom and Russell, 1999). Figure 8.1 also shows that: (i) a decrease in ω from 1.0 to 0.9 over a dark vegetation surface with values of R_s close to 0.2 can reduce the change in the ToA-level flux by about 50%, with ΔF_{ToA} varying from −4.3 to −2.3 W/m^2 for $R_s = 0.2$; and (ii) the same variation in ω over desert areas or snow fields with values of $R_s > 0.4$ can cause a particularly marked decrease in ΔF_{ToA}, leading in most cases to changes from cooling to warming effects, such as for instance: (a) from −2.4 to +1.0 W/m^2, for $R_s = 0.4$; (b) from −1.1 to +3.8 W/m^2, for $R_s = 0.6$; and (c) from −0.3 to +6.1 W/m^2, for $R_s = 0.8$. For columnar atmospheric loadings of continental-anthropogenic aerosol particles yielding mean values of ω close to 0.9, such as those monitored over Sagres (southern Portugal) in summer 1997 on days characterized by transport of air masses from the European area (Vitale et al., 2000b), it is estimated in Fig. 8.1 that: (i) cooling effects usually occur over oceanic and vegetation-covered areas with values of $R_s < 0.25$; (ii) very weak cooling or only slight warming effects are generally induced by these airborne aerosols over areas where R_s varies between 0.25 and 0.35; and (iii) pronounced warming effects are caused over desert regions. In particular, Fig. 8.1 shows that very marked warming effects may be induced by aerosols over ice- or snow covered areas with $R_s > 0.35$, especially when the particles contain relatively large fractions of soot substances, causing rather low SSA values.

Ramanathan et al. (2001a, 2001b) found that significant variations in SSA due to changes in the chemical composition and, hence, in the optical parameters of airborne aerosol particles can induce important changes in the radiation fluxes within and below the aerosol layers. Such changes induce thermodynamic forcing effects that do not modify the net energy budget of the surface–atmosphere system, but rather cause an internal redistribution of the energy surplus, modifying the amount of latent heat released by aerosol-induced changes in clouds and precipitations. Such exchange mechanisms may considerably alter the atmospheric stability conditions, thus influencing heating rates, surface temperatures, cloud formation, and cloud persistence, all of which contribute to changing local cooling and warming patterns in the atmosphere. Within this perspective, it is important to note that the aerosol optical parameters may vary considerably in cases where the chemical composition of particulate matter is modified as a result of changes in the transport

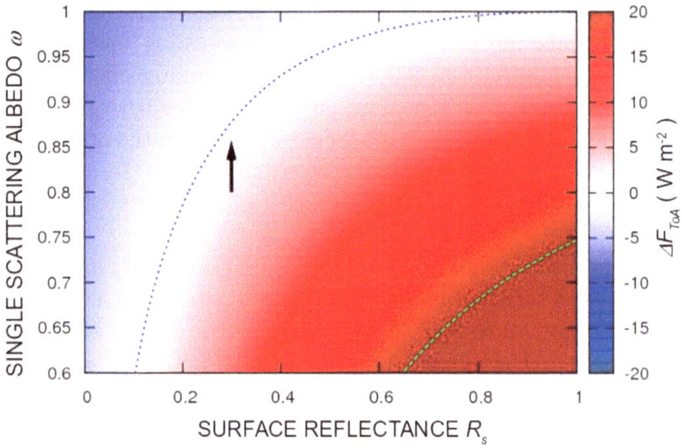

Fig. 8.1. Columnar aerosol single-scattering albedo ω as a function of surface reflectance R_s. The color scale reported on the right indicates the intensity of the DARF effects induced by columnar aerosol and provides the values of the aerosol-induced change ΔF_{ToA} (measured in W/m^2) caused by columnar aerosol in the upwelling flux of short-wave radiation at the ToA-level of the atmosphere. The calculations were made using a similar graph published by Russell et al. (2002) and evaluated using Eq. (8.4) for $\tau_a = 0.10$, $\beta = 0.17$, $A_c = 0$ (cloudless sky), and atmospheric transmittance $T = 0.75$. The blue dotted curve marks the critical values of ω, dividing the graph into two domains: (i) the upper one, where the aerosol loading produces cooling effects; and (ii) the lower one, where warming effects are induced by aerosol particles. The green dashed curve marks the domain of ω, where the changes in the upwelling flux of short-wave radiation at the ToA-level are higher than $+20$ W/m^2, thus inducing very strong warming effects. The vertical black arrow indicates the variation in ω for $R_s = 0.30$, which is mentioned in the text in section 8.3, where the results provided by Eq. (8.4) are presented.

mechanisms carrying aerosol particles from different sources. In this regard, it is reasonable to think that the atmospheric aerosol content generally maintains stable optical characteristics only over short periods of a few days, since it is subsequently influenced and altered by the dynamic patterns of the atmosphere that regulate the particulate matter transport over the observation site. In order to attain more meaningful evaluations of the DARF effects, it seems more appropriate to calculate the 'instantaneous' values of aerosol radiative forcing $\Delta F(t)$ at the ToA- and BoA-levels, for (i) daily sets of aerosol radiative parameters determined from field measurements continuously performed during the whole day, yielding the spectral values of $\tau_a(\lambda)$ and $\omega(\lambda)$, and (ii) daily sets of aerosol scattering, absorption, and extinction coefficients derived from combined measurements of direct solar irradiance and sky-brightness in the almucantar over the entire daily insolation period (Vitale et al., 2000a). As pointed out in section 8.2.1, the instantaneous radiative forcing terms $\Delta F_{ToA}(t)$ and $\Delta F_{BoA}(t)$ can be evaluated by measuring or calculating the instantaneous upwelling and downwelling fluxes of short-wave radiation at the ToA- and BoA-levels, respectively, for an atmosphere without aerosols and an atmosphere with aerosols, each net flux being thus calculated as the difference between the downwelling and upwelling flux. The calculations of instantaneous upwelling fluxes at the ToA-level and the instantaneous upward and downward fluxes at the BoA-level can be determined knowing all the thermal and pressure parameters of the atmosphere, the apparent solar zenith angle, and the optical parameters of the columnar aerosol load. Applying the radiative transfer procedure to the time series of all such parameters, the time-patterns of $\Delta F_{ToA}(t)$ and $\Delta F_{BoA}(t)$ can be determined over the entire insolation period. From them, realistic estimates of the diurnally averaged DARF terms ΔDF_{ToA} and ΔDF_{BoA} can be obtained on each day by calculating the integrals given in Eqs (8.1) and (8.2), respectively.

In the DARF calculations, it is important to take into account that all the atmospheric aerosol radiative parameters, such as $\tau_a(\lambda)$, $\omega(\lambda)$, and the volume scattering and absorption coefficients, which are all subject to appreciable variations during the day over the entire wavelength range. These variations are substantially due to changes in the number and mass concentrations of columnar aerosol, associated with the growth and removal processes of aerosols, and therefore related to the changes in the number and volume size-distribution curves of the columnar particle polydispersion, and in the mass density size-distributions of the various aerosol components. In addition, significant time variations in the radiative transfer processes occur regularly during the day as a function of θ_o, causing some slight time variations in the relative optical air mass m, which is used in the radiative transfer calculations to define the actual length of the optical slant path described by the incoming direct solar irradiance (Tomasi et al., 1998a).

Variations in the atmospheric field of diffuse solar radiation occur usually during the day as a function of the pressure, temperature, and moisture conditions of the atmosphere at various altitudes. This is due to the fact that the diffuse atmospheric radiation is mainly produced by Rayleigh scattering on air molecules and aerosol scattering, although its field is also not negligibly influenced by absorption processes caused by water vapor molecules and airborne aerosols. Such changes in the thermodynamic parameters of air usually lead to appreciable time variations in the upwelling fraction of scattered solar radiation and in atmospheric transmittance characteristics. In addition, the changes in the spectral and angular features of surface reflectance exert strong effects on the DARF processes, which can considerably vary with the surface type and the incidence angle of the solar beam, producing notable effects on the upwelling flux of short-wave radiation at both ToA- and BoA-levels.

Bearing the above considerations in mind, the time series of spectral measurements of $\tau_a(\lambda)$ performed at visible and near-infrared wavelengths during different sun-photometer measurement campaigns are examined in the following subsections. They are studied together with evaluations of columnar aerosol optical parameters derived from sky-brightness measurements in the almucantar taken using sun-/sky-radiometers or from *in situ* nephelometer, aethalometer, and absorption-photometer measurements and/or chemical analyses of aerosol samples. To calculate the DARF effects on the basis of these optical data, realistic non-Lambertian models of surface reflectance were used, as defined by taking into account the coverage characteristics of the areas where the field experiments were conducted. It is fair to point out that not negligible uncertainties can affect both the clear-sky and all-sky DARF evaluations based on field measurements, as evidenced by Loeb and Su (2010), who estimated the total DARF uncertainty arising from the measurements of aerosol optical parameters to be of 0.5–1.0 W m^{-2}—that is, a factor of two to four times greater than the most updated IPCC's assumptions. To reduce such uncertainties, it was recently recommended by Kahn (2012) to combine space-based and targeted suborbital measurements of the aerosol amount, type, and distribution derived from space-borne satellite data along with field measurements, since, by themselves, the satellite-based observations cannot provide sufficient quantitative details on the aerosol microphysical properties. In fact, ground-based measurements of direct solar irradiance and sky-brightness taken at different wavelengths with the sun-/sky-radiometers routinely employed in the AERONET (Holben et al., 1998) and SKYNET (Nakajima et al., 2007) networks can be very useful for obtaining accurate evaluations of the DARF effects, and then validating the satellite-based estimates of such effects acting on the radiative budget of the surface–atmosphere system.

Thus, the present chapter illustrates DARF evaluations obtained by us from sets of field measurements of aerosol radiative parameters carried out in different areas of Earth. The detailed spectral features of the field observations and the DARF calculations derived from them are provided in the following subsections, for various types of natural and anthropogenic aerosol loads monitored in different regions of Europe (Algarve in southern Portugal, Puglia in southern Italy, and Po Valley in northern Italy), the Himalayan region (Nepal), numerous Arctic and Antarctic remote sites, some sectors of the Atlantic Ocean, and north-central Oklahoma (US).

8.3.1 DARF evaluations from the CLEARCOLUMN (ACE-2) field measurements in southern Portugal

The CLEARCOLUMN project was one of the six Aerosol Characterization Experiment (ACE-2) activities planned as parts of a general clear-sky column closure experiment conducted in June/July 1997 over a large area of the eastern Atlantic Ocean including the south-western corner of Portugal, the Canary Islands, and the Azores (Russell and Heintzenberg, 2000). Spectral measurements of direct solar irradiance were regularly performed by Vitale et al. (2000a, 2000b) at Sagres (37° 19′ N, 8° 34′ W, 50 m above mean sea level (a.m.s.l.)) (Algarve, Portugal) from June 16 to July 25, 1997, using the IR-RAD sun-photometer manufactured at the ISAC-CNR Institute (Bologna, Italy). The instrument had a circular field-of-view with angular diameter of 1° 16′, and was equipped with a series of 13 interference filters chosen to define narrow spectral bands within the main transparency windows of the atmospheric transmittance spectrum from 0.4 to 3.7 μm. The spectral channels were selected with peak-wavelengths λ_c equal to 0.401, 0.443, 0.501, 0.550, 0.610, 0.667, 0.780, 0.873, 1.025, 1.224, 1.587, 2.245, and 3.676 μm and half-bandwidths varying between 9 and 14 nm over the $\lambda_c < 1.05$-μm range and increasing from 30 to 156 nm at the four wavelengths longer than 1.05 μm. Each spectral series of direct solar irradiance required an overall time of seven minutes. Therefore, a variable number of 50–130 spectral series of output voltages were recorded on each measurement day. The measurements were analyzed in terms of the Lambert–Beer law using the IR-RAD calibration constants determined at the high-altitude Schneefernerhaus Observatory (2665 m a.m.s.l.) in the Bavarian Alps (Germany), during the RAD-I-CAL 96 (ACE-2) intercomparison campaign of October 1996 (Tomasi et al., 1998b; Vitale et al., 2000b), obtaining as many spectral series of $\tau_a(\lambda)$ at the above 13 wavelengths on each of the 21 clear-sky measurement days of the campaign (Vitale et al., 2000a), and collecting an overall number of more than 2100 spectral series of $\tau_a(\lambda)$ during the CLEARCOLUMN campaign. The values of $\tau_a(0.55\ \mu\text{m})$ were found to mainly vary between 0.05 (for pure maritime and background (hereinafter referred to as BG) continental aerosol particle loads) and more than 0.35 (for polluted aerosol from anthropogenic sources), with overall instrumental errors smaller than 0.01. Such good accuracy of the $\tau_a(\lambda)$ measurements was mainly achieved thanks to the very stable responsivity of the IR-RAD sun-photometer, obtained by (i) housing the thermopile sensor and all electronic and optical components in thermostated boxes regulated to maintain an internal temperature of $15 \pm 0.5°\text{C}$ during the measurements and (ii) mounting the instrument on an alt-azimuth automatic tracker, which assured its pointment to the solar disk with an angular accuracy better than $1'$ for clear-sky conditions.

Among the 21 cloud-free sky measurement days, 10 days were chosen by Vitale et al. (2000b) as 'golden days', presenting well diversified and stable conditions of atmospheric turbidity parameters, associated with columnar aerosol loads of different origin. The spectral series of $\tau_a(\lambda)$ measured over the 0.400–1.025-μm spectral range were examined in terms of the Ångström (1964) formula to determine the atmospheric turbidity parameters α and β: (i) exponent α was obtained with an average error of ± 0.02, providing a reliable measure of the dependence of $\tau_a(\lambda)$ on wavelength, which is closely related to the columnar aerosol size-distribution shape-parameters, and (ii) β was obtained with an accuracy of ± 0.01, in practice giving an average measure of $\tau_a(1.00\ \mu\text{m})$.

In keeping with the concepts and definitions of the DARF terms given in sections 8.2.1–8.2.3, a rigorous procedure subdivided into seven steps (and hereinafter referred to as DARF-PROC) was adopted to analyze the experimental data collected at Sagres during the CLEARCOLUMN campaign. The steps involved were as follows:

(1) *Analysis of field data to determine the columnar aerosol extinction parameters*

The daily mean values of $\tau_a(0.55\ \mu\text{m})$, α and β obtained from the above analysis for the 10 golden days are given in Table 8.1a, which also contains concise information on the origins of the air masses transported over Sagres on the 10 days, as obtained from the NOAA/HYSPLIT (National Oceanic and Atmospheric Administration/Hybrid Single Particle Lagrangian Integrated Trajectory model) (Draxler and Hess, 1998) backward trajectories drawn at 12:00 UTC/GMT of each golden day. The data in Table 8.1a show that, among the 10 chosen days, only June 19, 1997, was characterized by a columnar aerosol load consisting almost totally of pure maritime particles, while the three following days (June 20, 23, and 25) presented mixed compositions, with maritime aerosol particles suspended within the boundary layer, and continental polluted aerosol suspended at middle troposphere altitudes, having been transported from north-western Europe or from the Mediterranean Sea. On the remaining six golden days (i.e. July 5, 7, 10, 11, 18, and 20), the columnar aerosol load substantially consisted of anthropogenic/continental aerosol mainly originating from northern Spain and France, which was concentrated within the low troposphere below 2.0–2.5 km. Correspondingly, the daily mean values of $\tau_a(0.55\ \mu\text{m})$ were found to range between 0.05 (on June 19,

Table 8.1a. Daily mean values of aerosol optical thickness $\tau_a(0.55\ \mu m)$, Ångström's (1964) exponent α calculated over the 0.401–1.025-μm wavelength range, real part $n(0.55\ \mu m)$ and imaginary part $k(0.55\ \mu m)$ of the columnar aerosol refractive index (as derived from the sky-brightness measurements in the almucantar performed following the procedure of von Hoyningen-Huene et al. (1998)), and columnar aerosol single-scattering albedo $\omega(0.55\ \mu m)$, as determined from the measurements performed on the 10 golden days of the CLEARCOLUMN (ACE–2) campaign conducted at Sagres (Algarve, southern Portugal) in summer 1997 (Vitale et al., 2000a, 2000b; Tomasi et al., 2003).

Measurement day	Daily mean $\tau_a(0.55\ \mu m)$	Daily mean Ångström's exponent α	Daily mean values of refractive index parts		Daily mean $\omega(0.55 \mu m)$	Type and origin of columnar aerosol load, as derived from the NOAA/HYSPLIT backward trajectories passing over Sagres at the 12:00 UTC/GMT of each measurement day
			Real part $n(0.55\ \mu m)$	Imaginary part $k(0.55\ \mu m)$		
June 19	0.053	0.63	1.43	3.7×10^{-3}	0.96	Pure maritime aerosol from northern Atlantic Ocean
June 20	0.067	0.14	1.49	8.2×10^{-3}	0.81	maritime aerosol from northern Atlantic Ocean, and continental polluted aerosol from North American coasts of the Atlantic Ocean
June 23	0.065	0.14	1.47	6.7×10^{-3}	0.87	maritime aerosol from northern Atlantic Ocean, and continental aerosol from British Isles, Gulf of Biscay and Portugal
June 25	0.086	0.57	1.49	1.3×10^{-2}	0.90	maritime aerosol from northern Atlantic Ocean, and anthropogenic/continental aerosol from British Isles and north Europe
July 5	0.051	0.35	1.47	6.7×10^{-3}	0.91	Anthropogenic/continental aerosol from Portugal and western Spain
July 7	0.132	0.83	1.50	1.4×10^{-2}	0.80	Anthropogenic aerosol from Iberian Peninsula and western Mediterranean
July 10	0.233	0.91	1.49	1.3×10^{-2}	0.85	Anthropogenic/continental aerosol from British Isles, Gulf of Biscay and north-western part of the Iberian peninsula
July 11	0.217	0.22	1.49	8.2×10^{-3}	0.89	Anthropogenic/continental aerosol from France, Gulf of Biscay and north-western Portugal
July 18	0.298	1.03	1.48	1.2×10^{-2}	0.90	Anthropogenic/continental aerosol from Europe, Iberian peninsula and western Mediterranean
July 20	0.291	0.82	1.50	1.4×10^{-2}	0.84	Anthropogenic polluted aerosol from France, Bay of Biscay and Iberian peninsula

Table 8.1b. Daily values of the diurnally averaged aerosol forcing terms ΔDF_{ToA} at the ToA-level, ΔDF_{BoA} at the BoA-level, and ΔDF_{Atm} within the atmosphere, diurnal average aerosol fractional forcing AFF_{ToA} at the ToA-level (given by the ratio between flux change $\Delta FToA$ at the ToA-level and the incoming flux $I_S \downarrow$ of solar radiation at the ToA-level), and the diurnal average DARF efficiencies E_{ToA}, E_{BoA}, and E_{Atm} giving the rates at which the surface–atmosphere system is forced per unit $\tau_a(0.55\ \mu m)$, as obtained for the 10 golden days of the CLEARCOLUMN (ACE–2) experiment conducted in summer 1997 at Sagres (southern Portugal).

Measurement day	Surface albedo model	Diurnal average (DARF) terms (W/m^2)			Diurnal average	Diurnal average values of DARF efficiencies (W/m^2)		
		ΔDF_{ToA}	ΔDF_{BoA}	ΔDF_{Atm}	AFF_{ToA}	E_{ToA}	E_{BoA}	E_{Atm}
June 19	OS2	−5.2	−1.1	−4.2	-1.2×10^{-2}	−98.5	−19.9	−78.6
	VS1	−2.7	−4.5	+1.7	-6.0×10^{-3}	−51.8	−84.7	+32.9
	VS4	−1.7	−3.5	+1.8	-4.0×10^{-3}	−31.6	−66.1	+34.6
June 20	OS2	−4.9	−4.0	−0.8	-1.1×10^{-2}	−72.4	−60.3	−12.1
	VS1	−2.0	−6.9	+4.9	-5.0×10^{-3}	−30.3	−103.6	+73.2
	VS4	−0.5	−5.7	+5.3	-1.0×10^{-3}	−6.8	−85.3	+78.4
June 23	OS2	−4.7	−4.2	−0.5	-1.1×10^{-2}	−72.5	−64.8	−7.7
	VS1	−1.0	−7.7	+6.7	-2.0×10^{-3}	−14.9	−118.1	+103.2
	VS4	+1.8	−5.8	+7.6	$+4.0 \times 10^{-3}$	+27.9	−89.5	+117.4
June 25	OS2	−7.3	−10.3	+3.0	-1.7×10^{-2}	−85.1	−120.3	+35.2
	VS1	−3.1	−12.8	+9.7	-7.0×10^{-3}	−35.6	−148.4	+112.9
	VS4	−0.5	−11.0	+10.5	-1.0×10^{-3}	−6.1	−127.9	+121.8
July 5	OS2	−3.7	−3.2	−0.5	-9.0×10^{-3}	−73.1	−63.0	−10.1
	VS1	−1.0	−5.6	+4.6	-2.0×10^{-3}	−19.1	−110.1	+91.0
	VS4	+1.0	−4.3	+5.3	$+2.0 \times 10^{-3}$	+18.9	−84.8	+103.7
July 7	OS2	−7.0	−9.3	+2.3	-1.6×10^{-2}	−52.8	−70.3	+17.5
	VS1	−2.3	−13.2	+10.8	-5.0×10^{-3}	−17.6	−99.7	+82.1
	VS4	+1.0	−11.0	+12.0	$+2.0 \times 10^{-3}$	+7.5	−83.3	+90.8
July 10	OS2	−12.2	−19.0	+6.8	-2.9×10^{-3}	−52.5	−81.7	+29.1
	VS1	−5.8	−21.4	+15.6	-1.4×10^{-2}	−24.9	−92.0	+67.1
	VS4	−1.9	−18.8	+16.9	-4.0×10^{-3}	−8.1	−80.5	+72.4
July 11	OS2	−5.5	−5.9	+0.4	-1.3×10^{-2}	−25.4	−27.1	+1.7
	VS1	−2.5	−7.9	+5.4	-6.0×10^{-3}	−11.7	−36.5	+24.8
	VS4	−0.9	−6.7	+5.8	-2.0×10^{-3}	−4.3	−31.0	+26.7
July 18	OS2	−12.0	−13.5	+1.5	-2.9×10^{-2}	−40.1	−45.3	+5.2
	VS1	−5.7	−20.5	+14.8	-1.4×10^{-2}	−19.2	−68.9	+49.7
	VS4	−1.6	−17.5	+15.9	-4.0×10^{-3}	−5.5	−58.7	+53.2
July 20	OS2	−11.3	−14.9	+3.5	-2.7×10^{-3}	−39.0	−51.0	+12.0
	VS1	−5.4	−19.6	+14.2	-1.3×10^{-2}	−18.5	−67.4	+48.9
	VS4	−1.6	−16.9	+15.3	-4.0×10^{-3}	−5.5	−58.1	+52.6

1997, in the presence of pure maritime aerosols) to nearly 0.30 (on July 18, 1997, for anthropogenic/continental polluted aerosol), with daily mean values of α ranging between 0.14 (June 20 and 23, 1997) and more than 1.0 (July 18, 1997), the lowest values of the Ångström's exponent being due to the more significant extinction effects of coarse particles.

Following the above procedure, the daily time-patterns of $\tau_a(0.55\ \mu\text{m})$ and α were also defined for each golden day. Four examples of the daily time-patterns of $\tau_a(0.55\ \mu\text{m})$ are shown in Fig. 8.2, as derived from the IR-RAD measurements performed on (i) June 19, 1997 (pure maritime aerosol), with a daily mean value of $\alpha = 0.63$; (ii) June 23, 1997 (mixed maritime/continental aerosol) with $\alpha = 0.14$; (iii) July 5, 1997 (anthropogenic/continental aerosol) with $\alpha = 0.35$; and (iv) July 7, 1997 (anthropogenic aerosol) with $\alpha = 0.83$. It can be observed that $\tau_a(0.55\ \mu\text{m})$ decreased slowly during the morning hours of the first two golden days, from 0.10 to 0.15 in the early morning to about 0.03–0.05 at noon, after which it slowly increased in the afternoon to reach values of 0.08–0.10 at sunset. On July 5, 1997, $\tau_a(0.55\ \mu\text{m})$ decreased during the morning, from about 0.10 to 0.03 at noon, and increased from 0.03 to 0.08 in the afternoon, while, on July 7, 1997, parameter $\tau_a(0.55\ \mu\text{m})$ slowly decreased from about 0.18 to 0.12 during the morning and gradually increased in the afternoon until reaching a value of around 0.18 at sunset. Figure 8.2 also shows the time-patterns of columnar aerosol $\omega(0.55\ \mu\text{m})$ determined on the same four days considered in the upper part of Fig. 8.2. These were derived for the real and imaginary parts of columnar aerosol refractive index evaluated at step (2), and the multimodal size-distribution curves inferred at step (4), as described below.

(2) *Determination of the columnar aerosol refractive index*

The second step was devoted to the analysis of the simultaneous measurements of sky-brightness in the almucantar and short-wave radiation flux at ground level (von Hoyningen-Huene et al., 1998) performed at the Sagres and Mt. Foia sites during the CLEARCOLUMN campaign, using a pair of ASP sun-/sky-radiometers manufactured at the Leipzig University (Germany). The measurements were examined following the procedure described by Posse and von Hoyningen-Huene (1996), obtaining daily mean estimates of complex refractive index $n(\lambda) - ik(\lambda)$ of columnar aerosol load on all the 10 days. The monochromatic values of $n(0.55\ \mu\text{m})$ and $k(0.55\ \mu\text{m})$ are given in Table 8.1a, indicating that $n(0.55\ \mu\text{m})$ varied during the campaign between 1.43 (maritime aerosol) and 1.50 (mixed anthropogenic and continental aerosol), and $k(0.55\ \mu\text{m})$ between 4.0×10^{-3} and 1.4×10^{-2}.

(3) *Determination of the size-distribution curves of columnar aerosol*

The spectral series of $\tau_a(\lambda)$ recorded for different hours of the 10 golden days were analyzed following the King (1982) inversion method for the daily mean values of complex refractive index given in Table 8.1a. Using the said procedure, the columnar particle size-distribution curves were retrieved, obtaining the sequences of multimodal size-distribution curves of columnar aerosol over the particle radius range from 0.08 to about 12 μm (Vitale et al., 2000a; Tomasi et al., 2003). Examples of the size-distribution curves derived at various hours of three golden days are shown in Fig. 8.3. The columnar aerosol size-distributions of particle number density $N(r)$ and particle volume $V(r)$ turned out to be clearly characterized by bimodal features in all the three cases, with the fine particle mode centered at radii lower than 0.08 μm, and the coarse particle mode peaked at radii equal to ~5 μm on June 19, 1997 (pure maritime aerosol), ~1.2 μm on July 5, 1997 (anthropogenic/continental aerosol), and ~1.8 μm on July 20, 1997 (anthropogenic polluted aerosol). Figure 8.3 also provides evidence of the predominant contribution made by coarse particles

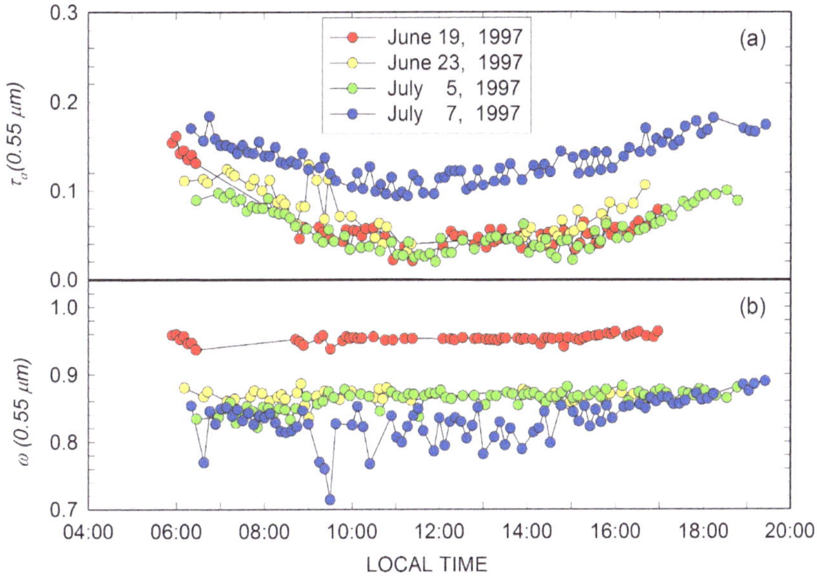

Fig. 8.2. (a) Time-patterns of aerosol optical thickness $\tau_a(0.55\ \mu\text{m})$ derived from the IR-RAD direct solar irradiance measurements performed at Sagres on: (i) June 19, 1997, for pure maritime aerosol (red circles), giving a daily mean value of $\alpha = 0.63$; (ii) June 23, 1997, for mixed maritime/continental aerosol (yellow circles), yielding $\alpha = 0.14$; (iii) July 5, 1997, for anthropogenic/continental aerosol (green circles), with $\alpha = 0.35$; and (iv) July 7, 1997, for anthropogenic aerosol (blue circles), with $\alpha = 0.83$. (b) Time-patterns of columnar aerosol single-scattering albedo $\omega(0.55\ \mu\text{m})$ calculated for (i) the values of the real and imaginary parts of columnar aerosol refractive index derived from the sky-brightness measurements performed by the Leipzig University group (Germany) using the ASP sun-/sky-radiometer and (ii) the size-distribution curves of columnar aerosol particles inferred from the spectral series of $\tau_a(\lambda)$ measured with the IR-RAD sun-photometer using the King (1982) inversion procedure.

to the particulate volume occupying the atmospheric column in the three cases, explaining the fact that the corresponding daily mean values of α were all found to be lower than 1.

(4) *Determination of the columnar aerosol single-scattering albedo*

Using the multimodal size-distribution curves retrieved at various hours of each golden day together with the evaluations of $n(\lambda)$ and $k(\lambda)$ given in Table 8.1a for the columnar particulate matter, calculations of SSA parameter $\omega(\lambda)$ were made at the various visible and near-IR wavelengths, from which the daily mean values of $\omega(0.55\ \mu\text{m})$ were obtained, as given in Table 8.1a for each golden day. They were found to (i) be equal to 0.96 on June 19, 1997, for a pure maritime aerosol content, (ii) vary between 0.81 and 0.90 on the three golden days characterized by mixed contents of maritime/continental aerosols, (iii) range between 0.85 and 0.91 on the four golden days characterized by the presence of mixed anthropogenic/continental particle loads, and (iv) be equal to 0.80 and 0.84 on the two days of July 7, 1997, and July 20, 1997, respectively, when the columnar contents mainly consisted of

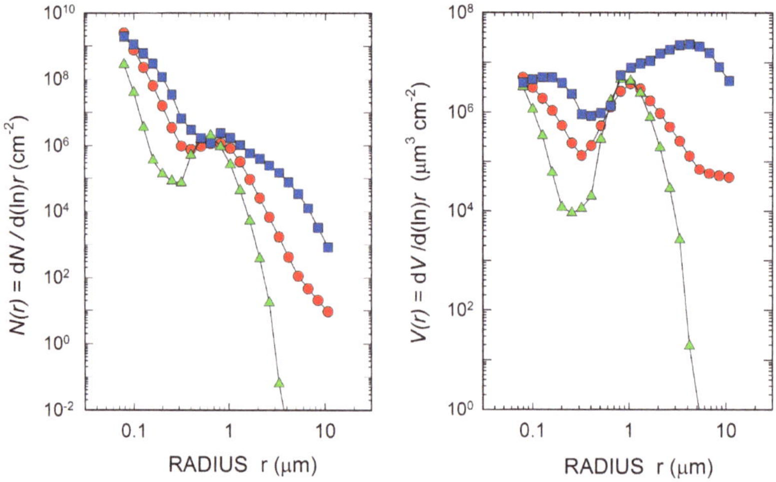

Fig. 8.3. Examples of bimodal size-distribution curves of the columnar aerosol particle number density $N(r) = \mathrm{d}N/\mathrm{d}(\ln r)$ measured per cm^2 (left) and columnar aerosol particle volume $V(r) = \mathrm{d}V/\mathrm{d}(\ln r)$ measured in μm^3/cm^2 (right), as retrieved by applying the King (1982) inversion method to the IR-RAD spectral series of $\tau_a(\lambda)$ determined for the following three cases: (i) June 19, 1997 (09:52 UTC/GMT), for pure maritime aerosol (red circles); (ii) July 5, 1997 (17:14 UTC/GMT) for mixed anthropogenic/continental aerosol (green triangles); and (iii) July 20, 1997 (11:23 UTC/GMT) for anthropogenic aerosol (blue squares). The retrieval procedure was used in all cases for the daily mean values of complex refractive index given in Table 8.1a.

anthropogenic polluted aerosol. Some examples of time-patterns of $\omega(0.55\ \mu\mathrm{m})$ determined on four golden days for different columnar aerosol loads are shown in Fig. 8.2.

(5) *Definition of local surface albedo models*

The fifth step involved the definition of a set of realistic non-Lambertian models of BRDF surface albedo to represent the average characteristics of the ocean and land surfaces in the different areas of southern Portugal, since these data were necessarily used to calculate the instantaneous values of terms $\Delta F_{ToA}(t)$ and $\Delta F_{BoA}(t)$ for the various solar zenith angles recorded during the sunlit periods of the 10 golden days. To define the main spectral characteristics of the Atlantic Ocean surface albedo over the area off-shore of the Sagres coastal station, we calculated the average surface-level wind velocity V_w recorded at Sagres during the campaign. For this purpose, we analyzed the ground-level meteorological data recorded during the campaign (Verver et al., 2000) and the climatological data set of wind velocity downloaded over the sector of Atlantic Ocean in proximity of southern Portugal coasts from the NOAA/OAR/ESRL PSD website *www.esrl.noaa.gov/psd/* (Kalnay et al., 1996). On examining such data, the daily average value of V_w over the ocean surface was found to be close to 5 m/s during the period from mid-June to mid-July. We therefore decided to use the OS2 model of Tomasi et al. (2013) to represent as realistically as possible the average BRDF surface reflectance characteristics of the ocean surface during the CLEARCOLUMN campaign in the DARF calculations,

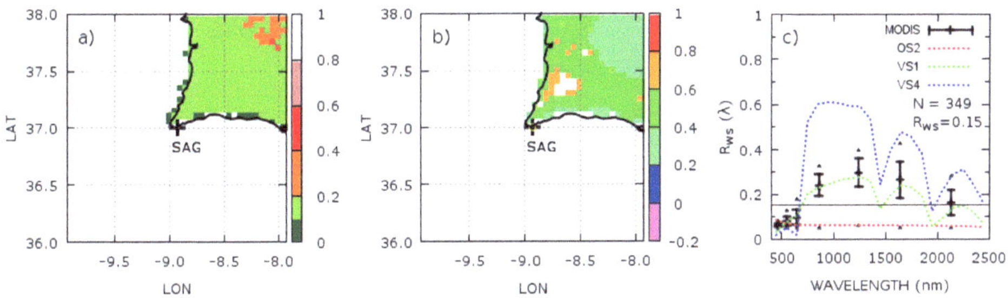

Fig. 8.4. Maps of the average land surface albedo (a) and Normalized Difference Vegetation Index (NDVI) (b) obtained over the Algarve region (southern Portugal) from the MODIS Level 3.0 surface albedo data (MCD43C3 products) recorded during the second half of June. The data recorded during the first half of July do not appreciably differ from those of June. The cross symbol labelled SAG in both graphs indicates the geographical position of Sagres, where the sun-photometer measurements were performed during the CLEARCOLUMN field experiment. (c) Spectral values (black vertical bars) of the white-sky albedo R_{ws}, averaged over a land area of around 1° latitude × 1° longitude, giving a mean value of $R_{ws} = 0.15$, with standard deviations of around 0.1 determined within all the MODIS spectral channels, and minimum and maximum values represented with small triangles over the spectral range from 0.50 to 2.50 μm. Also shown are the spectral curves of $R_{ws}(\lambda)$ relative to (i) the ocean surface albedo model OS2 (dotted red curve) and (ii) the vegetation-covered surface albedo models VS1 (dotted green curve) and VS4 (dotted blue curve) proposed by Tomasi et al. (2013), as obtained applying the best-fit procedure to the data sets recorded within all the seven MODIS spectral channels.

model OS2 being determined for $V_w = 5$ m/s and sea-water pigment concentration of 34.3 ppt. Analyzing the set of MCD43C3 products derived from the MODIS Level 3.0 surface albedo data recorded over the southern Portugal from mid-June to mid-July of 2009, we obtained the surface reflectance maps shown in Fig. 8.4. They clearly indicate that the surface reflectance characteristics over land were best represented throughout the campaign by the vegetation-covered surface albedo models VS1 and VS4 of Tomasi et al. (2013). The said models were found to provide the lowest values of Root Mean Square Error (RMSE) on applying the best-fit procedure to the surface albedo data measured within the seven spectral channels of the MODIS sensor: (i) model VS1 was determined for a low-depth canopy coverage with Leaf Area Index (LAI) = 0.10, and is therefore typical of the sparsely vegetated land areas encountered in proximity to the Atlantic Ocean coasts; and (ii) model VS4 was defined for a vegetation surface covered by high-depth canopies with LAI = 5.0, and is therefore suitable for representing the surface reflectance features of the hill and mountain region between the Atlantic coast and Mt. Foia, which is covered by extensive woods of eucalyptus and cork oaks. Thus, models VS1 and VS4 were used to carry out the most representative DARF calculations over such land area.

The spectral curves of models OS2, VS1, and VS4 are shown in Fig. 8.5 for nine values of θ_o taken in steps of 10° from 0° to 80°. It can be seen that the spectral curves of the OS2 model are flat and exhibit surface albedo values lower than 0.10 for $\theta_o \leq 50°$, which then gradually increase to around 0.15 at $\theta_o = 60°$, until exceeding 0.40 at $\theta_o = 70°$ for $\lambda > 0.7$ μm, and becoming nearly unit at $\theta_o = 80°$ for

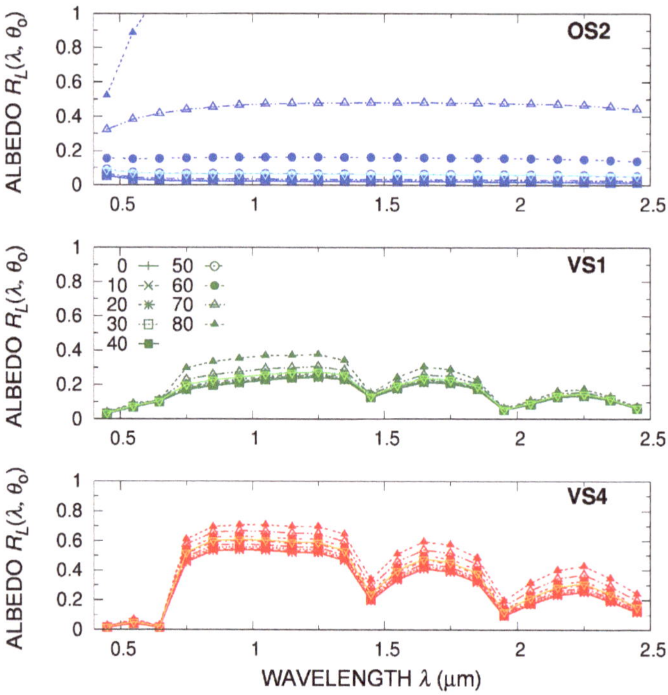

Fig. 8.5. Spectral curves of the BRDF surface albedo $R_L(\lambda, \theta_o)$ defined by Tomasi et al. (2013) for the ocean surface albedo model OS2, and the vegetation-covered surface albedo models VS1 and VS4, as evaluated for nine values of solar zenith angle θ_o, taken in steps of 10° over the 0°–80° range. Model OS2 refers to an oceanic surface with surface-level wind velocity $V_w = 5$ m/s; model VS1 to a vegetation-covered surface, like that of the land areas surrounding Sagres, covered by low-depth canopies (with Leaf Area Index LAI = 0.10); and VS4 to a vegetation-covered surface, like those of the Algarve hill and mountain areas covered by woods (with LAI = 5.0, for high-depth canopies). The spectral curves of the white-sky albedo $R_{ws}(\lambda)$ are also shown in each graph using a lighter color.

$\lambda > 0.6$ μm. Tomasi et al. (2013) estimated that (i) model OS2 presents black-sky albedo $R_{bs}(\theta_o = 0°) = 0.026$, bi-hemispherical reflectance (also known as white-sky albedo) $R_{ws} = 0.069$, and broadband albedo $A(\theta_o = 60°) = 0.158$; (ii) model VS1 has $R_{bs}(\theta_o = 0°) = 0.134$, $R_{ws} = 0.153$, and $A(\theta_o = 60°) = 0.155$; and (iii) model VS4 has $R_{bs}(\theta_o = 0°) = 0.250$, $R_{ws} = 0.292$, and $A(\theta_o = 60°) = 0.306$.

(6) *Calculations of the daily time-patterns of instantaneous DARF terms and diurnally averaged DARF effects*

The time-patterns of $\Delta F_{ToA}(t)$ and $\Delta F_{BoA}(t)$ were calculated for all 10 days by employing the procedure of Tomasi et al. (2013), applied to (i) the size-distribution curves retrieved at step (3) from the spectral series of $\tau_a(\lambda)$ determined at step (1) for various solar zenith angles θ_o; (ii) the complex refractive index data defined at step (2), together with the main aerosol optical parameters calculated at step (4); and (iii) the OS2, VS1, and VS4 surface albedo models defined at step (5). The time-patterns of $\Delta F_{ToA}(t)$ and $\Delta F_{BoA}(t)$ were integrated over the sunlit period of each golden day, according to Eqs (8.1) and (8.2) to calculate the daily mean values of the diurnally averaged DARF terms ΔDF_{ToA} and ΔDF_{BoA}, for each of the OS2, VS1, and VS4 models. The evaluations are provided in Table 8.1b, together with estimates of the daily mean values of ΔDF_{Atm}, calculated as differences between the corresponding ΔDF_{ToA} and ΔDF_{BoA}, according to Eq. (8.3).

The time-patterns of the daily mean values of ΔDF_{ToA}, ΔDF_{BoA}, and ΔDF_{Atm} obtained on the 10 golden days for the OS2, VS1, and VS4 surface albedo models are shown in Fig. 8.6, together with those of $\tau_a(0.55\ \mu\text{m})$ and columnar aerosol SSA parameter $\omega(0.55\ \mu\text{m})$. As can be seen, the time-patterns of ΔDF_{ToA} are mainly negative for all three surface models, differing by a few W/m^2 from one model to another, and hence inducing cooling effects at the ToA-level. Similarly, the time-patterns of ΔDF_{BoA} are rather similar on all the golden days, with differences of a few W/m^2 passing from one surface albedo model to another. Consequently, the time-patterns of ΔDF_{Atm} calculated for the VS1 and VS4 models were found to assume very similar values, which are all positive (heating effects) and ranging mainly between +5 and +15 W/m^2, in spite of the considerable spectral discrepancies between the two surface albedo models. Appreciably lower values of ΔDF_{Atm} were obtained for the OS2 model, with both positive and negative signs. Their absolute values are in general smaller than those achieved for the VS1 and VS4 models, by 5–10 W/m^2 on the various days, giving a measure of the change in the DARF effects that may arise from the different surface albedo conditions of sea and land surfaces. The most noteworthy outcome of the calculations of ΔDF_{ToA}, ΔDF_{BoA}, and ΔDF_{Atm} is that the above DARF terms are roughly proportional to $\tau_a(0.55\ \mu\text{m})$, with proportionality coefficients varying according to aerosol origin and surface albedo. The different dependence features can be observed in Fig. 8.7, which presents the scatter plots of the daily mean values of ΔDF_{ToA}, ΔDF_{BoA}, and ΔDF_{Atm} versus the corresponding daily mean values of $\tau_a(0.55\ \mu\text{m})$, as obtained on the 10 golden days for the OS2 and VS4 surface albedo models. The data dispersion appears to be at least partly caused by the day-to-day variations in $\omega(0.55\ \mu\text{m})$, which ranged between 0.80 and 0.96 on the 10 golden days.

The daily mean values of ΔDF_{ToA} were found to vary between -3 and 13 W/m^2 over sea, and between -6 and $+2$ W/m^2 over land, depending closely on $\tau_a(\lambda)$ at visible wavelengths, and more weakly on the columnar aerosol optical characteristics. At the same time: (i) daily mean values of ΔDF_{BoA} varied between -1 and -22 W/m^2 over sea, and between -3 and -23 W/m^2 over land, while (ii) those of ΔDF_{Atm} between -5 and $+7$ W/m^2 over sea, and between $+2$ and $+17$ W/m^2 over land. In spite of such variable results, it is evident that ΔDF_{ToA}, ΔDF_{BoA}, and ΔDF_{Atm} decrease almost linearly as $\tau_a(0.55\ \mu\text{m})$ gradually increases, although these evaluations differ significantly from one surface albedo model to another, with considerable dispersion due to changes in the columnar aerosol SSA.

Calculations of AFF_{ToA} were also made for the daily mean values of ΔDF_{ToA} given in Table 8.1b and the daily average calculations of incoming solar radiation flux $I_S \downarrow$ at the ToA-level averaged over the whole sunlit period of each day, as obtained for the 10 golden days. The daily estimates of AFF_{ToA} are shown in Table 8.1b, assuming negative values mainly ranging between 10^{-3} and 10^{-1} for all the three surface albedo models, and a few positive values for the high-surface model VS4 only.

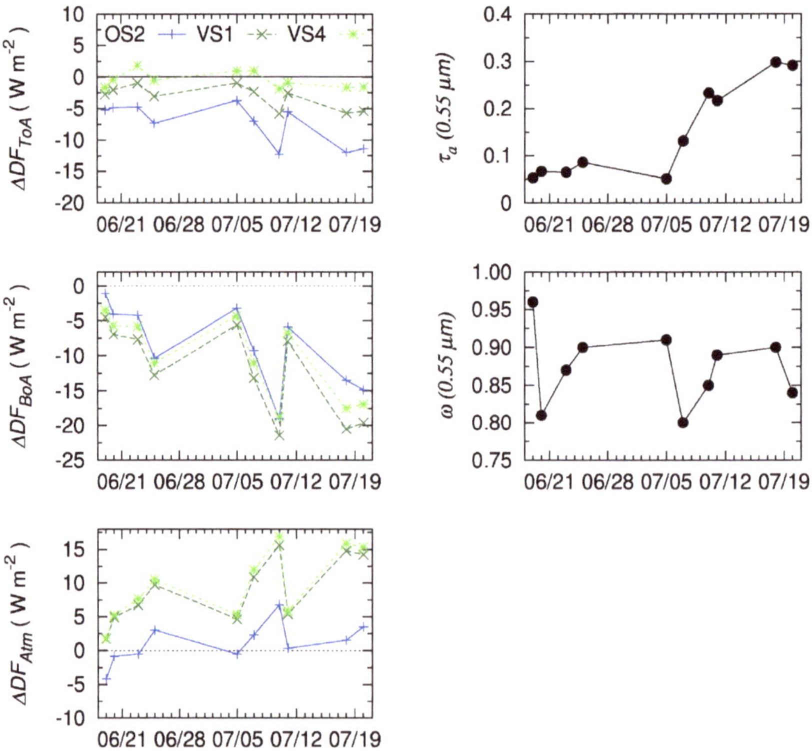

Fig. 8.6. Left: Time-patterns of the daily mean values of diurnally averaged DARF terms ΔDF_{ToA} at the ToA-level, ΔDF_{BoA} at the BoA-level, and ΔDF_{Atm} within the atmosphere, as calculated on the 10 golden days of the CLEARCOLUMN field campaign conducted at Sagres (southern Portugal) from June 16 to July 25, 1997, using the surface albedo models OS2, VS1, and VS4 of Tomasi et al. (2013). Right: Time-patterns of the daily mean values of aerosol optical thickness $\tau_a(0.55\ \mu\text{m})$ and columnar aerosol single-scattering albedo $\omega(0.55\ \mu\text{m})$, as determined for the columnar aerosol size-distribution curves and complex refractive index measured on the 10 golden days of the CLEARCOLUMN field campaign.

(7) *Calculations of the daily mean values of DARF efficiencies*

In spite of the data dispersion evident in Fig. 8.7, the slope coefficients of the regression lines drawn in these scatter plots of ΔDF_{ToA}, ΔDF_{BoA}, and ΔDF_{Atm} versus $\tau_a(0.55\ \mu\text{m})$ provide reliable average estimates of the corresponding efficiency terms. It is clear that such efficiency variations arise principally from the different SSA features, which are closely related to the composition of columnar particulate matter, and the diverse surface albedo characteristics considered in the DARF calculations. The daily mean values of DARF efficiencies E_{ToA} at the ToA-level, E_{BoA} at the BoA-level, and E_{Atm} in the atmosphere were calculated on the 10 golden days by dividing the daily values of ΔDF_{ToA}, ΔDF_{BoA}, and ΔDF_{Atm} by the corresponding values of $\tau_a(0.55\ \mu\text{m})$ given in Table 8.1a. The estimates are plotted in Fig. 8.8 versus the corresponding daily mean values of columnar aerosol $\omega(0.55\ \mu\text{m})$, separately for the OS2, VS1, and VS4 surface albedo models.

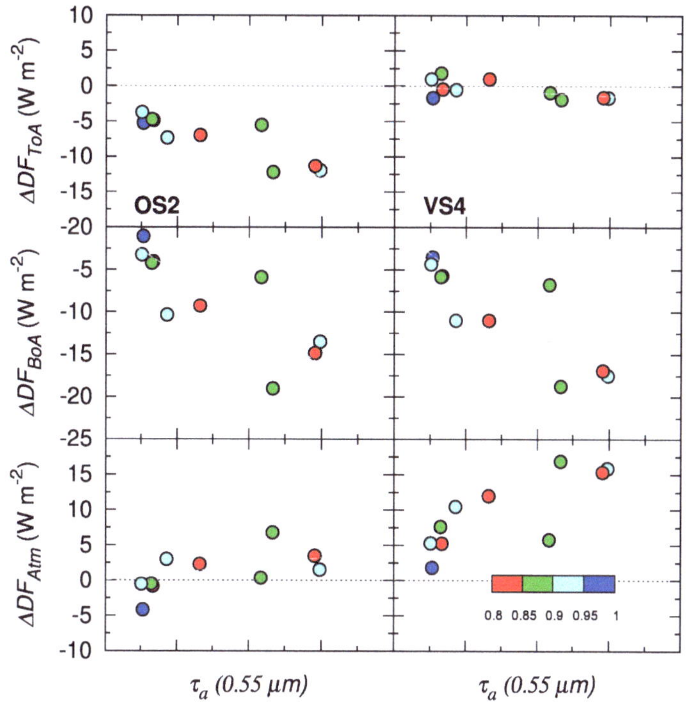

Fig. 8.7. Scatter plots of the daily mean values of DARF terms ΔDF_{ToA}, ΔDF_{BoA}, and ΔDF_{Atm} versus the corresponding daily mean values of aerosol optical thickness $\tau_a(0.55\ \mu m)$ determined from the CLEARCOLUMN field measurements performed on the 10 golden days, as calculated for the oceanic surface albedo model OS2 (left) and the vegetation-covered surface albedo model VS4 (right), and for different values of columnar aerosol single-scattering albedo $\omega(0.55\ \mu m)$, indicated using differently colored circles according to the color scale reported on the right.

The scatter plots confirm that the DARF efficiency terms vary considerably with changing surface albedo model. In particular: (i) E_{ToA} decreases in general as $\omega(0.55\ \mu m)$ increases, presenting the highest (both positive and negative) values for model VS4 (i.e. for a high-canopy covering the surface), moderately negative values for the low-canopy coverage represented by model VS1, and very negative values over the sea surface represented by model OS2; (ii) E_{BoA} assumes very scattered values as a function of $\omega(0.55\ \mu m)$, without appreciable differences for the three surface reflectance models; and (iii) E_{Atm} exhibits mainly positive values, which are higher for the VS1 and VS4 surface models and lower but more scattered for the OS2 model, evidencing a generally decreasing trend from positive to negative values with increasing $\omega(0.55\ \mu m)$—that is, as the mass fraction of anthropogenic polluted aerosol gradually decreases.

The results achieved through the analysis of the CLEARCOLUMN measurements made for different surface albedo conditions are evidence of the complexity of the radiative transfer processes occurring in the atmosphere for different combinations of aerosol scattering characteristics and surface reflectance properties, both of which

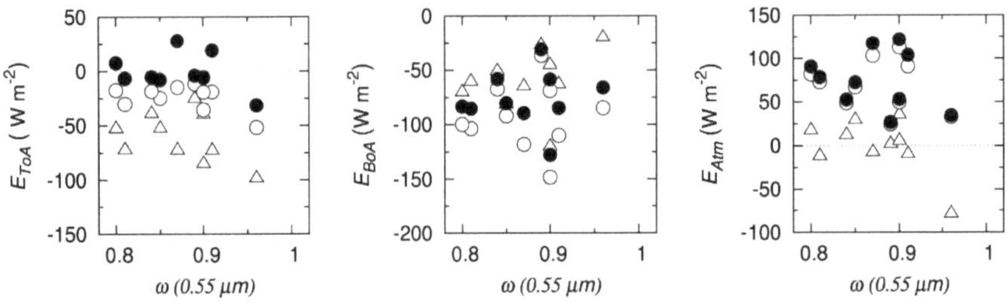

Fig. 8.8. Scatter plots of the daily mean values of DARF efficiency E_{ToA} at the ToA-level, DARF efficiency E_{BoA} at the BoA-level, and DARF efficiency E_{Atm} within the atmosphere versus the columnar aerosol single-scattering albedo $\omega(0.55\ \mu m)$, as obtained on the 10 golden days of the CLEARCOLUMN experiment conducted at Sagres (Portugal) in summer 1997, by assuming the surface albedo characteristics represented by ocean surface model OS2 (open triangles), vegetation-covered surface model VS1 (open circles), and vegetation-covered surface model VS4 (solid circles).

contribute to modulating the upwelling flux of short-wave radiation in the surface–atmosphere system.

8.3.2 DARF evaluations from the PRIN-2004 project measurements in southern Italy

The PRIN-2004 project was one of the research projects supported by the Italian Ministry of the University and Scientific Research (MiUR) during the triennium 2003–2005. It specifically aimed to (i) characterize the optical and microphysical properties of aerosols, (ii) calculate the DARF effects, and (iii) provide a first picture of the natural and anthropogenic aerosol species present over the southern part of Puglia (Salento) in southern Italy. The project was developed by the Physics Department at the University of Salento (Lecce), in cooperation with two research groups of the Universities of Basilicata (Potenza) and Ferrara. The ISAC-CNR (Bologna) group participated in the project as an external partner, its main goal being to apply a new procedure for calculating the DARF effects at the ToA- and BoA-levels, testing the resulting DARF evaluations through a comparison with the estimates made by Tafuro et al. (2007) on the basis of the field measurements performed by Perrone et al. (2005). The field measurements were regularly recorded by (i) the AERONET Cimel CE-318 sun-photometer, providing regular measurements of all the columnar aerosol parameters; (ii) the EARLINET Lidar routinely operated at the Lecce University site to measure vertical profiles of the aerosol backscattering coefficient at the 0.532-μm wavelength; and (iii) various *in situ* sampling instruments, used to measure the microphysical characteristics and chemical composition of ground-level aerosol particles. These measurements were analyzed by Lanconelli (2007) in his Ph.D. thesis, to infer the aerosol optical characteristics and determine the mean scattering and absorption parameters of columnar aerosol during the field campaign conducted from March 2003 to January 2004. The backward trajectories passing over Lecce were downloaded from the NOAA/HYSPLIT data archive to define the last 96-hour transport patterns of industrial, continental, and desert aerosol particles over south-eastern Italy. In addition, satellite data downloaded from the Sea-viewing Wide Field-of-view Sensor 1 (SeaWiFS-1) and MODIS/Terra and MODIS/Aqua archives were also examined by Lanconelli (2007) to better characterize the Saharan dust transport episodes over Lecce. Through this exhaustive analysis, aerosols were estimated to be transported over the southern Puglia from a number of different sources: (i) anthropogenic/continental aerosols from central and eastern Europe; (ii) urban/industrial aerosols from north-western Europe and northern Italy; (iii) mineral dust from the Saharan region and North African coasts; and (iv) maritime particles from the central Atlantic Ocean and Mediterranean Sea. Eighty-seven clear-sky days were selected by Perrone et al. (2005), who found daily mean values of $\tau_a(0.55\ \mu m)$ mainly varying between 0.05 and 0.80, and those of Ångström's exponent α between 0.05 and 2.0. Parameter α was calculated by applying the Ångström (1964) best-fit procedure to the spectral series of $\tau_a(\lambda)$ measured over the 0.44–0.87-μm wavelength range. Correspondingly, the columnar parameter $\omega(0.44\ \mu m)$ was estimated to assume daily mean values mainly varying between 0.83 and 0.99. The daily mean values of parameter $n(0.55\ \mu m)$ were estimated by Lanconelli (2007) mainly to range between 1.33 and 1.60 during the PRIN-2004 field campaign, and those of $k(0.55\ \mu m)$ between 5×10^{-4} and 10^{-1}. Analyzing the set of measurements taken during the most massive Saharan dust transport episodes, $n(0.55\ \mu m)$ was found to vary between 1.45 and 1.60, and $k(0.55\ \mu m)$ from 2×10^{-3} to 2×10^{-2}. These variable characteristics are associated with the diverse origins of the aerosol particle loads, as can be clearly seen in Fig. 8.9, which shows the scatter plots of the daily mean values of exponent $\alpha(0.44–0.87\ \mu m)$, and $\omega(0.44\ \mu m)$ versus $\tau_a(0.55\ \mu m)$, as obtained analyzing the AERONET data of Perrone et al. (2005). Bearing in mind that $\omega(0.44\ \mu m)$ is closely related to the scattering and absorbtion characteristics of columnar aerosols, a rough classification of the aerosol optical features is offered by the scatter plots of Fig. 8.9, since parameter α closely depends on the size-distribution shape parameters and gives a measure of the optical weights of the fine and coarse particle loads in extinguishing the incoming solar radiation. In particular, the scatter plot shown in the right part of Fig. 8.9 highlights that continental particles produce very different extinction and scattering effects from those caused by mineral dust, since they contain in general prevailing fractions of fine particles on those of coarse particles, even though the SSA features of the two particle size classes are very similar. Some noticeable seasonal differences can be seen in Fig. 8.9, which gives evidence that higher values of $\tau_a(0.44\mu m)$ were generally found in spring and summer, and lower aerosol particle loads were observed over Lecce during the colder period of the year, with features more frequently associated with maritime particles.

Using the seven-step DARF-PROC procedure adopted in the previous subsection for analyzing the CLEARCOLUMN measurements, we examined a limited set of measurements carried out at the Lecce University station on 12 selected golden days, obtaining the following results in the various steps:

(1) *Analysis of field data to determine the columnar aerosol extinction parameters*

The AERONET and EARLINET measurements performed at Lecce University (40° 20′ N, 18° 06′ E, 53 m a.m.s.l.) from March 2003 to January 2004 were analyzed to find 12 golden days, for which the spectral characteristics and the main extinction parameters of columnar aerosol were determined. The 12 golden days

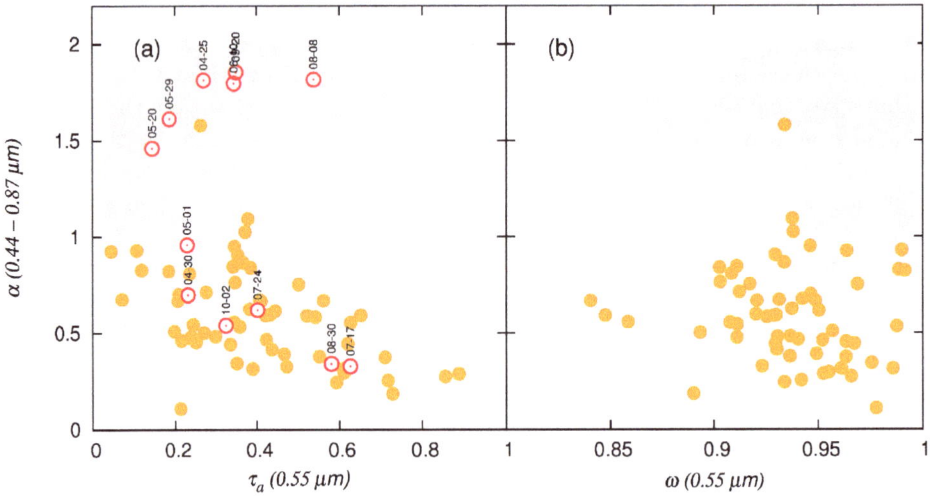

Fig. 8.9. Scatter plots of the daily mean values of Ångström's exponent $\omega(0.44–0.87\ \mu m)$ versus aerosol optical thickness $\tau_a(0.55\ \mu m)$ (left) and columnar aerosol single-scattering albedo $\omega(0.44\ \mu m)$ (right), as obtained from the set of AERONET measurements performed by Perrone et al. (2005) over the 0.44–0.87-μm wavelength range on 87 clear-sky days of the AERONET campaign conducted from March 1, 2003, to January 17, 2004, at Lecce (Salento, Puglia, southern Italy). Gray circles refer to the days characterized by predominant extinction due to mixed anthropogenic/continental aerosol, and light brown circles to the days presenting predominant extinction effects produced by Saharan dust particles. Red-edged circles shown in the left part refer to the 12 golden days of the PRIN-2004 experiment, subdivided into two sets each consisting of six golden days, the upper set regarding anthropogenic/continental aerosol and the lower set pertaining to Saharan dust transport episodes.

were chosen among the 87 clear-sky days studied by Perrone et al. (2005). They are listed in Table 8.2a, together with the origins of the aerosol loads indicated by the NOAA/HYSPLIT data. As can be seen, only one day (April 25, 2003) was characterized by the presence of pure continental aerosol from northern and eastern Europe. The remaining 11 days presented the following features: (i) April 30, 2003, with anthropogenic polluted aerosol transported from the southern Mediterranean, mixed with marine aerosol from the central Atlantic Ocean at levels >1 km; (ii) May 1 and July 17, 2003 with transport of Saharan dust at low levels (<1 km and <3 km, respectively) and marine/continental aerosol from Atlantic Ocean and Iberian peninsula at the upper levels; (iii) May 20 and August 30, 2003, with anthropogenic/continental aerosol transported from southern France, northern Italy, and the Balkans; (iv) May 29, June 10, August 8, and September 20, 2003, with fine continental aerosol transported from eastern Europe and the southern Balkans; and (v) July 24 and October 2, 2003, with continental aerosol transported from southern Italy and the Balkans, mixed with Saharan dust at middle and high levels. Table 8.2a also provides the daily mean values of $\tau_a(0.50\ \mu m)$ derived from the AERONET Level 2.0 data recorded on the 12 golden days, which were estimated to vary between 0.14 (May 20, 2003, for anthropogenic/continental aerosol) and 0.63 (July 17, 2003, with Saharan dust transport).

The spectral series of $\tau_a(\lambda)$ obtained on the 12 golden days were examined in terms of the Ångström (1964) formula to determine the best-fit values of exponent $\alpha(0.44\text{–}0.87\ \mu\text{m})$. Three examples of the best-fit procedure are shown in Fig. 8.10, as applied to different atmospheric turbidity conditions associated with (i) anthropogenic/continental aerosol transported from southern France and northern Italy (May 20, 2003, 09:45 UTC/GMT, $\alpha = 1.496$); (ii) fine continental aerosol from the Black Sea region and southern Balkans (June 10, 2003, 10:03 UTC/GMT, $\alpha = 1.803$); and (iii) anthropogenic/continental aerosol from central and southern Italy, mixed with a significant load of mineral dust transported from the Saharan region (August 30, 2003, 07:30 UTC/GMT, $\alpha = 0.342$).

Examining the overall set of daily mean values of $\alpha(0.44\text{–}0.87\ \mu\text{m})$ given in Table 8.2a, it can be noted that exponent α varied between 0.33 (measured on July 17, 2003, for optically predominant Saharan dust particles) and 1.86 (September 20, 2003, for main extinction effects produced by continental aerosol suspended within the first kilometre of the atmosphere over southern Italy, and weaker extinction by

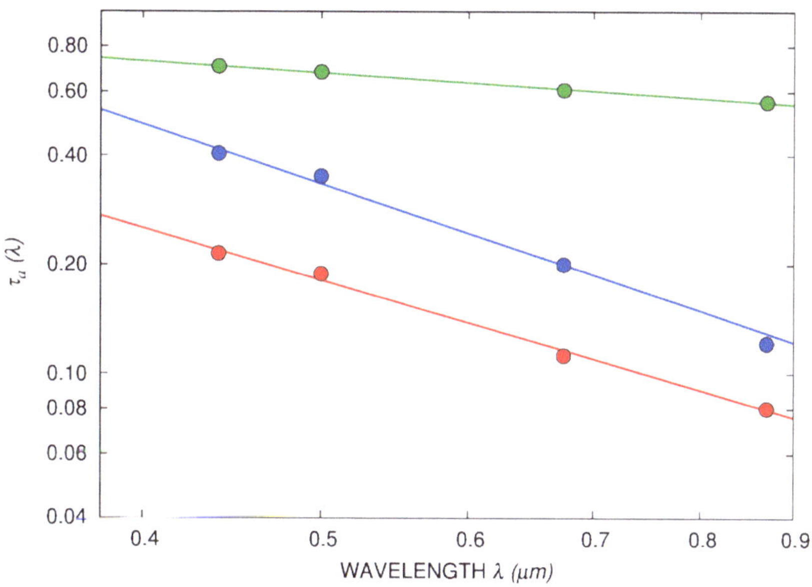

Fig. 8.10. Spectral series of aerosol optical thickness $\tau_a(\lambda)$ obtained from the AERONET measurements performed at Lecce (Italy) on three golden days for different atmospheric turbidity conditions, and examined in terms of the Ångström (1964) formula to determine the best-fit values of exponent α over the 0.44–0.87-μm wavelength range: (i) May 20, 2003 (09:45 UTC/GMT) (red circles), for mixed anthropogenic/continental aerosol from southern France and northern Italy ($\alpha = 1.496$); (ii) June 10, 2003 (10:03 UTC/GMT) (blue circles), for fine continental aerosol from the Black Sea and southern Balkans ($\alpha = 1.803$); and (iii) August 30, 2003 (07:30 UTC/GMT) (dark green circles), for mixed anthropogenic/continental aerosol from central and southern Italy at levels <1 km, and Saharan dust from North Africa at upper levels ($\alpha = 0.342$).

Table 8.2a. Daily mean values of aerosol optical thickness $\tau_a(0.50\ \mu m)$, Ångström's (1964) exponent α measured over the 0.441–0.873-μm wavelength range, real part $n(0.50\ \mu m)$ and imaginary part $k(0.50\ \mu m)$ of the complex refractive index of columnar aerosol determined from sky-brightness measurements in the almucantar, together with the daily mean values of columnar aerosol single-scattering albedo $\omega(0.55\ \mu m)$, as determined from the AERONET Cimel sun-/sky-radiometer data recorded on the 12 golden days selected among those recorded during the PRIN–2004 project, over the period from March 2003 to January 2004 (Lanconelli, 2007).

Measurement day	Daily mean $\tau_a(0.50\ \mu m)$	Daily mean α	Daily mean values of refractive index parts		Daily mean	Aerosol types and origins, as derived from the NOAA/ HYSPLIT backward trajectories passing over Lecce at
			$n(0.50\ \mu m)$	$k(0.50\ \mu m)$	$\omega(0.55 \mu m)$	the 12:00 UTC/GMT
April 25	0.271	1.814	1.435	1.09×10^{-2}	0.919	Pure continental aerosol from Scandinavian and Baltic area, transported over eastern Europe and Balkans
April 30	0.232	0.700	1.505	9.04×10^{-2}	0.901	Anthropogenic polluted aerosol from local circulation over the south Mediterranean at levels <1 km, and mixed marine/continental aerosol from central Atlantic Ocean, Iberian Peninsula, and central Italy at levels >1 km
May 1	0.230	0.595	1.554	3.42×10^{-3}	0.954	Saharan dust from the Libyan coasts, circulating over the Ionic Sea and central Mediterranean sector at lower levels, and marine/continental mixed aerosol from Atlantic Ocean, Iberian peninsula, and central/southern Italy at levels from 1 to 4 km
May 20	0.144	1.462	1.530	1.57×10^{-2}	0.874	Anthropogenic/continental aerosol from Gulf of Biscay, southern France, northern Italy, and Balkans at all levels
May 29	0.186	1.614	1.372	5.27×10^{-3}	0.938	Fine continental aerosol from Caucasian region, Black Sea, southern Balkans, and eastern Mediterranean at all levels
June 10	0.345	1.798	1.421	2.80×10^{-3}	0.977	Fine continental aerosol from Black Sea and southern Balkans at all levels

Table 8.2a. Continued.

Measurement day	Daily mean $\tau_a(0.50\ \mu m)$	Daily mean α	Daily mean values of refractive index parts		Daily mean	Aerosol types and origins, as derived from the NOAA/HYSPLIT backward trajectories passing over Lecce at
			$n(0.50\ \mu m)$	$k(0.50\ \mu m)$	$\omega(0.55 \mu m)$	the 12:00 UTC/GMT
July 17	0.627	0.328	1.483	2.76×10^{-3}	0.940	Saharan dust from North African coasts at levels <1 km, from the north-western Saharan regions at the 1–3 km levels, and from Sahara, Canary Islands, and Atlantic Ocean at altitudes of around 4 km, with marked atmospheric content variations throughout the day
July 24	0.402	0.621	1.453	5.28×10^{-3}	0.919	Continental aerosol from central and southern Italy and Balkans area at levels <1 km, and Saharan dust from Morocco, western Mediterranean, and southern part of Iberian peninsula at middle and high levels
Aug. 8	0.540	1.817	1.455	7.59×10^{-3}	0.939	Fine continental aerosol from northern Atlantic, North Sea, central-eastern Europe, and Balkans region at all levels
Aug. 30	0.582	0.341	1.469	2.94×10^{-3}	0.938	Anthropogenic/continental aerosol from local circulation over central and southern Italy at levels <1 km, and Saharan dust from North Africa at levels >1.5 km
Sept. 20	0.350	1.856	1.391	5.65×10^{-3}	0.949	Fine continental aerosol from long-time circulation over the Balkans at low and middle levels, and from central Europe and Balkans at levels >1.5 km
Oct. 2	0.325	0.540	1.506	5.04×10^{-3}	0.920	Continental aerosol from local circulation over southern Italy and central Mediterranean at low levels, mixed with Saharan dust from the north-western Saharan region (Morocco) and southern Spain at middle and high levels

thin layers of Saharan dust between levels of 1–4 km). The results in Table 8.2a are evidence of the close relationship between exponent α and the origins of particulate matter, which define the mass fractions of fine and coarse particles that contribute to attenuating the short-wave radiation, the fine particles generally producing optically predominant effects in continental particles that do not contain anthropogenic substances, and coarse particles causing prevailing extinction effects when they mainly consist of marine and/or desert dust particles.

Figure 8.11 shows the time-patterns of $\tau_a(0.55\ \mu m)$ and $\alpha(0.44–0.87\ \mu m)$ measured on four golden days for (i) pure continental aerosol transported from Scandinavia and eastern Europe on April 25, 2003; (ii) anthropogenic polluted aerosol from the southern Mediterranean and the Iberian peninsula on April 30, 2003; (iii) fine continental aerosol from the Black Sea and southern Balkans on June 10, 2003; and (iv) anthropogenic/continental aerosol from central and southern Italy, mixed with Saharan dust from North Africa, as observed over the Lecce University station on August 30, 2003. The values of $\tau_a(0.55\ \mu m)$ measured during the sunlit periods of the first three golden days were found to be very stable at the various measurement hours, while those recorded on the fourth day, when a significant load of Saharan dust was transported over Lecce, showed decreasing patterns during

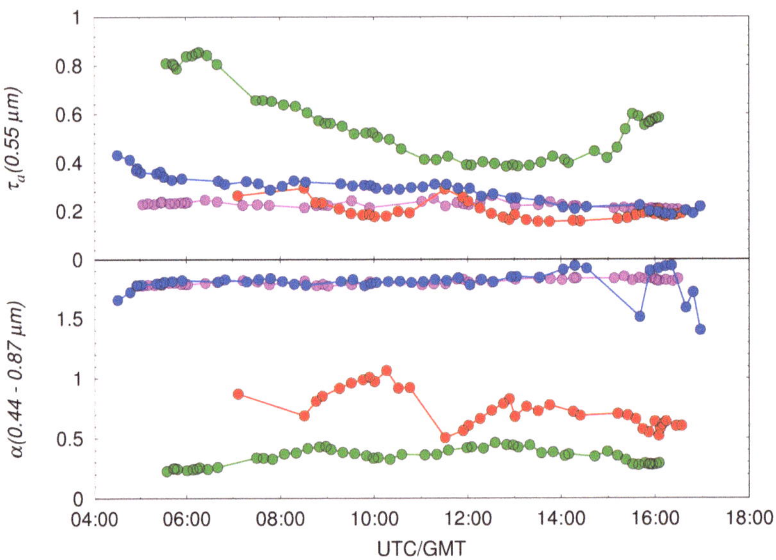

Fig. 8.11. Time-patterns of aerosol optical thickness $\tau_a(0.55\ \mu m)$ (upper part) and Ångström's exponent $\alpha(0.44–0.87\ \mu m)$ (lower part), as derived from the AERONET measurements performed at Lecce (Puglia, Italy) during the PRIN-2004 experimental campaign on the following four golden days: (i) April 25, 2003, for pure continental aerosol from Scandinavian area and eastern Europe (fuchsia circles); (ii) April 30, 2003, for anthropogenic polluted aerosol from southern Mediterranean and Iberian peninsula (red circles); (iii) June 10, 2003, for fine continental aerosol from Black Sea and southern Balkans (blue circles); and (iv) August 30, 2003, for anthropogenic/continental aerosol from central and southern Italy, mixed with Saharan dust transported from North Africa (dark green circles).

the morning and around midday, followed by a pronounced increase in the late afternoon. Correspondingly, the time-patterns of $\alpha(0.44\text{–}0.87\ \mu m)$ determined on the first and third golden days for pure continental aerosol (April 25, 2003) and fine continental aerosol (June 10, 2003), respectively, were rather stable in both cases, apart from the last hours of June 10, 2003. Those of April 30, 2003, associated with anthropogenic polluted aerosol, varied greatly between about 0.5 and 1.0 during the day, suggesting the occurrence of major extinction effects by the coarse particle mass component. Those of August 30, 2003, due to a mixed load of anthropogenic/continental and Saharan dust particles, presented values mainly varying between 0.2 and 0.4, clearly associated with predominant extinction effects produced by coarse particles mainly containing anthropogenic substances and mineral dust.

(2) *Determination of the refractive index of columnar aerosol*

The AERONET Level 2.0 data recorded on the 12 golden days were examined to determine the values of real part $n(\lambda)$ and imaginary part $k(\lambda)$ of the refractive index retrieved as inversion products at various hours of each golden day (Holben et al., 1998). The daily mean values of $n(0.50\ \mu m)$ and $k(0.50\ \mu m)$ are reported in Table 8.2a. They show that (i) $n(0.50\ \mu m)$ varied between 1.372 (May 29, 2003, for fine continental aerosol from eastern Europe and the eastern Mediterranean) and 1.554 (May 1, 2003, for Saharan dust mixed with marine/continental aerosol from the Atlantic Ocean, Spain, and Italy), while (ii) $k(0.50\ \mu m)$ varied between 2.76×10^{-3} (July 17, 2003, for Saharan dust) and 9.04×10^{-2} (April 30, for anthropogenic polluted aerosol). The spectral series of $n(\lambda)$ and $k(\lambda)$ were also retrieved at the various hours of the 12 golden days, to complete the picture of the complex refractive index characteristics.

(3) *Determination of the columnar aerosol size-distribution curves*

The spectral series of $\tau_a(\lambda)$ for the various hours of the 12 golden days were examined together with the spectral curves of $n(\lambda)$ and $k(\lambda)$ obtained above, to retrieve the number density and volume size-distribution curves of columnar aerosol particles. Some examples of the resulting size-distribution curves are shown in Fig. 8.12, as obtained at different hours on May 20, 2003 (for anthropogenic/continental aerosol from southern France and northern Italy), June 10, 2003 (for fine continental aerosol from the Black Sea and southern Balkans), and August 30, 2003 (for mixed anthropogenic/continental aerosol from central and southern Italy, mixed with Saharan dust transported from North Africa). The curves of number density $N(r)$ determined in the three study cases reveal different patterns, with (i) gradually decreasing concentrations of fine particles, passing from the August 30, 2007, case (anthropogenic, continental, Saharan particles) to the May 20, 2003, case (anthropogenic and continental particles, only) and (ii) comparable contents of coarse particles in the two cases of May 20 and June 10, 2003, without Saharan dust loads, and generally higher concentrations in the mixed case of August 30, 2003, with anthropogenic/continental aerosol from central and southern Italy, and Saharan dust from North Africa. The multimodal curves of total columnar particle volume $V(r)$ yield a more accurate measure of the differences in the fine and accumulation particle volumes within the vertical column of the atmosphere, which characterized the

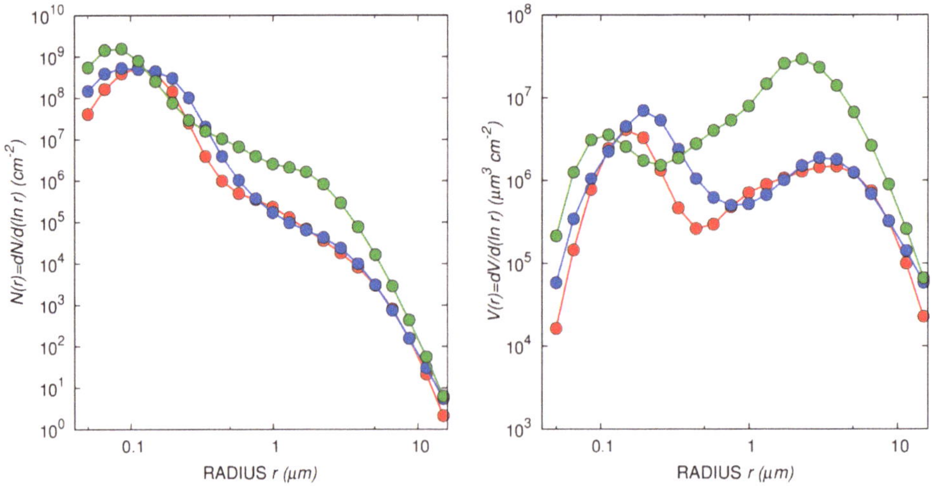

Fig. 8.12. Examples of multimodal size-distribution curves of columnar total particle number density $N(r) = \mathrm{d}N/\mathrm{d}(\ln r)$ measured per cm^2 (left) and columnar total particle volume $V(r) = \mathrm{d}V/\mathrm{d}(\ln r)$ measured in $\mu\mathrm{m}^3/\mathrm{cm}^2$ (right), as retrieved from the AERONET spectral series of $\tau_a(\lambda)$ determined at Lecce (Puglia, Italy) on: (i) May 20, 2003 (09:45 UTC/GMT) for mixed anthropogenic/continental aerosol from southern France and northern Italy ($\alpha = 1.496$) (red circles); (ii) June 10, 2003 (10:03 UTC/GMT) for fine continental aerosol from Black Sea and southern Balkans ($\alpha = 1.803$) (blue circles); and (iii) August 30, 2003 (07:30 UTC/GMT) for mixed anthropogenic/continental aerosol from central and southern Italy at levels <1 km, and Saharan dust from North Africa at the upper levels ($\alpha = 0.342$) (dark green circles).

three study cases. They indicate (i) a marked predominance of Saharan coarse particles on August 30, 2003 (mixed anthropogenic/continental aerosol and Saharan dust) and (ii) comparable volumes of fine and coarse particles on June 10, 2003.

(4) *Determination of the columnar aerosol single-scattering albedo*

For the various multimodal size-distribution curves of columnar aerosol determined above, and the correspondingly variable complex values of columnar aerosol refractive index, the calculations of $\omega(0.55\ \mu\mathrm{m})$ were made at various hours of each golden day, using the 6S code of Vermote et al. (1997). The daily mean values of this parameter were then calculated for the above instantaneous data, obtaining the daily mean values given in Table 8.2a, found to range between 0.874 (on May 20, 2003, for anthropogenic/continental aerosol) and 0.977 (on June 10, 2003, for fine continental aerosol). This clearly indicated that $\omega(0.55\ \mu\mathrm{m})$ increased assuming gradually higher values as the columnar aerosol extinction by anthropogenic particulate matter decreased and that of non-absorbing dust and background (hereinafter referred to as BG) continental aerosol became optically more significant.

(5) *Definition of local surface albedo models*

The land surface albedo characteristics of the Salento area were estimated on the 12 golden days of the PRIN-2004 project by analyzing the MCD43C3 products derived over land from the MODIS Level 3.0 surface albedo data recorded in the winter and summer periods of 2009, obtaining the surface albedo maps shown in Fig. 8.13. Applying the best-fit procedure to all the seven MODIS spectral channels, it was found that the surface albedo characteristics of this agricultural area are well represented by the VS1 model in winter and by the VS4 model in spring and summer, as defined for LAI equal to 0.1 and 5.0, respectively. The values of albedo parameters $R_{bs}(\theta_o = 0°)$, R_{ws}, and $A(\theta_o = 60°)$ of the two VS models are those defined in section 8.3.1, where the DARF calculations for the CLEARCOLUMN project are illustrated. To define the most suitable BRDF surface albedo models for representing the surface reflectance characteristics of the Adriatic Sea and Ionic Sea in proximity to the Salento peninsula coasts, an analysis of the surface-level wind velocity was undertaken for various multi-year seasonal periods, using the climatological data downloaded from the NOAA/OAR/ESRL PSD website (*www.esrl.noaa.gov/psd/*; Kalnay et al., 1996) over this Mediterranean Sea sector. Examining the data, it was found that the most appropriate sea surface albedo models for obtaining realistic estimates of DARF effects on the 12 golden days were the OS3 and OS4 surface albedo models of Tomasi et al. (2013), defined for surface-level wind velocity V_w equal to 10 m/s and 20 m/s, respectively, and for sea-water pigment concentration of 34.3 ppt in both models. The spectral characteristics of models OS3 and OS4 are shown in Fig. 8.14. They both present almost neutral surface albedo values as a function of wavelength, from 0.5 to 2.5 μm, but with model OS3 yielding considerably higher spectral albedo at all the selected values of θ_o from 0° to 80°. In fact, model OS3 was evaluated to provide $R_{bs}(0°) = 0.028$, $R_{ws} = 0.069$, and $A(60°) = 0.127$, and model OS4 to give $R_{bs}(0°) = 0.044$, $R_{ws} = 0.081$, and $A(60°) = 0.105$.

(6) *Calculation of the daily time-patterns of instantaneous DARF terms and diurnally averaged DARF effects*

The instantaneous values of DARF terms $\Delta F_{ToA}(t)$ and $\Delta F_{BoA}(t)$ were calculated for the OS3, OS4, VS1, and VS4 models and the data sets obtained from the field measurements performed on the 12 golden days. The calculations were made by using the 6S code for the various values of θ_o observed during the sunlit period. It was assumed that the time-patterns of the columnar aerosol optical parameters were those calculated in the previous steps: namely those of $\tau_a(\lambda)$ and α in step (1), those of $n(\lambda)$ and $k(\lambda)$ in step (2), those of curves $N(r)$ and $V(r)$ in step (3), and the evaluations of $\omega(\lambda)$ in step (4). The daily time-patterns of $\Delta F_{ToA}(t)$ and $\Delta F_{BoA}(t)$ were subsequently integrated over their corresponding sunlit period according to Eqs (8.1) and (8.2), to determine (i) the daily mean values of ΔDF_{ToA} and ΔDF_{BoA} on the 12 golden days, separately for the four surface albedo models OS3, OS4, VS1, and VS4, and (ii) the corresponding values of ΔDF_{Atm} as differences between ΔDF_{ToA} and ΔDF_{BoA}. The time-patterns of the daily mean values of ΔDF_{ToA}, ΔDF_{BoA}, and ΔDF_{Atm} are shown in Fig. 8.15, together with those of columnar aerosol parameters $\tau_a(0.55\ \mu m)$ and $\omega(0.55\ \mu m)$. The values of ΔDF_{ToA} are all negative but their absolute values increase appreciably passing from the VS4 model to the VS1 one, to become even higher for the OS4 and OS3 models, reaching the highest cooling values in correspondence with the maximum of $\tau_a(0.55\ \mu m)$ measured on July 17, 2003. More similar values of ΔDF_{BoA} were obtained using

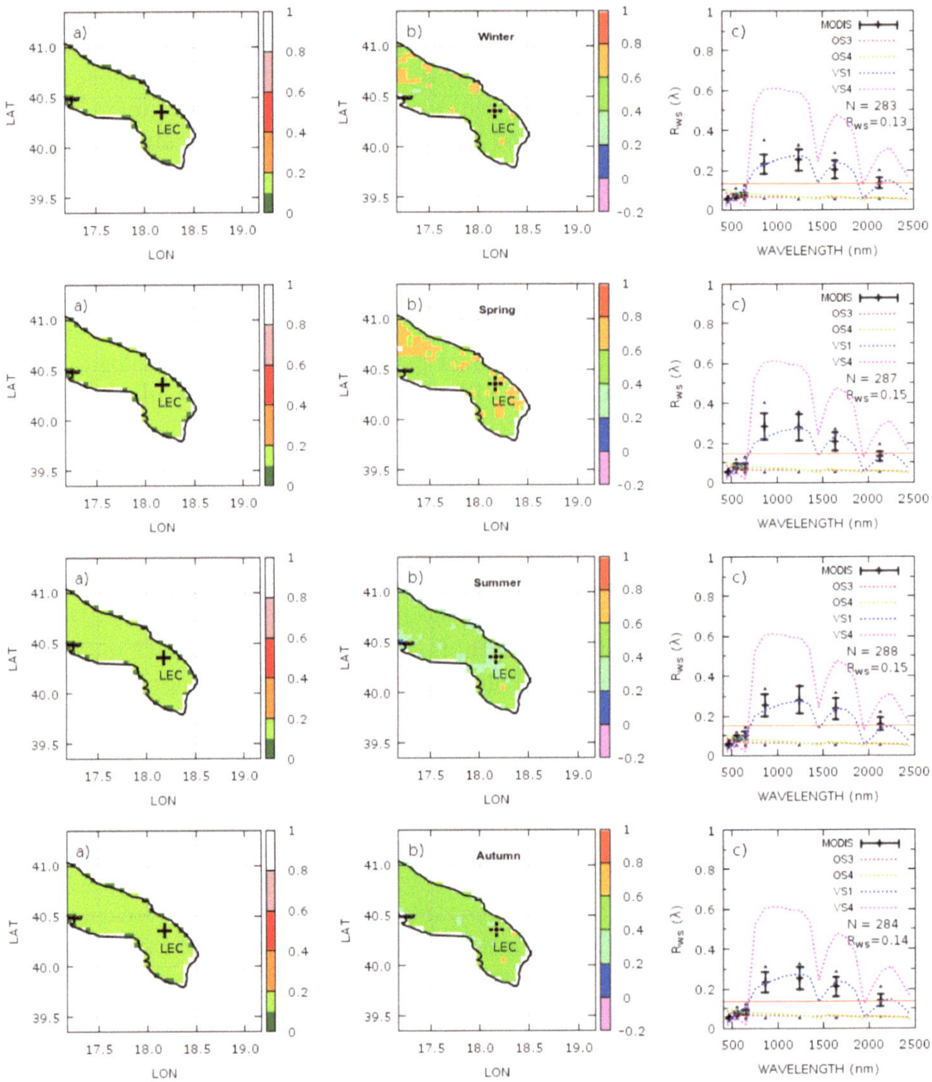

Fig. 8.13. Average seasonal maps of the land surface albedo ((a) in the left column) and Normalized Difference Vegetation Index (NDVI) ((b) in the central column) obtained over the southern Puglia (Salento, Italy) from the MODIS Level 3.0 surface albedo data (MCD43C3 products) recorded during the four seasons of 2009. The crosses labelled LEC in the graphs indicate the geographical position of Lecce, where the sun-photometer measurements were regularly performed during the PRIN-2004 field experiment. The graphs shown in the third column provide the spectral values (black vertical bars) of the white-sky albedo R_{ws} determined over the land area, yielding mean values of 0.13 in winter, 0.15 in spring and summer, and 0.14 in autumn, all found with standard deviations of around 0.1 within the seven MODIS channels, and with spectral values of minimum and maximum represented by small triangles. Also shown are the spectral curves of $R_{ws}(\lambda)$ obtained as best-fit solutions for the ocean surface albedo models OS3 (dotted red curve) and OS4 (dotted green curve), and the vegetation-covered surface albedo models VS1 (dotted blue curve) and VS4 (dotted fuchsia curve).

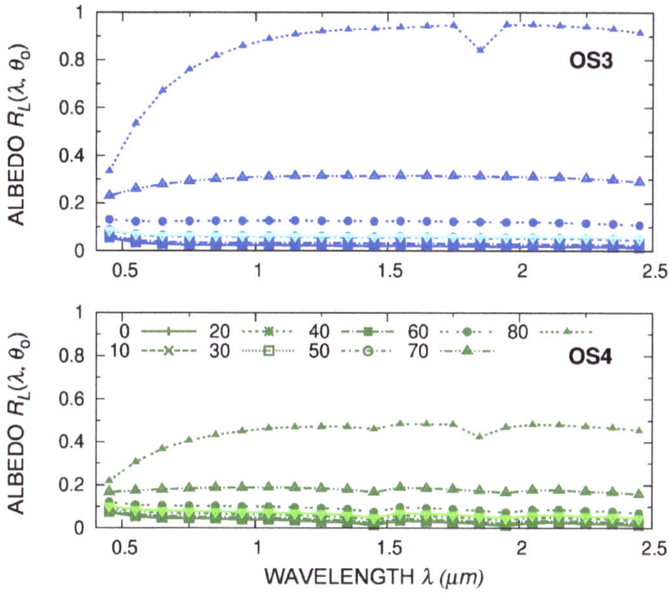

Fig. 8.14. Spectral curves of the BRDF surface albedo $R_L(\lambda, \theta_o)$ defined by Tomasi et al. (2013) for the OS3 and OS4 surface albedo models, as evaluated for nine values of solar zenith angle θ_o, taken in steps of 10° over the 0°–80° range, and used to represent the sea surface albedo characteristics of the southern Adriatic Sea and Ionic Sea surrounding the coasts of the Salento peninsula (Italy). Model OS3 refers to a sea surface with surface-level wind velocity V_w = 10 m/s; and model OS4 to a sea surface with V_w = 20 m/s. The spectral curves of the white-sky albedo $R_{ws}(\lambda)$ are also shown in each graph using a different lighter color.

the four surface albedo models, with discrepancies within ±10 W/m^2, presenting the most negative value on July 17, 2003. Because of the said differences, the values of ΔDF_{Atm} obtained on the 12 golden days were mostly positive and lower than 20 W/m^2, with a maximum of around 35 W/m^2 on July 17, 2003, obtained using the VS4 model.

The daily mean values of diurnally averaged ΔDF_{ToA}, ΔDF_{BoA}, and ΔDF_{Atm} terms are given in Table 8.2b for all the 12 golden days. They are plotted in Fig. 8.16 as a function of $\tau_a(0.55\ \mu m)$, separately for the four surface albedo models adopted in the observation area around Lecce. Both ΔDF_{ToA} and ΔDF_{BoA} decrease on average as $\tau_a(0.55\ \mu m)$ increases, with similar slope coefficients, which do not vary greatly as $\omega(0.55\ \mu m)$ assumes different values. Conversely, the values of ΔDF_{Atm}appreciably increase as $\tau_a(0.55\ \mu m)$ increases, with rates varying only slightly as a function of the columnar aerosol SSA characteristics.

In particular, it is worth noting that three of the 12 golden days listed in Table 8.2a were characterized by Saharan dust transport over Puglia on (i) May 1, 2003, when mineral dust was mixed with marine/continental aerosol; (ii) July 17, 2003, with a predominant load of Saharan dust, mixed with minor fractions of marine and continental aerosol particles; and (iii) October 2, 2003, with a prevailing content of continental aerosol, but with a minor load of Saharan dust at

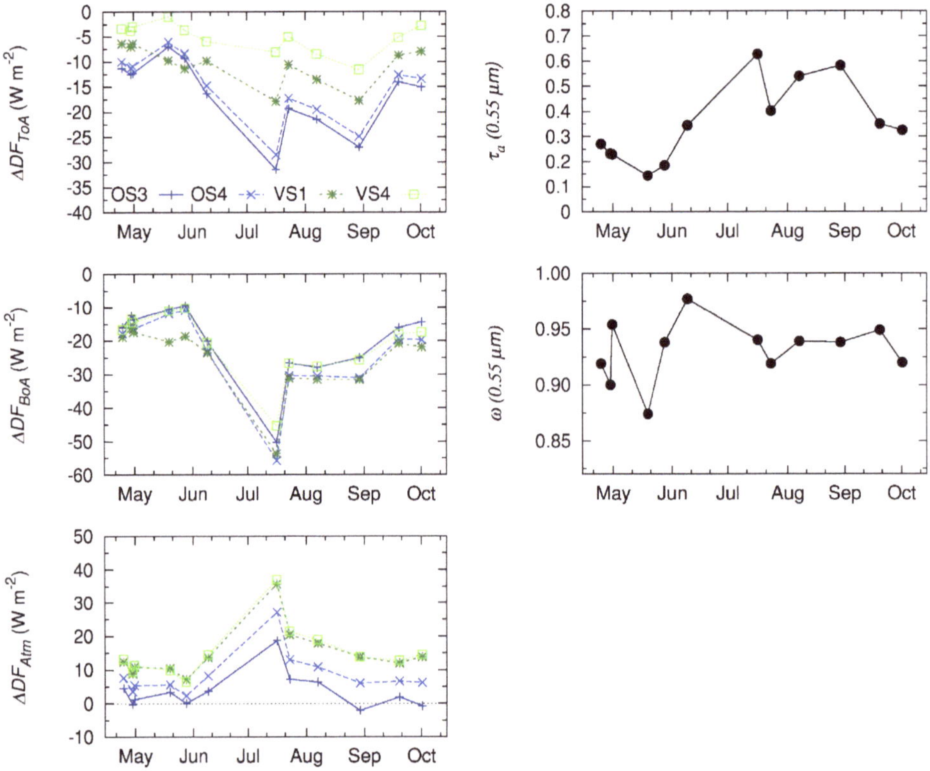

Fig. 8.15. Left: Time-patterns of the daily mean values of diurnally averaged DARF terms ΔDF_{ToA} at the ToA-level, ΔDF_{BoA} at the BoA-level, and ΔDF_{Atm} in the atmosphere, calculated for the 12 golden days of the PRIN-2004 experiment at Lecce (Italy) from April 25 to October 2, 2003, using the surface albedo models OS3, OS4, VS1, and VS4 proposed by Tomasi et al. (2013). Right: Time-patterns of the daily mean values of aerosol optical thickness $\tau_a(0.55\ \mu\text{m})$ and columnar aerosol single-scattering albedo $\omega(0.55\ \mu\text{m})$, as determined for the columnar aerosol size-distribution features and complex refractive index measured on the 12 golden days.

relatively high tropospheric levels. Examining the time-patterns of DARF terms and aerosol parameters shown in Fig. 8.15, it can be noted that the dust transport episode of May 1, 2003, at levels <1 km caused DARF effects, leading to aerosol optical thickness features very similar to those recorded on April 25 and 30, 2003, when continental and anthropogenic aerosol loads predominated, yielding a columnar value of $\omega(0.55\ \mu\text{m}) = 0.954$. The dust transport episode of July 17, 2003, was characterized by the most negative values of ΔDF_{ToA} and ΔDF_{BoA} among those evaluated on the 12 golden days, associated with the highest positive value of ΔDF_{Atm} and $\tau_a(0.55\ \mu\text{m})$ measured during the field campaign, for $\omega(0.55\ \mu\text{m}) = 0.940$, clearly indicating that intense transport of Saharan dust over southern Italy can produce very strong DARF effects. In addition, the weak transport of Saharan dust observed on October 2, 2003, gave daily average values of $\tau_a(0.55\ \mu\text{m}) = 0.35$ and $\omega(0.55\ \mu\text{m}) = 0.949$ due to a predominant content of

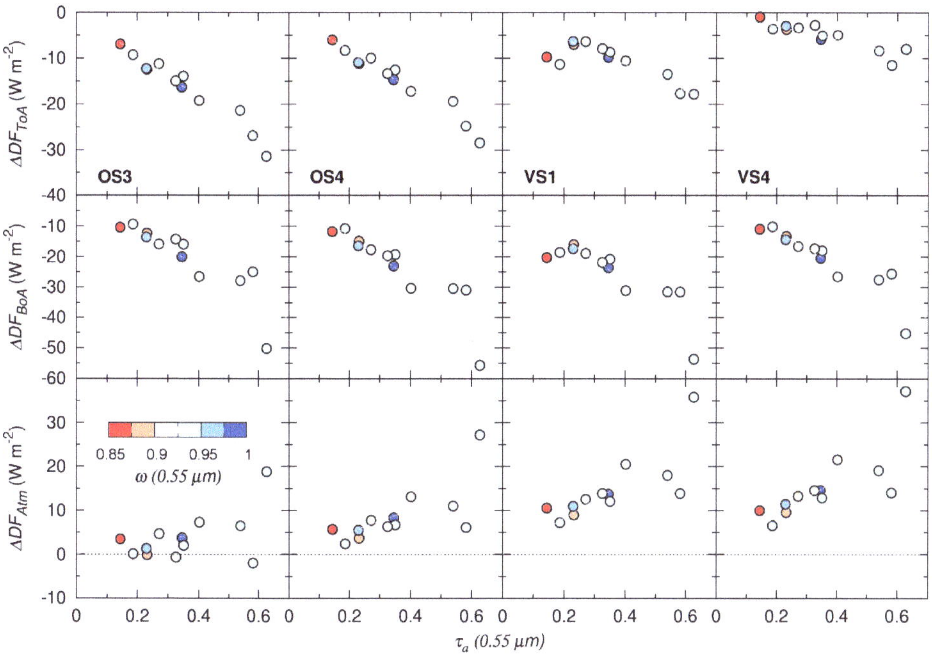

Fig. 8.16. Scatter plots of the daily mean values of DARF terms ΔDF_{ToA}, ΔDF_{BoA}, and ΔDF_{Atm} versus the corresponding daily mean values of aerosol optical thickness $\tau_a(0.55\ \mu m)$, as determined from the PRIN-2004 field measurements performed on the 12 golden days using the oceanic surface albedo models OS3 and OS4, and the vegetation-covered surface albedo models VS1 and VS4, for different values of columnar aerosol single-scattering albedo $\omega(0.55\ \mu m)$ indicated by differently colored circles chosen according to the color scale reported on the left.

continental aerosol that induced DARF effects comparable to those evaluated on September 20, 2003, for a mean columnar content of continental aerosol yielding similar values of $\tau_a(0.55\ \mu m)$ and $\omega(0.55\ \mu m)$.

(7) *Calculations of the daily mean values of DARF efficiencies*

The daily mean values of ΔDF_{ToA}, ΔDF_{BoA}, and ΔDF_{Atm} given in Table 8.2b were divided by the corresponding values of $\tau_a(0.55\ \mu m)$ given in Table 8.2a, to determine the daily mean values of DARF efficiency parameters E_{ToA} at the ToA-level, E_{BoA} at the BoA-level, and E_{Atm} within the atmosphere. The estimates are reported in Table 8.2b for all 12 golden days, indicating that E_{ToA} and E_{BoA} did not exhibit relevant variations on passing from the anthropogenic and continental cases (April 25, May 20, May 29, July 24, August 8, August 30, and September 20, 2003) to cases where continental particles were mixed with Saharan dust in the atmospheric column (on May 1, July 17, and October 2, 2003).

All the evaluations of the three efficiency parameters given in Table 8.2b are plotted in Fig. 8.17 versus the columnar aerosol $\omega(0.55\ \mu m)$, over the range from 0.87 to nearly 0.98. The first graph shows that E_{ToA} tends to decrease slightly on average with $\omega(0.55\ \mu m)$, for all the four surface albedo models considered in

Table 8.2b. Daily values of the diurnally averaged aerosol forcing terms ΔDF_{ToA} at the ToA-level, ΔDF_{BoA} at the BoA-level, and ΔDF_{Atm} within the atmosphere, diurnal average aerosol fractional forcing AFF_{ToA} at the ToA-level (given by the ratio between flux change ΔF_{ToA} at the ToA-level and the incoming flux $I_S \downarrow$ of solar radiation at the ToA-level), and the diurnal average DARF efficiencies E_{ToA}, E_{BoA}, and E_{Atm} giving the rates at which the surface–atmosphere system is forced per unit $\tau_a(0.55\ \mu m)$, as obtained for the 12 golden days selected among those of the PRIN–2004 campaign conducted from April 25 to October 2, 2003, at Lecce (Salento, Puglia, southern Italy) (Lanconelli, 2007).

Measurement day	Surface albedo model	Diurnal average DARF terms (W/m^2)			Diurnal average	Diurnal average values of DARF efficiencies (W/m^2)		
		ΔDF_{ToA}	ΔDF_{BoA}	ΔDF_{Atm}	AFF_{ToA}	E_{ToA}	E_{BoA}	E_{Atm}
April 25	OS3	−11.2	−15.9	+4.6	-3.0×10^{-2}	−41.4	−58.6	+17.2
	OS4	−10.0	−17.7	+7.7	-2.6×10^{-2}	−36.7	−65.3	+28.6
	VS1	−6.4	−18.9	+12.5	-1.7×10^{-2}	−23.6	−69.7	+46.1
	VS4	−3.3	−16.6	+13.3	-9.0×10^{-3}	−12.2	−61.1	+48.9
April 30	OS3	−12.5	−12.3	−0.1	-3.2×10^{-2}	−53.7	−53.1	−0.6
	OS4	−11.2	−14.9	+3.7	-2.9×10^{-2}	−48.1	−64.1	+16.0
	VS1	−7.0	−16.0	+9.0	-1.8×10^{-2}	−30.2	−68.9	+38.7
	VS4	−3.7	−13.3	+9.5	-1.0×10^{-2}	−16.1	−57.2	+41.1
May 1	OS3	−12.3	−13.6	+1.3	-3.1×10^{-2}	−53.4	−59.0	+5.6
	OS4	−10.9	−16.4	+5.5	-2.8×10^{-2}	−47.3	−71.2	+23.9
	VS1	−6.4	−17.3	+10.9	-1.6×10^{-2}	−27.7	−75.3	+47.6
	VS4	−2.9	−14.4	+11.4	-8.0×10^{-3}	−12.7	−62.4	+49.7
May 20	OS3	−6.9	−10.4	+3.5	-1.7×10^{-2}	−48.2	−72.3	+24.1
	OS4	−6.0	−11.7	+5.7	-1.5×10^{-2}	−42.0	−81.5	+39.5
	VS1	−3.3	−12.6	+9.3	-8.0×10^{-3}	−23.0	−87.5	+64.5
	VS4	−1.0	−10.9	+10.0	-2.0×10^{-3}	−6.6	−76.0	+69.4
May 29	OS3	−9.3	−9.4	+0.1	-2.2×10^{-2}	−49.8	−50.4	+0.6
	OS4	−8.3	−10.7	+2.4	-2.0×10^{-2}	−44.7	−57.8	+13.1
	VS1	−5.5	−11.7	+6.2	-1.3×10^{-2}	−29.7	−63.1	+33.4
	VS4	−3.6	−10.1	+6.6	-8.0×10^{-3}	−19.2	−54.5	+35.3
June 10	OS3	−16.3	−20.1	+3.8	-3.8×10^{-2}	−47.3	−58.2	+10.9
	OS4	−14.7	−23.0	+8.3	-3.4×10^{-2}	−42.6	−66.7	+24.1
	VS1	−9.8	−23.5	+13.7	-2.3×10^{-2}	−28.5	−68.3	+39.8
	VS4	−5.9	−20.5	+14.6	-1.4×10^{-2}	−17.1	−59.4	+42.3
July 17	OS3	−31.4	−50.2	+18.8	-7.5×10^{-2}	−50.0	−80.0	+30.0
	OS4	−28.4	−55.7	+27.2	-6.8×10^{-2}	−45.3	−88.8	+43.5
	VS1	−17.8	−53.6	+35.8	-4.2×10^{-2}	−28.4	−85.5	+57.1
	VS4	−8.0	−45.2	+37.2	-1.9×10^{-2}	−12.8	−72.1	+59.3
July 24	OS3	−19.2	−26.5	+7.3	-4.7×10^{-2}	−47.9	−66.0	+18.1
	OS4	−17.2	−30.3	+13.1	-4.2×10^{-2}	−42.8	−75.4	+32.6
	VS1	−10.6	−31.0	+20.5	-2.6×10^{-2}	−26.3	−77.2	+50.9
	VS4	−4.9	−26.5	+21.6	-1.2×10^{-2}	−12.2	−65.9	+53.7
Aug. 8	OS3	−21.4	−27.8	+6.5	-5.4×10^{-2}	−39.6	−51.6	+12.0
	OS4	−19.4	−30.4	+11.0	-4.9×10^{-2}	−35.9	−56.3	+20.4
	VS1	−13.5	−31.5	+18.0	-3.4×10^{-2}	−25.0	−58.3	+33.3
	VS4	−8.4	−27.5	+19.1	-2.1×10^{-2}	−15.5	−50.9	+35.4

Table 8.2b. Continued.

Measurement day	Surface albedo model	Diurnal average DARF terms (W/m^2)			Diurnal average	Diurnal average values of DARF efficiencies (W/m^2)		
		ΔDF_{ToA}	ΔDF_{BoA}	ΔDF_{Atm}	AFF_{ToA}	E_{ToA}	E_{BoA}	E_{Atm}
Aug. 30	OS3	−26.9	−24.9	−2.0	-7.7×10^{-2}	−46.2	−42.8	−3.4
	OS4	−24.8	−30.9	+6.1	-7.1×10^{-2}	−42.6	−53.1	+10.5
	VS1	−17.7	−31.5	+13.9	-5.0×10^{-2}	−30.4	−54.2	+23.8
	VS4	−11.5	−25.5	+14.0	-3.3×10^{-2}	−19.8	−43.9	+24.1
Sept. 20	OS3	−13.9	−15.9	+2.0	-4.6×10^{-2}	−39.8	−45.6	+5.8
	OS4	−12.6	−19.3	+6.7	-4.2×10^{-2}	−35.9	−55.1	+19.2
	VS1	−8.7	−20.8	+12.1	-2.9×10^{-2}	−24.9	−59.4	+34.5
	VS4	−5.1	−18.0	+12.9	-1.7×10^{-2}	−14.6	−51.3	+36.7
Oct. 2	OS3	−15.0	−14.3	−0.7	-5.5×10^{-2}	−46.1	−44.0	−2.1
	OS4	−13.3	−19.7	+6.3	-4.9×10^{-2}	−41.0	−60.5	+19.5
	VS1	−8.0	−21.8	+13.9	-2.9×10^{-2}	−24.5	−67.1	+42.6
	VS4	−2.7	−17.3	+14.6	-1.0×10^{-2}	−8.4	−53.2	+44.8

the present analysis, as evidenced by the regression lines drawn separately for the various surface albedo classes. Conversely, the scatter plots of E_{BoA} and E_{Atm} in the two other graphs are observed to be rather dispersed. They indicate an average trend of E_{Boa} to increase as a function of $\omega(0.55\ \mu m)$ until assuming values of about -40 W/m^2 for nearly unit $\omega(0.55\ \mu m)$, and a general trend of E_{Atm} to decrease slightly with $\omega(0.55\ \mu m)$, until assuming values extrapolated to unit $\omega(0.55\ \mu m)$ that vary between a few W/m^2 for the OS3 model and around +40 W/m^2 for the VS1 and VS4 models. In particular, the efficiency values of E_{ToA}, E_{BoA}, and E_{Atm} shown in Fig. 8.17 for the three days characterized by Saharan dust transport over Puglia, relative to $\omega(0.55\ \mu m)$ ranging between 0.94 and 0.96, were found not to differ appreciably from those evaluated on the other golden days for marine/continental and continental/anthropogenic aerosol columnar loads.

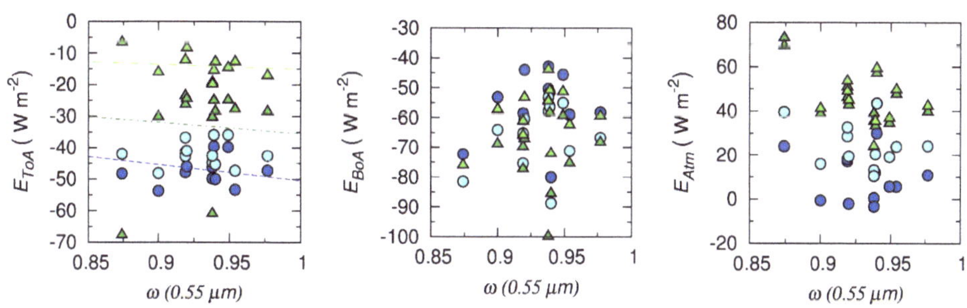

Fig. 8.17. Scatter plots of the daily mean values of DARF efficiency E_{ToA} at the ToA-level, DARF efficiency E_{BoA} at the BoA-level, and DARF efficiency E_{Atm} within the atmosphere versus the columnar aerosol single-scattering albedo $\omega(0.55\ \mu m)$, as obtained on the 12 golden days of the PRIN-2004 experiment conducted at Lecce (Italy) from March to October of 2003, by assuming the surface albedo characteristics represented by the oceanic surface models OS3 (dark blue circles) and OS4 (cyan circles), and the vegetation-covered surface models VS1 (dark green triangles)and VS4 (light green triangles).

8.3.3 DARF evaluations obtained from the AEROCLOUDS project measurements in northern Italy

AEROCLOUDS is the acronym of a national project supported by the Italian MiUR, which was organized in cooperation with the Italian National Council of Research (CNR) under the title 'Study of the direct and indirect effects of aerosols and clouds on climate'. Numerous ground-based aerosol remote sensing measurements were performed from spring 2007 to spring 2010, along with particulate matter chemical and microphysical studies made at numerous sites over the Italian peninsula. In particular, regular measurements of direct solar irradiance and sky-brightness in the almucantar were performed from May 2007 to March 2008 at the ISAC-CNR meteorological station at SPC, 44° 37′ N, 17° 57′ E, 10 m a.m.s.l.) in the center of the Po Valley (northern Italy), about 25 km north-east of Bologna. Simultaneous measurements were taken at SPC with (i) the portable ISAC-CNR/ENEA tropospheric Lidar, to define each day the vertical profile of volume backscattering coefficient within the first 4 km of the atmosphere; (ii) the Radiance Research (RR) nephelometer (M903 model) and the RR Particle Soot Absorption Photometer (PSAP), to carry out regular ground-level measurements of the volume scattering and absorption coefficients $\beta_{sca}(0.530\ \mu\text{m})$ and $\beta_{abs}(0.573\ \mu\text{m})$, respectively, from which average values of volume extinction coefficient $\beta_{ext}(0.550\ \mu\text{m})$ and ground-level SSA parameter $\omega_o(0.550\ \mu\text{m})$ were derived; and (iii) different particulate sampling techniques, to define the aerosol chemical composition and microphysical properties at the surface. Instantaneous DARF evaluations were made using these optical, microphysical, and chemical data, following the seven-step DARF-PROC procedure adopted for calculating the DARF effects during the CLEARCOLUMN and PRIN-2004 field campaigns:

(1) *Analysis of field data to determine the columnar aerosol extinction parameters*

The columnar aerosol optical parameters were determined at SPC using the PREDE POM-02L sun-/sky-radiometer of the SKYNET network, equipped with 11 narrow-band interference filters centered at the 0.315, 0.340, 0.380, 0.400, 0.500, 0.675, 0.870, 0.940 (water vapor band), 1.020, 1.600, and 2.200-μm wavelengths, with half-bandwidths varying from 5 to 12 nm. The instrument was carefully calibrated before the campaign, following the accurate procedure defined by Campanelli et al. (2007), and after the campaign through intercomparison tests performed at the Izaña Observatory (Tenerife, Canary Islands, Spain) with other PREDE and AERONET sun-/sky-radiometers (Mazzola et al., 2012).

It was found over the period from May 2007 to March 2008 that:

(a) The daily values of $\tau_a(0.50\ \mu m)$ covered quite different ranges in the various seasons, equal to (i) 0.09–0.63 in winter, with a mean value of 0.22 ± 0.16 and a median value of 0.26; (ii) 0.12–0.54 in spring, with a mean value of 0.27 ± 0.16 and a median value of 0.32; (iii) 0.10–0.44 in summer, with a mean value of 0.24 ± 0.14 and a median value of 0.25; and (iv) 0.09–0.46 in autumn, with a mean value of 0.24 ± 0.16 and a median value of 0.27.
(b) The daily mean values of Ångström's exponent $\alpha(0.40\text{–}0.87\ \mu\text{m})$ varied mainly between: (i) 0.76 and 1.57 in winter, with an average seasonal value of 1.37 ± 0.30, and a median value of 1.48; (ii) 0.52 and 1.58 in spring, with an average value

of 1.18 ± 0.36 and a median value of 1.36; (iii) 0.56 and 1.65 in summer, with an average value of 1.22 ± 0.37 and a median value of 1.44; and (iv) 0.68 and 1.47 in autumn, with an average value of 1.13 ± 0.25 and a median value of 1.34. An overall average value of α(0.40–0.87 μm) equal to 1.34 was found for the set of clear-sky days without Saharan dust transport, while an average value of 1.00 was obtained for the whole set, which included the measurements taken with Saharan dust loads in the vertical atmospheric column, the lowering of α being due to the high load of coarse desert dust particles.

A number of 18 golden days were chosen among those of the AEROCLOUDS campaign conducted from May 2007 to March 2008. They are listed in Table 8.3a, together with short daily descriptions of the NOAA/HYSPLIT backward trajectories found at the 12:00 UTC/GMT of each day and during the previous 96 hours: seven days were chosen for marine/continental aerosol transported from the northern Atlantic Ocean and north-western/central Europe; five days for continental aerosol transported from north-western and central Europe; five days for transport episodes of continental aerosol from Scandinavian and Baltic regions, north-eastern Europe, eastern Europe, and central Italy (in the last case mixed with Saharan dust transported from North Africa at levels of around 4 km); and one day with transport of anthropogenic/continental aerosol from southern Italy and the Mediterranean Sea. Table 8.3a provides the daily mean values of τ_a(0.50 μm) that vary between about 0.07 (determined on January 24, 2008, for marine/continental aerosol from the north Atlantic and northern Europe) and 0.37 (measured on August 2, 2007, for continental aerosol from north-western and central Europe).

The spectral series of aerosol optical thickness $\tau_a(\lambda)$ recorded at SPC on the 18 golden days were examined to determine the Ångström (1964) best-fit values of exponent α over the 0.40–0.87-μm wavelength range. Figure 8.18 shows four examples of application of the best-fit procedure for the spectral series of $\tau_a(\lambda)$ measured at different hours of the following four golden days: (i) July 17, 2007, when continental aerosol circulated over the Po Valley, central Europe, and the Adriatic Sea, mixed with maritime aerosol from the Mediterranean Sea and Saharan dust from Northern Africa, suspended at levels of $\sim$4 km, giving $\alpha = 0.424$; (ii) October 15, 2007, when continental aerosol was transported from the Arctic Ocean, Scandinavian peninsula, and subsequently over central Europe, yielding $\alpha = 1.107$; (iii) December 19, 2007, when continental aerosol particles were transported mainly from Russia and the Balkans, giving $\alpha = 1.694$; and (iv) March 13, 2008, when marine/continental aerosol moved from the northern Atlantic and western Europe, providing $\alpha = 1.337$.

The time-patterns of aerosol optical thickness τ_a(0.55 μm) and exponent α(0.40–0.87 μm) measured on three golden days are reported in Fig. 8.19, as obtained for columnar aerosol loads of different origins, namely for (i) anthropogenic aerosol mixed with dust from North Africa on June 21, 2007; (ii) continental aerosol from north-western and central Europe on August 19, 2007; and (iii) continental aerosol from the Arctic Ocean, Scandinavian peninsula, and central Europe on October 15, 2007. The values of τ_a(0.55 μm) were very stable throughout the measurement period, apart from a peak in the early afternoon of June 21, 2007, and a sharp increase in the late afternoon of August 19, 2007. Different ranges of α(0.40–0.87 μm) were determined for the different types of aerosol particles, the highest

Table 8.3a. Daily mean values of aerosol optical thickness $\tau_a(0.50\ \mu m)$, Ångström's (1964) exponent $\alpha(0.40–0.87\ \mu m)$, real part $n(0.50\ \mu m)$, and imaginary part $k(0.50\ \mu m)$ of the complex refractive index of columnar aerosol, and columnar aerosol single-scattering albedo $\omega(0.55\ \mu m)$, as determined from the SKYNET measurements taken with the PREDE POM–02L sun-/sky-radiometer on 18 golden days selected among those of the two AEROCLOUDS project campaigns performed at San Pietro Capofiume (Po Valley, Italy) from May 2007 to March 2008 (Mazzola et al., 2010).

Measurement day	Daily mean $\tau_a(0.50\ \mu m)$	Daily mean	Daily mean values of complex refractive index		Daily mean	Types and origins of aerosols, as derived from the NOAA/HYSPLIT backward trajectories passing over San Pietro Capofiume (SPC, Po Valley, Italy) at the 12:00 UTC/GMT
		α	Real part $n(0.50\ \mu m)$	Imaginary part $k(0.50\ \mu m)$	$\omega(0.55 \mu m)$	
May 18, 2007	0.115	1.174	1.515	5.55 × 10–4	0.986	Marine/continental aerosol from northern Atlantic Ocean and north-western Europe at all levels $z < 4$ km
May 24, 2007	0.290	1.402	1.427	1.35×10^{-3}	0.980	Continental aerosol from north-eastern Europe at all levels $z < 4$ km
June 21, 2007	0.349	0.880	1.453	5.74×10^{-3}	0.902	Anthropogenic/continental aerosol from southern Italy at levels <1 km, and mixed aerosol (marine/continental/dust) from the Mediterranean Sea and North Africa at levels from 1 to 4 km
July 1, 2007	0.318	1.321	1.395	3.14×10^{-3}	0.952	Marine/continental aerosol from Atlantic Ocean, southern France, and Ligurian Sea at all levels $z < 4$ km
July 8, 2007	0.156	1.272	1.426	2.09×10^{-3}	0.968	Marine/continental aerosol from Atlantic Ocean, southern France, and Ligurian Sea at all levels $z < 4$ km
July 17, 2007	0.292	0.424	1.566	4.27×10^{-3}	0.901	Continental aerosol circulating over central Italy and Adriatic Sea at levels $z < 1$ km, over central Europe and Adriatic Sea at levels of 1–2 km, and mixed marine/dust aerosol at the 4 km level, transported from Atlantic Ocean and north-western Sahara region
July 26, 2007	0.153	1.191	1.433	4.30×10^{-3}	0.932	Marine/continental aerosol from Atlantic Ocean and north-western and central Europe at all levels $z < 4$ km
Aug. 2, 2007	0.374	0.694	1.521	2.95×10^{-3}	0.940	Continental aerosol from north-western and central Europe at levels $z < 1$ km, and mixed marine/continental aerosol from Atlantic Ocean, southern France and Iberian peninsula at levels $z < 1$ km

Table 8.3a. Continued.

Measurement day	Daily mean $\tau_a(0.50\ \mu m)$	Daily mean α	Daily mean values of complex refractive index: Real part $n(0.50\ \mu m)$	Daily mean values of complex refractive index: Imaginary part $k(0.50\ \mu m)$	Daily mean $\omega(0.55 \mu m)$	Types and origins of aerosols, as derived from the NOAA/HYSPLIT backward trajectories passing over San Pietro Capofiume (SPC, Po Valley, Italy) at the 12:00 UTC/GMT
Aug. 5, 2007	0.132	1.373	1.430	1.80×10^{-3}	0.976	Continental aerosol from north-western and central Europe at levels $z < 2$ km, and mixed marine/continental aerosol from Atlantic Ocean, northern France, and Alpine region at around 4 km level
Aug. 19 2007	0.283	1.381	1.409	1.80×10^{-3}	0.974	Continental aerosol from north-western Europe and circulating the last two days over Austria and Slovenia before reaching Po Valley at levels $z < 1.2$ km, and marine/continental aerosol from Atlantic Ocean and southern France at the upper levels
Aug. 27, 2007	0.256	0.502	1.515	2.53×10^{-3}	0.935	Continental aerosol from north-western Europe and circulating for two days over southern Germany and Austria at all levels, before reaching the Po Valley
Sept. 9, 2007	0.207	1.283	1.429	2.80×10^{-3}	0.963	Marine/continental aerosol from northern Atlantic Ocean and central Europe at all levels
Sept. 13, 2007	0.084	0.927	1.417	5.28×10^{-3}	0.910	Marine/continental aerosol from northern Atlantic Ocean and central Europe at all levels
Oct. 15, 2007	0.296	1.102	1.462	4.18×10^{-3}	0.928	Continental aerosol from the Arctic Ocean and Scandinavian regions, subsequently transported across the central Europe at all levels $z < 4$ km
Dec. 19, 2007	0.147	1.568	1.450	8.66×10^{-3}	0.937	Continental aerosol from eastern Europe (including Russia, Belarus, and northern part of Balkans area) at all levels $z < 4$ km
Jan. 24, 2008	0.071	1.264	1.450	6.71×10^{-3}	0.915	Continental aerosol from northern Europe at levels $z < 1$ km, and marine/continental aerosol from northern Atlantic Ocean and western Europe at the upper levels
Feb.12, 2008	0.245	1.498	1.443	6.51×10^{-3}	0.945	Continental aerosol from eastern Europe and the northern Balkans area at all levels
March 13, 2008	0.080	1.327	1.522	5.51×10^{-3}	0.948	Marine/continental aerosol from northern Atlantic Ocean and western Europe at all levels

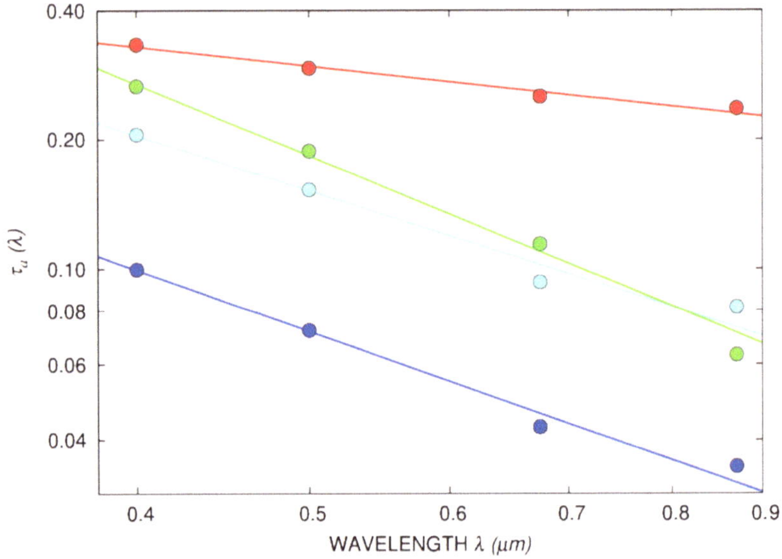

Fig. 8.18. Four examples of the Ångström (1964) best-fit procedure applied to spectral series of aerosol optical thickness $\tau_a(\lambda)$ measured using the PREDE POM-02L sun-/sky-radiometer at San Pietro Capofiume (Po Valley, Italy) at various hours of four AERO-CLOUDS golden days, to determine the best-fit values of exponent α over the 0.40–0.87-μm wavelength range. The four examples refer to: (i) July 17, 2007 (11:15 UTC/GMT), for continental aerosol circulated during the previous 4 days over central Italy, central Europe and Adriatic Sea at levels <2 km, and mixed with Saharan dust transported from North Africa at upper levels (red circles), giving $\alpha = 0.424$; (ii) October 15, 2007 (11:45 UTC/GMT), for continental aerosol from Arctic Ocean, Scandinavian peninsula, and central Europe (cyan circles), giving $\alpha = 1.107$; (iii) December 19, 2007 (08:22 UTC/GMT), for continental aerosol from eastern Europe (green circles), with $\alpha = 1.694$; and (iv) March 13, 2008 (14:30 UTC/GMT) for mixed marine/continental aerosol from northern Atlantic Ocean and western Europe (blue circles), with $\alpha = 1.337$.

values pertaining to continental aerosol and the lowest ones being associated with anthropogenic aerosol mixed with Saharan dust.

In addition to the PREDE POM-01L sun-/sky-radiometer measurements, regular Lidar measurements were performed at SPC during the AEROCLOUS campaign, using the ISAC-CNR tropospheric Lidar model to measure the vertical profiles of backscattering ratio and depolarization degree at the 0.532-μm wavelength up to tropospheric levels of 4–6 km. The measurements highlighted that aerosol particles were suspended above SPC, presenting multi-layered features at different levels: (i) within the first 1.5 km of the atmosphere, on winter days characterized by marked stability conditions below strong thermal inversions; (ii) in the first 3.5 km of the troposphere on clear-sky summer days, when the convective vertical mixing of aerosol particles was particularly strong; and (iii) at altitudes varying between 2 and 4 km during the most significant advective transport episodes of dense Saharan dust layers from the North African mobilization areas, most frequently observed in spring and early summer.

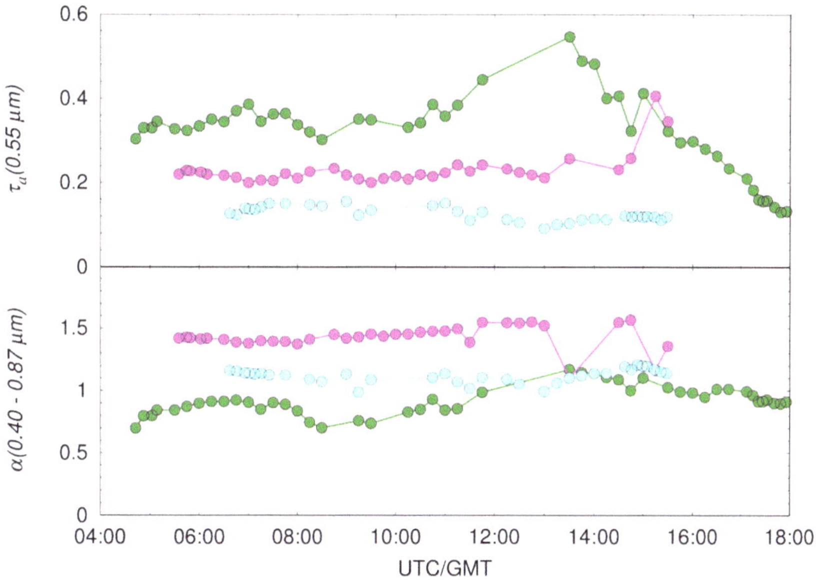

Fig. 8.19. Time-patterns of aerosol optical thickness $\tau_a(0.55\ \mu\text{m})$ (upper part) and Ångström's exponent $\alpha(0.40\text{–}0.87\ \mu\text{m})$ (lower part), as derived from the PREDE POM-02L sun-/sky-radiometer measurements performed at San Pietro Capofiume (Po Valley, Italy) during the AEROCLOUDS experimental campaign conducted from May 2007 to March 2008, on the following three golden days: (i) June 21, 2007, for anthropogenic aerosol transported from southern Italy and mixed with aerosol transported from Mediterranean Sea and North Africa at levels >1 km (dark green circles); (ii) August 19, 2007, for continental aerosol from north-western and central Europe (fuchsia circles); and (iii) October 15, 2007, for continental aerosol from Arctic Ocean, Scandinavian peninsula, and central Europe (cyan circles).

(2) *Determination of the columnar aerosol refractive index*

The PREDE POM-02L sky-brightness measurements in the almucantar were regularly performed at some selected wavelengths from 0.40 to 1.02 μm, for the purpose of deriving estimates of the complex refractive index $n(\lambda) - ik(\lambda)$ of columnar aerosol, and then inferring those of single-scattering albedo $\omega(\lambda)$, asymmetry factor $g(\lambda)$. The retrieval of these measurements provided columnar values of the complex refractive index, with real part $n(0.50\ \mu\text{m})$ ranging mainly from 1.40 to 1.52 in all seasons, giving an overall average value of 1.44 ± 0.06, and values of imaginary part $k(0.50\ \mu\text{m})$ of from 10^{-4} to 10^{-2}. It was found that the daily values of $n(0.50\ \mu\text{m})$ provided seasonal averages of 1.415 in winter, 1.428 in spring, 1.422 in summer, and 1.423 in autumn, while those of $k(0.50\ \mu\text{m})$ yielded seasonal averages of 2×10^{-3} in spring, 5×10^{-3} in winter and summer, and 1.5×10^{-2} in autumn. The daily mean values of $n(0.50\ \mu\text{m})$ determined on the 18 golden days reported in Table 8.3a were found to range between 1.395 (July 1, 2007, for marine/continental aerosol) and 1.566 (July 17, 2007, for continental aerosol mixed with Saharan dust), while the daily mean values of $k(0.50\ \mu\text{m})$ varied between 5.6×10^{-4} (May 18, 2007, for marine/continental aerosol from the northern Atlantic Ocean and north-western

Europe) and 8.7×10^{-3} (December 19, 2007, for continental aerosol from eastern Europe).

(3) *Determination of the size-distribution curves of columnar aerosol*

For the spectral series of $\tau_a(\lambda)$ recorded on the different hours of the 18 golden days and the corresponding values of $n(\lambda)$ and $k(\lambda)$ determined in the previous step, the size-distribution curves of columnar aerosol particles were retrieved for each SKYNET scanning recorded at SPC during the AEROCLOUDS campaign. Figure 8.20 shows some examples of bimodal size-distribution curves of columnar total particle number density $N(r)$ and columnar total particle volume $V(r)$, calculated for various measurement times on July 1, 2007 (with marine/continental aerosol), October 15, 2007 (with continental aerosol from the Arctic and Scandinavia), and December 19, 2007 (with continental aerosol from eastern Europe). Pronounced modes of fine and coarse particles are present in the first case, presumably due to continental and maritime aerosol, respectively, while the size-distribution curves of continental aerosol from unpolluted regions exhibit a mode of accumulation particles combined with a mode of coarse particles having larger sizes than in the previous case. The third case represents a columnar load of continental aerosol particles giving a value of $\alpha(0.40–0.87\ \mu m)$ appreciably greater than those of the first two examples: the first mode exhibits features similar to those of the Arctic particles, but the coarse particle mode is considerably less marked than in the other two cases.

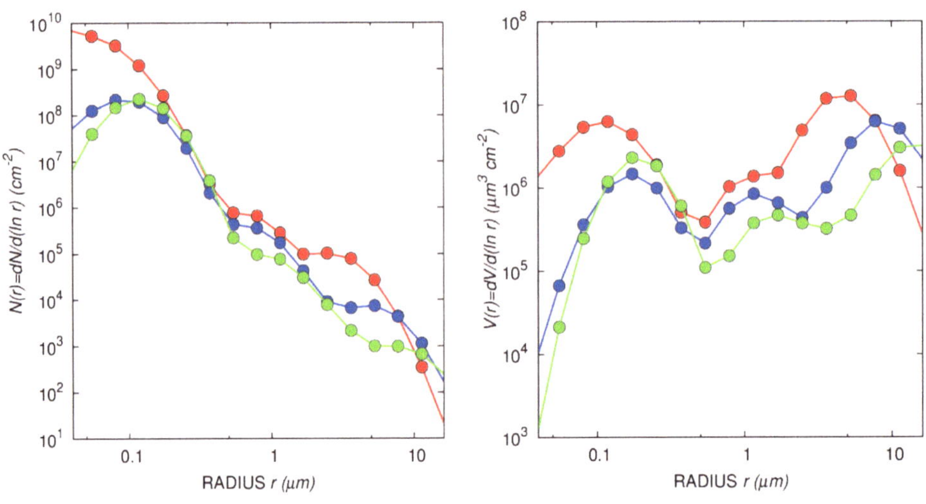

Fig. 8.20. Examples of multimodal size-distribution curves of columnar particle number density $N(r) = \mathrm{d}N/\mathrm{d}(\ln r)$ measured per cm^2 (left) and columnar particle volume $V(r) = \mathrm{d}V/\mathrm{d}(\ln r)$ measured in $\mu\mathrm{m}^3/\mathrm{cm}^2$ (right), retrieved from the SKYNET spectral series of $\tau_a(\lambda)$ determined at San Pietro Capofiume (Po Valley, Italy) on the following three AEROCLOUDS golden days: (i) July 1, 2007 (07:45 UTC/GMT), for mixed marine/continental aerosol from Atlantic Ocean, southern France, and Ligurian Sea, giving $\alpha = 1.347$ (red circles); (ii) October 15, 2007 (11:45 UTC/GMT), for continental aerosol from Arctic Ocean, Scandinavian peninsula, and central Europe (blue circles), yielding $\alpha = 1.107$; and (iii) December 19, 2007 (08:22 UTC/GMT), for continental aerosol from eastern Europe (green circles), providing $\alpha = 1.694$.

(4) *Determination of the columnar aerosol single-scattering albedo*

The daily mean values of $\omega(0.50\ \mu m)$ were derived through a Skyrad inversion procedure from the sky-brightness measurements taken with the PREDE POM-02L sun-/sky-radiometer. They were found to range (i) between 0.92 and 0.99 in winter, with an average seasonal value of 0.953; (ii) between 0.93 and 0.99 in spring, with an average seasonal value of 0.966; (iii) between 0.87 and 0.98 in summer, with an average seasonal value of 0.935; and (iv) between 0.85 and 0.98 in autumn, with an average seasonal value of 0.897. The measurements substantially agree with the corresponding daily mean values of ground-level SSA parameter $\omega_o(0.55\ \mu m)$ obtained by means of simultaneous nephelometer and PSAP measurements, which are more scattered than the columnar ones because they are more strongly influenced by the trapping mechanisms of aerosol particles occurring near the ground for thermal inversion conditions. Because of such larger variability, rather wide seasonal ranges of $\omega_o(0.55\ \mu m)$ were found, equal to (i) 0.80–0.94 in winter, with an average seasonal value of 0.88 ± 0.06; (ii) 0.78–0.94 in spring, with an average seasonal value of 0.86 ± 0.07; (iii) 0.71–0.96 in summer, with an average seasonal value of 0.85 ± 0.06; and (iv) 0.88–0.93 in autumn, with an average seasonal value of 0.90 ± 0.04. Together with the daily mean values of $\omega(0.55\ \mu m)$ derived in Table 8.3a through calculations made with the 6S code of Vermote et al. (1997), the asymmetry factor $g(0.55\ \mu m)$ relative to the vertical atmospheric column was found to assume average seasonal values equal to 0.67 ± 0.04 in winter and spring, 0.66 ± 0.04 in summer, and 0.69 ± 0.04 in autumn.

(5) *Definition of the local surface albedo models*

For all the retrieved bimodal size-distribution curves and the corresponding values of particulate matter refractive index obtained on the 18 golden days, the present procedure requires the use of realistic non-Lambertian models of BRDF surface albedo. Examining the set of MCD43C3 products obtained from the MODIS Level 3.0 surface albedo observations for the winter and summer months of the two-year period 2007–2008, over the south-eastern part of the Po Valley (northern Italy), the seasonal average maps of surface albedo shown in Fig. 8.21 were determined. The findings indicate that the surface reflectance characteristics over the Po Valley area can be realistically simulated using the VS1, VS2, VS3, and VS4 surface albedo models defined by Tomasi et al. (2013), for obtaining correct evaluations of the DARF effects. The four VS-type models give values of white-sky albedo R_{ws} varying from 0.153 to 0.292, broadband albedo $A(\theta_o = 60°)$ varying from 0.149 to 0.289, and LAI parameter varying from 0.10 to 5.0. In particular, comparing the surface albedo evaluations derived from satellite-borne observations with those of the VS1, VS2, VS3, and VS4 non-Lambertian surface reflectance models of Tomasi et al. (2013), it can be stated that (i) model VS1 can be most reliably used to represent the surface albedo characteristics observed in winter; (ii) model VS2 can be preferably used in late winter and late autumn; (iii) model VS3 in early spring and late summer; and (iv) model VS4 in late spring and early summer. The spectral curves of the BRDF surface albedo $R_L(\lambda, \theta_o)$ calculated for models VS1 and VS4 have been presented in section 8.3.1 and shown in Fig. 8.5, where the DARF calculations for the CLEARCOLUMN project are reported. The spectral curves of the BRDF surface albedo $R_L(\lambda, \theta_o)$ defined for models VS2 and VS3 at

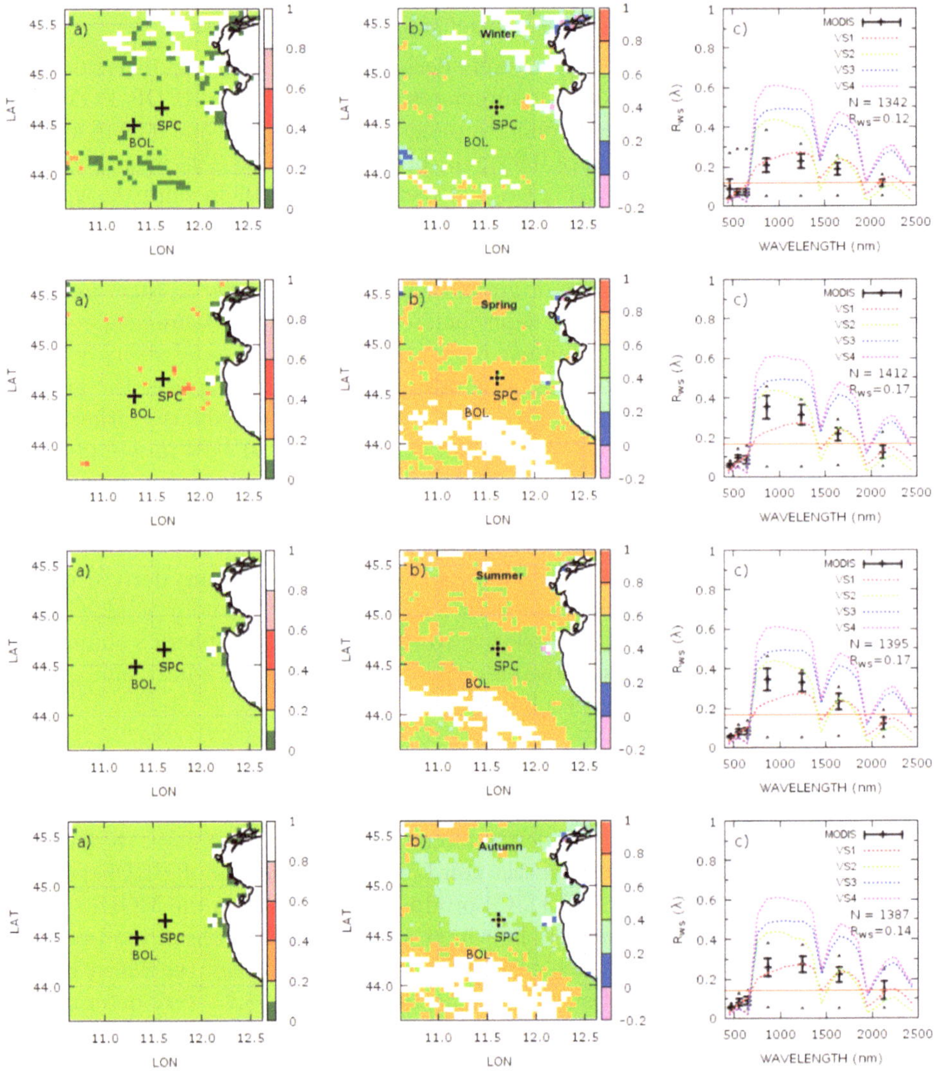

Fig. 8.21. Seasonal maps of the average land surface albedo ((a) in the left column) and Normalized Difference Vegetation Index (NDVI) ((b) in the central column) obtained over the eastern part of the Po Valley area (northern Italy) from the MODIS Level 3.0 surface albedo data (MCD43C3 products) recorded during the four seasonal periods of 2010. The crosses labelled BOL and SPC indicate the geographical positions of Bologna and San Pietro Capofiume stations, where MFRSR shadow-band radiometer and PREDE POM-02L sun-/sky-radiometer measurements were regularly performed during the two AEROCLOUDS field campaigns, respectively. The graphs shown in the third column present the spectral values (black vertical bars) of the white-sky albedo R_{ws} determined over the land area, yielding mean values of 0.12 in winter, 0.17 in spring and summer, and 0.14 in autumn, all found with standard deviations of around 0.1 within the seven MODIS channels, and with spectral values of minimum and maximum represented by small triangles. The graphs also show the spectral curves of $R_{ws}(\lambda)$ obtained as best-fit solutions for the vegetation-covered surface albedo models VS1 (dotted red curve), VS2 (dotted green curve), VS3 (dotted blue curve), and VS4 (dotted fuchsia curve) of Tomasi et al. (2013).

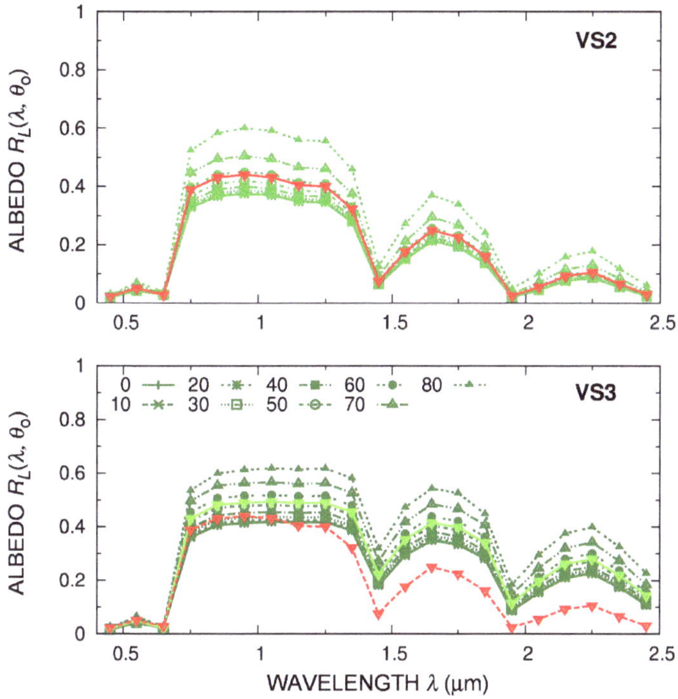

Fig. 8.22. Spectral curves of the BRDF surface albedo $R_L(\lambda, \theta_o)$ defined by Tomasi et al. (2013) for the vegetation-covered surface albedo models VS2 and VS3, as evaluated for nine values of solar zenith angle θ_o, taken in steps of 10° over the 0°–80° range, over the eastern Po Valley area (northern Italy) around San Pietro Capofiume station. Model VS2 refers to a vegetation-covered surface with LAI = 1.5, and model VS3 to a vegetation-covered surface with LAI = 2.5. The spectral curve of white-sky albedo $R_{ws}(\lambda)$ for the VS2 model is marked in red in both graphs, while that of model VS3 is marked in light green in the lower graph, for an easier comparison with the red VS2 curve of $R_{ws}(\lambda)$.

nine values of solar zenith angle θ_o, taken in steps of 10° over the $0° \leq \theta_o \leq 80°$ range, are shown in Fig. 8.22. Model VS2 defined for LAI = 1.0 presents values of $R_{bs}(\theta_o = 0°) = 0.170$, $R_{ws} = 0.203$, and $A(\theta_o = 60°) = 0.210$, and model VS3 defined for LAI = 2.5 yields values $R_{bs}(\theta_o = 0°) = 0.201$, $R_{ws} = 0.243$ m and $A(\theta_o = 60°) = 0.258$. It is evident that model VS3 exhibits appreciably higher spectral values of BRDF surface albedo $R_L(\lambda, \theta_o)$ than model VS2, and therefore also gives noticeably higher spectral values of white-sky albedo $R_{ws}(\lambda)$ than those of model VS2, as can be seen in the comparison made in Fig. 8.22.

(6) *Calculations of the daily time-patterns of instantaneous DARF terms and diurnally averaged DARF effects*

Adopting the procedure used in the PRIN-2004 analysis described in section 8.3.2, the instantaneous values of DARF forcing terms $\Delta F_{ToA}(t)$ and $\Delta F_{BoA}(t)$ were calculated for the VS1, VS2, VS3, and VS4 surface albedo models assumed in the previous step (5), and the data sets of the aerosol optical parameters obtained at steps (1)–(4) from the field measurements performed at SPC on the 18 selected golden days listed in Table 8.3a. As in the CLEARCOLUMN and PRIN-2004 analyses, the calculations were made by employing the 6S code for the various values of θ_o in the insolation period, and the time-patterns of the columnar aerosol optical parameters calculated in the previous steps: they were determined from (i) the sets of $\tau_a(\lambda)$ and α (calculated in step (1)), (ii) the sets of refractive index parts $n(\lambda)$ and $k(\lambda)$ (obtained in step (2)), (iii) the $N(r)$ and $V(r)$ size-distribution curves (determined in step (3)), and (iv) the daily mean spectral patterns of $\omega(\lambda)$ (determined in step (4)). The daily time-patterns of $\Delta F_{ToA}(t)$ and $\Delta F_{BoA}(t)$ for the 18 golden days were integrated over the corresponding sunlit periods to calculate (i) the daily values of ΔDF_{ToA} and ΔDF_{BoA} on the 18 golden days, for the four surface albedo models VS1, VS2, VS3, and VS4, and (ii) the corresponding values of ΔDF_{Atm} as differences between ΔDF_{ToA} and ΔDF_{BoA}. The resulting daily values of ΔDF_{ToA}, ΔDF_{BoA}, and ΔDF_{Atm} are given in Table 8.3b for the 18 golden days, while their time-patterns are shown in Fig. 8.23, where they are compared with those of the corresponding daily mean values of columnar aerosol parameters $\tau_a(0.55\ \mu\text{m})$ and $\omega(0.55\ \mu\text{m})$ given in Table 8.3a. The values of ΔDF_{ToA} are all negative, but gradually decrease in absolute value on passing from model VS1 to VS4. In other words, the cooling effects become progressively more moderate as the daily mean values of $\tau_a(0.55\ \mu\text{m})$ gradually increase, while weaker dependence features can be attributed to the vegetation-covered surface albedo and the variations in $\omega(0.55\ \mu\text{m})$, found to range between around 0.93 and nearly 1.00 during the experiment. It can be also seen in Fig. 8.23 that small differences characterized each set of four daily values of ΔDF_{BoA} determined for the four models VS1–VS4 on each golden day, while even more limited differences were found for each set of four daily values of ΔDF_{Atm}. These findings clearly indicate that the daily values of the three DARF terms depend closely on $\tau_a(0.55\ \mu\text{m})$ and to a lesser extent on the other aerosol optical parameters. In addition, the effects due to the surface albedo characteristics are considerable in the diurnally averaged ΔDF_{ToA} terms, of appreciable intensity in the diurnally averaged ΔDF_{BoA} term, and very low in the diurnally averaged ΔDF_{Atm} term.

The daily values of diurnally averaged ΔDF_{ToA}, ΔDF_{BoA}, and ΔDF_{Atm} terms are given in Table 8.3b for all 18 golden days of the AEROCLOUDS campaign, showing that ΔDF_{ToA} was estimated to vary between -0.1 W/m^2 (on September 13, 2007, for the VS4 model) and -13.5 W/m^2 (on August 2, 2007, for the VS1 model), both cases pertaining to mixed marine/continental aerosol. DARF term ΔDF_{BoA} was found to vary between -3.8 W/m^2 (on January 24, 2008, for the VS4 model and marine/continental aerosol) and -29.0 W/m^2 (on June 21, 2007, for the VS1 model and anthropogenic/continental aerosol), while ΔDF_{Atm} ranged between $+1.1$ W/m^2 (on December 19, 2007, for the VS1 model and mixed anthropogenic and continental aerosol) and $+18.8$ W/m^2 (on June 21, 2007, for the VS4 model and continental aerosol).

The daily values of diurnally averaged ΔDF_{ToA}, ΔDF_{BoA}, and ΔDF_{Atm} terms given in Table 8.3b for the 18 AEROCLOUDS golden days are plotted in Fig. 8.24 as a function of $\tau_a(0.55\ \mu\text{m})$, separately for the four VS surface albedo models assumed in the Po Valley area around SPC. The scatter plots show that ΔDF_{ToA} and ΔDF_{BoA} decrease on average as $\tau_a(0.55\ \mu\text{m})$ increases, with slope coefficients that vary appreciably with the different VS models, and only slightly with $\omega(0.55\ \mu\text{m})$.

Table 8.3b. Daily values of the diurnally averaged aerosol forcing terms ΔDF_{ToA} at the ToA-level, ΔDF_{BoA} at the BoA-level, and ΔDF_{Atm} within the atmosphere, diurnal average aerosol fractional forcing AFF_{ToA} at the ToA-level (given by the ratio between flux change ΔF_{ToA} at the ToA-level and the incoming flux $I_S \downarrow$ of solar radiation at the ToA-level), and the diurnal average DARF efficiencies E_{ToA}, E_{BoA}, and E_{Atm} giving the rates at which the surface–atmosphere system is forced per unit $\tau_a(0.55\ \mu m)$, as obtained for the 18 golden days selected among those of the two AEROCLOUDS project campaigns performed at San Pietro Capofiume (Po Valley, Italy) from May 2007 to March 2008 (Mazzola et al., 2010).

Measurement day	Surface albedo model	Diurnal average DARF terms (W/m^2)			Diurnal average	Diurnal average values of DARF efficiencies (W/m^2)		
		ΔDF_{ToA}	ΔDF_{BoA}	ΔDF_{Atm}	AFF_{ToA}	E_{ToA}	E_{BoA}	E_{Atm}
May 18, 2007	VS1	−5.1	−7.6	+2.5	-1.2×10^{-2}	−43.7	−65.2	+21.5
	VS2	−4.3	−7.0	+2.7	-1.1×10^{-2}	−37.1	−59.9	+22.8
	VS3	−4.0	−6.8	+2.8	-1.0×10^{-2}	−34.2	−58.3	+24.1
	VS4	−3.5	−6.4	+2.9	-8.0×10^{-3}	−29.7	−54.8	+25.1
May 24, 2007	VS1	−11.9	−15.6	+3.7	-2.8×10^{-2}	−42.7	−56.1	+13.4
	VS2	−10.9	−14.5	+3.6	-2.6×10^{-2}	−39.2	−52.3	+13.1
	VS3	−10.3	−14.1	+3.8	-2.5×10^{-2}	−37.1	−50.6	+13.5
	VS4	−9.6	−13.3	+3.7	-2.3×10^{-2}	−34.4	−47.8	+13.4
June 21, 2007	VS1	−11.5	−29.0	+17.5	-2.7×10^{-2}	−33.6	−84.6	+51.0
	VS2	−9.5	−27.3	+17.9	-2.2×10^{-2}	−27.6	−79.7	+52.1
	VS3	−7.9	−26.3	+18.3	-1.8×10^{-2}	−23.1	−76.6	+53.5
	VS4	−6.0	−24.8	+18.8	-1.4×10^{-2}	−17.4	−72.2	+54.8
July 1, 2007	VS1	−12.0	−17.3	+5.3	-2.8×10^{-2}	−41.4	−59.8	+18.4
	VS2	−10.5	−16.0	+5.5	-2.4×10^{-2}	−36.1	−55.2	+19.1
	VS3	−9.8	−15.5	+5.7	-2.3×10^{-2}	−33.8	−53.6	+19.8
	VS4	−8.7	−14.6	+5.9	-2.0×10^{-2}	−30.2	−50.6	+20.4
July 8, 2007	VS1	−7.4	−10.5	+3.1	-1.7×10^{-2}	−47.8	−67.9	+20.1
	VS2	−6.5	−9.8	+3.3	-1.5×10^{-2}	−42.4	−63.7	+21.3
	VS3	−6.2	−9.6	+3.4	-1.4×10^{-2}	−40.0	−62.3	+22.3
	VS4	−5.5	−9.1	+3.6	-1.3×10^{-2}	−35.9	−59.0	+23.1
July 17, 2007	VS1	−9.2	−22.3	+13.1	-2.2×10^{-2}	−32.0	−77.3	+45.3
	VS2	−7.3	−20.7	+13.4	-1.7×10^{-2}	−25.3	−71.7	+46.4
	VS3	−5.8	−19.6	+13.8	-1.4×10^{-2}	−20.3	−68.0	+47.7
	VS4	−4.3	−18.4	+14.1	-1.0×10^{-2}	−14.8	−63.6	+48.8
July 26, 2007	VS1	−5.6	−11.2	+5.7	-1.4×10^{-2}	−36.7	−74.0	+37.3
	VS2	−4.4	−10.4	+6.1	-1.1×10^{-2}	−28.8	−68.9	+40.1
	VS3	−3.9	−10.2	+6.3	-9.0×10^{-3}	−25.4	−67.1	+41.7
	VS4	−3.0	−9.6	+6.6	-7.0×10^{-3}	−19.6	−63.1	+43.5
Aug. 2, 2007	VS1	−13.5	−21.0	+7.4	-3.4×10^{-2}	−39.5	−61.2	+21.7
	VS2	−11.7	−19.4	+7.6	-3.0×10^{-2}	−34.2	−56.4	+22.2
	VS3	−10.5	−18.4	+7.8	-2.7×10^{-2}	−30.7	−53.5	+22.8
	VS4	−9.2	−17.2	+7.9	-2.3×10^{-2}	−27.0	−50.1	+23.1
Aug. 5, 2007	VS1	−5.2	−9.2	+4.0	-1.3×10^{-2}	−38.2	−67.3	+29.1
	VS2	−4.2	−8.5	+4.3	-1.1×10^{-2}	−30.8	−61.9	+31.1
	VS3	−3.8	−8.2	+4.4	-1.0×10^{-2}	−27.8	−60.1	+32.3
	VS4	−3.1	−7.7	+4.6	-8.0×10^{-3}	−22.9	−56.4	+33.5

Table 8.3b. Continued.

Measurement day	Surface albedo model	Diurnal average DARF terms (W/m^2)			Diurnal average	Diurnal average values of DARF efficiencies (W/m^2)		
		ΔDF_{ToA}	ΔDF_{BoA}	ΔDF_{Atm}	AFF_{ToA}	E_{ToA}	E_{BoA}	E_{Atm}
Aug. 19, 2007	VS1	−9.2	−15.9	+6.8	-2.5×10^{-2}	−27.8	−48.4	+20.6
	VS2	−7.7	−14.6	+6.9	-2.1×10^{-2}	−23.3	−44.4	+21.1
	VS3	−7.0	−14.1	+7.1	-1.9×10^{-2}	−21.2	−42.8	+21.6
	VS4	−6.0	−13.3	+7.3	-1.7×10^{-2}	−18.3	−40.4	+22.1
Aug. 27, 2007	VS1	−9.0	−15.5	+6.4	-2.6×10^{-2}	−34.9	−59.8	+24.9
	VS2	−7.6	−14.1	+6.5	-2.2×10^{-2}	−29.2	−54.4	+25.2
	VS3	−6.5	−13.2	+6.7	-1.9×10^{-2}	−25.3	−51.2	+25.9
	VS4	−5.6	−12.4	+6.8	-1.6×10^{-2}	−21.6	−47.9	+26.3
Sept. 9, 2007	VS1	−7.3	−9.9	+2.6	-2.3×10^{-2}	−36.6	−49.8	+13.2
	VS2	−6.4	−9.0	+2.7	-2.0×10^{-2}	−32.2	−45.5	+13.3
	VS3	−6.0	−8.8	+2.8	-1.9×10^{-2}	−30.1	−44.2	+14.1
	VS4	−5.5	−8.3	+2.9	-1.8×10^{-2}	−27.5	−41.9	+14.4
Sept. 13, 2007	VS1	−2.7	−8.0	+5.3	-9.0×10^{-3}	−35.5	−105.4	+69.9
	VS2	−1.5	−7.3	+5.8	-5.0×10^{-3}	−19.9	−96.5	+76.6
	VS3	−0.9	−7.0	+6.1	-3.0×10^{-3}	−12.2	−92.5	+80.3
	VS4	−0.1	−6.5	+6.4	0.0 to -1.0×10^{-3}	−0.8	−85.7	+84.9
Oct. 15, 2007	VS1	−3.6	−8.8	+5.2	-1.7×10^{-2}	−24.9	−60.4	+35.5
	VS2	−2.5	−8.0	+5.4	-1.2×10^{-2}	−17.5	−54.7	+37.2
	VS3	−2.1	−7.7	+5.6	-1.0×10^{-2}	−14.2	−52.7	+38.5
	VS4	−1.5	−7.2	+5.8	-7.0×10^{-3}	−10.2	−49.8	+39.6
Dec. 19, 2007	VS1	−3.6	−4.7	+1.1	-3.2×10^{-2}	−26.9	−35.1	+8.2
	VS2	−3.0	−4.1	+1.1	-2.7×10^{-2}	−22.7	−30.6	+7.9
	VS3	−2.9	−4.1	+1.2	-2.6×10^{-2}	−21.7	−30.4	+8.7
	VS4	−2.7	−3.9	+1.2	-2.5×10^{-2}	−20.4	−29.2	+8.8
Jan. 24, 2008	VS1	−1.2	−4.5	+3.3	-9.0×10^{-3}	−18.2	−66.8	+48.6
	VS2	−0.7	−4.1	+3.3	-5.0×10^{-3}	−11.0	−60.4	+49.4
	VS3	−0.6	−4.0	+3.4	-4.0×10^{-3}	−8.3	−59.1	+50.8
	VS4	−0.3	−3.8	+3.5	-2.0×10^{-3}	−4.8	−56.4	+51.6
Feb. 12, 2008	VS1	−6.9	−10.4	+3.4	-3.9×10^{-2}	−28.9	−43.2	+14.3
	VS2	−5.9	−9.3	+3.4	-3.3×10^{-2}	−24.4	−38.6	+14.2
	VS3	−5.4	−9.0	+3.6	-3.0×10^{-2}	−22.5	−37.3	+14.8
	VS4	−4.9	−8.5	+3.6	-2.8×10^{-2}	−20.5	−35.4	+14.9
March 13, 2008	VS1	−2.8	−4.8	+2.0	-1.1×10^{-2}	−31.4	−54.3	+22.9
	VS2	−2.3	−4.4	+2.1	-9.0×10^{-3}	−25.8	−49.6	+23.8
	VS3	−2.1	−4.3	+2.2	-8.0×10^{-3}	−23.5	−48.5	+25.0
	VS4	−1.8	−4.1	+2.3	-7.0×10^{-3}	−20.4	−46.0	+25.6

Conversely, the scatter plot of ΔDF_{Atm} versus $\tau_a(0.55\ \mu\text{m})$ indicates that heating effects within the atmosphere (i) increase as $\tau_a(0.55\ \mu\text{m})$ assumes gradually higher values, to an extent that slowly increases passing from the VS1 to the VS4 model, and (ii) vary only slightly as a function of $\omega(0.55\ \mu\text{m})$.

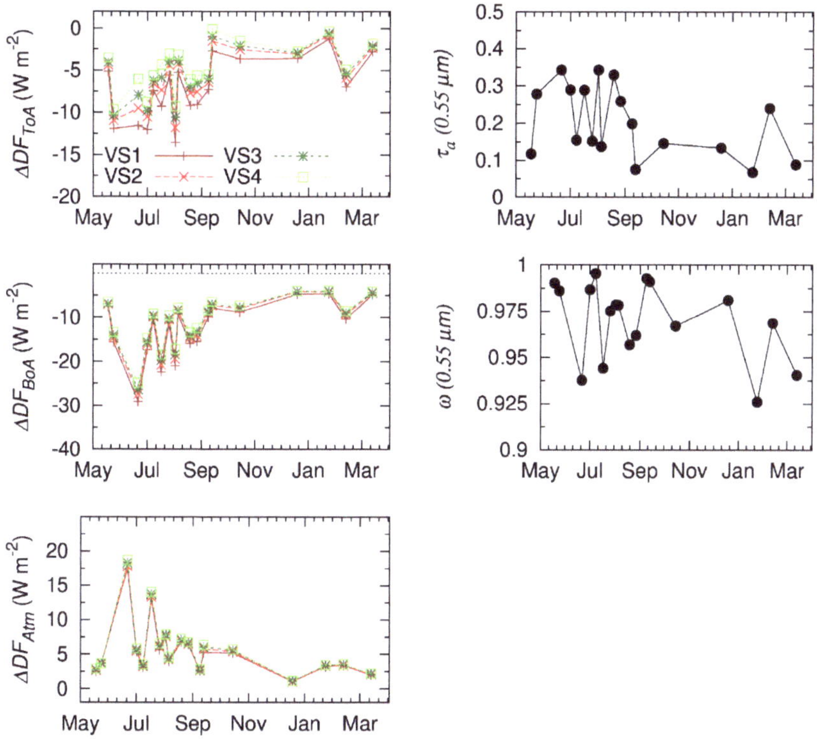

Fig. 8.23. Left: Time-patterns of the daily values of diurnally averaged DARF terms ΔDF_{ToA} at the ToA-level, ΔDF_{BoA} at the BoA-level, and ΔDF_{Atm} within the atmosphere, calculated on the 18 golden days of the AEROCLOUDS experiment conducted at San Pietro Capofiume (Po Valley, northern Italy) from May 2007 to March 2008, using the vegetation-covered surface albedo models VS1, VS2, VS3, and VS4 of Tomasi et al. (2013). Right: Time-patterns of the daily mean values of aerosol optical thickness $\tau_a(0.55\ \mu\text{m})$ and columnar aerosol single-scattering albedo $\omega(0.55\ \mu\text{m})$, determined for the columnar aerosol size-distributions and complex refractive index measured on the 18 AEROCLOUDS golden days.

(7) *Calculations of the daily values of DARF efficiencies*

The scatter plots in Fig. 8.24 showing the daily values of diurnally averaged ΔDF_{ToA}, ΔDF_{BoA}, and ΔDF_{Atm} represented as a function of $\tau_a(0.55\ \mu\text{m})$, provide evidence of the strong dependence of these DARF terms on $\tau_a(0.55\ \mu\text{m})$, in spite of the dispersion features linked to the choice of different surface albedo models and aerosol optical characteristics. Assuming that a rough proportionality exits between each DARF term and $\tau_a(0.55\ \mu\text{m})$, with proportionality coefficients varying as a function of surface albedo and aerosol SSA, the daily mean values of ΔDF_{ToA}, ΔDF_{BoA}, and ΔDF_{Atm} in Table 8.3b were divided by the corresponding values of $\tau_a(0.55\ \mu\text{m})$ in Table 8.3a, in order to determine the daily mean values of DARF efficiency parameters E_{ToA} at the ToA-level, E_{BoA} at the BoA-level, and E_{Atm} in the atmosphere, for all 18 selected golden days of the AEROCLOUDS campaign. It can be observed that the daily values of DARF efficiency E_{ToA} were estimated

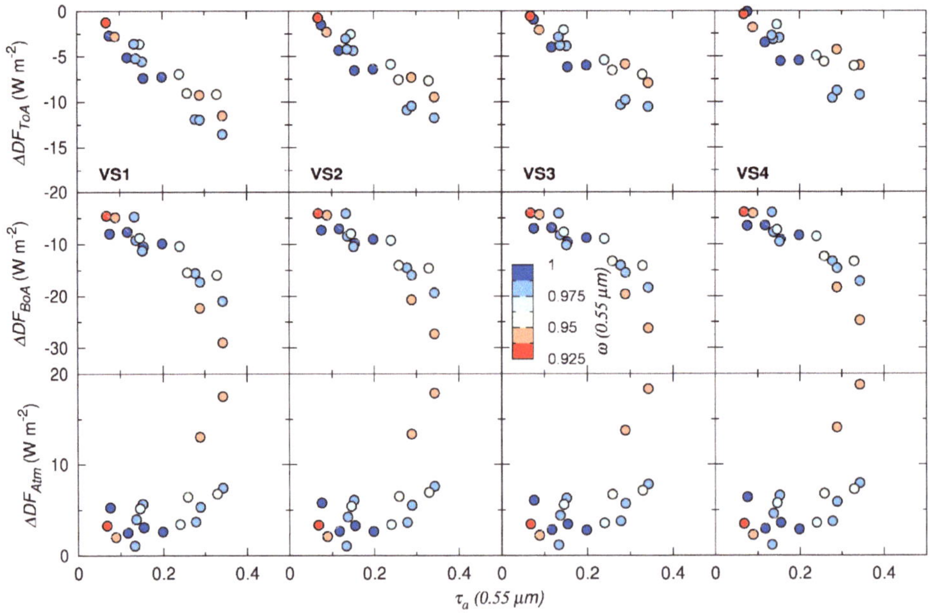

Fig. 8.24. Scatter plots of the daily mean values of DARF terms ΔDF_{ToA}, ΔDF_{BoA}, and ΔDF_{Atm} versus the corresponding daily mean values of aerosol optical thickness $\tau_a(0.55\ \mu m)$, determined from the AEROCLOUDS field measurements performed on the 18 golden days, for the vegetation-covered surface albedo models VS1 (first column), VS2 (second column), VS3 (third column), and VS4 (fourth column), and for different values of columnar aerosol single-scattering albedo $\omega(0.55\ \mu m)$, as indicated by the different colors of the symbols, chosen according to the color scale in the third column.

to vary between −2 and −48 W/m^2, while those of E_{BoA} varied between −30 and −98 W/m^2, and those of E_{Atm} between +8 and +78 W/m^2 for daily mean values of $\omega(0.55\ \mu m)$ of between 0.92 and 0.99. The three sets of daily mean values of ΔDF_{ToA}, ΔDF_{BoA}, and ΔDF_{Atm} are plotted in Fig. 8.25 versus the corresponding daily mean values of $\omega(0.55\ \mu m)$. The figure shows that E_{ToA} ranged between −47.8 W/m^2 and nearly null values, found for maritime/continental aerosol characterized by values of $\omega(0.55\ \mu m)$ very close to unit. The values of E_{ToA} slightly decrease on average as $\omega(0.55\ \mu m)$ increases from 0.93 to 0.99, presenting similarly sloping trends for the four subsets determined for the four VS surface albedo models. It is also worth noting that the most negative values of E_{ToA} were obtained for the VS1 model characterized by the lowest reflectance features, while the less negative ones were found for the VS4 model, which presents the highest reflectance characteristics among the four VS models. The diurnal average values of E_{BoA} were found to vary between −29.2 and −105.4 W/m^2, being widely scattered throughout the range of $\omega(0.55\ \mu m)$ and showing a generally increasing trend as $\omega(0.55\ \mu m)$ increases. The diurnal average values of E_{Atm} were found to vary between +7.9 and +84.8 W/m^2, presenting large dispersion features and an increasing trend with $\omega(0.55\ \mu m)$.

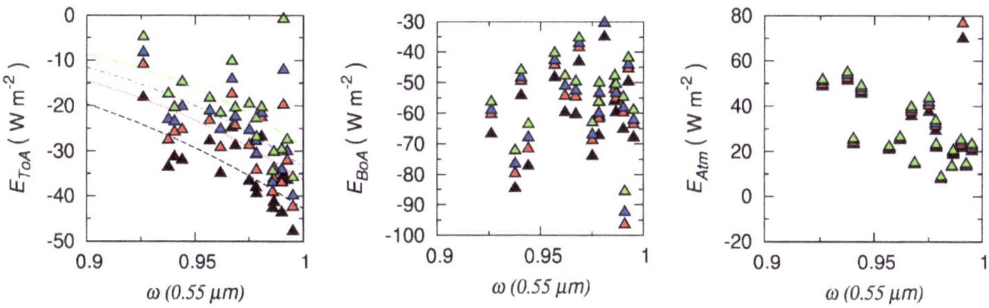

Fig. 8.25. Scatter plots of the daily mean values of DARF efficiency E_{ToA} at the ToA-level, DARF efficiency E_{BoA} at the BoA-level, and DARF efficiency E_{Atm} within the atmosphere versus the columnar aerosol single-scattering albedo $\omega(0.55\ \mu m)$, as obtained on the 18 AEROCLOUDS golden days chosen among the measurements performed at San Pietro Capofiume (Po Valley, northern Italy) from May 2007 to March 2008, by assuming the surface albedo characteristics represented by the vegetation-covered surface albedo models VS1 (black triangles), VS2 (red triangles), VS3 (blue triangles), and VS4 (green triangles).

8.3.4 DARF evaluations from the Ev-K2-CNR project measurements in Himalaya (Nepal)

Regular measurements of direct solar irradiance were performed at the CNR Pyramid Laboratory (27° 57′ N, 86° 49′ E, 5050 m a.m.s.l.), located at the foot of Mt. Everest (Himalaya, Nepal), during the Ev-K2-CNR project field campaigns conducted in the summer months of 1991 and 1992 (Tomasi et al., 1997). The measurements were taken using two portable Volz (1974) sun-photometers, of which the first was employed in 1991 during the two periods from July 24 to August 5, and from September 19 to October 2, and the second in 1992 from July 23 to August 11. The main purpose of the measurements was to (i) evaluate the mean atmospheric turbidity characteristics before and after the appearance of the Mt. Pinatubo stratospheric particle layer and (ii) measure the atmospheric turbidity variations caused by the volcanic particles a few months and about 1 year after the Pinatubo eruption in the Philippines on June 15, 1991. The sun-photometer measurements were undertaken during the 'monsoon' seasons of 1991 and 1992, when rather low concentrations of PM1, PM10, and black carbon (BC) are usually monitored at this high-altitude mountain site, as shown by the *in situ* particulate sampling measurements carried out by Marinoni et al. (2010).

To analyze the above data sets, we used the DARF-PROC procedure already followed to analyse the previous field data sets. It consisted of the seven steps described above:

(1) *Analysis of field sun-photometer measurements and determination of the columnar aerosol extinction parameters*

The series of sun-photometer measurements performed at the CNR Pyramid Laboratory were examined to determine the instantaneous values of $\tau_a(0.380\ \mu m)$, $\tau_a(0.500\ \mu m)$, and $\tau_a(0.875\ \mu m)$ at various hours of each clear-sky day (Tomasi et al., 1997). Examining such daily measurement sets, the daily mean values of

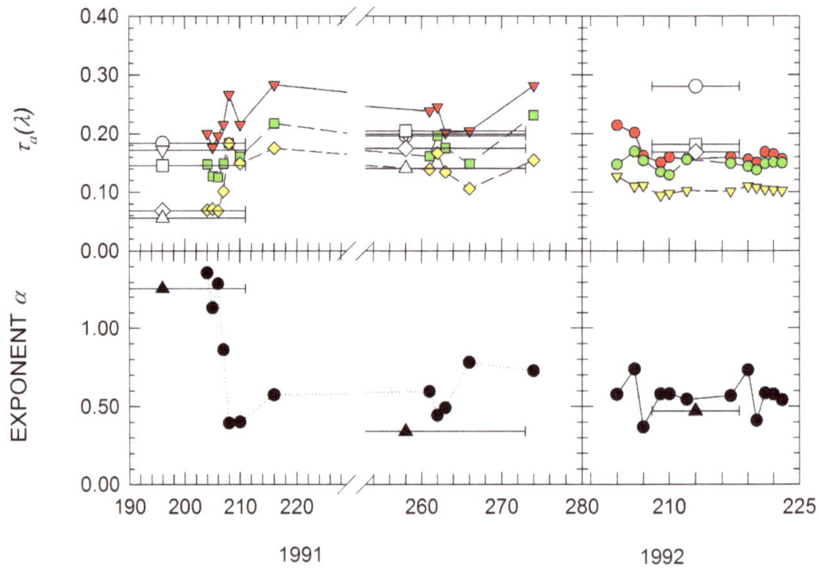

Fig. 8.26. Upper part: Time-patterns of the daily mean values of aerosol optical thickness $\tau_a(\lambda)$ measured at wavelengths $\lambda = 0.380$ μm (red symbols), $\lambda = 0.500$ μm (green symbols), and $\lambda = 0.875$ μm (yellow symbols) by the ISAC-CNR group (Tomasi et al., 1997) at the CNR Pyramid Laboratory (5050 m a.m.s.l.) (Himalaya, Nepal), during the three periods (A), (B), and (C) of summer 1991 (left) and period (D) in summer 1992 (right). The results are compared with the monthly mean values of $\tau_a(\lambda)$ measured by Pueschel et al. (1993) at the Mauna Loa Observatory (Hawaii) in July and September 1991 and in July and August 1992 at wavelengths $\lambda = 0.382$ μm (open circles), $\lambda = 0.451$ μm (open down triangles), $\lambda = 0.528$ μm (open squares), $\lambda = 0.865$ μm (open up triangles), and $\lambda = 1.060$ μm (open diamonds). Lower part: Time-patterns of the daily mean values of the Ångström (1964) exponent α (solid circles) determined at the Pyramid Laboratory during summer 1991 (left) and summer 1992 (right), and compared with the monthly mean values of α (solid triangles) determined by Pueschel et al. (1993) at the Mauna Loa Observatory in July and September 1991 (left) and in August 1992 (right).

$\tau_a(\lambda)$ were calculated at the three wavelengths, for which the corresponding daily mean values of Ångström (1964) parameters α and β were determined over the 0.380–0.875-μm spectral range. The time-patterns of the daily mean values of τ_a(0.380 μm), τ_a(0.500 μm), τ_a(0.875 μm), and α(0.380–0.875 μm) obtained during the two summer campaigns are shown in Fig. 8.26. Due to the greatly differing spectral characteristics of such parameters, the overall data set was subdivided into four subsets, each pertaining to one of the four periods listed in Table 8.4a, as follows:

(A) *Period from July 24 to 26, 1991*
Relatively low daily mean values of $\tau_a(\lambda)$ were obtained from the measurements taken during the three clear-sky days from July 24 to 26, 1991, which immediately preceded the appearance of the Mt. Pinatubo stratospheric particles over the Himalayan chain, giving values of α and β typical of high atmospheric transparency conditions. These background (BG) atmospheric turbid-

ity conditions are often observed at the high-altitude Himalayan sites during summer, when particulate matter is only weakly transported from the Indo-Gangetic regions, and the optical extinction effects are mainly produced by fine particles. During period (A), the daily values of $\tau_a(0.380\ \mu m)$, $\tau_a(0.500\ \mu m)$, and $\tau_a(0.875\ \mu m)$ were rather stable, with average values equal to 0.19 ± 0.04, 0.13 ± 0.04, and 0.07 ± 0.02, respectively. Correspondingly, Ångström's exponent α varied between 1.13 and 1.36, yielding an average value of 1.21 ± 0.10, and β varied between 0.057 and 0.061. The average values were found to agree closely with the monthly mean values of $\tau_a(\lambda)$ evaluated by Pueschel et al. (1993) at five wavelengths from 0.382 to 1.060 μm, examining the sun-photometer measurements performed at the Mauna Loa Observatory (Hawaii) in July 1991, as can be seen in Fig. 8.26. The atmospheric turbidity measurements performed at the CNR Pyramid Laboratory during period (A) clearly indicate that the columnar aerosol load mainly consisted of BG tropospheric aerosol particles, with a prevailing content of fine particles and a relatively low content of large stratospheric particles.

(B) *Period from July 28 to August 5, 1991*
As shown by Tomasi et al. (1997), the values of $\tau_a(\lambda)$ suddenly increased on July 27, 1991, about 40 days after the Mt. Pinatubo eruption. The sharp increase was caused by the massive transport of Mt. Pinatubo volcanic aerosol at stratospheric levels over the Himalayan region. It was followed by a further pronounced increase of $\tau_a(\lambda)$ on the subsequent day, with an overall two-day increase in the monochromatic values of $\tau_a(\lambda)$ from about 0.21 to 0.28 at $\lambda = 0.380\ \mu m$, from 0.13 to more than 0.18 at $\lambda = 0.500\ \mu m$, and from 0.06 to more than 0.12 at $\lambda = 0.875\ \mu m$. Average values of $\tau_a(0.380\ \mu m) = 0.24 \pm 0.04$, $\tau_a(0.500\ \mu m) = 0.18 \pm 0.03$, and $\tau_a(0.875\ \mu m) = 0.15 \pm 0.04$ were determined over period (B). As can be seen in Table 8.4a, the corresponding daily values of α ranged between 0.39 and 0.57, giving an average value of 0.46 ± 0.10, while β varied between around 0.14 and less than 0.17, with an average value of 0.15 ± 0.01. Such marked variations in $\tau_a(\lambda)$, α, and β were clearly due to the extinction effects caused by the new stratospheric particles of volcanic origins, which caused considerably more pronounced extinction effects than those due to BG tropospheric particles.

(C) *Period from September 19 to October 2, 1991*
The sun-photometer measurements of period (B) were suspended on August 5, 1991, and subsequently resumed in period (C), namely about three months after the Mt. Pinatubo eruption. During this period, the daily mean values of $\tau_a(\lambda)$ presented considerable day-to-day variations, with $\tau_a(0.380\ \mu m)$ ranging between 0.20 and 0.28, $\tau_a(0.500\ \mu m)$ between 0.15 and 0.23, and $\tau_a(0.875\ \mu m)$ between 0.10 and 0.17, yielding average daily values of $\tau_a(0.380\ \mu m) = 0.23 \pm 0.03$, $\tau_a(0.500\ \mu m) = 0.18 \pm 0.03$, and $\tau_a(0.875\ \mu m) = 0.14 \pm 0.02$. The comparison made in Table 8.4a indicates that these average values of $\tau_a(\lambda)$ were similar to those obtained in period (B), while (i) the daily mean values of α varied between 0.47 and 0.80, yielding an average value of 0.61 ± 0.14, which is only slightly higher than that recorded in period (B); and (ii) the daily mean values of β ranged between 0.10 and 0.15, giving an average value of 0.13 ± 0.02, which is about 20% lower than that recorded in period (B).

Table 8.4a. Average values of aerosol optical thickness $\tau_a(\lambda)$ measured at the three wavelengths $\lambda = 0.380$ μm, $\lambda = 0.500$ μm, and $\lambda = 0.875$ μm during the EV-K2-CNR campaigns of summer 1991 and summer 1992 employing two Volz sun-photometers at the CNR Pyramid Laboratory (5050 m a.m.s.l.), in Himalaya (Nepal), together with the average values of Ångström's (1964) atmospheric turbidity parameters α and β and their ranges, as determined by Tomasi et al. (1997) for the four periods (A)–(D), into which the two field campaigns of summer 1991 and summer 1992 were subdivided.

Measurement period	Average values of aerosol optical thickness $\tau_a(\lambda)$			Range of α	Average value of α(0.380–0.875 μm)	Range of β	Average value of β	Interpolated value of $\tau_a(0.55\ \mu m)$
	$\tau_a(0.380\ \mu m)$	$\tau_a(0.500\ \mu m)$	$\tau_a(0.875\ \mu m)$					
Period (A) from July 24 to July 26, 1991	0.191 ± 0.042	0.134 ± 0.035	0.069 ± 0.023	1.13–1.36	1.21 ± 0.10	0.057–0.061	0.059 ± 0.002	0.116
Period (B) from July 28 to August 5, 1991	0.245 ± 0.035	0.178 ± 0.030	0.152 ± 0.037	0.39–0.57	0.46 ± 0.10	0.138–0.165	0.154 ± 0.010	0.173
Period (C) from September 19 to October 2, 1991	0.234 ± 0.033	0.182 ± 0.032	0.140 ± 0.023	0.47–0.80	0.61 ± 0.14	0.102–0.154	0.126 ± 0.020	0.174
Period (D) from July 23 to August 11, 1992	0.160 ± 0.020	0.141 ± 0.009	0.102 ± 0.010	0.41–0.74	0.55 ± 0.09	0.087–0.112	0.099 ± 0.008	0.138

(D) *Period from July 23 to August 11, 1992*
Volz sun-photometer measurements were also performed in summer 1992 at the Pyramid Laboratory, about 13–14 months after the Mt. Pinatubo eruption in mid-June 1991. During this period, the daily mean values of $\tau_a(0.380\ \mu m)$ varied between 0.13 and 21, those of $\tau_a(0.500\ \mu m)$ between 0.12 and 0.16, and those of $\tau_a(0.875\ \mu m)$ between 0.08 and 0.13, showing that an appreciable decrease in $\tau_a(\lambda)$ occurred about one year after the first field measurements. Average values of $\tau_a(0.380\ \mu m) = 0.16 \pm 0.02$, $\tau_a(0.500\ \mu m) = 0.14 \pm 0.01$, and $\tau_a(0.875\ \mu m) = 0.10 \pm 0.01$ were measured in period (D), indicating that $\tau_a(\lambda)$ decreased by 25–30% compared to the average values obtained in period (C). The daily mean values of α varied between 0.41 and 0.74, yielding an average value of $\alpha = 0.55 \pm 0.09$, which is slightly lower than that found in period (C). Such variations clearly indicate that the strong extinction effects produced by accumulation and coarse particles of volcanic origin continued to play a significant role in determining the atmospheric turbidity conditions of the atmosphere during summer 1992, as a result of the 11-month aging processes that appreciably modified the microphysical properties of volcanic particles in the low stratosphere. This is also clearly confirmed by the variations in β, which ranged mainly between 0.09 and 0.11 during period (D), giving an average value of 0.10 ± 0.01, which was about 20% lower than that measured in period (C).

(2) *Determination of the columnar aerosol refractive index*

Taking into account the spectral features of the daily mean values of $\tau_a(\lambda)$, α, and β measured at the CNR Pyramid Laboratory in 1991 and 1992, we decided to represent the columnar aerosol optical characteristics assuming that the columnar aerosol size-distribution curves are given in the four periods (A)–(D) by different linear combinations of a fine particle size-distribution and a coarse particle size-distribution, chosen among those defined by Tomasi et al. (2013) to represent the extinction features of BG aerosol particles and stratospheric particles of volcanic origin, respectively. Bearing in mind that the value of α closely depends on the shape-parameters of the aerosol size-distribution curves and the particulate optical characteristics, the number concentrations of fine and coarse particle modes were made to vary until fitting the average values of α obtained in the four periods (A)–(D). More precisely, the following assumptions were made to define the complex refractive index of columnar aerosol during the four periods:

(A) In the first period (A), characterized by predominant extinction due to BG tropospheric aerosol and the absence of an important volcanic particle load in the stratosphere, it was assumed that the overall columnar aerosol size-distribution is given by an appropriate linear combination of the bimodal rural aerosol (SF-R) model and the monomodal stratospheric BG aerosol (PV-1) model defined by Tomasi et al. (2013):

(i) The SF-R model was defined by Shettle and Fenn (1979) for representing the average rural particle load in the troposphere, composed of mass fractions equal to 70% water-soluble substances (ammonium and calcium sulfates, with organic compounds), about 28% dust-like particulate, and about 2% liquid water, since such calculations were made for air relative humidity RH = 50%. The optical characteristics of these tropospheric particles were determined using the refractive index evaluations of Volz (1972a, 1972b, 1973) for water-soluble substances and dust samples, which realistically represent the optical features of tropospheric aerosol particles above the CNR Pyramid Laboratory on summer clear-sky days. The spectral values of parameters $n(\lambda)$ and $k(\lambda)$ calculated for RH = 50% are given in Table 8.4b at the most significant visible and near-infrared wavelengths.

(ii) The PV-1 model was defined by Tomasi et al. (2013) to represent a polydispersion of BG aerosol particles suspended in the stratosphere, presenting features similar to those defined by Pueschel et al. (1989) from *in situ* observations made during a long volcanic quiescence period. The mass composition of the PV-1 particles was assumed to consist of 72% sulfuric acid, 24% liquid water, and 4% water-soluble (nitrate) substances. For these mass percentages, the spectral values of $n(\lambda)$ and $k(\lambda)$ were calculated over the 0.36–3.75-μm wavelength range for the various components, according to the evaluations of Palmer and Williams (1975), Hummel et al. (1988), Hale and Querry (1973), and Vermote et al. (1997) for dry air conditions, giving the results shown in Table 8.4b.

(B) In period (B), in the presence of fresh Mt. Pinatubo volcanic particles, the optical characteristics of columnar aerosol were represented in terms of a linear combination of (i) the SF-R model proposed by Shettle and Fenn (1979) to simulate the optical parameters of BG tropospheric aerosol and, (ii) the bimodal stratospheric volcanic aerosol extinction model (PV-2) defined by Tomasi et al. (2013) on the basis of the *in situ* sampling measurements performed by Pueschel et al. (1993) at an altitude of around 16.5 km in the stratosphere, about two months after the Mt. Pinatubo eruption of mid-June 1991, finding a bimodal volcanic particle load. The values of $n(\lambda)$ and $k(\lambda)$ relating to these stratospheric particles are given in Table 8.4b, as obtained according to the spectral evaluations made by Pueschel et al. (1993), who used the data of Palmer and Williams (1975) at wavelengths greater than 0.70 μm for an aqueous solution consisting of 75% sulfuric acid and 25% liquid water mass percentages. The lack of PV-2 particulate optical parameters over the 0.36–0.70-μm wavelength range was overcome by us assuming the values of $n(\lambda)$ and $k(\lambda)$ proposed by Hummel et al. (1988).

(C) In period (C), the optical characteristics of columnar aerosol were represented in terms of a linear combination of the SF-R bimodal model and the bimodal stratospheric volcanic aerosol model PV-2 used during period (B). Therefore, the spectral patterns of $n(\lambda)$ and $k(\lambda)$ were assumed to be the same as given in Table 8.4b for period (B).

(D) In period (D), the optical characteristics of columnar aerosol were represented in terms of a linear combination of the SF-R model used above to simulate the BG tropospheric aerosol optical characteristics in summer, and the bimodal stratospheric volcanic aerosol model PV-3, defined by Tomasi et al. (2013) on the basis of the evaluations obtained by Pueschel et al. (1993) through the analysis of a set of mid-latitude *in situ* sampling measurements performed at an altitude of around 12.5 km, about nine months after the Mt. Pinatubo

Table 8.4b. Average values of the real part $n(\lambda)$ and imaginary part $k(\lambda)$ of the complex refractive index determined at the three Volz sun-photometer wavelengths, and some other supplementary wavelengths, for the four multimodal size-distribution curves of the aerosol extinction models SF-R (Shettle and Fenn, 1979) defined for relative humidity RH = 50%, and of the stratospheric aerosol models PV-1, PV-2, and PV-3 proposed by Tomasi et al. (2013) and used to define the best-fit linear combinations of fine and coarse particle size-distributions giving the average values of $\alpha(0.380–0.875\ \mu m)$ obtained during the four selected periods. The spectral values of $n(\lambda)$ and $k(\lambda)$ for the volcanic particle loads in the low stratosphere were evaluated according to the estimates of Kinne and Pueschel (2001) and using the Palmer and Williams (1975) measurements.

Aerosol extinction model	Spectral values of refractive index real part $n(\lambda)$							Spectral values of refractive index imaginary part $k(\lambda)$						
	Wavelength (μm)							Wavelength (μm)						
	0.360	0.500	0.694	0.860	1.060	2.250	3.750	0.360	0.500	0.694	0.860	1.060	2.250	3.750
Rural SF-R, first mode	1.520	1.520	1.520	1.510	1.510	1.257	1.396	5.6×10^{-3}	5.6×10^{-3}	6.9×10^{-3}	1.0×10^{-2}	1.4×10^{-2}	9.2×10^{-3}	6.0×10^{-3}
Rural SF-R, second mode	1.520	1.520	1.520	1.510	1.510	1.356	1.396	5.6×10^{-3}	5.6×10^{-3}	6.9×10^{-3}	1.0×10^{-2}	1.4×10^{-2}	9.2×10^{-3}	6.0×10^{-3}
BG stratospheric PV-1 aerosol unimodal model	1.455	1.436	1.433	1.429	1.426	1.370	1.398	2.0×10^{-4}	2.0×10^{-4}	2.8×10^{-4}	4.8×10^{-4}	5.7×10^{-4}	2.1×10^{-3}	1.3×10^{-3}
Volcanic stratospheric PV-2 (bimodal) and PV-3 (trimodal) aerosol models	1.452	1.432	1.429	1.426	1.422	1.368	1.396	1.1×10^{-8}	1.1×10^{-8}	2.1×10^{-8}	1.8×10^{-7}	1.5×10^{-6}	1.8×10^{-3}	1.3×10^{-1}

eruption. Thus, model PV-3 seems to be particularly suitable for representing the radiative parameters of a volcanic particle load of 11-month age, like the one observed above the CNR Pyramid Laboratory in summer 1992. In practice, due to a composition consisting of mass percentages of 75% sulfuric acid and 25% liquid water, the spectral values of $n(\lambda)$ and $k(\lambda)$ determined by Tomasi et al. (2013) were closely similar to those assumed for the PV-2 model, according to the Palmer and Williams (1975) and Hummel et al. (1988) evaluations. They are provided in Table 8.4b.

(3) *Determination of the size-distribution curves of columnar aerosol*

Bearing the above assumptions and remarks in mind, the multimodal size-distribution curves of columnar aerosol were defined for the four periods as linear combinations of fine and coarse particle size-distributions: (i) in period (A), as a linear combination of the SF-R rural aerosol model and the PV-1 stratospheric aerosol model; (ii) in periods (B) and (C), as linear combinations of the SF-R rural aerosol model and the PV-2 stratospheric volcanic aerosol model; and (iii) in period (D), as a linear combination of the SF-R rural aerosol model and the PV-3 stratospheric nine-month-old volcanic aerosol model.

The shape-parameters of the log-normal modes giving form to the bimodal SF-R, monomodal PV-1, bimodal PV-2, and trimodal PV-3 size-distribution curves are given in Table 8.4c. The total particle number density size-distribution curves $N(r) = \mathrm{d}N/\mathrm{d}(\ln r)$ and the total particle volume size-distribution curves $V(r) = \mathrm{d}V/\mathrm{d}(\ln r)$ of tropospheric aerosol model SR-R and stratospheric aerosol models PV-1, PV-2, and PV-3, and the corresponding spectral curves of $n(\lambda)$, $k(\lambda)$, $\beta_{ext}(\lambda)$, and $\omega(\lambda)$ were assumed to be those determined by Tomasi et al. (2013). Using the 6S radiative transfer code of Vermote et al. (1997) for the spectral values of $n(\lambda)$ and $k(\lambda)$ given in Table 8.4b and the log-normal size-distribution shape-parameters given in Table 8.4c, the values of volume extinction coefficients $\beta_{ext}(0.380\ \mu\mathrm{m})$, $\beta_{ext}(0.500\ \mu\mathrm{m})$, and $\beta_{ext}(0.860\ \mu\mathrm{m})$ were then calculated for all the four aerosol models, obtaining the values given in Table 8.4c, which were determined for the overall columnar number contents obtained for the average values of $\tau_a(\lambda)$ measured during periods (A)–(D) (see Table 8.4a). It is interesting to note that the best-fit values of exponent α calculated for the spectral triplets of $\beta_{ext}(\lambda)$ in Table 8.4b are equal to 1.193 for the SF-R aerosol model and 1.688 for the PV-1 aerosol model, suggesting that the extinction effects simulated by the linear combination of the two models are predominantly due to fine particles. The best-fit values of exponent α obtained for models PV-2 and PV-3 are rather low, being equal to 0.186 in the first case and 0.054 in the second case, clearly indicating that the spectral extinction features simulated using these two aerosol models are nearly neutral, being predominantly produced by coarse particles. Considering that exponent α closely depends on the shape-parameters of the columnar aerosol size-distributions and to a lesser extent on the particulate optical characteristics, the columnar aerosol size-distributions were realistically represented by appropriate linear combinations of a tropospheric aerosol model with a prevailing content of fine particles (SF-R model) and one of the stratospheric aerosol models PV-1, PV-2, and PV-3, all presenting predominant contents of coarse particles. The average estimate of α determined

Table 8.4c. Values of the shape-parameters of the unimodal log-normal size-distribution curves and the main physical and optical parameters giving form to the four SF-R, PV-1, PV-2, and PV-3 aerosol models proposed in the present study, where N_j is number density concentration of the j-th particle mode, σ is the geometric standard deviation of the log-normal curve, and r_c is the mode radius. The values of volume extinction coefficients $\beta_{ext}(0.380\ \mu m)$, $\beta_{ext}(0.500\ \mu m)$, and $\beta_{ext}(0.875\ \mu m)$ and Ångström's exponent α determined over the 0.40–0.86-μm wavelength range are given in the last four columns for overall number density $N = 1000$/cm in all the four aerosol models.

Aerosol models	Shape parameters of the log-normal monomodal size-distributions									β_{ext}	β_{ext}	β_{ext}	Expo-
	First mode			Second mode			Third mode			(0.380 μm)	(0.500 μm)	(0.875 μm)	nent
	N_1 (per cm^3)	σ	r_c (μm)	N_2 (per cm^3)	σ	r_c (μm)	N_3 (per cm^3)	σ	r_c (μm)	(km^{-1})	(km^{-1})	(km^{-1})	α
(SF-R) Rural (mixture of water-soluble, and dust-like aerosols)	999.875	2.239	0.02748	0.125	2.500	0.4377	–	–	–	1.571×10^{-2}	1.132×10^{-2}	5.810×10^{-3}	1.19
PV-1 (BG stratospheric, unimodal)	1000	1.80	0.070	–	–	–	–	–	–	6.701×10^{-2}	4.216×10^{-2}	1.639×10^{-2}	1.69
PV-2 (Post-Pinatubo 2-month volcanic stratospheric, bimodal)	586.207	1.50	0.090	413.83	1.50	0.31	–	–	–	6.613×10^{-1}	5.826×10^{-1}	5.250×10^{-1}	1.86×10^{-1}
PV-3 (Post-Pinatubo 9-month volcanic stratospheric, trimodal)	757.785	1.40	0.060	214.53	1.50	0.18	27.682	1.20	0.75	1.962×10^{-1}	1.933×10^{-1}	1.875×10^{-1}	5.40×10^{-2}

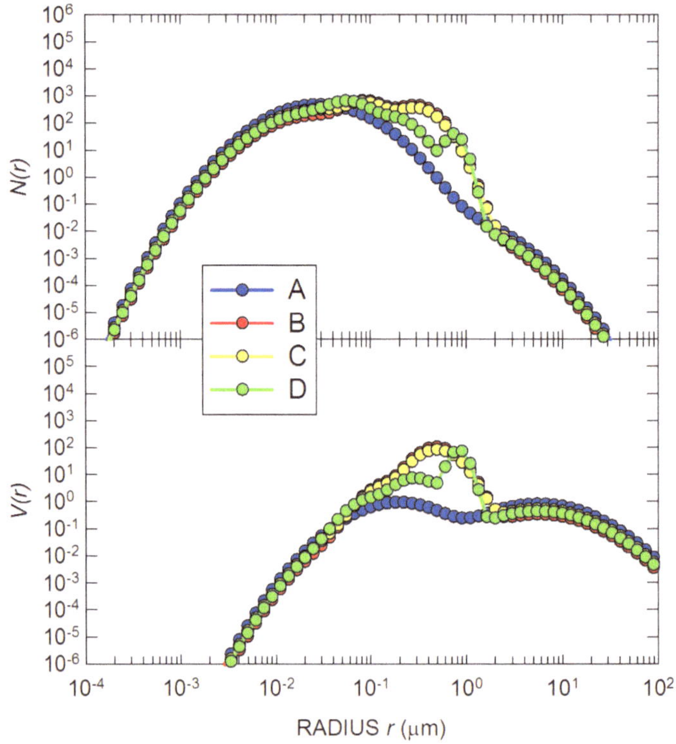

Fig. 8.27. Multimodal size-distribution curves of columnar particle number density $N(r) = \mathrm{d}N/\mathrm{d}(\ln r)$ measured per cm^2 (upper part) and columnar particle volume $V(r) = \mathrm{d}V/\mathrm{d}(\ln r)$ measured in cm (lower part), obtained as best-fit linear combinations of the SF-R, PV-1, PV-2, and PV-3 size-distribution curves (Tomasi et al., 2013) for the four periods (A) (blue circles), (B) (red circles), (C) (yellow circles), and (D) (green circles).

during period (A) was equal to 1.21 ± 0.10 and hence placed intermediately between the values of 1.193 and 1.688 found for the SF-R and PV-1 aerosol extinction models, respectively. Thus, a best-fit procedure was adopted to define the linear combination of models SF-R and PV-1 giving $\alpha = 1.21$. The multimodal best-fit size-distribution curve obtained is shown in Fig. 8.27, presenting a bimodal curve with a columnar number content $N_{SF-R} = 1.029 \times 10^4$ per cm^2 of SF-R tropospheric aerosol particles, combined with a third mode of stratospheric particles with $N_{PV-1} = 4.288 \times 10^2$ per cm^2.

Following a similar procedure, the best-fit linear combination of period (B) was determined for the average value of $\alpha = 0.46$, which was intermediate between the values of 1.193 and 0.186 given in Table 8.4c for the SF-R and PV-2 aerosol models, respectively. Here, the number density concentrations of the PV-2 and SF-R particles was changed step by step until obtaining a spectral series of the overall coefficient $\beta_{ext}(\lambda)$, giving the value of $\alpha = 0.46$. The multimodal best-fit size-distribution curve thus obtained is shown in Fig. 8.27, presenting a bimodal curve of tropospheric particles with $N_{SF-R} = 1.093 \times 10^2$ per cm^2, and a bimodal curve of stratospheric particles with $N_{PV-2} = 2.957 \times 10^2$ per cm^2.

Considering that the columnar aerosol extinction features did not substantially change on passing from period (B) to period (C), the size-distributions and radiative parameters of models SF-R and PV-2 were used to determine the best-fit solution for the average value of $\alpha = 0.61$ obtained in period (C). Applying the same best-fit procedure adopted in the previous period, the overall columnar number concentrations of SF-R and PV-2 aerosol particles were changed until obtaining a best-fit value of $\alpha = 0.61$, for $N_{SF-R} = 2.145 \times 10^2$ per cm^2 and $N_{PV-2} = 2.962 \times 10^2$ per cm^2. The overall four-modal best-fit size-distribution curves obtained using the procedure is shown in Fig. 8.27, for comparison with those achieved for the two previous periods.

During period (D), an average value of $\alpha = 0.55$ was found, which is slightly lower than that evaluated for period (C) and intermediate between the values of 1.193 and 0.054 estimated for the aerosol models SF-R and PV-3, respectively. This rather low value of α indicates that, in period (D), the overall columnar aerosol extinction predominantly came from stratospheric particles, gradually growing as a result of the aging processes occurring over the 11 months that separated period (C) from period (D). It is worth noting that the PV-3 size-distribution curve of these aged stratospheric particles exhibits three modes, with mode radii equal to 0.060 μm for fine particles, 0.18 μm for accumulation particles, and 0.75 μm for coarse particles. Applying the above best-fit procedure to the spectral series of $\beta_{ext}(\lambda)$ given in Table 8.4c for models SF-R and PV-3, the overall values of $N_{SF-R} = 5.248 \times 10^2$ per cm^2 and $N_{PV-3} = 6.680 \times 10^2$ per cm^2 were obtained. The overall five-modal size-distribution curve determined with the best-fit procedure is shown in Fig. 8.27, for comparison with the multimodal curves defined in the previous three observation periods. It consists of two modes of SF-R tropospheric particles and three modes of PV-3 aged volcanic stratospheric particles. Figure 8.27 also shows that a limited content of coarse particles was present in the vertical atmospheric column during period (A), while the coarse particle fraction was far more pronounced in periods (B) and (C), presenting clearly multimodal features, gradually becoming more marked during period (D), with three distinct modes over the 0.5–2.0-μm radius range.

The spectral curves of parameters $n(\lambda)$, $k(\lambda)$, $\beta_{ext}(\lambda)$, and $\omega(\lambda)$ determined for the four best-fit multimodal size-distribution curves described above are shown in Fig. 8.28 over the 0.36–3.75-μm wavelength range, providing a measure of the mean variations that characterized the spectral features of columnar aerosol extinction over the four periods. It can be seen that $n(\lambda)$ was found to have higher values in period (A) than in periods (B)–(D), during which it assumed comparable values over the whole spectral range. Correspondingly, higher values of $k(\lambda)$ were determined in period (A) than in the other periods, due to the fact that stratospheric volcanic particles scatter strongly and only slightly absorb short-wave radiation. As a result of such variations, as well as the considerable changes in the columnar aerosol size-distribution curves, coefficient $\beta_{ext}(\lambda)$ was estimated to increase appreciably at all wavelengths when passing from period (A) to the subsequent ones, owing to the gradual growth of volcanic particles in the low stratosphere.

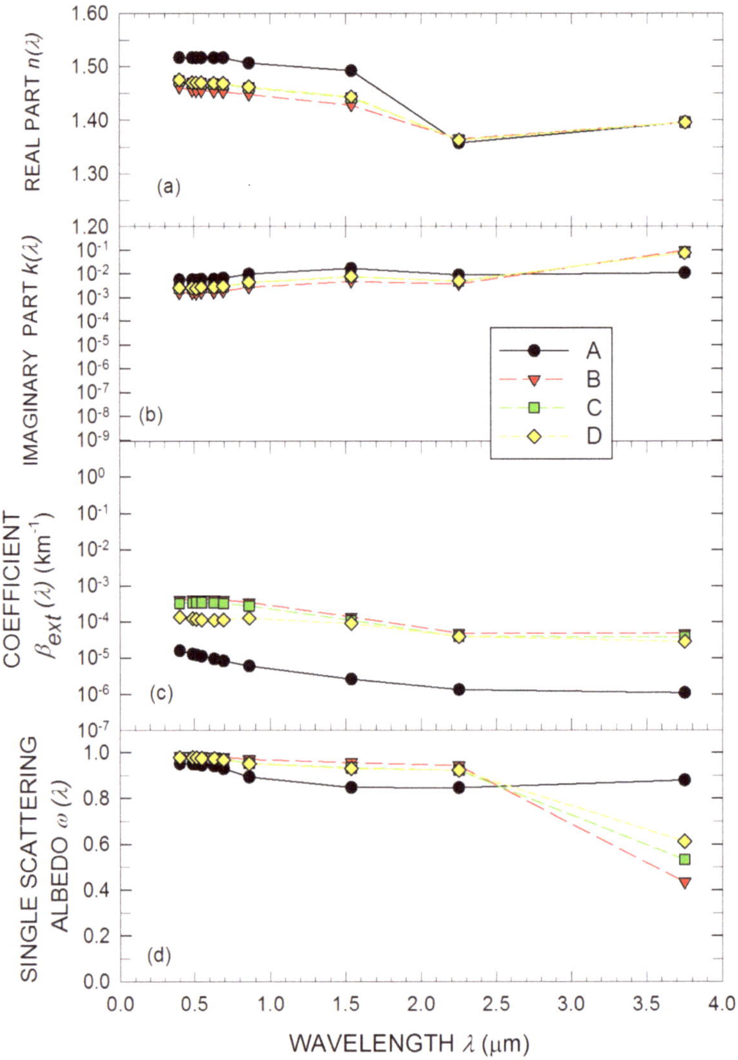

Fig. 8.28. Spectral curves of real part $n(\lambda)$ and imaginary part $k(\lambda)$ of particulate refractive index, volume extinction coefficient $\beta_{ext}(\lambda)$, and columnar aerosol single-scattering albedo $\omega(\lambda)$ determined for the multimodal best-fit aerosol extinction models obtained in the four measurement periods (A)–(D) of the Ev-K2-CNR campaigns conducted in summer 1991 and summer 1992, as defined in Table 8.4a.

(4) *Determination of the columnar aerosol single-scattering albedo*

In spite of the significant variations in the size-distribution curves of fine and coarse particles modes shown in Fig. 8.27, and the variations in $n(\lambda)$ and $k(\lambda)$ presented in Fig. 8.28, the columnar aerosol $\omega(\lambda)$ was found to assume very similar values on passing from one measurement period to another, with an average value of $\omega(0.55\ \mu m) = 0.945$ in period (A), for atmospheric turbidity conditions typical of optically predominant BG tropospheric fine particle loads, and with values of

Table 8.4d. Average values of the volume extinction coefficients $\beta_{ext}(0.380\ \mu m)$, $\beta_{ext}(0.500\ \mu m)$, and $\beta_{ext}(0.875\ \mu m)$ calculated for the overall columnar particle number contents derived from the values of $\tau_a(\lambda)$ given in Table 8.4a, and columnar aerosol single-scattering albedo $\omega(0.55\ \mu m)$, as determined for the four selected periods of the EV-K2-CNR campaigns conducted in 1991 and 1992, using (i) the complex refractive index parts $n(\lambda)$ and $k(\lambda)$ given in Table 8.4b, and (ii) the four best-fit multimodal size-distribution curves of aerosol particles obtained as linear combinations of the unimodal aerosol extinction models presented in Table 8.4c.

Measurement period	Average columnar volume extinction coefficients (per km)			Average columnar
	$\beta_{ext}(0.380\ \mu m)$	$\beta_{ext}(0.500\ \mu m)$	$\beta_{ext}(0.875\ \mu m)$	$\omega(0.55\ \mu m)$
Period (A) (July 24–26, 1991)	1.742×10^{-2}	1.250×10^{-2}	6.350×10^{-3}	0.945
Period (B) (July 28–August 5, 1991)	4.985×10^{-1}	4.394×10^{-1}	3.396×10^{-1}	0.984
Period (C) (September 19–October 2, 1991)	4.742×10^{-1}	3.564×10^{-1}	2.253×10^{-1}	0.975
Period (D) (July 23–August 11, 1992)	1.374×10^{-1}	1.182×10^{-1}	8.680×10^{-2}	0.975

$\omega(0.55\ \mu m)$ decreasing from 0.984 to 0.975 for volcanic coarse particles in periods (B)–(D). The spectral variations of $\omega(\lambda)$ are presented in Fig. 8.28 over the whole 0.40–3.75-μm wavelength range, showing that the spectral values of columnar SSA were lower during period (A) at all the visible and near-infrared wavelengths than those measured during periods (B) and (C), since the volcanic coarse particles had gradually grown with age.

(5) *Definition of the local surface albedo models*

Examining the MCD43C3 products derived from the MODIS Level 3.0 surface albedo data for summer over the Himalayan region and the surrounding areas, very different surface reflectance characteristics emerged over three typical areas. In fact, this mountainous region can be subdivided according to the different coverage conditions of the surface, associated with diverse altitudes and precipitation regimes, causing very different reflection effects on the short-wave radiation. Such areas are listed below:

(i) The Himalayan Mountain Region (HMR), which includes the entire Himalayan chain, whose southern part rises abruptly into the zone of perpetual snow, north of which the Himalayas soar upward to constitute a virtual wall beyond the snowline at 5000–5500 m, with numerous peaks over 7000 m and eight peaks over 8000 m. The HMR exhibits different vegetation types covering the surface at relatively low altitudes, mainly consisting of shrublands, mountain grasslands and meadows, and rocky terrains, snow fields, and glaciers at the highest elevations, also during summer. The surface albedo evaluations derived from the MCD43C3 products recorded in September over a 2° latitude × 2° longitude pixel centered at the CNR Pyramid Laboratory (covering roughly the 27°–29° N latitude and the 86°–88° E longitude intervals) allowed us to define the average surface albedo map shown in Fig. 8.29.

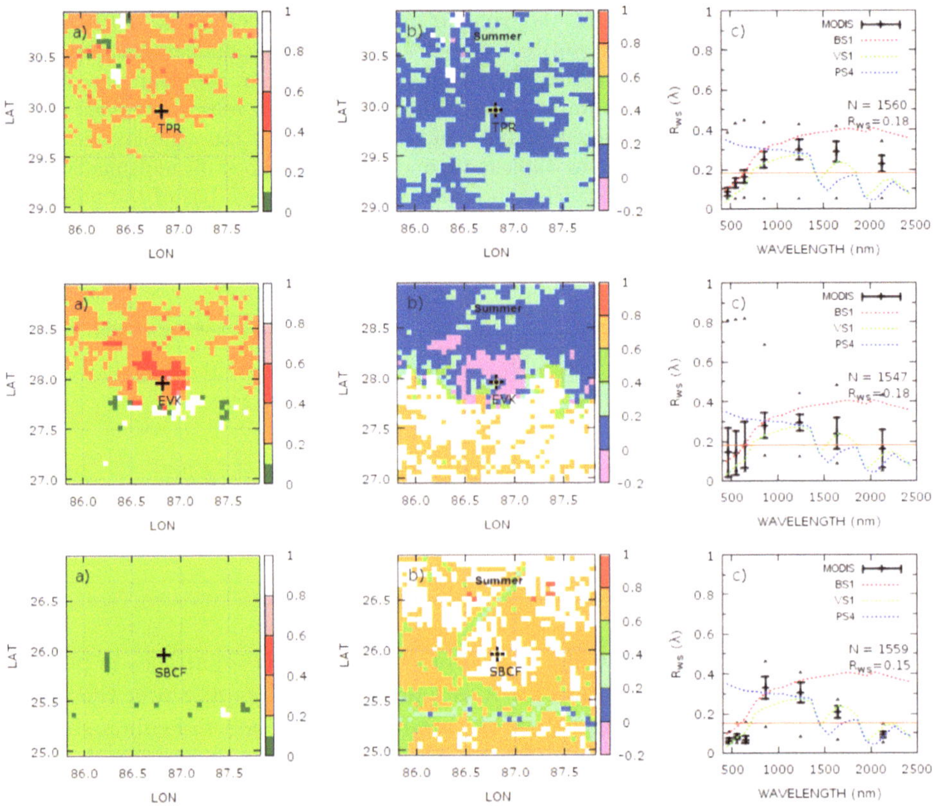

Fig. 8.29. Maps of the average land surface albedo ((a) in the left column) and Normalized Difference Vegetation Index (NDVI) ((b) in the central column) obtained from the MODIS Level 3.0 surface albedo data (MCD43C3 products) recorded in late summer 2009 (from the 269th to 285th day of year) over pixels of 2° latitude × 2° longitude, including the Tibetan Plateau Region (TPR) (upper part), the Himalayan Mountain Region (HMR) (middle part), and the Subtropical Broadleaf and Coniferous Forests area (SBCF) (lower part). The crosses labelled TPR, EVK, and SBCF indicate the geographical positions of the central sites of the three regions under study. The graphs reported in the right column provide the spectral values (black vertical bars) of the average white-sky albedo R_{ws}, found over the TPR, HMR, and SBCF regions to give mean values of 0.18, 0.18, and 0.15, respectively, retrieved for best-fit solutions with standard deviation values of around 0.1, and minimum and maximum values indicated by small triangles at each of the central wavelengths of the seven MODIS channels. The three best-fit spectral curves of $R_{ws}(\lambda)$ are also shown, as obtained for the surface albedo model BS1 (dotted red curve), chosen for the TPR area, model PS4 (dotted blue curve) chosen for the HMR area, and model VS1 (dotted green curve) chosen for the SBDF area.

(ii) The Subtropical Broadleaf and Coniferous Forests (SBCF) area, which occupies the southern part of Nepal and is situated south of the HMR area, mostly between 700 and 4000 m altitudes. These steep southern Himalayan slopes are nearly uninhabited, and exhibit different vegetation coverages depending on altitude: (a) tropical savannas along the Indian border, (b) subtropical broadleaf species growing between 500 and 1000 m, coniferous forests growing in the hills, including pine forests between 1000 and 2000 m altitudes, and subtropical forests dominating the lower elevations in summer, constitute a patchwork running east–west across Nepal, and (c) temperate broadleaf forests at elevations >3000 m, and coniferous forests on the lower Himalayan slopes. Examining the MCD43C3 products derived from the MODIS Level 3.0 surface data for summer, the surface albedo map in Fig. 8.29 was achieved to characterize the surface reflectance over such an area covered by high-depth canopies.

(iii) The Tibetan Plateau Region (TPR) is located north of the Himalayan range, and is characterized by very arid terrains covered by a high-altitude steppe interspersed with mountain reliefs and large brackish lakes, caused by low annual precipitation, usually ranging between 100 and 300 mm. The main vegetation coverage of the steppe consists of grasslands, which sustainably support populations of nomadic herdsmen, although frost occurs for six months of the year. Analyzing the MCD43C3 products, derived from the MODIS Level 3.0 surface data, the surface albedo map in Fig. 8.29 was obtained for this high-altitude area in summer.

Examining the above results, we found that (i) the non-Lambertian surface reflectance model PS4 is the most realistic model among those determined by Tomasi et al. (2013) for representing the surface albedo characteristics of the HMR region, yielding the spectral curve shown in Fig. 8.30, associated with snow fields, rocky slopes, and glaciers; (ii) the non-Lambertian surface reflectance model VS1 represents very well the average surface albedo characteristics of the SBCF region in summer, yielding the spectral curve of surface albedo already shown in Fig. 8.5 for representing the canopy features monitored over southern Portugal (CLEAR-COLUMN experiment); and (iii) the non-Lambertian surface albedo model BS1 defined by Tomasi et al. (2013) for a sandy area can be correctly used to represent the average TPR surface reflectance characteristics of the high-altitude desert and arid area of the Tibetan Plateau.

The spectral curves of BRDF surface albedo $R_L(\lambda, \theta_o)$ of models PS4 and BS1 are shown in Fig. 8.30, defined for nine values of solar zenith angle θ_o, taken in steps of 10° over the $0° \leq \theta_o \leq 80°$ range: (i) model PS4 yields the values of white-sky albedo $R_{ws} = 0.296$, black-sky albedo $R_{bs}(\theta_o = 0°) = 0.214$, and $A(\theta_o = 60°) = 0.329$, which are typical of mixed snow-covered and rocky terrain; and (ii) model BS1 pertains to dry-sand surface reflectance characteristics, giving $R_{ws} = 0.237$, $R_{bs}(\theta_o = 0°) = 0.225$, and $A(\theta_o = 60°) = 0.240$, which are typical of the highly reflecting surfaces of the TPR area.

(6) *Calculations of the daily time-patterns of instantaneous DARF terms and diurnally averaged DARF effects*

The sun-photometers clear-sky measurements performed at the CNR Pyramid Laboratory did not revealed strong variations in $\tau_a(\lambda)$ during the sunlit period of each day. Therefore, in order to calculate the daily time-patterns of instantaneous DARF effects over the sunlit period of four typical clear-sky days of the four periods (A)–(D), it was assumed that:

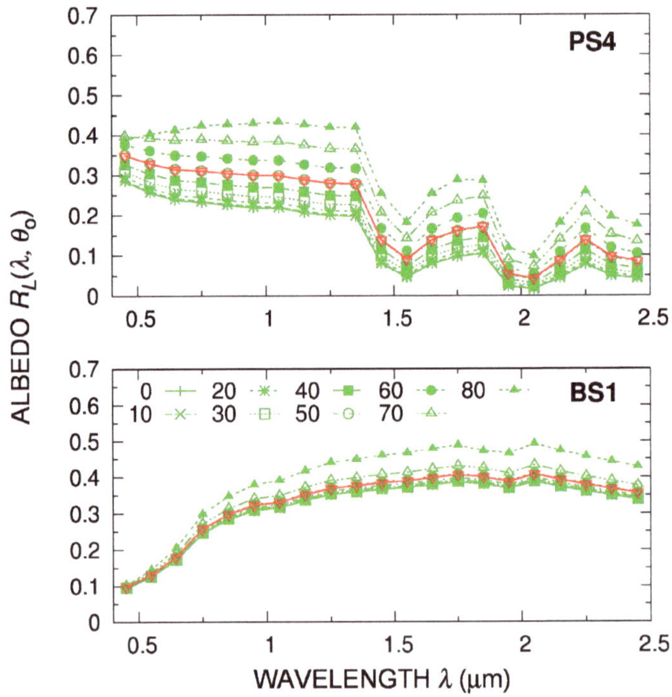

Fig. 8.30. Spectral curves of the BRDF surface albedo $R_L(\lambda, \theta_o)$ determined for the non-Lambertian surface albedo models PS4 (upper part) and BS1 (lower part) of Tomasi et al. (2013), and nine values of solar zenith angle θ_o, taken in steps of 10° over the 0°–80° range. Model PS4 was used to represent the surface reflectance characteristics of the Himalayan Mountain Region (HMR), and model BS1 those of the Tibetan Plateau Region (TPR) located north of the Himalayan chain, obtained as best-fit solutions examining the MCD43C3 products derived from the MODIS Level 3.0 data recorded in September 2009 over the two regions. The red solid curves give the spectral values of the white-sky albedo $R_{ws}(\lambda)$.

(i) The values of $\tau_a(\lambda)$ at the visible and near-IR wavelengths were constant during the sunlit period and equal to the daily average values given in Table 8.4a.

(ii) The time-patterns of solar zenith angle θ_o varied throughout each day, being calculated for the geographical coordinates (27° 57′ N, 86° 49′ E, 5050 m a.m.s.l.) of the CNR Pyramid Laboratory on July 25, 1991, for period (A); August 1, 1991, for period (B); September 26, 1991, for period (C); and August 3, 1992, for period (D).

(iii) The average multimodal size-distribution curves of columnar aerosol particles were equal to those determined at step (3) and shown in Fig. 8.27 for the four measurement periods, as determined for the shape-parameters of the various monomodal curves given in Table 8.4c.

(iv) The spectral curves of the radiative parameters of the multimodal size-distribution curves defined at step (3) were equal to those given in Table 8.4b and shown in Fig. 8.28.

(v) The spectral and geometrical characteristics of surface albedo were represented by using surface albedo model BS1 over the TPR area, model PS4 over the HMR area, and model VS1 over the SBCF area. The calculations of the instantaneous DARF effects were made during typical days of the four periods (A)–(D) employing the DARF-PROC procedure for the time-patterns of $\tau_a(\lambda)$ and solar zenith angle θ_o defined above during each typical day. The values of the diurnally averaged terms ΔDF_{ToA} at the ToA-level, ΔDF_{BoA} at the BoA-level, and ΔDF_{Atm} within the atmosphere, and those of aerosol fractional forcing AFF_{ToA} at the ToA-level were determined by a simple integration of the instantaneous DARF terms calculated over the various sunlit periods. The results are given in Table 8.4e, and shown in Fig. 8.31, to describe the sequence of the diurnally averaged DARF terms ΔDF_{ToA}, ΔDF_{BoA}, and ΔDF_{Atm} calculated for the four periods, separately for the surface albedo characteristics of models PS4, VS1, and BS1. It can be seen that ΔDF_{ToA} was found to decrease slowly from 1991 to 1992, while ΔDF_{BoA} assumes substantially stable values, leading to a gradual decrease in ΔDF_{Atm}. The estimates of ΔDF_{ToA}, ΔDF_{BoA}, and ΔDF_{Atm} are shown in Fig. 8.32 as a function of the average values of $\tau_a(0.55\ \mu m)$, to give a measure of the significant variations induced by the changes in $\tau_a(\lambda)$ and in the other aerosol optical parameters, for the different surface reflectance characteristics of the HMR, SBCF, and TPR areas.

Examining the airborne measurements of the optical parameters of Mt. Pinatubo volcanic aerosol particles at stratospheric altitudes of 20–25 km in March 1992 over the mid-latitude regions, Kinne and Pueschel (2001) estimated a mean value of effective mode radius $r_{eff} = 0.50\ \mu m$, a mean value of $\tau_a(0.55\ \mu m) = 0.20\ \mu m$, and a mean value of stratospheric SSA $\omega_S(0.55\ \mu m) = 1.00$. They obtained values of $\Delta DF_{ToA} = -4.7\ W/m^2$, $\Delta DF_{BoA} = -8.5\ W/m^2$, and $\Delta DF_{Atm} = +3.8\ W/m^2$, which are also given in Table 8.4e for comparison with the results found by us in periods (A)–(D). Comparing the estimates obtained in periods (C) and (D) with those made by Kinne and Pueschel (2001) in March 1992 (i.e. in a period that is intermediate between periods (C) and (D)), it can be noted that a close agreement exists between the present evaluations of the three DARF terms and those of Kinne and Pueschel (2001).

(7) *Calculations of the daily mean values of DARF efficiencies*

The values of DARF efficiencies E_{ToA}, E_{BoA}, and E_{Atm}, give in practice the rates at which the surface–atmosphere system is forced by unit $\tau_a(0.55\ \mu m)$ at the two extreme levels of the atmosphere. They were simply calculated by dividing the values of the diurnally averaged DARF terms ΔDF_{ToA}, and ΔDF_{BoA}, given in Table 8.4e, by the corresponding daily mean values of $\tau_a(0.55\ \mu m)$ shown in Table 8.4a. Similar efficiency values were obtained as ratios between the estimates of ΔDF_{ToA}, and ΔDF_{BoA} determined by Kinne and Pueschel (2001) and the mean value of $\tau_a(0.55\ \mu m) = 0.20$. The average values of efficiency E_{Atm} within the atmosphere were then calculated as differences between E_{ToA} and E_{BoA}, according to Eq. (8.4). All the above evaluations of E_{ToA}, E_{BoA}, and E_{Atm} are given in Table 8.4e, together with those determined over land by Kinne and Pueschel (2001), which are $E_{ToA} = -23.5\ W/m^2$, $E_{BoA} = -42.5\ W/m^2$, and $E_{Atm} = +19.0\ W/m^2$. All the average values of E_{ToA}, E_{BoA}, and E_{Atm} obtained over the four periods (A)–(D)

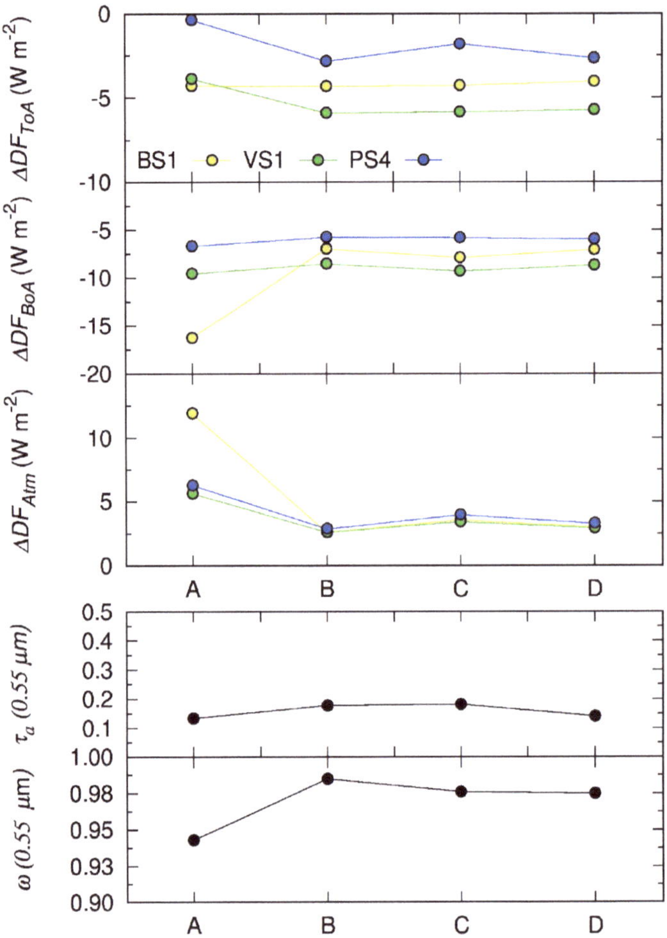

Fig. 8.31. Upper part: Daily values of diurnally averaged DARF terms ΔDF_{ToA} at the ToA-level, ΔDF_{BoA} at the BoA-level, and ΔDF_{Atm} within the atmosphere, calculated for the average multimodal aerosol size-distribution curves and aerosol radiative parameters determined in Tables 8.4a and 8.4c during the four periods (A)–(D) of the Ev-K2-CNR campaign conducted in summer 1991 and summer 1992. The values of ΔDF_{ToA}, ΔDF_{BoA}, and ΔDF_{Atm} were obtained for the three surface albedo models PS4 over the Himalayan Mountain Region (blue circles), VS1 over the Subtropical Broadleaf and Coniferous Forests area (dark green circles), and BS1 over the Tibetan Plateau Region (mustard yellow circles). Lower part: Time-patterns of the daily mean values of aerosol optical thickness $\tau_a(0.55\ \mu m)$ and columnar aerosol single-scattering albedo $\omega(0.55\ \mu m)$, determined in periods (A)–(D) on the basis of the average multimodal aerosol size-distribution curves shown in Fig. 8.28 and the aerosol radiative properties given in Tables 8.4b–d.

Table 8.4e. Daily values of the diurnally averaged aerosol forcing terms ΔDF_{ToA} at the ToA-level, ΔDF_{BoA} at the BoA-level, and ΔDF_{Atm} within the atmosphere, diurnal average aerosol fractional forcing AFF_{ToA} at the ToA-level (given by the ratio between flux change ΔF_{ToA} at the ToA-level and the incoming flux $I_S\downarrow$ of solar radiation at the ToA-level), and the diurnal average DARF efficiencies E_{ToA}, E_{BoA}, and E_{Atm} giving the rates at which the surface–atmosphere system is forced per unit $\tau_a(0.55\ \mu m)$, as obtained for the four median values of the aerosol radiative parameters defined in Table 8.4a during the four periods of the EV-K2-CNR project measurement campaign, using the aerosol optical data measured by Tomasi et al. (1997) for the surface albedo models BS1 (over the TPR area), VS1 (over the SBCF area), and PS4 (over the HMR area). The results are compared with those obtained for the Mt. Pinatubo volcanic particle extinction model defined by Kinne and Pueschel (2001) in March 1992 (i.e. for 9-month-aged stratospheric volcanic aerosol, after the Mt. Pinatubo eruption in mid-June 1991) and calculated for surface albedo characteristics similar to those of the vegetation-covered model VS2 of Tomasi et al. (2013).

Measurement period	Surface albedo model	Diurnal average DARF terms (W/m^2)			Diurnal average	Diurnal average DARF efficiencies (W/m^2)		
		ΔDF_{ToA}	ΔDF_{BoA}	ΔDF_{Atm}	AFF_{ToA}	E_{ToA}	E_{BoA}	E_{Atm}
Period (A) (July 24–26, 1991)	BS1	−4.3	−16.2	+12.0	-5.2×10^{-3}	−37.1	−139.7	+103.4
	VS1	−3.9	−9.5	+5.7	-9.4×10^{-3}	−33.6	−81.9	+49.1
	PS4	−0.4	−6.6	+6.3	-8.6×10^{-4}	−3.4	−56.9	+54.3
Period (B) (July 28–August 5, 1991)	BS1	−4.3	−6.9	+2.6	-1.1×10^{-2}	−24.9	−39.9	+15.1
	VS1	−5.9	−8.5	+2.6	-1.4×10^{-2}	−34.1	−49.1	+15.0
	PS4	−2.8	−5.7	+2.9	-6.9×10^{-3}	−16.2	−32.9	+16.8
Period (C) (September 19–October 2, 1991)	BS1	−4.3	−7.9	+3.6	-1.3×10^{-2}	−24.7	−45.4	+20.7
	VS1	−5.8	−9.3	+3.4	-1.7×10^{-2}	−33.3	−53.4	+19.5
	PS4	−1.8	−5.8	+4.0	-5.3×10^{-3}	−10.3	−33.3	+23.0
Period (D) (July 23–August 11, 1992)	BS1	−4.0	−6.8	+2.9	-1.1×10^{-2}	−29.0	−49.3	+21.0
	VS1	−5.5	−8.4	+2.8	-1.5×10^{-2}	−39.9	−60.9	+20.2
	PS4	−2.3	−5.5	+3.2	-6.3×10^{-3}	−16.7	−39.9	+23.2
March 1992	Similar to VS2 model	−4.7	−8.5	+3.8	Not given	−23.5	−42.5	+19.0

and in March 1992 are plotted versus $\omega(0.55\ \mu m)$ in Fig. 8.33, to provide evidence of the dependence patterns of these climate-forcing parameters on the SSA characteristics of columnar aerosol, for different surface albedo conditions and during various growth phases of stratospheric volcanic particles. The results clearly indicate that (i) E_{ToA}, assumes the highest values over the HMR area and the lowest over the SBCF area; (ii) E_{BoA} is higher over the HMR area than over the TPR and SBCF areas, where comparable values were obtained; and (iii) E_{Atm} assumes values decreasing on average as $\omega(0.55\ \mu m)$ increases, over both the HMR and SBCF areas, and clearly higher values over the TPR. Interestingly, considerably higher efficiency estimates were obtained in period (A) for BG summer aerosol particles at both tropospheric and stratospheric levels than in the subsequent three periods

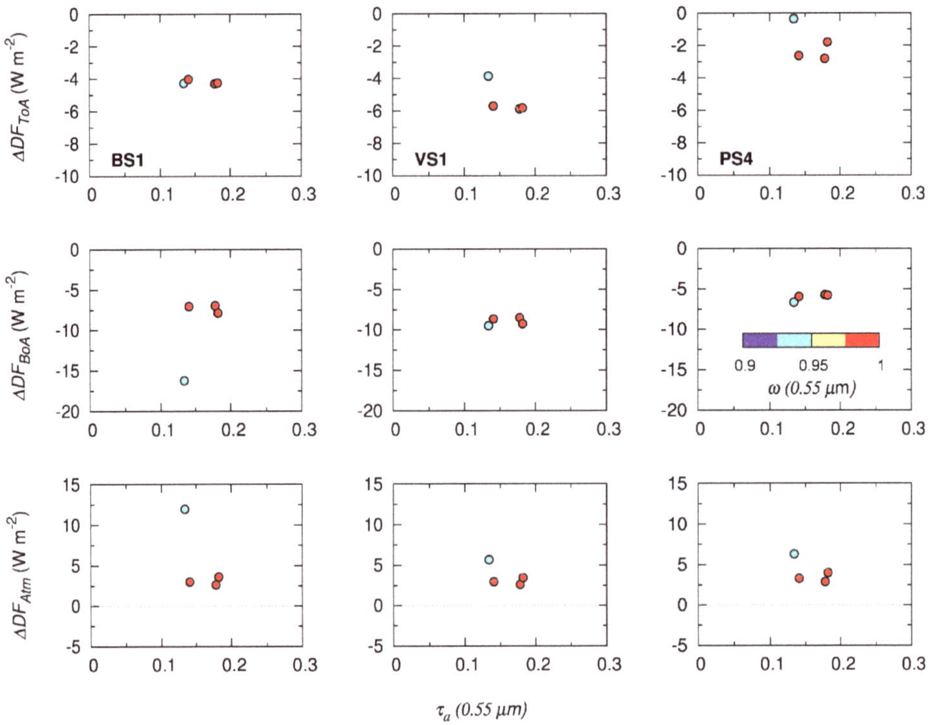

Fig. 8.32. Scatter plots of the daily values of DARF terms ΔDF_{ToA} (upper part), ΔDF_{BoA} (middle part), and ΔDF_{Atm} (lower part) plotted versus the daily mean values of aerosol optical thickness $\tau_a(0.55\ \mu\mathrm{m})$, determined from the Volz sun-photometer measurements performed during the four periods (A)–(D) of the two Ev-K2-CNR field campaigns at the Pyramid Laboratory (5050 m a.m.s.l.) (Himalaya, Nepal), and calculated for the surface albedo models BS1 over the TPR area (first column), VS1 over the SBCF area (second column), and PS4 over the HMR area (third column). The values shown in the three columns were obtained for different values of columnar aerosol single-scattering albedo $\omega(0.55\ \mu\mathrm{m})$, which are indicated using differently colored circles, the colors being chosen according to those shown in the color scale reported on the right.

characterized by the optical predominance of stratospheric volcanic particles. This indicates that the formation of volcanic aerosol layers in the low stratosphere can produce an increase in the spectral values of $\tau_a(\lambda)$, because of the more pronounced scattering effects of the stratospheric coarse particles, although with more moderate efficiency effects. Is is also worth noting that the DARF efficiency evaluations made in periods (C) and (D) closely agree, at the ToA- and BoA-levels and in the atmosphere, with those made by Kinne and Pueschel (2001) in March 1992, with relative discrepancies not exceeding ±10% for all three surface albedo models adopted in the present calculations. In addition, the present DARF efficiency evaluations in Table 8.4c show that all three DARF efficiency parameters slowly increased as the stratospheric volcanic particles gradually aged from summer 1991 to summer 1992.

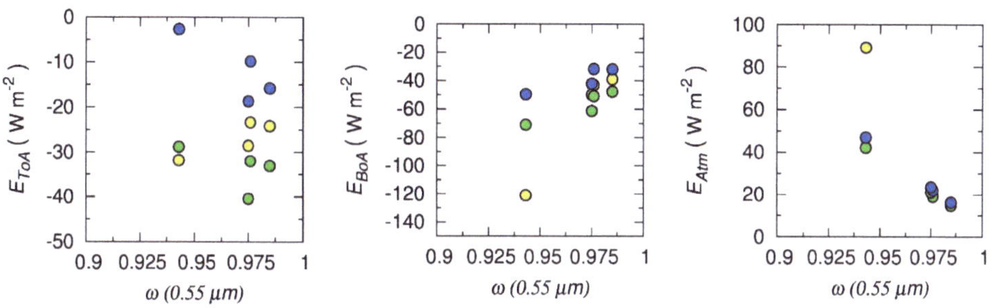

Fig. 8.33. Scatter plots of the daily mean values of DARF efficiency parameters E_{ToA} at the ToA-level, E_{BoA} at the BoA-level, and E_{Atm} within the atmosphere versus the columnar aerosol single-scattering albedo $\omega(0.55\ \mu m)$, as obtained for the four measurement periods (A)–(D) at the CNR Pyramid Laboratory in 1991 and 1992, by using the surface albedo models PS4 over the Himalayan Mountain Region (HMR) area (blue circles), BS1 over the Tibetan Plateau Region (TPR) area (mustard yellow circles), and VS1 over the Sub-tropical Broadleaf Coniferous Forest (SBCF) area (dark green circles).

(8) *Calculations of the DARF effects at the Nepal Climate Observatory-Pyramid (NCOP) site over the 2006–2009 period*

The values of the DARF efficiency terms given in Table 8.4e for period (A) were determined for the columnar BG aerosol without the Mt. Pinatubo volcanic particles. To give a picture of the DARF effects induced by the BG tropospheric aerosol over the Himalayan region for the entire year, the measurements performed by Marcq et al. (2010) are briefly presented here. Accurate observations of aerosol optical features were carried out by Marcq et al. (2010) from November 2006 to March 2009 at the renovated station of the CNR Pyramid Laboratory, currently named with the acronym NCOP. Regular measurements of $\tau_a(\lambda)$ have been taken at visible and near-IR wavelengths by Gobbi et al. (2010) with the AERONET sun-photometer model since 2006, while regular measurements of the light absorption and scattering coefficients of aerosol particles were derived by Marcq et al. (2010) from ground-level measurements taken using an integrating nephelometer (model TSI 3563) and a Multi-Angle Absorption Photometer (MAAP 5012), respectively, deriving from them evaluations of the ground-level SSA ω_o in the visible. The measurements were subdivided by Marcq et al. (2010) into three seasonal sets: (i) a pre-monsoon data set, collected in April and May; (ii) a monsoon data set, recorded from June to September; and (iii) a post-monsoon data set, collected in October and November, while very poor data were achieved during the winter season from December to February. A classification of the aerosol transport events was made for each annual observational period from April to November, subdividing the experimental data sets into the following three subsets: (i) the BG aerosol transport episodes, with air masses generally flowing from the Tibetan Plateau, characterized by very low particle concentrations; (ii) the regional pollution (RP) episodes associated with air masses moving from the Nepal plains and driven by the timing of thermal-induced ventilation from the valley towards higher altitudes; and (iii) special transport events (SE) often associated with long-range transport of polluted aerosols, not necessarily linked to the mountain breeze circulation. For

Table 8.4f. Seasonal average values of the diurnally averaged aerosol forcing terms ΔDF_{ToA} at the ToA-level, ΔDF_{BoA} at the BoA-level, and ΔDF_{Atm} within the atmosphere, seasonal average values of aerosol optical thickness $\tau_a(0.55\ \mu m)$, seasonal average values of ground-level single-scattering albedo ω_o in the visible, and diurnal average DARF efficiencies E_{ToA}, E_{BoA}, and E_{Atm} giving the rates at which the surface–atmosphere system is forced per unit $\tau_a(0.55\ \mu m)$, as obtained by Marcq et al. (2010) for snow-covered surface (S) and rocky surface (R) albedo conditions during the pre-monsoon (PRE), monsoon (MON), and post-monsoon (POS) seasonal periods from November 2006 to March 2009, for the aerosol loads observed during the background aerosol (BG) events, regional polluted aerosol (RP) events and special pollution aerosol (SP) events.

Measurement period	Surface albedo type	Average DARF terms (W/m^2)			Average τ_a(0.55 μm)	Average ground-level ω_o	Diurnal average DARF efficiencies (W/m^2) calculated per unit τ_a(0.55 μm)		
		ΔDF_{ToA}	ΔDF_{BoA}	ΔDF_{Atm}			E_{ToA}	E_{BoA}	E_{Atm}
PRE–BG	S	+2.1	−1.6	+3.7	0.010	0.89	+210.0	−16.0	+226.0
	R	+0.2	−1.7	+1.9	0.010	0.89	+20.0	−17.0	+37.0
PRE–RP	S	+5.9	−4.4	+10.2	0.025	0.84	+236.0	−176.0	+412.0
	R	+0.7	−4.4	+5.2	0.025	0.84	+28.0	−176.0	+204.0
PRE–SP	S	+22.6	−19.5	+42.1	0.093	0.82	+243.0	−209.7	+452.7
	R	+3.3	−19.0	+22.3	0.093	0.82	+35.5	−204.3	+239.8
MON–BG	R	+0.9	−6.2	+7.1	0.031	0.76	+29.0	−200.0	+229.0
MON–RP	R	+1.3	−8.8	+10.1	0.043	0.81	+30.2	−204.7	+234.9
MON–SP	R	+3.8	−19.4	+23.2	0.077	0.88	+49.4	−251.9	+301.3
POS–BG	R	0.0	−1.0	+1.0	0.008	0.85	0.0	−125.0	+125.0
POS–RP	R	+0.1	−3.4	+3.5	0.023	0.85	+4.3	−147.8	+152.1
POS–SP	R	+1.3	−17.2	+18.5	0.098	0.84	+13.3	−175.5	+188.8

the three classes of transport conditions, Marcq et al. (2010) calculated the diurnally averaged DARF terms ΔDF_{ToA}, ΔDF_{BoA}, and ΔDF_{Atm} by assuming two surface albedo scenarios around the NCOP site, characterized by (i) snow-covered mountain surface (S model) and (ii) a terrain covered by rocks without snow (R model). Their results are reported in Table 8.4f for all the combinations of seasonal periods, aerosol transport events, and surface albedo characteristics. The average daily mean values of $\tau_a(0.55\ \mu m)$ and average ground-level SSA ω_o assumed by Marcq et al. (2010) are also given in Table 8.4f. Comparing such results with those obtained by us in Table 8.4a during period (A), namely at the end of July 1991 for BG summer aerosol only, it can be noted that the values of $\tau_a(0.55\ \mu m)$ obtained in late July 1991 on three measurement days only are considerably higher than the average values of $\tau_a(0.55\ \mu m)$ reported by Marcq et al. (2010). In view of these great discrepancies, the evaluations of the DARF efficiency parameters obtained by dividing the Marcq et al. (2010) values of ΔDF_{ToA}, ΔDF_{BoA}, and ΔDF_{Atm} by the corresponding seasonal average values of $\tau_a(0.55\ \mu m)$ turned out to be surprisingly high. In fact, (i) the diurnal average values of E_{ToA} were found to vary between a nearly null value determined for BG aerosol over surface R in the post-monsoon period, and a value of around 240 W/m^2 determined for SE aerosol over surface S during the pre-monsoon season; (ii) the diurnal average values of E_{BoA} were found to be all negative, varying between -16 W/m^2 for BG aerosol conditions over surface S in the pre-monsoon period, and around -252 W/m^2 for SE aerosol conditions over surface R during the monsoon season; and (iii) the diurnal average values of E_{Atm} were found to be mainly positive in all cases, suggesting the occurrence of intense heating processes within the troposphere above the NCOP site, varying between a minimum of $+125$ W/m^2 for BG aerosol over surface R in the post-monsoon period, and a value of around $+453$ W/m^2 obtained for SE aerosol conditions over surface S in the pre-monsoon season. Such very high efficiency values can be plausibly ascribed to the use of the very low average values of $\tau_a(0.55\ \mu m)$ of Marcq et al. (2010) compared to those measured by Tomasi et al. (1997), the discrepancies being mostly comparable with the experimental errors usually made when using sun-photometers to measure $\tau_a(\lambda)$ at such high altitudes.

8.3.5 DARF evaluations from the POLAR-AOD project measurements performed at Arctic and Antarctic sites

The POLAR-AOD IPY project was one of the International Polar Year (IPY) projects endorsed by the IPY Committee in November 2005. It was designated as one of the lead projects in the cluster devoted to 'Cloud, aerosol and atmospheric chemistry', with the main objective of 'characterising the means, variability and trends of the climate-forcing effects induced by aerosols in Polar regions'. Regular and precise measurements of $\tau_a(\lambda)$ were performed at several visible and near-IR wavelengths during the POLAR-AOD campaigns conducted at various Arctic and Antarctic sites, from which measurements of Ångström's exponent α were obtained, together with additional measurements of aerosol physical, chemical, and optical characteristics.

The DARF effects induced by polar aerosols were evaluated from these measurements by following the seven-step DARF-PROC procedure described in the previous mid-latitude experiments. The main purpose was to determine the daily time-patterns of instantaneous DARF terms and the diurnally averaged DARF effects for various polar surface albedo models, and to evaluate the daily mean values of DARF efficiency at the various levels. The analysis of experimental data was made as follows:

(1) *Analysis of field data to determine the columnar aerosol extinction parameters*
The daily mean values of $\tau_a(\lambda)$ at visible and near-IR wavelengths were determined by examining the daily sets of sun-photometer measurements performed by the various POLAR-AOD research groups over the past 10 years (Tomasi et al., 2007, 2012). The most significant estimates of $\tau_a(0.50\ \mu m)$ and exponent α were carried out at the following 10 polar sites:

(i) Barrow, in northern Alaska, where measurements of $\tau_a(\lambda)$ and α were carried out by NOAA/GMD (Boulder, Colorado, US) employing the SP02 and SP022 Carter Scott sun-photometers from February 2002 to October 2010 (Tomasi et al., 2012). Four different types of atmospheric turbidity conditions were studied during the various seasons by determining the daily mean values of $\tau_a(0.50\ \mu m)$ and $\alpha(0.368–0.862\ \mu m)$ for: (a) Arctic Haze (AH), mainly from

January to May, obtaining the average values of $\tau_a(0.50\ \mu\text{m}) = 0.116$ and $\alpha = 1.28$; (b) BG summer aerosol, finding the average values of $\tau_a(0.50\ \mu\text{m}) = 0.078$ and $\alpha = 1.40$; (c) Asian Dust (AD), observed during some transport episodes monitored most frequently in spring 2002, for which a peak value of $\tau_a(0.50\ \mu\text{m}) = 0.20$ was estimated by Stone et al. (2007), with a mean value of $\alpha = 0.80$, clearly due to prevailing extinction effects by coarse particles; and (d) Boreal Forest Fire smoke (hereinafter referred to as BFF), for which the average value of $\tau_a(0.50\ \mu\text{m}) = 0.30$ was measured in July 2004, with an average value of $\alpha = 1.20$, which clearly indicates that the most pronounced optical effects were produced by fine particles (Tomasi et al., 2007).

(ii) Ny-Ålesund, in Spitsbergen (Svalbard), where measurements of $\tau_a(\lambda)$ and α were performed by AWI (Bremerhaven, Germany) using the SP1A and SP2H sun-photometers and the STAR01 star-photometer from May 2006 to September 2010 (Herber et al., 2002). Three different sets of atmospheric turbidity conditions were defined for: (a) AH in winter and spring, with average values of $\tau_a(0.50\ \mu\text{m}) = 0.080$ and $\alpha(0.380\text{–}0.864\ \mu\text{m}) = 1.25$; (b) Arctic Dense Summer (ADS) aerosol, with average values of $\tau_a(0.50\ \mu\text{m}) = 0.12$ and $\alpha = 1.00$; and (c) BG summer aerosol, with average values of $\tau_a(0.50\ \mu\text{m}) = 0.041$ and $\alpha = 1.20$.

(iii) Summit, on the central Greenland ice sheet, where measurements of $\tau_a(\lambda)$ and α have been performed since 2002 by PMOD/WRC (Davos, Switzerland) using PFR sun-photometers, yielding the average values of $\tau_a(0.50\ \mu\text{m}) = 0.039$ and $\alpha = 1.48$ for BG summer aerosol.

(iv) Sodankylä, in northern Finland, where measurements of $\tau_a(\lambda)$ and α were performed by FMI (Helsinki, Finland) during various campaigns from 2004 to 2012, using PFR sun-photometers, finding average values of $\tau_a(0.50\ \mu\text{m}) = 0.066$ and $\alpha = 1.25$ for AH episodes observed in winter and early spring, and of $\tau_a(0.50\ \mu\text{m}) = 0.060$ and $\alpha = 1.42$ for BG summer aerosol.

(v) Tiksi, in north-central Siberia, where measurements of $\tau_a(\lambda)$ and α were taken in summer 2010 with the AERONET Cimel CE-318 sun-photometer of NASA/GSFC (Holben et al., 1998), obtaining the seasonal average values of $\tau_a(0.50\ \mu\text{m}) = 0.085$ and $\alpha = 1.60$ for BG summer aerosol.

(vi) Mario Zucchelli, on the Terra Nova Bay (Ross Sea, Victoria Land) in Antarctica, where regular measurements were performed by Tomasi et al. (2007) using various sun-photometer models (UVISIR-2, FISBAT, PREDE POM-01L, and ASP-15WL), having the technical characteristics described by Tomasi et al. (2012). The measurements were carried out during various austral summer campaigns from December 1987 to January 2005, from which the average values of $\tau_a(0.50\ \mu\text{m}) = 0.030$ and $\alpha = 0.90$ were obtained for BG austral summer aerosol.

(vii) Neumayer, on the Weddel Sea coast, where regular sun-photometer measurements were carried out by AWI (Bremerhaven, Germany) from January 1991 to April 2004 using the same sun-photometer models employed at Ny-Ålesund (Herber et al., 1993, 2002). Average values of $\tau_a(0.50\ \mu\text{m}) = 0.043$ and $\alpha = 0.68$ were determined for BG austral summer aerosol.

(viii) Mirny, on the Davis Sea coast, where routine sun-photometer measurements were performed by AARI (St. Petersburg, Russia) during six field campaigns conducted in the austral summer period from January 1994 to December 2009, using sun-photometer models manufactured at AARI (Radionov et al., 2002). The average values of $\tau_a(0.50\ \mu m) = 0.023$ and $\alpha = 1.35$ were obtained from these measurements, to characterize the radiative properties of BG austral summer aerosol.

(ix) Dome Concordia (Dome C) on the East Antarctic Plateau, where sun-photometer measurements were performed in January 2003 and throughout austral summer 2006 by NOAA/GMD (Boulder, Colorado, US) using the SP01-A and SP02 Carter Scott sun-photometers. The average values of $\tau_a(0.50\ \mu m) = 0.019$ and $\alpha = 1.77$ were determined for the BG austral summer aerosol at high altitudes.

(x) South Pole (Amundsen Scott Station, US), where sun-photometer measurements were carried out by NOAA/GMD (Boulder, Colorado, US) from November 2001 to March 2010, using the SP02 Carter Scott sun-photometer. Average values of $\tau_a(0.50\ \mu m) = 0.018$ and $\alpha = 1.49$ were obtained for the BG austral summer aerosol.

The geographical coordinates and altitudes of the five Arctic and five Antarctic sun-photometer stations are listed in Table 8.5a, together with the average values of $\tau_a(0.50\ \mu m)$ and α determined at the various polar sites for different aerosol loads. It is worth mentioning that the average values of $\tau_a(0.50\ \mu m)$ were usually obtained with standard deviations of around 0.05 at the Arctic coastal sites, no more than 0.02 at Summit and Antarctic coastal sites, and 0.01 at the two Antarctic Plateau sites, while the average values of α were determined with standard deviations of around 0.20 at the Arctic and Antarctic coastal sites, and 0.10 at the high-altitude sites (Summit, Dome C and South Pole).

(2) *Determination of the columnar aerosol refractive index*

Due to the great variety of both natural and anthropogenic particle sources in the Arctic (Hirdman et al., 2010) and Antarctica (Bodhaine, 1995; Weller and Wagenbach, 2007; Weller et al., 2008), the different aerosol types observed at the Arctic and Antarctic sites listed in Table 8.5a were assumed to consist of linear combinations of a fine/accumulation particle polydispersion and an accumulation/coarse particle polydispersion. The columnar aerosol number contents of the two particle classes have been then varied until fitting the average median values of $\tau_a(0.50\ \mu m)$ and exponent $\alpha(0.40\text{–}0.87\ \mu m)$ given in Table 8.5a. According to the above-mentioned measurements of particulate chemical composition, the fine/accumulation particles were assumed to consist mainly of water-soluble, soot, sea salt, and insoluble substances, and the accumulation/coarse particles of variable mass fractions of water-soluble, sea salt, insoluble matter, and mineral dust transported from remote mid-latitude regions. For these mixed characteristics of the multimodal size-distribution curves, it was assumed that the fine/accumulation and accumulation/coarse particles consist of the mass mixing percentages given in Table 8.5b. Thus, parameters $n(\lambda)$ and $k(\lambda)$ were calculated in terms of the mass composition percentages assumed by Hess et al. (1998) for the OPAC components calculated at RH = 50%, which present different chemical composition features for the size classes of the various particle types. On the basis of these assumptions, the monochromatic values of $n(\lambda)$ and $k(\lambda)$ were separately determined for

Table 8.5a. Daily median values of aerosol optical thickness $\tau_a(0.50\ \mu m)$ and Ångström's (1964) exponent α determined over the 0.40–0.87-μm wavelength range from the sun-photometer measurements performed at five Arctic sites and five Antarctic sites during the POLAR-AOD campaigns conducted for different atmospheric turbidity conditions.

Arctic or Antarctic site	Geographical coordinates and altitude	Aerosol type and season	Average median value of $\tau_a(0.50\ \mu m)$	Average median value of exponent $\alpha(0.40–0.87\ \mu m)$
Barrow (Alaska)	71° 19′ N, 156° 36′ W, 8 m a.m.s.l.	Arctic haze (winter–early spring)	0.116	1.28
		Asian dust	0.20	0.80
		Boreal forest fire smoke	0.30	1.20
		BG summer aerosol	0.078	1.40
Ny-Ålesund (Spitsbergen, Svalbard)	78° 54′ N, 11° 53′ E, 5 m a.m.s.l.	Arctic haze (winter–early spring)	0.080	1.29
		Arctic dense summer aerosol	0.12	1.00
		BG summer aerosol	0.041	1.20
Summit (Greenland ice sheet)	72° 20′ N, 38° 45′ W, 3270 m a.m.s.l.	BG summer aerosol	0.039	1.48
Sodankylä (Finland)	67° 22′ N, 26° 38′ E, 184 m a.m.s.l.	Arctic haze (winter–early spring)	0.066	1.25
		BG summer aerosol	0.060	1.42
Tiksi (northern-central Siberia)	71° 35′ N, 128° 47′ E, 40 a.m.s.l.	BG summer aerosol	0.085	1.60
Mario Zucchelli (Ross Sea coast, Victoria Land, Antarctica)	74° 42′ S, 164° 07′ E, 15 m a.m.s.l.	BG austral summer aerosol	0.030	0.90
Neumayer (Weddell Sea coast, Antarctica)	70° 39′ S, 08° 15′ W, 40 m a.m.s.l.	BG austral summer aerosol	0.043	0.68
Mirny (Davis Sea coast, Antarctica)	66° 33′ S, 93° 01′ E, 40 m a.m.s.l.	BG austral summer aerosol	0.023	1.35
Dome C (East Antarctic Plateau)	75° 06'S, 123° 21'E, 3233 m a.m.s.l.	BG austral summer aerosol	0.019	1.77
South Pole (Antarctic Plateau)	89° 59′ S, 139° 16′ E, 2835 m a.m.s.l.	BG austral summer aerosol	0.018	1.49

the fine/accumulation and accumulation/coarse particle components, by calculating the weighted averages of such values for the best-fit evaluations of the number concentrations of the fine/accumulation and accumulation/coarse particles, yielding the median values of $\tau_a(0.50\ \mu m)$ and $\alpha(0.40\text{–}0.87\ \mu m)$ given in Table 8.5a. The overall values of $n(0.50\ \mu m)$ and $k(0.50\ \mu m)$ obtained as linear combinations of such optical parameters calculated separately for the fine/accumulation and the accumulation/coarse particle fractions given in Table 8.5b are provided in Table 8.5c. The overall spectral curves of $n(\lambda)$ and $k(\lambda)$ are shown in Fig. 8.34 for the 11 Arctic aerosol types, and in Fig. 8.35 for the five types of BG austral summer aerosol monitored in Antarctica. It can be seen in Fig. 8.34 that $n(\lambda)$ in the visible varies between about 1.40 (for AH mainly consisting of polluted aerosol at Sodankylä) and about 1.53 (for AD monitored at Barrow), while $k(\lambda)$ was correspondingly estimated to range between 3.7×10^{-3} (ADS aerosol at Ny-Ålesund) and 6.3×10^{-2} (AD at Barrow). More limited variations of these optical parameters were found for the BG austral summer aerosol types monitored at the Antarctic sites, since $n(\lambda)$ at visible wavelengths was estimated to vary between about 1.44 (Mirny, Dome C, and South Pole) and 1.47 (Mario Zucchelli), as shown in Fig. 8.35, and $k(\lambda)$ ranged between about 9×10^{-4} (Mirny) and 3×10^{-3} (South Pole).

(3) *Determination of the size-distribution curves of columnar aerosol*

The multimodal size-distribution curves of the BG summer aerosol, ADS aerosol, AH, AD, and BFF smoke particles observed at the Arctic and Antarctic sites were determined using the same procedure adopted in the previous step to define the values of $n(\lambda)$ and $k(\lambda)$. Here, the best-fit linear combinations of the fine/accumulation and accumulation/coarse particle modes and their columnar number concentrations were made to vary until fitting the median values of $\tau(0.50\ \mu m)$ and $\alpha(0.40\text{–}0.87\ \mu m)$ given in Table 8.5a. It is worth noting that the composition of the fine/accumulation particles was assumed to be given by (i) most significant mass fractions of water-soluble (W-S) and insoluble substances, and sea salt, together with small fractions of soots in the AH and BG summer aerosol cases; (ii) a prevailing mass fraction of insoluble substances and mineral dust in the AD case; (iii) relevant mass fractions of water-soluble and insoluble substances, and a minor fraction of small sea salt aerosol, together with a rather marked fraction of soot substances in the BFF smoke case; and (iv) a predominant mass fraction of water-soluble substances in the BG austral summer aerosol cases monitored at the five Antarctic sites, combined with smaller contents of sea salt and small soot particles. The accumulation/coarse particles were assumed to consist mainly of water-soluble, sea-salt coarse, and insoluble particles at Arctic and Antarctic sites. For the mass mixing percentages given in Table 8.5b, the overall particle size-distribution curves of various types were found to consist in general of two to four modes of fine/accumulation particles, and of one to four modes of accumulation/coarse particles, which are in part superimposed, as can be seen in Fig. 8.36 for the Arctic aerosol cases, and in Fig. 8.37 for the Antarctic aerosol cases.

Figure 8.36 shows the size-distribution curves of columnar total particle number density $N(r)$ determined for (i) the AH case observed at Barrow, (ii) the BFF smoke case observed at Barrow, (iii) the ADS aerosol case monitored at Ny-Ålesund, and (iv) the AH case measured at Sodankylä. They mainly consist of fine particle modes,

Table 8.5b. Mass mixing percentages of fine and coarse particle loads defined in terms of the OPAC (RH = 50%) components (Hess et al., 1998), as assumed at the various POLAR-AOD sites in the Arctic and Antarctic regions for the different atmospheric turbidity conditions defined in Table 5a, on the basis of the data sets collected by Tomasi et al. (2007, 2012). The Antarctic data were assumed according to the *in situ* sampling measurements performed by Hillamo et al. (1998) and Fattori et al. (2005) at Mario Zucchelli; Weller and Wagenbach (2007), Weller et al. (2008), and Weller and Lampert (2008) at Neumayer; and Tuncel et al. (1989) and Arimoto et al. (2004) at the South Pole.

Arctic/ Antarctic site	Aerosol type and season	Mass mixing percentages of the OPAC (RH = 50%) components separately assumed for the fine/accumulation particles and the accumulation/coarse particles defined by Hess et al. (1998) to determine the composition features of the two aerosol components					
		W–S (Fine)	Sea salt (Accum.)	Sea salt (Coarse)	Insol. (Accum.)	Soot (Fine)	Mineral transp. (Coarse)
Barrow	Arctic haze (winter–early spring):						
	Fine/accum. particle fraction composition	41.3%	34.9%	–	21.8%	2.0%	–
	Accum./coarse particle fraction composition	4.1%	–	57.6%	38.3%	–	–
	Asian dust:						
	Fine/accum. particle fraction composition	24.0%	10.0%	–	66.0%	–	–
	Accum./coarse particle fraction composition	6.0%	–	4.0%	30.0%	–	60.0%
	Boreal forest fire smoke:						
	Fine/accum. particle fraction composition	34.0%	29.0%	–	35.0%	2.0%	–
	Accum./coarse particle fraction composition	29.0%	–	35.0%	35.0%	1.0%	–
	BG summer aerosol (4-modal):						
	Fine/accum. particle fraction composition	35.0%	17.8%	–	46.0%	1.2%	–
	Accum./coarse particle fraction composition	1.3%	–	70.9%	27.8%	–	–
Ny-Ålesund	Arctic haze (winter–early spring):						
	Fine particle fraction composition	98.0%	–	–	–	2.0%	–
	Accum./coarse particle fraction composition	–	–	90.9%	9.1%	–	–
	Arctic dense summer aerosol:						
	Fine/accum. particle fraction composition	35.0%	10.0%	–	54.0%	1.0%	–
	Accum./coarse particle fraction composition	5.0%	–	95.0%	–	–	–
	BG summer aerosol:						
	Fine particle fraction composition	98.0%	–	–	–	2.0%	–
	Accum./coarse particle fraction composition	–	–	50.0%	50.0%	–	–

Table 8.5b. Continued.

Arctic/ Antarctic site	Aerosol type and season	Mass mixing percentages of the OPAC (RH = 50%) components separately assumed for the fine/accumulation particles and the accumulation/coarse particles defined by Hess et al. (1998) to determine the composition features of the two aerosol components					
		W–S (Fine)	Sea salt (Accum.)	Sea salt (Coarse)	Insol. (Accum.)	Soot (Fine)	Mineral transp. (Coarse)
Summit	BG summer aerosol:						
	Fine particle fraction composition	98.0%	–	–	–	2.0%	–
	Accum./coarse particle fraction composition	–	–	50.0%	50.0%	–	–
Sodankylä	Arctic haze (winter–early spring):						
	Arctic (AR) OPAC model, with RH = 50%	40.6%	54.4%	–	3.0%	2.0%	–
	BG summer aerosol:						
	Fine particle fraction composition	98.0%	–	–	–	2.0%	–
	Accum./coarse particle fraction composition	–	–	50.0%	50.0%	–	–
Tiksi	BG summer aerosol:						
	Fine particle fraction composition	98.0%	–	–	–	2.0%	–
	Accum./coarse particle fraction composition	–	–	50.0%	50.0%	–	–
Mario Zucchelli	BG austral summer aerosol:						
	Fine particle fraction composition	99.5%	–	–	–	0.5%	–
	Accum./coarse particle fraction composition	–	–	73.9%	26.1%	–	–
Neumayer	BG austral summer aerosol:						
	Fine particle fraction composition	99.7%	–	–	–	0.3%	–
	Accum./coarse particle fraction composition	–	–	85.6%	14.4%	–	–
Mirny	BG austral summer aerosol:						
	Fine/accum. fraction composition	79.0%	19.0%	–	1.9%	0.1%	–
	Accum./coarse particle fraction composition	83.0%	–	13.0%	3.9%	0.1%	–
Dome C	BG austral summer aerosol:						
	Fine/accum. particle fraction composition	95.3%	4.7%	–	–	–	–
	Coarse particle fraction composition	–	–	–	–	–	100.0%
South Pole	BG austral summer aerosol:						
	Fine particle fraction composition	99.6%	–	–	–	0.4%	–
	Accum./coarse particle fraction composition	–	–	83.6%	16.4%	–	–

Table 8.5c. Daily mean values of the real part $n(0.50\ \mu m)$ and imaginary part $k(0.50\ \mu m)$ of the complex refractive index, the volume extinction coefficient $\beta_{ext}(0.50\ \mu m)$ (obtained for the overall columnar particle number contents derived for the median values of $\tau_a(\lambda)$ given in Table 8.5a), and the average values of columnar aerosol single-scattering albedo $\omega(0.55\ \mu m)$ determined for the various polar aerosol types listed in Table 8.5a at the various POLAR-AOD sites, and for the radiative characteristics of columnar aerosol particles defined in Table 8.5b.

polar aerosol type	Measurement site	Average values of columnar aerosol refractive index at the 0.50-μm wavelength		Average columnar volume extinction coefficient	Columnar aerosol single-scattering albedo
		Real part $n(0.50\ \mu m)$	Imaginary part $k(0.50\ \mu m)$	$\beta_{ext}(0.50\ \mu m)$ (per km)	$\omega(0.55\ \mu m)$
BG summer aerosol	Barrow	1.444	9.2×10^{-3}	7.44×10^{-3}	0.978
	Ny-Ålesund	1.461	7.9×10^{-3}	3.92×10^{-3}	0.966
	Summit	1.449	8.8×10^{-3}	3.27×10^{-3}	0.969
	Sodankylä	1.452	8.6×10^{-3}	5.71×10^{-3}	0.965
	Tiksi	1.444	9.2×10^{-3}	8.04×10^{-3}	0.977
BG austral summer aerosol	Mario Zucchelli	1.468	2.3×10^{-3}	2.93×10^{-3}	0.964
	Neumayer	1.457	1.3×10^{-3}	4.22×10^{-3}	0.975
	Mirny	1.442	8.9×10^{-3}	2.17×10^{-3}	0.985
	Dome C	1.441	2.0×10^{-3}	1.81×10^{-3}	0.999
	South Pole	1.445	3.2×10^{-3}	1.17×10^{-3}	0.988
Arctic haze (winter–early spring)	Barrow	1.399	3.3×10^{-3}	1.11×10^{-2}	0.840
	Ny-Ålesund	1.424	7.3×10^{-3}	7.63×10^{-3}	0.949
	Sodankylä	1.399	3.9×10^{-3}	6.26×10^{-3}	0.840
Arctic dense summer aerosol	Ny-Ålesund	1.437	3.7×10^{-3}	1.16×10^{-2}	0.852
Asian dust	Barrow	1.527	6.3×10^{-3}	1.99×10^{-2}	0.858
Boreal forest fire smoke	Barrow	1.469	2.5×10^{-3}	2.89×10^{-2}	0.758

with the volume fractions of fine particles prevailing over those of coarse particles by at least one order of magnitude. In the AD case monitored at Barrow, the size-distribution curve exhibits clearly bimodal features, with a largely predominant contribution of coarse particles. In the other cases, pertaining to the AH case at Ny-Ålesund, and the five BG summer aerosol cases monitored at Barrow, Ny-Ålesund, Summit, Sodankylä, and Tiksi, the size-distribution curves of $N(r)$ were found to show substantially bimodal characteristics, yielding a predominant volume content of coarse particles in the atmospheric column by at least seven to eight orders of magnitude over the overall fine particle volume content.

The multimodal size-distribution curves of columnar total particle number density $N(r)$ determined for the BG austral summer cases observed at the five Antarctic sites are shown in Fig. 8.37. They provide evidence of the multimodal characteristics of the columnar Antarctic aerosol number concentration, with important contributions made by the accumulation particle modes centered at radii of around 0.08 μm at Mirny, Dome C, and South Pole, and by coarse particle modes centered

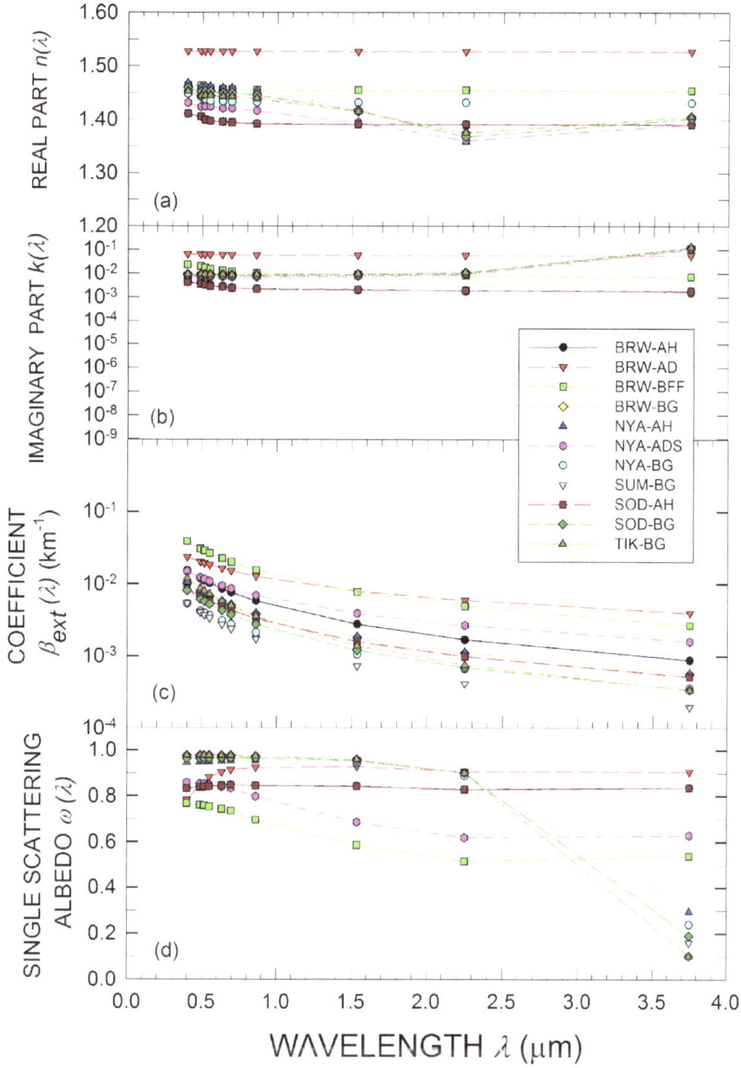

Fig. 8.34. Spectral curves of the real part $n(\lambda)$ and imaginary part $k(\lambda)$ of the particulate matter refractive index, volume extinction coefficient $\beta_{ext}(\lambda)$, and columnar aerosol SSA $\omega(\lambda)$. The four sets of curves were determined for the size-distribution curves of aerosol particles shown in Fig. 8.36 to represent the average atmospheric turbidity conditions associated with the 11 aerosol types monitored at five Arctic sites by means of the sun-photometer measurements reported in Table 8.5a, and the linear combinations of fine and coarse particle modes having the mass fractions of the OPAC components given in Table 8.5b.

at radii of around 2–4 μm at Mario Zucchelli and Neumayer. Correspondingly, the size-distribution curves of columnar total particle volume content $V(r)$ yielded contributions of accumulation particles prevailing over those of coarse particles at all

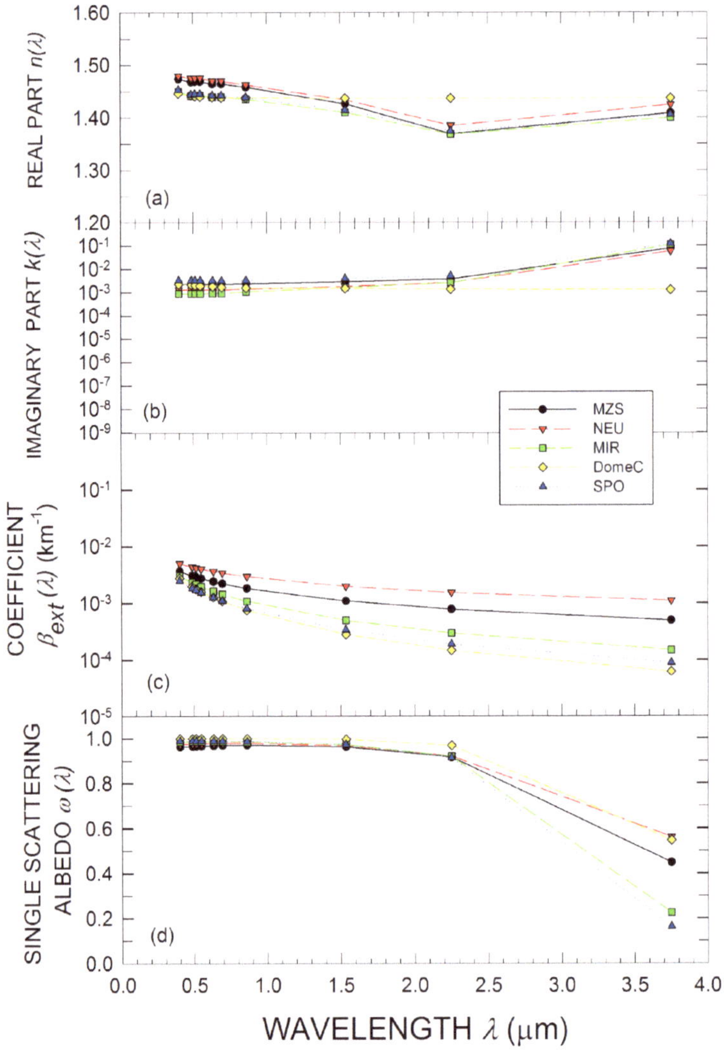

Fig. 8.35. As in Fig. 8.34, for the average columnar loads of BG austral summer aerosol particles monitored at five Antarctic sites for the sun-photometer measurements of atmospheric turbidity parameters given in Table 8.5a, and the linear combinations of fine and coarse particle modes having the mass fractions of the OPAC components given in Table 8.5b.

sites, which were peaked at radii of around 10 μm in the three coastal cases and at the South Pole, and a radius of around 4 μm at Dome C.

(4) *Determination of the columnar aerosol single-scattering albedo*

The spectral values of volume coefficients $\beta_{sca}(\lambda)$, $\beta_{abs}(\lambda)$, and $\beta_{ext}(\lambda)$ were calculated for all the Arctic and Antarctic aerosol types considered in the present analysis in order to determine the spectral characteristics of columnar $\omega(\lambda)$ for the various polar aerosol types. Figure 8.34 shows the spectral curves of $\beta_{ext}(\lambda)$ calculated for

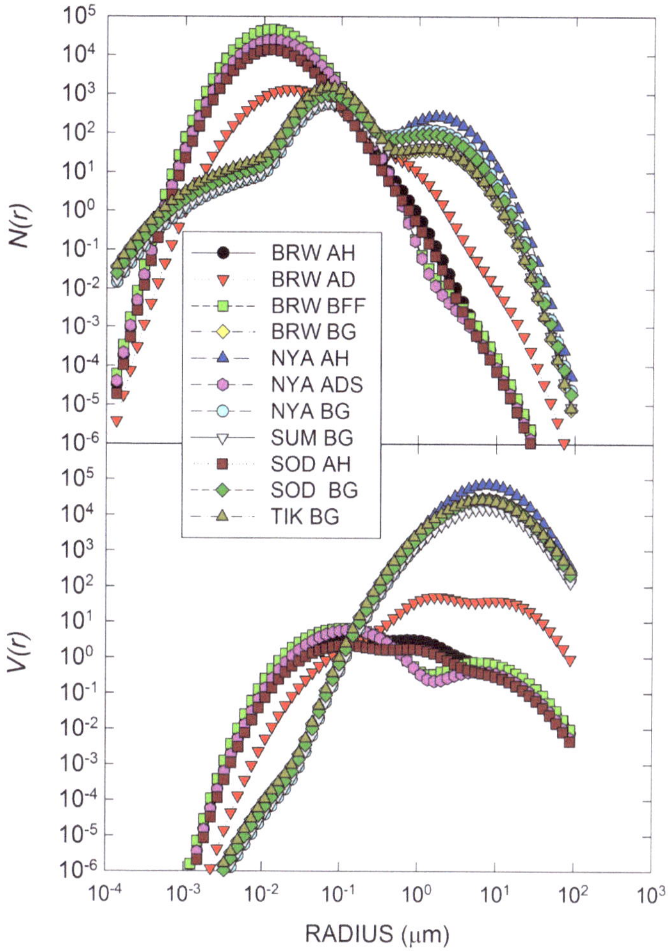

Fig. 8.36. Multimodal size-distribution curves of columnar particle number density $N(r) = \mathrm{d}N/\mathrm{d}(\ln r)$ measured per cm^2 (upper part) and columnar particle volume $V(r) = \mathrm{d}V/\mathrm{d}(\ln r)$ measured in $\mu\mathrm{m}^3/\mathrm{cm}^2$ (lower part), obtained as linear combinations of the OPAC components defined for RH = 50% and weighted using the mass percentages given in Table 8.5b, to represent the 11 Arctic aerosol types monitored at Barrow (BRW), Ny-Ålesund (NYA), Summit (SUM), Sodankylä (SOD), and Tiksi (TIK), for the atmospheric turbidity parameters given in Table 8.5a.

the overall columnar number contents of aerosol particles equal to those derived from the median values of $\tau_a(\lambda)$ given in Table 8.5a. They present similar spectral patterns for all the aerosol types monitored at the five Arctic sites, except for the AD case observed at Barrow, which was found to exhibit more slowly decreasing features, due to the significant extinction effects of coarse desert particles.

The spectral values of average columnar $\omega(\lambda)$ were calculated for all the multimodal size-distribution curves defined at the Arctic and Antarctic sites and shown in Figs 8.36 and 8.37, as determined for the spectral values of $n(\lambda)$ and $k(\lambda)$ shown in Figs 8.34 and 8.35, respectively. The values of $\omega(\lambda)$ at visible wavelengths vary

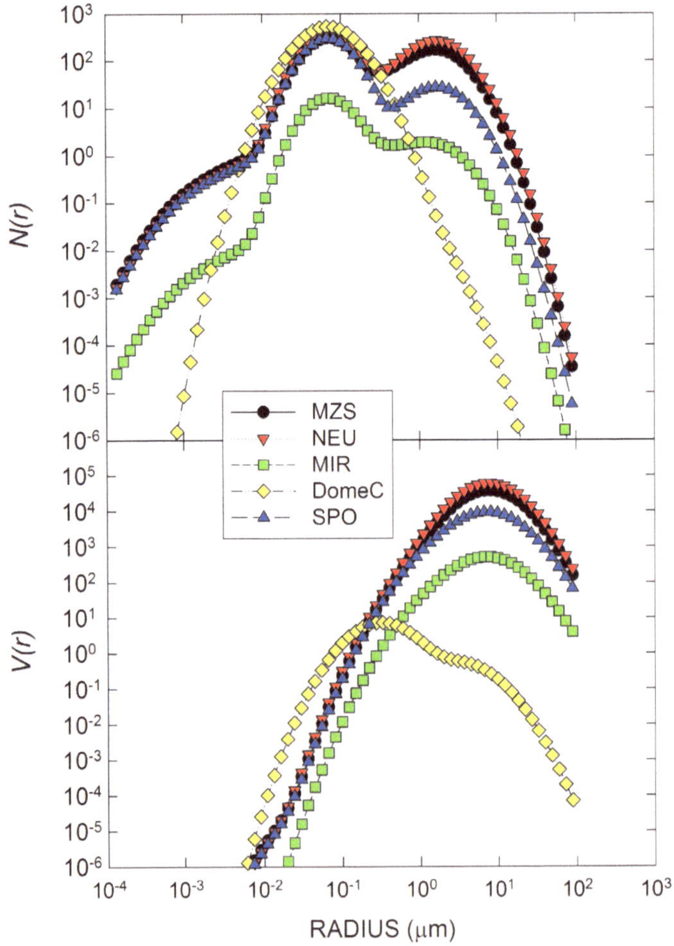

Fig. 8.37. As in Fig. 8.36, for the five Antarctic BG austral summer aerosol types monitored at Mario Zucchelli (MZS), Neumayer (NEU), Mirny (MIR), Dome C, and South Pole (SPO) sites in Antarctica, for the atmospheric turbidity parameters given in Table 8.5a.

between about 0.76 (BFF at Barrow) and nearly 0.98 (BG summer aerosol at Barrow and Tiksi). Interestingly, $\omega(\lambda)$ exhibits slowly decreasing spectral patterns from the visible to the near-IR, except for the AD case observed at Barrow, where $\omega(\lambda)$ increased from less than 0.8 to more than 0.9 over the 0.40–1.00 wavelength range. Similar spectral features of $\beta_{ext}(\lambda)$ were also found for all the BG austral summer aerosol cases monitored at the five Antarctic sites, shown in Fig. 8.35, which present values ranging between less than 1.2×10^{-3} (South Pole) and 4.2×10^{-3} per km (Neumayer) at the 0.50-μm wavelength. Correspondingly, the spectral curves of $\omega(\lambda)$ calculated over the 0.40–3.75-μm wavelength range assumes values that turn out to be very close to unity in the visible, especially those obtained at Dome C and South Pole.

(5) *Definition of the local surface albedo models*

The surface reflectance characteristics over sea and land are known to vary considerably in the polar regions, closely depending on the local seasonal climatic conditions. Ocean surfaces are usually frosted in the winter months, presenting high-reflectance characteristics for both ice and snow coverages. Conversely, the ice-free sea surface usually observed near the coast in late spring and summer yields relatively low surface reflectance values, which can vary considerably as a function of wind velocity at sea level and wave motion. The ocean surface reflectance characteristics observed in summer at the Arctic and Antarctic sites listed in Table 8.5d (i.e. Barrow, Ny-Ålesund, Sodankylä, and Tiksi on clear-sky days of boreal late spring and summer, and Mario Zucchelli and Neumayer on clear-sky days of austral summer) were assumed to be most realistically represented by the OS1 surface albedo model defined by Tomasi et al. (2013) for wind velocity of 2 m/s at sea level, giving values of white-sky albedo $R_{ws} = 0.069$, black-sky albedo $R_{bs}(\theta_o = 0°) = 0.026$, and broadband albedo $A(\theta_o = 60°) = 0.193$.

Land areas can present very different surface reflectance features in the Arctic, since they depend on the type of vegetation coverage and on the snow precipitation regime, therefore varying with season and latitude. In the Arctic region, there are two main types of vegetation coverage that are most commonly referred to as 'taiga' and 'tundra'. Taiga covers extended areas of coniferous forests, mostly consisting of pines, spruces, and larches, which grow in inland Canada and Alaska in North America, some lowland/coastal areas of Iceland, most of the Scandinavian peninsula, and a large area of Russia, from Karelia in the west to the eastern Pacific Ocean coast, thus including much of northern Siberia. Tundra vegetation consists of dwarf shrubs, sedges and grasses, mosses, lichens, and sometimes scattered trees, because the tree growth is hindered by low temperatures and short growing seasons. For this reason, Arctic tundra occurs north of the taiga belt, in remote areas where the subsoil is permafrost, or permanently frozen soil.

Examining the sets of MCD43C3 products derived from the MODIS Level 3.0 surface albedo data recorded over land, within all the seven MODIS spectral bands, the average white-sky albedo maps shown in Fig. 8.38 were determined for the Arctic and Antarctic regions during local summer periods. The data obtained in the Arctic region were recorded during the first two weeks of July 2009, indicating that white-sky albedo assumes values mainly ranging between 0.15 and 0.30 at Barrow and Sodankylä, around 0.40 at Tiksi, around 0.60 at Ny-Ålesund, and around 0.80 at Summit. On the basis of these results, it can be stated that the taiga surface reflectance characteristics near Barrow, Sodankylä, and Tiksi can be realistically represented in this season using the VS1 surface albedo model of Tomasi et al. (2013), which provides the set of spectral reflectance curves shown in Fig. 8.5, yielding values of $R_{ws} = 0.153$, $R_{bs}(\theta_o = 0°) = 0.134$, and $A(\theta_o = 60°) = 0.155$. Conversely, the surface albedo conditions at Summit, located in the middle of the high-altitude Greenland ice sheet, can be well represented in summer using the PS1 surface albedo model, and those around Ny-Ålesund (Spitsbergen, Svalbard) over land using model PS4. The austral summer average white-sky map over Antarctica shows that white-sky albedo ranges between 0.65 and 0.75 at the three coastal sites of Mirny, Neumayer, and Mario Zucchelli, and is higher than 0.75 over the Antarctic Plateau, at Dome C, and South Pole. Therefore, as reported in Table 8.5d, the

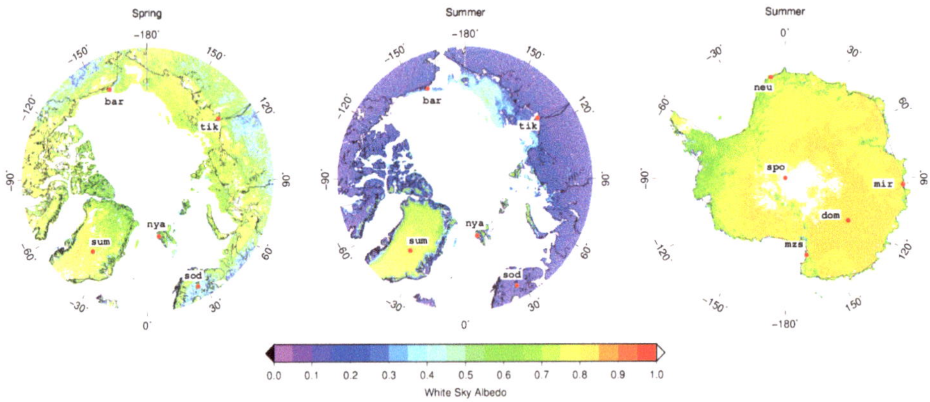

Fig. 8.38. Maps of white-sky albedo R_{ws} over land determined from the MODIS Level 3.0 surface albedo data (MCD43C3 products) taken (a) over the Arctic region in the first half of April 2009 (boreal spring) (left part), and in the first half of July 2009 (middle part), and (b) over Antarctica in the first half of January 2009 (austral summer) (right part). Red circles indicate the geographical positions of the five sun-photometer stations in Arctic and the five sun-photometer stations in Antarctica, where measurements of aerosol radiative properties were also performed (see Table 8.5a).

most appropriate models for representing the surface albedo characteristics at the Antarctic sites are the non-Lambertian models PS1 and PS2 defined by Tomasi et al. (2013) for snow- and ice-covered surfaces of the coastal regions, and model PS1 at the two high-altitude stations on the Antarctic Plateau. Model PS1 yields average values of white-sky albedo $R_{ws} = 0.847$, black-sky albedo $R_{bs}(0°) = 0.824$, and broadband albedo $A(\theta_o = 60°) = 0.854$, while the PS2 model gives average values of $R_{ws} = 0.761$, $R_{bs}(0°) = 0.720$, and $A(\theta_o = 60°) = 0.775$. It is worth noting that the spectral features of the four models PS proposed by Tomasi et al. (2013) agree very closely with those determined by Lupi et al. (2001) from some sets of surface reflectance measurements carried out with high-resolution spectrometers, employed in some coastal areas near the Mario Zucchelli station (Campo Icaro, Nansen Ice Sheet, Ross Ice Shelf) as well as over some high-altitude glaciers nearby.

The analysis of the MODIS Level 3.0 surface albedo data recorded over land in the Arctic region, during the early two-week period of April 2009, provided the third average white-sky albedo map shown in Fig. 8.38, charted for spring conditions in the Arctic. It clearly indicates that the sun-photometer stations are surrounded by areas covered by snow fields and glaciers in that season, presenting rather high surface albedo conditions, varying mainly between 0.50 and 0.75. They can be well represented by the non-Lambertian polar surface models PS1, PS2, and PS3 defined by Tomasi et al. (2013) for snow- and ice-covered surfaces. Model PS3 was estimated to give average values of $R_{ws} = 0.536$, $R_{bs}(0°) = 0.461$, and $A(\theta_o = 60°) = 0.564$, which are appreciably lower than those obtained for models PS1 and PS2 used above to represent surface albedo conditions in Antarctica. Therefore, they were deemed very appropriate for representing the land surface reflectance features observed around Barrow in winter–spring, over both ice-covered sea and snow-covered land areas. In particular, (i) model PS1 was chosen to represent the inner parts of Greenland, which are covered by snow fields and glaciers throughout the year, including the high-altitude area of Summit; (ii) model PS3 was correctly employed in summer to simulate the land area characteristics at Ny-Ålesund; (iii) model PS2 was selected to represent the surface albedo characteristics of the glaciers and snow fields not far from Sodankylä; and (iv) model PS4 was used to represent realistically the mixed ice-covered and ice-free areas in Spitsbergen, not far from Ny-Ålesund, since it yields average values of $R_{ws} = 0.296$, $R_{bs}(0°) = 0.214$, and $A(\theta_o = 60°) = 0.329$.

The spectral curves of white-sky albedo $R_{ws}(\lambda)$ determined for models PS1, PS2, PS3, and PS4 are shown in Fig. 8.39, to give an idea of the surface albedo characteristics that can be reliably used to calculate the instantaneous DARF values at the ToA- and BoA-levels. The calculations of the instantaneous DARF terms at the Arctic and Antarctic sites were made using the spectral curves of surface albedo $R_L(\lambda, \theta_o)$ calculated at the various solar zenith angles for the four PS models shown in Fig. 8.39. The most representative PS model was chosen at each site and for each aerosol type, to simulate the white-sky albedo characteristics derived from the MCD43C3 products, as indicated in Table 8.5d for the various aerosol types.

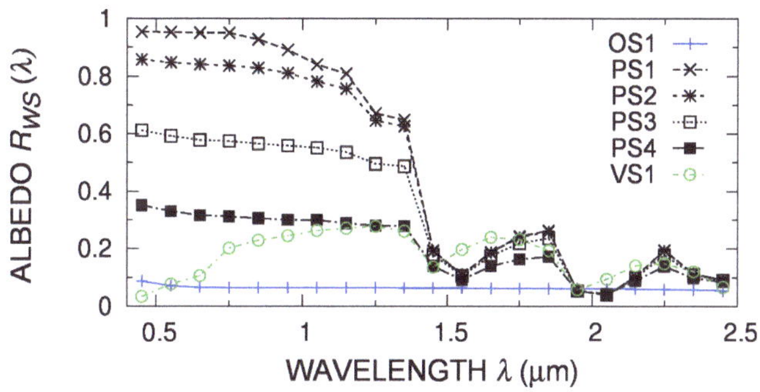

Fig. 8.39. Spectral curves of the white-sky albedo $R_{ws}(\lambda)$ determined for the surface albedo models PS1, PS2, PS3, and PS4 models (black symbols) of Tomasi et al. (2013), and the spectral curves of $R_{ws}(\lambda)$ determined for the oceanic surface albedo model OS1 (blue crosses) and vegetation-covered surface albedo model VS1 (green open circles), employed to carry out the DARF calculations of polar aerosols for the various polar surface albedo characteristics, the oceanic surface albedo characteristics in summer, and the taiga surface albedo characteristics in summer, respectively.

(6) *Calculations of the daily time-patterns of instantaneous DARF terms and diurnally averaged DARF effects*

As pointed out in the previous sub-paragraph, polar regions have extensive areas characterized by high-reflectance features, which strongly reflect the incoming solar radiation upward. For the rather low solar elevation angles usually observed at polar sites, aerosols are expected to induce significant climatic effects, which can vary appreciably according to the variable absorption features of particulate matter. In fact, for particles containing not negligible fractions of soot substances (and, in particular, BC), the changes in the SSA characteristics can cause pronounced

Table 8.5d. Daily values of the diurnally averaged aerosol forcing terms ΔDF_{ToA} at the ToA-level, ΔDF_{BoA} at the BoA-level, and ΔDF_{Atm} within the atmosphere, diurnal average aerosol fractional forcing AFF_{ToA} at the ToA-level (given by the ratio between flux change ΔF_{ToA} at the ToA-level and the incoming flux $I_S\downarrow$ of solar radiation at the ToA-level), and the diurnal average DARF efficiencies E_{ToA}, E_{BoA}, and E_{Atm} giving the rates at which the surface–atmosphere system is forced per unit $\tau_a(0.55\ \mu m)$, as obtained for the polar aerosol types defined in Table 8.5a from the sun-photometer measurements performed at the various POLAR-AOD sites and for the radiative characteristics of columnar aerosol given in Table 8.5b.

polar aerosol	Measurement	Surface albedo	Diurnal average DARF terms (W/m^2)			Diurnal	Diurnal average DARF efficiencies (W/m^2)		
type	site	model	ΔDF_{ToA}	ΔDF_{BoA}	ΔDF_{Atm}	AFF_{ToA}	E_{ToA}	E_{BoA}	E_{Atm}
BG summer aerosol	Barrow	OS1	−9.2	−1.9	−7.3	-2.3×10^{-2}	−118.0	−23.8	−94.2
		VS1	−4.2	−6.5	+2.3	-1.1×10^{-2}	−53.6	−82.9	+29.3
	Ny-Ålesund	OS1	−7.2	+1.5	−8.6	-1.8×10^{-2}	−174.6	+36.3	−210.9
		PS4	+1.0	−2.3	+3.4	$+3.0\times10^{-3}$	+25.5	−56.6	+82.0
	Summit	PS1	+2.2	−0.2	+2.5	$+6.0\times10^{-3}$	+57.3	−6.3	+63.5
	Sodankylä	OS1	−6.5	−1.6	−4.9	-1.7×10^{-2}	−108.6	−26.9	−81.7
		VS1	−3.0	−5.5	+2.4	-8.0×10^{-3}	−50.0	−90.8	+40.8
		PS2	+3.0	−0.4	+3.4	$+8.0\times10^{-3}$	+50.3	−6.5	+56.9
	Tiksi	OS1	−10.3	−1.6	−8.7	-2.6×10^{-2}	−121.5	−18.7	+102.8
		VS1	−5.6	−9.0	+3.3	-1.4×10^{-2}	−66.3	−105.3	+39.0
BG austral summer aerosol	Mario Zucchelli	OS1	−6.2	+0.9	−7.1	-1.4×10^{-2}	−207.7	+29.4	−237.1
		PS2	+2.0	−0.1	+2.1	$+4.0\times10^{-3}$	+65.0	−3.7	+68.7
	Neumayer	OS1	−8.2	+1.4	−9.6	-1.8×10^{-2}	−190.3	+33.0	−223.3
		PS1	+1.9	−0.2	+2.1	$+4.0\times10^{-3}$	+45.2	−3.9	+49.1
		PS2	+2.1	+0.0	+2.1	$+5.0\times10^{-3}$	+49.8	+0.7	+49.1
	Mirny	PS1	+0.2	−0.0	+0.3	$+5.0\times10^{-4}$	+9.6	−1.7	+11.3
		PS2	+0.6	−0.1	+0.6	$+1.0\times10^{-3}$	+24.2	−3.0	+27.2
	Dome C	PS1	+0.4	+0.2	+0.2	$+1.0\times10^{-3}$	+19.2	+8.9	+10.3
		PS2	+0.6	+0.4	+0.2	$+1.0\times10^{-3}$	+30.8	+20.3	+10.6
	South Pole	PS1	+0.6	−0.1	+0.7	$+1.0\times10^{-3}$	+35.9	−5.8	+41.7
		PS2	+0.7	−0.1	+0.8	$+2.0\times10^{-3}$	+40.5	−2.8	+43.3

Table 8.5d. Continued.

polar aerosol type	Measurement site	Surface albedo model	Diurnal average DARF terms (W/m^2) ΔDF_{ToA}	ΔDF_{BoA}	ΔDF_{Atm}	Diurnal AFF_{ToA}	Diurnal average DARF efficiencies (W/m^2) E_{ToA}	E_{BoA}	E_{Atm}
Arctic haze (winter–early spring)	Barrow	OS1	−5.5	+6.5	−12.0	-4.1×10^{-2}	−47.1	+55.9	−103.0
		PS1	+9.0	+0.0	+8.9	$+6.8 \times 10^{-2}$	+77.2	+0.2	+77.1
		PS2	+8.8	+0.1	+8.7	$+6.7 \times 10^{-2}$	+76.2	+0.9	+75.3
	Ny-Ålesund	OS1	−5.3	−0.6	−4.7	-6.7×10^{-2}	−66.2	−7.3	−58.9
		PS3	+1.0	−0.4	+1.4	$+1.3 \times 10^{-2}$	+13.1	−4.5	+17.6
	Sodankylä	OS1	−7.7	−1.1	−6.5	-2.0×10^{-2}	−116.2	−17.3	−99.0
		VS1	−3.5	−5.3	+1.8	-9.0×10^{-3}	−52.9	−79.6	+26.7
		PS2	+2.2	−0.3	+2.5	$+6.0 \times 10^{-3}$	+33.1	−4.5	37.7
Arctic dense summer aerosol	Ny-Ålesund	OS1	−10.4	+4.2	−14.6	-2.6×10^{-2}	−86.7	+35.3	−122.0
		PS4	+13.0	−5.2	+18.2	$+3.2 \times 10^{-2}$	+108.0	−43.1	+151.2
Asian dust	Barrow	OS1	−9.3	+16.0	−25.3	-3.9×10^{-2}	−46.5	+80.0	−126.5
		PS2	+18.5	+0.3	+18.2	$+7.7 \times 10^{-2}$	+92.3	+1.3	+91.0
Boreal forest fire smoke	Barrow	OS1	−32.9	−6.7	−26.2	-8.6×10^{-2}	−109.8	−22.3	−87.5
		VS1	−20.2	−28.6	+8.3	-5.2×10^{-2}	−67.5	−95.3	+27.8

variations in the solar radiation flux passing through the atmosphere, thus causing a significant change in the overall albedo of the climate system. In addition, polar aerosols can induce changes in cloud microphysical properties, thus perturbing the climate of polar regions through indirect effects (Stone, 2002). Such effects on climate can be very strong, since sudden pronounced variations in aerosol mass concentration may be caused by the transport of anthropogenic polluted aerosol from the densely populated mid-latitude areas, biomass-burning and forest fire smokes from the mid-latitude and temperate polar regions, and dust mobilized in the mid-latitude desert and agricultural regions. These poleward transport episodes of aerosol particles are often observed in the Arctic regions, involving airborne aerosols of industrial and urban origin, desert dust, and biomass-burning particulate matter (Shaw, 1995). The transport episodes of biomass-burning smoke particles can be so intense towards Antarctica as to influence considerably the optical characteristics of aerosol particles sampled at the Antarctic coastal sites, estimated to cause some slight decreases in SSA (Wolff and Cachier, 1998).

The mean time-patterns of instantaneous DARF terms were determined at the polar sites, for all the aerosol types listed in Table 8.5a, by using the 6S radiative transfer code (Vermote et al., 1997) for (i) the median values of $\tau_a(0.50\ \mu\mathrm{m})$ and $\alpha(0.40–0.87\ \mu\mathrm{m})$ given in Table 8.5a, which were assumed not to vary during the sunlit period of clear-sky days; (ii) the values of solar zenith angle θ_o calculated at the various hours of the day for the longitude and latitude coordinates of the various stations, as listed in Table 8.5a; (iii) the multimodal size-distribution curves determined as linear combinations of the OPAC (RH = 50%) components defined in Table 8.5b, according to the calculations made at step (3) for the various aerosol types; (iv) the optical characteristics of the multimodal size-distribution curves determined at step (2); and (v) the surface albedo models chosen at step (5) for the various Arctic and Antarctic sites.

All the time-patterns of instantaneous DARF terms $\Delta F_{ToA}(t)$, $\Delta F_{BoA}(t)$, and $\Delta F_{Atm}(t)$ determined above were then integrated over the sunlit periods of the various polar sites to calculate the corresponding values of diurnally averaged DARF terms ΔDF_{ToA}, ΔDF_{BoA}, and ΔDF_{Atm}. The results are shown in Table 8.5d for all the polar aerosol types monitored at the 10 POLAR-AOD sites. The corresponding evaluations of diurnal average AFF_{ToA} were also calculated, obtaining the values given in Table 8.5d. The scatter plots of the daily mean values of ΔDF_{ToA}, ΔDF_{BoA}, and ΔDF_{Atm} are shown in Fig. 8.40 versus the corresponding daily mean values of $\tau_a(0.55\ \mu\mathrm{m})$, for the five POLAR-AOD Arctic sites, as determined for the corresponding 11 aerosol types listed in Table 8.5d and the surface albedo models adopted at step (5). It can be seen that ΔDF_{ToA} varies appreciably at Barrow in the AD and BFF aerosol cases on passing from model OS1 to VS1 and from model OS1 to PS2, respectively, while only limited variations of ΔDF_{ToA} from about -10 W/m^2 to no more than $+10$ W/m^2 were evaluated in the BG summer aerosol and AH cases observed at the five Arctic sites. Similar features of ΔDF_{BoA} were also obtained, with rather large variations in the AD and BFF aerosol cases monitored at Barrow, and more limited estimates mainly ranging between -5 and $+5$ W/m^2 in all the BG summer aerosol and AH cases observed at the five Arctic sites and examined for the various surface albedo models chosen at step (5). Correspondingly, rather large variations in ΔDF_{Atm}were found not only for the AD

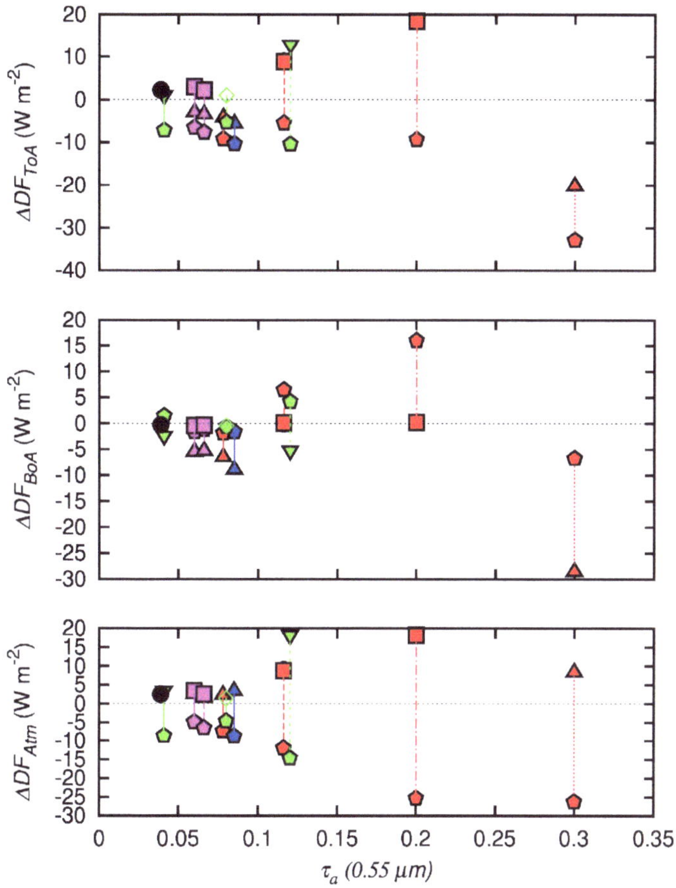

Fig. 8.40. Scatter plots of the daily mean values of DARF terms ΔDF_{ToA} (upper part), ΔDF_{BoA} (middle part), and ΔDF_{Atm} (lower part) as a function of the daily mean aerosol optical thickness $\tau_a(0.55\ \mu m)$, determined for the POLAR-AOD average atmospheric turbidity conditions indicated in Table 8.5a for the five Arctic sites, and associated with 11 seasonal different types of columnar aerosol. The colored symbols refer to Barrow (red), Ny-Ålesund (green), Summit (black), Sodankylä (fuchsia), and Tiksi (blue) sun-photometer stations. Symbols of different shape refer to surface albedo models OS1 (pentagons), PS1 (circles), PS2 (squares), PS3 (diamonds), PS4 (down triangles), and VS1 (up triangles) shown in Fig. 8.39. Different outlining types were adopted to draw the vertical bars for the various Arctic aerosol types: continuous bars refer to BG summer aerosol, long-dashed bars to Arctic haze formed in winter–early spring period, short-dashed bars to Arctic dense summer aerosol, dotted bars to BFF smoke, and dotted and dashed bars to Asian dust.

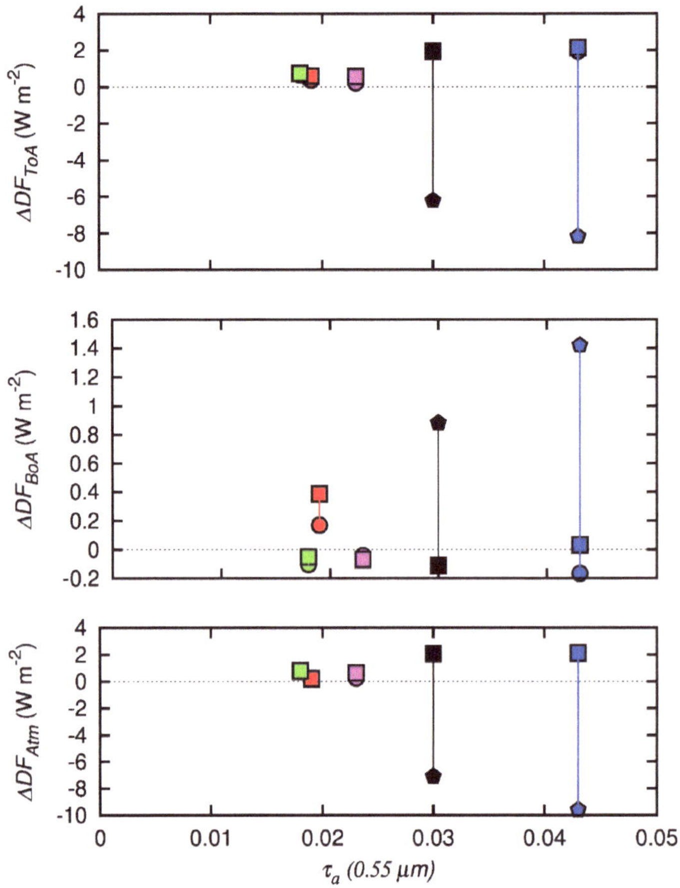

Fig. 8.41. As in Fig. 8.40, for the five Antarctic sites, and the corresponding BG austral summer aerosol types. The differently colored symbols refer to the sun-photometer stations Mario Zucchelli (black), Neumayer (blue), Mirny (fuchsia), Dome C (red), and South Pole (green). Symbols of different shape refer to surface albedo models OS1 (pentagons), PS1 (circles), and PS2 (squares) shown in Fig. 8.39. Solid vertical bars define the ranges of the DARF terms determined for different loads of BG austral summer aerosol.

and BFF aerosol cases at Barrow, but also for the AH cases observed at Barrow and Ny-Ålesund for values of $\tau_a(0.50\ \mu m)$ greater than 0.10.

The scatter plots of the daily mean values of ΔDF_{ToA}, ΔDF_{BoA}, and ΔDF_{Atm} are shown in Fig. 8.41 as a function of the corresponding daily mean values of $\tau_a(0.55\ \mu m)$, for the five Antarctic sites, where various types of BG austral summer aerosol were monitored, as shown in Table 8.5d, and different surface albedo models were considered in carrying out the DARF-PROC calculations. The scatter plot substantially shows that large variations in ΔDF_{ToA}, ΔDF_{BoA}, and ΔDF_{Atm} are usually obtained at the Mario Zucchelli and Neumayer stations when passing from the oceanic surface albedo model OS1 to the land model PS2 charaterizing the glacier-covered inner regions. By contrast, the results shown in Fig. 8.41 clearly indicate that very limited changes of the three DARF terms are obtained at all the

Antarctic sites when passing from one area to another of the Antarctic continent covered mainly by glaciers and snow fields, and seldom by ice-free rocky terrains.

(7) *Calculations of the daily mean values of DARF efficiencies*

The median values of $\tau_a(0.50\ \mu m)$ and α given in Table 8.5a were used to calculate the corresponding values of $\tau_a(0.55\ \mu m)$ using the well-known Ångström (1964) best-fit procedure. Thereupon, the daily mean values of DARF efficiencies were calculated for all the evaluations of diurnally averaged DARF terms ΔDF_{ToA}, ΔDF_{BoA}, and ΔDF_{Atm} obtained at the previous step and the corresponding average values of $\tau_a(0.55\ \mu m)$ obtained at the 10 Arctic and Antarctic sites for the various polar aerosol types. The results are given in Table 8.5d. The values of parameters E_{ToA}, E_{BoA}, and E_{Atm} calculated at the five Arctic sites for the 11 columnar aerosol types listed in Table 8.5a are plotted in Fig. 8.42 versus the corresponding estimates of columnar $\omega(0.55\ \mu m)$ ranging between 0.84 and more than 0.97. The results provide evidence of the large variations in the efficiency features that take place when passing from low surface albedo characteristics (like those represented with model OS1) to average surface albedo conditions (like those represented with model VS1) and to the high-reflectance conditions represented by models PS. It can be noted in Fig. 8.42 that (a) E_{ToA} varies from less than -150 W/m^2 (at Ny-Ålesund for BG summer aerosol over the sea surface) to more than $+100$ W/m^2 (at Ny-Ålesund for ADS aerosol over mixed rocky and ice-covered surfaces); (b) E_{BoA} varies from about -100 W/m^2 (at Tiksi for BG summer aerosol over the taiga surface) to around $+80$ W/m^2 (at Barrow for an AH load over a calm-wind oceanic surface); and (c) E_{Atm} varies from less than -200 W/m^2 (at Ny-Ålesund for BG summer aerosol over the sea surface) to more than $+150$ W/m^2 (at Ny-Ålesund for ADS aerosol over the ice-covered area of Spitsbergen (Svalbard)).

The calculations of E_{ToA}, E_{BoA}, and E_{Atm} made at the five Antarctic sites for the five BG austral summer aerosol types described in Table 8.5a are presented in Fig. 8.43 as a function of columnar $\omega(0.55\ \mu m)$ over the 0.96–1.00 range. They give evidence of the large variations in the three efficiency quantities that are obtained passing from the low surface albedo characteristics of the oceanic surface to the high surface albedo conditions of the Antarctic Plateau. For instance, (a) E_{ToA} increases from less than -200 W/m^2 (at Mario Zucchelli over the OS1 oceanic surface) to more than $+60$ W/m^2 (at the same site, over the PS2 snow- and ice-covered surface); (b) E_{BoA} increases from about -5 W/m^2 (at Neumayer over the PS1 ice-covered surface) to more than $+33$ W/m^2 (at the same site over the OS1 oceanic surface); and (c) E_{Atm} varies from nearly -250 W/m^2 (at Mario Zucchelli over the OS1 oceanic surface) to about $+70$ W/m^2 (at the same site over the PS2 ice-covered surface). It is interesting to note in Fig. 8.43 that the three efficiency parameters turn out to vary only slightly or to a more limited extent, at Mirny, Dome C, and South Pole on passing from one area of the Antarctic Plateau to another.

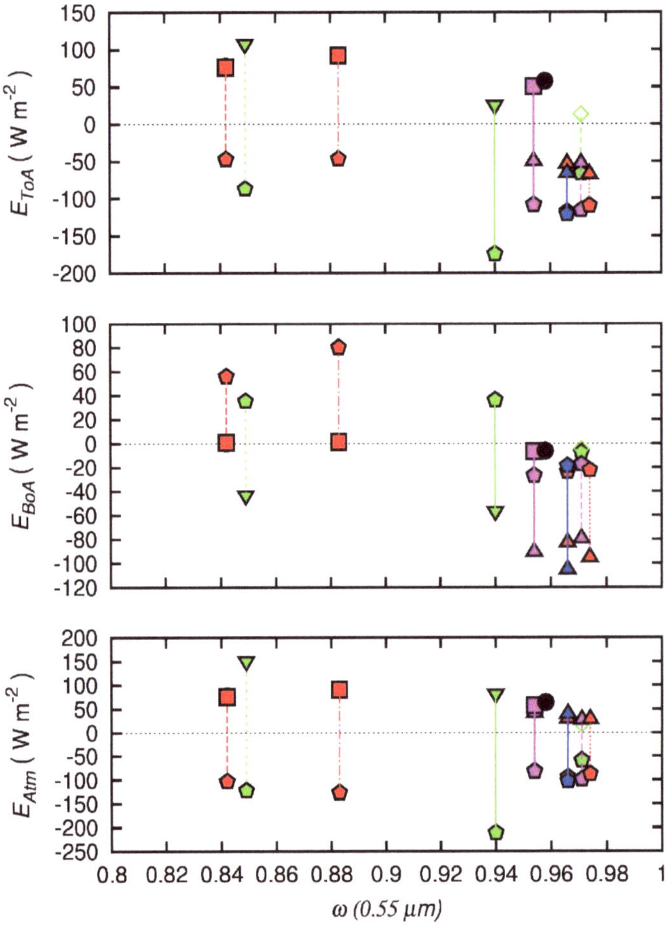

Fig. 8.42. Scatter plots of the daily mean values of DARF efficiency parameters E_{ToA} at the ToA-level, E_{BoA} at the BoA-level, and E_{Atm} within the atmosphere versus the columnar aerosol single-scattering albedo $\omega(0.55\ \mu m)$, determined for the average atmospheric turbidity characteristics associated with the 11 columnar aerosol types monitored at the five Arctic sites (and given in Table 8.5a). Differently colored symbols refer to Barrow (red), Ny-Ålesund (green), Summit (black), Sodankylä (fuchsia), and Tiksi (blue) sun-photometer stations. Symbols of different shape refer to surface albedo models OS1 (pentagons), PS1 (circles), PS2 (squares), PS3 (diamonds), PS4 (down triangles), and VS1 (up triangles) shown in Fig. 8.39. Different outlining types were adopted to draw the vertical bars for the various Arctic aerosol types: continuous bars refer to BG summer aerosol, long-dashed bars to Arctic haze formed in the winter–early spring period, short-dashed bars to Arctic dense summer aerosol, dotted bars to BFF smoke, and dotted and dashed bars to Asian dust.

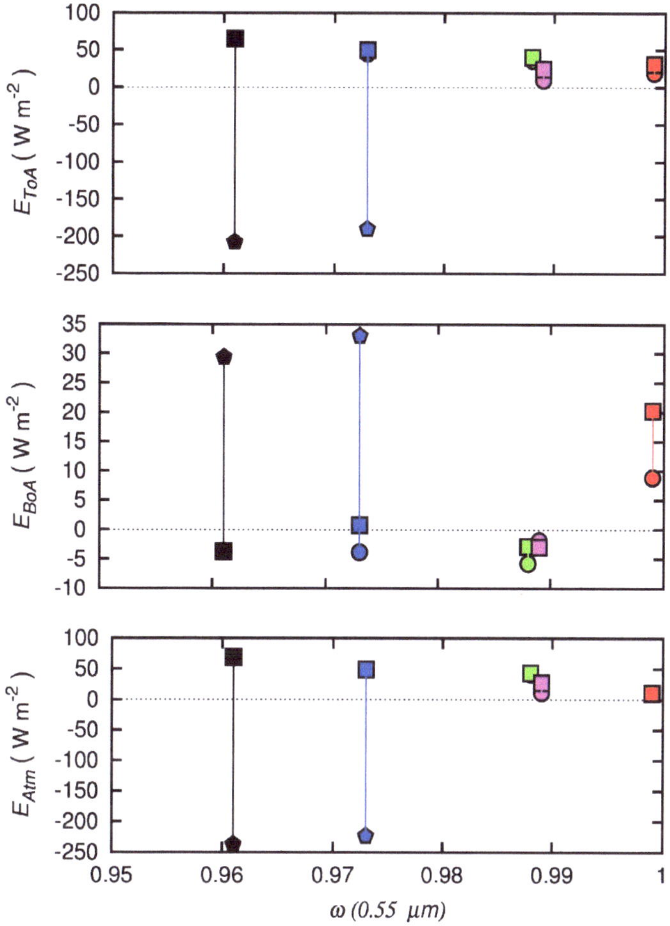

Fig. 8.43. As in Fig. 8.42, for the five different types of BG austral summer aerosol particle load monitored at the five Antarctic sites, as given in Table 8.5a. Differently colored symbols refer to the sun-photometer stations located at Mario Zucchelli (black), Neumayer (blue), Mirny (fuchsia), Dome C (red), and South Pole (green). Symbols of different shape refer to surface albedo models OS1 (pentagons), PS1 (circles), and PS2 (squares) shown in Fig. 8.39. Solid vertical bars define the ranges of DARF terms determined for atmospheric loads of BG austral summer aerosol.

8.3.6 DARF evaluations from the Aerosols99 measurements in the Atlantic Ocean

The Aerosols99 cruise crossed the Atlantic Ocean from Norfolk (Virginia, US) to Cape Town (South Africa) from January 14 to February 8, 1999. The main goal of the cruise was to study the chemical, physical, and optical characteristics of aerosol particles in the marine boundary layer (MBL). For this purpose, spectral ship-borne measurements of $\tau_a(\lambda)$ were carried out for clear-sky conditions, together with Lidar measurements of the vertical profiles of volume aerosol backscattering coefficient, and *in situ* measurements of ozone, CO, and peroxy-radical concentrations in the MBL. The sampling strategies were optimized to evaluate ocean color from the Sea-viewing Wide Field-of-view Sensor (SWFS) satellite-borne measurements, while the spectral values of $\tau_a(\lambda)$ were retrieved from the AVHRR measurements over the Atlantic Ocean. At the same time, evaluations of columnar ozone content were derived from the Total Ozone Mapping Spectrometer (TOMS) observations. An exhaustive overview of the results was presented by Bates et al. (2001).

Three types of columnar aerosol were monitored during the Aerosols99 cruise, consisting of different chemical substances and therefore presenting well-distinct size-distribution curves (Voss et al., 2001a): (i) pure maritime aerosol, monitored during the first part of the cruise; (ii) African dust mobilized from Saharan regions, which was observed during the middle part of the cruise; and (iii) biomass-burning smoke, transported from the southern Africa during the final part. Due to the different optical characteristics and size spectra of such aerosol types, the values of $\tau_a(\lambda)$ and exponent α were found to vary greatly passing from one regime to another: (i) in the northern part of the Atlantic Ocean, the measurements performed at the 35° N–30° N latitudes during the first week of the cruise provided the average values of $\tau_a(0.50\ \mu\text{m}) = 0.10 \pm 0.03$ and $\alpha = 0.3 \pm 0.3$; (b) African dust events observed in the central sector of Atlantic Ocean, during the second week of the cruise at the 20° N–25° N latitudes, were found to yield the average values of $\tau_a(0.50\ \mu\text{m}) = 0.29 \pm 0.05$ and $\alpha = 0.36 \pm 0.13$; and (c) no regular measurements were carried out within the Inter-Tropical Convergence Zone (ITCZ), because of prevailing cloudiness, while biomass-burning aerosol conditions were encountered after the ITCZ, at latitudes of around 20° S, for which a rather high average value of $\tau_a(0.50\ \mu\text{m}) = 0.36 \pm 0.13$ was measured with an average value of $\alpha = 0.88 \pm 0.30$ (Voss et al., 2001a). A micropulse Lidar system was employed almost continually during the cruise, to measure the profiles of aerosol backscattering coefficient, evidencing that clean maritime aerosols were capped at 1 km on days without dust advection from Africa, while the aerosol extinction profile showed its maximum at heights >2 km during dust advection episodes, and at around 4 km during some biomass-burning smoke transport episodes from southern Africa (Voss et al., 2001b).

The mass concentrations of the principal aerosol chemical components were measured at RH conditions close to 50–60%, finding that: (i) non-sea-salt (nss) sulfate aerosol particles provided a significant mass contribution to the submicron concentration in all regions, varying between 20% and 67%; (ii) maritime oceanic particles presented a mean submicron mass fraction of sea-salt particles ranging between 9% and 49%, with the maximum recorded in the northern oceanic regions and the minimum determined on the days characterized by biomass-burning particulate matter transport; (iii) supermicron marine particles presented a mean sea-salt mass fraction varying between 52% and 98%, with the highest values recorded in the middle of Atlantic Ocean; (iv) submicron and supermicron mass fractions of dust particles were estimated to be equal to around 22% and 26%, respectively; and (v) submicron mass fraction of particulate organic matter ranged from below the detection limits in the dust region to about 18% on average in the biomass-burning region, where the mean submicron mass fraction of BC was found to be of about 7%, giving an extinction contribution of 14% (Quinn et al., 2001).

On the basis of these evaluations, the DARF effects induced by the three above-mentioned columnar aerosol types were calculated using the DARF-PROC procedure subdivided into the usual seven steps as follows:

(1) Analyzing the field data collected by Voss et al. (2001a), the average values of $\tau_a(0.55\ \mu m)$ and α were separately determined for the three aerosol types, finding the results given in Table 8.6a.
(2) The refractive index of columnar aerosol was calculated for the three aerosol types by assuming that the particulate matter composition of the fine, accumulation, and coarse particle modes is given by the linear combinations of the mass percentages of the OPAC components defined by Hess et al. (1998), given in Table 8.6a, as evaluated for RH = 50%, according to the field observations of Quinn et al. (2001).
(3) The multimodal size-distribution curves of columnar aerosol were defined for the above assumptions of the OPAC percentage mass concentrations given in Table 8.6a, where: (i) the size-distributions of pure maritime aerosol particles were represented by the OPAC Maritime Clean model, in which the monomodal components of water-soluble and sea-salt accumulation particles were combined with the sea-salt coarse component until fitting the value of $\alpha = 0.30$; (ii) those of African dust were represented by the OPAC Desert model, in which the monomodal components of water-soluble particles, mineral dust nuclei, and mineral dust accumulation particles were combined with the mineral dust coarse particle component, until obtaining the best-fit value of $\alpha = 0.36$; and (iii) those of the biomass-burning smoke (hereinafter referred to as biomass-burning smoke (BBS)) were represented by the so-called BL trimodal model of Tomasi et al. (2013), in which the mass percentage concentrations of the Aitken nuclei and accumulation particle modes were kept stable and the mass concentration of coarse particles was suitably changed until obtaining the overall value of $\alpha = 0.88$. Figure 8.44 shows the multimodal size-distribution curves of total particle number density and total particle volume, obtained through the above best-fit procedure for the three multimodal size-distributions of columnar aerosol particles, found for pure maritime aerosol, African dust, and BBS particles. It can be clearly seen that the size-distribution curve of African dust exhibits a very pronounced mode of accumulation particles, together with a well-marked mode of coarse particles centered at around 1-μm radius, while the multimodal size-distribution curves of pure maritime and BBS particles show more continuous features over the whole radius range from 10^{-2} to around 20 μm.
(4) For the size-distribution curves evaluated at the previous step, presented in Fig. 8.44, the spectral curves of $n(\lambda)$ and $k(\lambda)$ and of $\beta_{ext}(\lambda)$ and $\omega(\lambda)$ were determined for the three Aerosols99 types. They are shown in Fig. 8.45, which indicates that $n(\lambda)$ assumes quite different values in the three cases, ranging between 1.50 and 1.55, with the highest value for the BBS particles and the lowest one for African dust, while $k(\lambda)$ decreases rapidly with wavelength in the visible for the pure maritime and African dust cases, assuming lower and more stable spectral values for the BBS particles. For the spectral characteristics of refractive index and the different size-distribution curves, coefficient $\beta_{ext}(\lambda)$ was found to assume higher and more stable spectral values in the pure maritime

Table 8.6a. Average values of aerosol optical thickness $\tau_a(0.55\ \mu m)$, Ångström exponent α, and columnar aerosol single-scattering albedo $\omega(0.55\ \mu m)$ determined for the three aerosol types monitored during the Aerosols99 cruise conducted in the Atlantic Ocean (Voss et al., 2001a), as obtained for the multimodal size-distribution curves determined using the partial mass mixing percentages given in brackets, to give form to the fine/accumulation and coarse particle components of the three aerosol types. The mass mixing percentages of the first two cases are calculated according to those given by Hess et al. (1998) for the wet (RH = 50%) OPAC aerosol components, and those of the third case according to those proposed by Tomasi et al. (2013) for the biomass-burning smoke BL trimodal aerosol extinction model.

Aerosol type	Average optical parameters			Mass mixing percentages of the unimodal fine, accumulation, and coarse particle components, as assumed for each mode to calculate the best-fit values of the overall columnar aerosol radiative parameters			
	$\tau_a(0.55\ \mu m)$	Exponent α	$\omega(0.55\ \mu m)$	Fine particle mode	Accumulation particle mode	Coarse particle mode	Coarse particle mode
Pure maritime aerosol	0.097	0.30	0.996	OPAC (RH = 50%) Water-soluble (98.68%)	OPAC (RH = 50%) Sea salt (Acc.) (1.32%)	OPAC (RH = 50%) Sea salt (Coarse) (100.00%)	–
African dust	0.280	0.36	0.883	OPAC (RH = 50%) Water-soluble (88.12%)	OPAC (RH = 50%) mineral dust (Nucl.) (11.88%)	OPAC (RH = 50%) mineral dust (Accum.) (99.54%)	OPAC (RH = 50%) mineral dust (Coarse) (0.46%)
Biomass-burning smoke	0.331	0.88	0.950	BL Aitken nuclei mode (40.46%)	BL accumulation mode (59.54%)	BL coarse mode (100.00%)	–

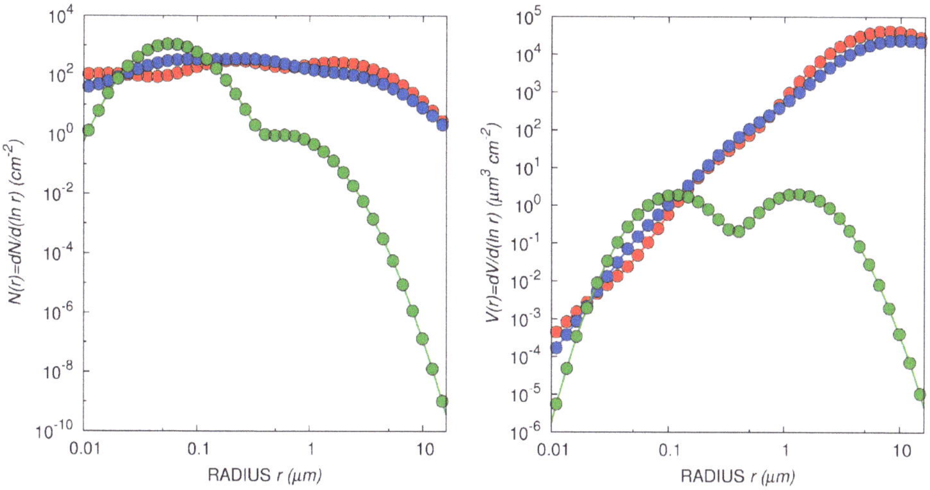

Fig. 8.44. Multimodal size-distribution curves of columnar particle number density $N(r) = \mathrm{d}N/\mathrm{d}(\ln r)$ measured per cm^2 (left) and columnar particle volume $V(r) = \mathrm{d}V/\mathrm{d}(\ln r)$ measured in μm^3/cm^2 (right), obtained for the three Aerosols99 cases: the first two cases pertain to pure maritime (blue circles) and African dust (dark green circles), both obtained as linear combinations of the OPAC (RH = 50%) components and the mass mixing percentages given in Table 8.6a. The third case pertains to BBS particles (red circles), obtained as a linear combination of the three modes consisting of Aitken nuclei, accumulation particles, and coarse particles defined by Tomasi et al. (2013) in the BL aerosol extinction model. The pure maritime aerosol was assumed to consist of water-soluble, sea-salt accumulation, and sea-salt coarse particle OPAC components; the African dust, of water-soluble, mineral dust nuclei, mineral dust accumulation, and mineral dust coarse particle OPAC components; and the BBS particles of 46% mass fraction of mineral dust, 35% water-soluble substances, 15% sea salt, and 4% soot matter.

aerosol case, and appreciably lower ones in the other two cases, by more than 50%. In particular, $\omega(\lambda)$ assumed spectral values (i) close to unity over the whole spectral range in the pure maritime aerosol case, with $\omega(0.55\ \mu\mathrm{m}) = 0.996$; (ii) increasing considerably with wavelength in the African dust case, with $\omega(0.55\ \mu\mathrm{m}) = 0.880$; and (iii) slightly decreasing with wavelength in the BBS case, with $\omega(0.55\ \mu\mathrm{m}) = 0.950$.

(5) The OS2 and OS4 surface albedo models defined by Tomasi et al. (2013) were adopted to represent the most frequently observed ocean surface reflectance characteristics on the clear-sky days of the cruise, model OS2 pertaining to calm-wind conditions with surface-level wind velocity $V_w = 5$ m/s, and OS4 determined for $V_w - 20$ m/s. The spectral and angular dependence features of model OS2 are shown in Fig. 8.38, and those of model OS4 in Fig. 8.14.

(6) The calculations of the daily time-patterns of instantaneous DARF terms and diurnally averaged DARF effects for the three Aerosols99 types were made using the 6S radiative transfer code of Vermote et al. (1997) for (i) spectral values of $\tau_a(\lambda)$ obtained at visible and near-IR wavelengths from the values of $\tau_a(0.55\ \mu\mathrm{m})$ and α given in Table 8.6a, which were assumed to remain

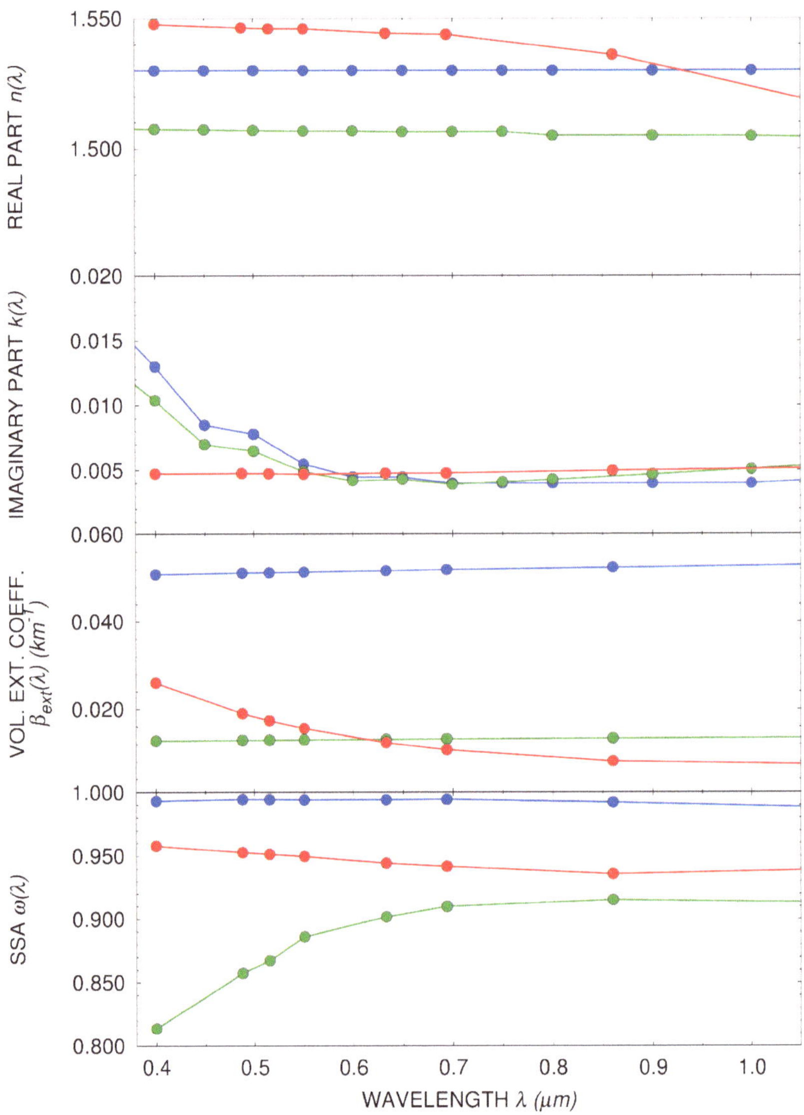

Fig. 8.45. Spectral curves of the real part $n(\lambda)$ and imaginary part $k(\lambda)$ of the particulate refractive index, volume extinction coefficient $\beta_{ext}(\lambda)$, and columnar aerosol single-scattering albedo $\omega(\lambda)$ of the multimodal aerosol particle size-distribution curves shown in Fig. 8.44 for the following aerosol extinction models determined in Table 8.6a: (i) pure maritime aerosol (blue circles), consisting of the OPAC (RH = 50%) components relative to water-soluble, sea-salt accumulation, and sea-salt coarse particle modes; (ii) African dust (dark green circles), consisting of the OPAC (RH = 50%) components relative to water-soluble, mineral dust nuclei, mineral dust accumulation, and mineral dust coarse particle modes; and (iii) biomass-burning smoke (red circles), consisting of the BL (RH = 50%) components given in Table 8.6a for the three modes of Aitken nuclei, accumulation particles, and coarse particles.

Table 8.6b. Daily values of the diurnally averaged aerosol forcing terms ΔDF_{ToA} at the ToA-level, ΔDF_{BoA} at the BoA-level, and ΔDF_{Atm} within the atmosphere, diurnal average aerosol fractional forcing AFF_{ToA} at the ToA-level (given by the ratio between flux change ΔF_{ToA} at the ToA-level and the incoming flux $I_S\downarrow$ of solar radiation at the ToA-level), and the diurnal average DARF efficiencies E_{ToA}, E_{BoA}, and E_{Atm} giving the rates at which the surface–atmosphere system is forced per unit $\tau_a(0.55\ \mu m)$, as obtained for the three study cases examined during the Aerosols99 cruise in the Atlantic Ocean (Voss et al., 2001a) by using the OS2 and OS4 oceanic surface albedo models.

Aerosol type	Surface albedo model	Diurnal average DARF terms (W/m^2)			Diurnal average	Diurnal average DARF efficiencies (W/m^2)		
		ΔDF_{ToA}	ΔDF_{BoA}	ΔDF_{Atm}	AFF_{ToA}	E_{ToA}	E_{BoA}	E_{Atm}
Pure maritime aerosol (January 18, 1999)	OS2	−4.4	+4.9	−9.3	-2.1×10^{-2}	−45.0	+50.1	−95.1
	OS4	−2.7	+3.0	−5.7	-1.3×10^{-2}	−28.1	+30.8	−58.9
African dust (January 25, 1999)	OS2	−4.0	−0.1	−3.9	-1.5×10^{-2}	−14.4	−0.4	−14.0
	OS4	−1.9	−7.8	+5.9	-7.0×10^{-3}	−6.7	−27.8	+21.1
Biomass-burning smoke (January 30, 1999)	OS2	−19.5	−18.1	−1.3	-4.6×10^{-2}	−58.8	−54.7	−4.1
	OS4	−17.3	−22.9	+5.6	-4.1×10^{-2}	−52.2	−69.0	+16.8

constant during the whole sunlit period; (ii) the values of solar zenith angle θ_o calculated at the various hours of the day for a site located at 35° N latitude and 45° W longitude in the pure maritime aerosol case, at 22° N latitude and 25° W longitude in the African dust transport event, and 20° S latitude and 10° E longitude in the BBS transport episode across the Atlantic Ocean; (iii) the multimodal size-distribution curves defined at step (3) for the three aerosol types and shown in Fig. 8.44; (iv) the optical characteristics of the three aerosol types determined at step (2) and shown in Fig. 8.45, and (v) the sea surface albedo models OS2 and OS4 chosen at step (5). The daily mean values of ΔDF_{ToA}, ΔDF_{BoA}, and ΔDF_{Atm} calculated with the DARF-PROC procedure are given in Table 8.6b, together with the corresponding evaluations of AFF_{ToA} obtained at the ToA-level. The diurnally averaged DARF terms are plotted in Fig. 8.46 versus the daily mean values of $\tau_a(0.55\ \mu\text{m})$ given in Table 8.6a for the three Aerosols99 types. It can be noted that the values of ΔDF_{ToA} determined over the OS1 oceanic surface were found not to differ much from those obtained for the OS4 oceanic surface model, since they present a variation smaller than 2 W/m^2 for the pure maritime aerosol and of about 2 W/m^2 for the African dust and BBS cases. Correspondingly, the daily mean values of ΔDF_{BoA} were estimated to decrease by less than 2 W/m^2 in the pure maritime aerosol case, by more than 7 W/m^2 in the African dust case, and by nearly 5 W/m^2 for the BBS particles. The DARF term ΔDF_{Atm} was found to increase from about −9 to nearly −6 W/m^2 for pure maritime aerosol, and more markedly in the other two cases, from about −4 to about +6 W/m^2 for the African dust, and from less than −1 to more than +5 W/m^2 for the BBS particles.

(7) The calculations of the daily mean values of DARF efficiencies were made for the sea surface albedo models OS2 and OS4 and the three aerosol types described in Table 8.6a, obtaining the average values of E_{ToA}, E_{BoA}, and E_{Atm} given in Table 8.6b. Furthermore, the scatter plots of these efficiency estimates versus $\omega(0.55\ \mu\text{m})$ are shown in Fig. 8.46 for the three Aerosols99 aerosol types, obtained for surface albedo models OS2 and OS4. It can be observed that (i) E_{ToA} assumes values ranging between −28 and −45 W/m^2 for pure maritime aerosol, values of about −10 W/m^2 for the African dust, and values of around −50 W/m^2 for the BBS particles; (ii) E_{BoA} exhibits relatively moderate variations for all the three aerosol types, varying from −50 to −31 W/m^2 for pure maritime aerosol, from −0.4 to −28 W/m^2 for African dust, and from about −55 to −69 W/m^2 for BBS particles; and (iii) E_{Atm} varies from −95 to −59 W/m^2 for pure maritime aerosol, from −14 to +21 W/m^2 for African dust, and from −4 to nearly +17 W/m^2 for BBS particles, on passing from the OS2 to the OS4 surface albedo conditions.

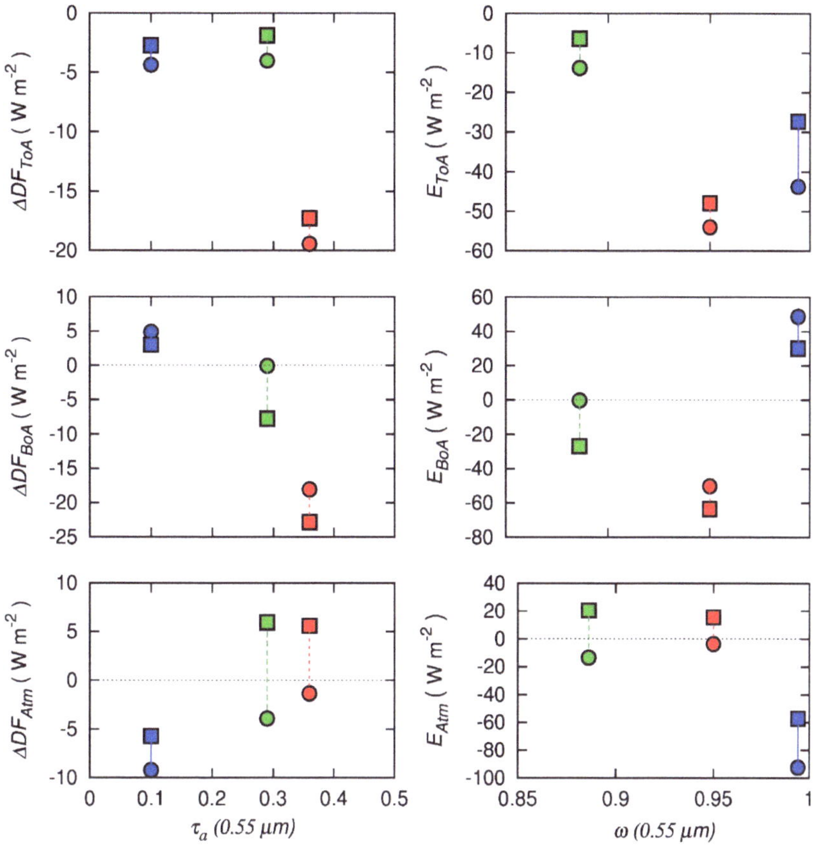

Fig. 8.46. Left part: Scatter plots of the daily values of DARF terms ΔDF_{ToA} (upper part), ΔDF_{BoA} (middle part), and ΔDF_{Atm} (lower part) plotted versus the daily mean aerosol optical thickness $\tau_a(0.55\ \mu m)$, determined for the three aerosol particle types monitored during the Aerosols99 cruise in the Atlantic Ocean (Voss et al., 2001a), and presenting the multimodal features and composition characteristics defined in Table 8.6a for (i) pure maritime aerosol (blue symbols); (ii) African dust (green symbols); and (iii) biomass-burning smoke (red symbols), suspended over the oceanic surfaces represented by the surface albedo models OS1 (circles) and OS4 (squares). Right part: Scatter plots of the daily mean values of DARF efficiency parameters E_{ToA} at the ToA-level, E_{BoA} at the BoA-level, and E_{Atm} within the atmosphere versus the columnar aerosol single-scattering albedo $\omega(0.55\ \mu m)$, represented using the same symbols adopted in the left part.

8.3.7 DARF evaluations from the DOE/ARM/AIOP project field measurements in north-central Oklahoma

Experimental studies were conducted by the DOE as part of the Atmospheric Radiation Measurement (ARM) program during the AIOP, with the main target of characterizing the aerosol optical properties and evaluating the radiative influence of aerosol particles on the energy budget of the surface–atmosphere system. Field measurements were performed at the ARM Center (Oklahoma, US) in May 2003, using ground-based and airborne instruments, as well as space-based remote sensing techniques (Ferrare et al., 2006). A numerous set of AERONET Cimel sun-photometer measurements (Holben et al., 1998; Eck et al., 1999) was recorded in May 2003 at the ARM Southern Great Plains (SGP) Climate Research Facility (CRF) site (36° 36′ N, 97° 29′ W, 315 m a.m.s.l.), obtaining a set of daily mean values of $\tau_a(0.355\ \mu m)$ ranging between 0.07 and 0.47, with $\alpha(0.34$–$0.87\ \mu m)$ correspondingly varying between 0.32 and 1.60. It can be noted in Table 8.7a that rather high values of $\tau_a(\lambda)$ were measured during the first and last weeks of the field campaign, with relatively low values of α in early May, due to the significant extinction produced by coarse dust particles, whereas the rather high values of α recorded on the last days of May were presumably due to predominant extinction effects by smoke particles generated by Siberian forest fires (hereinafter referred to as SFF) and transported over the north-central Oklahoma from May 25 to 27, 2003. In addition, small-size accumulation particles formed from anthropogenic pollution sources were observed over the ARM-SGP site from May 11 to 24, and from May 28 to 30, consisting of smoke, dust, and polluted particles transported from distant sources at altitudes varying between 2 and 5 km. These aerosols efficiently absorbed the solar radiation, presenting the relatively low SSA values given in Table 8.7b, as obtained from satellite images and backward trajectory analyses.

Table 8.7a. Daily values of aerosol optical thickness $\tau_a(0.355\ \mu m)$ and Ångström's parameters α and β derived from the sun-photometer measurements performed during the DOE/ARM/AIOP experiment conducted in May 2003 over north-central Oklahoma (US) (Ferrare et al., 2006), and corresponding values of aerosol optical thickness $\tau_a(0.55\ \mu m)$ calculated for the daily mean values of $\tau_a(0.355\ \mu m)$ and Ångström's exponent α for the aerosol types observed on the 19 clear-sky measurement days of the field campaign. Acronym SFF refers to Siberian Forest Fire smoke particles transported over long distances.

Day	Aerosol type	Daily mean $\tau_a(0.355\ \mu m)$	Daily mean Ångström's exponent α	Daily mean parameter β	Daily mean $\tau_a(0.55\ \mu m)$
May 5	Cont. clean + Dust	0.088	0.320	0.063	0.077
May 7	Cont. poll. + Soot	0.232	1.120	0.073	0.142
May 8	Cont. clean + Dust	0.600	0.544	0.342	0.473
May 9	Cont. clean + Dust	0.440	0.416	0.286	0.367
May 10	Cont. clean + Dust	0.560	0.560	0.314	0.438
May 11	Cont. poll. + Soot	0.120	1.296	0.031	0.068
May 12	Cont. poll. + Soot	0.160	0.848	0.067	0.110
May 14	Cont. poll. + Soot	0.480	0.960	0.178	0.315
May 18	Cont. poll. + Soot	0.548	0.880	0.220	0.373
May 20	Cont. poll. + Soot	0.340	1.280	0.090	0.194
May 21	Cont. poll. + Soot	0.400	1.120	0.125	0.245
May 22	Cont. poll. + Soot	0.420	1.320	0.107	0.219
May 24	Cont. poll. + Soot	0.220	1.216	0.062	0.129
May 25	Cont. poll. + Soot (SFF)	0.344	1.264	0.093	0.198
May 26	Cont. poll. + Soot (SFF)	0.320	1.600	0.061	0.159
May 27	Cont. poll. + Soot (SFF)	0.460	1.120	0.144	0.282
May 28	Cont. poll + Soot	0.680	1.312	0.175	0.383
May 29	Cont. poll + Soot	0.252	1.184	0.074	0.150
May 30	Cont. poll + Soot	0.280	1.280	0.074	0.160

Table 8.7b. Daily values of diurnally averaged aerosol forcing terms ΔDF_{ToA} at the ToA-level, ΔDF_{BoA} at the BoA-level, and ΔDF_{Atm} within the atmosphere, and the diurnal average DARF efficiencies E_{ToA}, E_{BoA}, and E_{Atm} giving the rates at which the surface–atmosphere system is forced per unit $\tau_a(0.55\ \mu m)$, as obtained for the VS2 and VS3 vegetation-covered surface albedo models assumed to represent realistically the surface reflectance characteristics over the north-central Oklahoma (US) in May 2003.

Measurement day of May 2003	Surface albedo model	Diurnal average DARF terms (W/m^2)			Diurnal average DARF efficiencies (W/m^2)		
		ΔDF_{ToA}	ΔDF_{BoA}	ΔDF_{Atm}	E_{ToA}	E_{BoA}	E_{Atm}
May 5	VS2	−3.2	−3.8	+0.4	−41.7	−47.3	+5.5
	VS3	−3.1	−3.5	+0.5	−39.6	−45.7	+6.1
May 7	VS2	−5.5	−6.4	+0.9	−38.6	−44.8	+6.2
	VS3	−5.2	−6.1	+0.9	−36.7	−43.2	+6.5
May 8	VS2	−17.0	−20.5	+3.4	−36.0	−43.3	+7.3
	VS3	−16.1	−19.6	+3.5	−34.1	−41.4	+7.4
May 9	VS2	−12.1	−14.9	+2.8	−33.0	−40.6	+7.5
	VS3	−11.3	−14.1	+2.8	−30.8	−38.4	+7.7
May 10	VS2	−14.5	−17.5	+3.1	−33.0	−40.1	+7.1
	VS3	−13.6	−16.7	+3.1	−31.0	−38.2	+7.2
May 11	VS2	−2.6	−3.0	+0.4	−37.9	−43.9	+6.0
	VS3	−2.5	−2.9	+0.4	−36.1	−42.7	+6.6
May 12	VS2	−4.3	−5.1	+0.7	−39.4	−46.2	+6.8
	VS3	−4.1	−4.9	+0.8	−37.3	−44.6	+7.3
May 14	VS2	−9.7	−11.3	+1.6	−30.7	−35.8	+5.1
	VS3	−9.2	−10.9	+1.6	−29.3	−34.6	+5.2
May 18	VS2	−11.7	−13.6	+1.9	−31.3	−36.5	+5.2
	VS3	−11.2	−13.1	+2.0	−29.9	−35.2	+5.3
May 20	VS2	−5.9	−7.0	+1.1	−30.4	−36.3	+5.9
	VS3	−5.6	−6.8	+1.2	−29.0	−35.2	+5.2
May 21	VS2	−8.1	−9.5	+1.4	−33.0	−38.9	+5.9
	VS3	−7.7	−9.2	+1.5	−31.4	−37.5	+6.1
May 22	VS2	−7.1	−8.4	+1.3	−32.6	−38.4	+5.8
	VS3	−6.8	−8.2	+1.3	−31.2	−37.3	+6.1
May24	VS2	−4.8	−5.4	+0.7	−37.0	−42.1	+5.1
	VS3	−4.6	−5.3	+0.7	−35.5	−40.9	+5.4
May 25	VS2	−7.4	−8.5	+1.0	−37.6	−42.7	+5.1
	VS3	−7.1	−8.2	+1.1	−36.1	−41.5	+5.4
May 26	VS2	−8.0	−6.8	+0.8	−37.9	−42.9	+5.0
	VS3	−5.8	−6.6	+0.8	−36.4	−41.7	+5.3
May 27	VS2	−9.1	−10.7	+1.6	−32.4	−37.9	+5.5
	VS3	−8.8	−10.4	+1.6	−31.1	−36.8	+5.7
May 28	VS2	−11.6	−13.9	+2.2	−30.4	−36.3	+5.9
	VS3	−11.2	−13.5	+2.3	−29.1	−35.2	+6.0
May 29	VS2	−5.4	−6.3	+0.9	−35.7	−41.9	+6.3
	VS3	−5.1	−6.1	+1.0	−34.1	−40.7	+6.6
May 30	VS2	−6.1	−7.3	+1.2	−37.9	−45.5	+7.6
	VS3	−5.8	−7.1	+1.3	−36.2	−44.1	+8.0

Ground-level measurements of volume coefficients $\beta_{ext}(0.675\ \mu m)$ and $\beta_{ext}(1.550\ \mu m)$ were performed during the field campaign using a new cavity ring-down (CRD) instrument, called Cadenza (NASA-ARC) (Strawa et al., 2006), from which the values of exponent α were derived. The said measurements of $\beta_{ext}(0.675\ \mu m)$ and simultaneous measurements of coefficient $\beta_{sca}(0.675\ \mu m)$ were employed to calculate the corresponding values of $\beta_{abs}(0.675\ \mu m)$, in terms of the differences between $\beta_{ext}(0.675\ \mu m)$ and $\beta_{sca}(0.675\ \mu m)$, from which values of $\omega(0.675\ \mu m)$ were obtained, mainly varying between 0.90 and 0.96. In addition, vertical profiles of $\beta_{sca}(\lambda)$ were determined at the 0.467, 0.530, and 0.675-μm wavelengths using airborne commercial nephelometers (Hallar et al., 2006). The measurements showed that strong extinction effects were caused by (i) coarse dust particle layers suspended at 2–4-km altitudes on May 8 and 9, and by SFF smoke particles transported over long distances on May 26 to 28, 2003.

The seven-step DARF-PROC procedure was used to analyze the DOE/ARM/AIOP field measurements and determine the DARF effects, as follows:

(1) The AERONET Level 1.5 sun-photometer measurements of direct solar irradiance were analyzed to determine the daily mean values of $\tau_a(0.355\ \mu m)$ and $\tau_a(0.550\ \mu m)$, and the corresponding values of parameters α and β calculated over the 0.38–0.87-μm wavelength range, as given in Table 8.7a for the 19 measurement days of the campaign.

(2) The complex values of columnar aerosol refractive index were defined on the 19 measurements days using the AERONET Level 1.5 data of sky-brightness in the almucantar. Figure 8.47 shows the daily mean spectral curves of $n(\lambda)$ and $k(\lambda)$ obtained on (i) May 8, 2003, for continental clean and dust particles, (ii) May 14, 2003, for continental polluted and soot particles, (iii) May 25, 2003 for continental polluted and soot/smoke SFF particles, and (iv) May 26, 2003 for continental polluted and soot/smoke SFF particles. The data indicate that $n(\lambda)$ varies between 1.42 and 1.53 at visible and near-IR wavelengths, in the extreme cases, both due to continental polluted aerosol associated with SFF smoke particles, while $k(\lambda)$ is close to 5×10^{-4} at all wavelengths for continental clean and dust aerosol on May 9, and close to 2.5×10^{-3} for continental polluted aerosol mixed with soot particulate matter on May 14.

(3) The size-distribution curves of columnar aerosol were directly retrieved from the spectral values of $\tau_a(\lambda)$, $n(\lambda)$, and $k(\lambda)$ determined at the previous step. Figure 8.48 shows some examples of number and volume multimodal size-distribution curves of columnar aerosol particles, as obtained on the same four days as considered in Fig. 8.47. The curves exhibit bimodal features in all cases, with a marked mode of coarse particles peaking at a radius of about 2 μm on May 9, presumably due to dust particles, and similar coarse particle modes on the three other days, presumably associated with soot particulate matter mainly arising from anthropogenic pollution or from SFF smoke particles. Similar bimodal size-distribution curves were also retrieved from the spectral series of $\tau_a(\lambda)$ measured on the other 15 measurement days of the campaign (see Table 8.7a).

(4) Daily mean values of columnar $\omega(0.55\ \mu m)$ were determined by analyzing the AERONET Level 1.5 products, collected on the 19 DOE/ARM/AIOP measurement days. As shown in Table 8.7b, they were found to range between

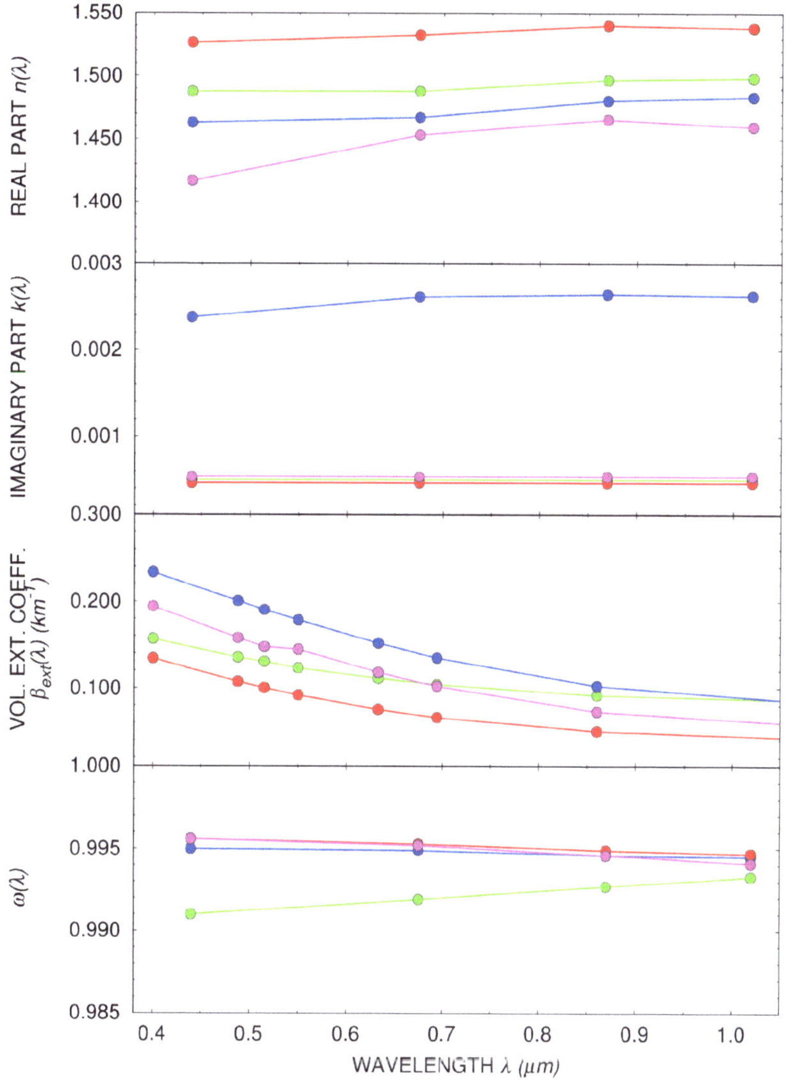

Fig. 8.47. Spectral curves of the real part $n(\lambda)$ and imaginary part $k(\lambda)$ of the particulate refractive index, volume extinction coefficient $\beta_{ext}(\lambda)$ (calculated for particle number density equal to 10^3 per cm^3), and columnar aerosol single-scattering albedo $\omega(\lambda)$, derived from the retrieved AERONET Level 1.5 products collected at the ARM-SGP-CRF site (north-central Oklahoma, US) in May 2003, on four of the 19 DOE/ARM/AIOP measurement days: (i) May 9, 2003 (14:08 UTC) (green circles), for continental clean aerosol and dust particles; (ii) May 14, 2003 (22:21 UTC) (blue circles), for continental polluted aerosol and soot particles; (iii) May 25, 2003 (22:58 UTC) (red circles), for continental polluted and soot containing Siberian Forest Fire smoke particles; and (iv) May 27, 2003 (16:28 UTC) (fuchsia circles), for continental polluted and soot containing SFF smoke particles.

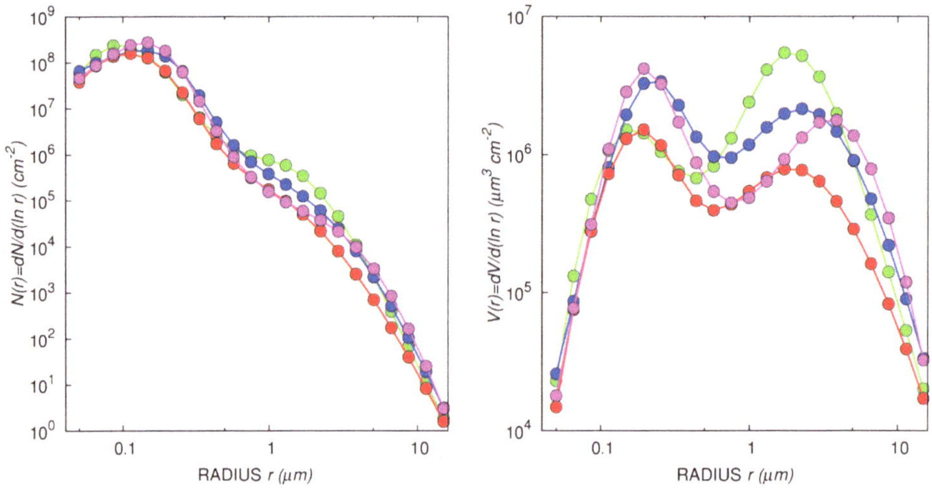

Fig. 8.48. Multimodal size-distribution curves of columnar particle number density $N(r) = \mathrm{d}N/\mathrm{d}(\ln r)$ measured per cm^2 (left) and columnar particle volume $V(r) = \mathrm{d}V/\mathrm{d}(\ln r)$ measured in $\mu\mathrm{m}^3/\mathrm{cm}^2$ (right), retrieved from the AERONET Level 1.5 sun-photometer measurements carried out at the ARM-SGP-CRF site (north-central Oklahoma, US) in May 2003, on four of the 19 DOE/ARM/AIOP measurement days: (i) May 9, 2003 (14:08 UTC) (green circles), for continental clean aerosol and dust particles; (ii) May 14, 2003 (22:21 UTC) (blue circles), for continental polluted aerosol and soot particles; (iii) May 25, 2003 (22:58 UTC) (red circles), for continental polluted and soot aerosols containing Siberian Forest Fire smoke particles; and (iv) May 27, 2003 (16:28 UTC) (fuchsia circles), for continental polluted and soot aerosols containing SFF smoke particles.

0.989 (May 12) and 0.996 (May 25), therefore varying only slightly around an average value of 0.993 ± 0.003 throughout the measurement period. The spectral curves of $\beta_{ext}(\lambda)$ and $\omega(\lambda)$ are shown in Fig. 8.47, obtained for the four measurement days selected above: (a) the spectral curves of $\beta_{ext}(\lambda)$ determined for an overall particle number density of 10^3 per cm^3 decrease as a function of wavelength, with differences due to the various size-distribution shape-parameters and complex refractive index, assuming values ranging between 10^{-1} and 2×10^{-1} per km, of which the highest was measured on May 14 for continental polluted aerosol with soot substances, and the lowest on May 25 for continental polluted aerosol with SFF smoke; (b) the spectral curves of $\omega(\lambda)$ are rather stable, assuming values of around 0.995 in the three cases of continental polluted aerosol mixed with soot or SFF smoke particles, and values slowly increasing with wavelength from about 0.991 to 0.993 in the case of May 9, for continental clean aerosol and dust.

(5) Examining the MCD43C3 products derived from the MODIS Level 3.0 surface albedo data recorded over north-central Oklahoma in May 2008, we obtained the two average maps of land surface albedo and Normalized Difference Vegetation Index (NDVI). On the basis of these surface reflectance data, the VS2 and VS3 vegetation-covered surface albedo models of Tomasi et al. (2013) were

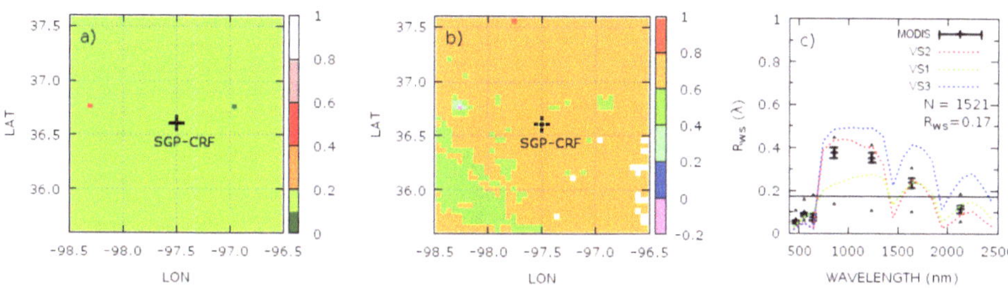

Fig. 8.49. Maps of the average land surface albedo ((a) in the left column) and Normalized Difference Vegetation Index (NDVI) ((b) in the central column) obtained from the MODIS Level 3.0 surface albedo data (MCD43C3 products) recorded over the north-central Oklahoma (US) in May 2008 over a pixel of 2° latitude × 2° longitude. The crosses labelled SGP-CRF indicate the geographical position of the ARM Southern Great Plains Climate Research Facility site. (c) Shows the spectral values (black vertical bars) of the average white-sky albedo $R_{ws}(\lambda)$ giving a mean value of 0.17 and determined for spectral standard deviations of around 0.1, and minimum and maximum values indicated by small triangles at each MODIS channel peak-wavelength. The three colored dotted curves are the best-fit solutions found by retrieving the MODIS data recorded within the seven spectral channels, for which the surface albedo models VS2 (dotted red curve), VS1 (dotted green curve), and VS3 (dotted blue curve) were found to represent the surface albedo characteristics of this agricultural area, presenting gradually worst approximation coefficients.

adopted to represent realistically the surface reflectance characteristics of this agricultural area, as can be seen in the right-hand graph of Fig. 8.49. The spectral and angular features of models VS2 and VS3 have already been shown in Fig. 8.22, when examining the AEROCLOUDS measurements over the Po Valley in northern Italy.

(6) The daily time-patterns of instantaneous DARF terms and diurnally averaged DARF effects were calculated for the DOE/ARM/AIOP 19 measurement days following the DARF-PROC procedure described for the other mid-latitude experiments. The calculations were made for (i) the AERONET daily mean values of $\tau_a(\lambda)$; (ii) the daily mean spectral series of $n(\lambda)$ and $k(\lambda)$; (iii) the aerosol extinction parameters obtained on each day, given in Table 8.7a, kept constant during each measurement day; (iv) the values of solar zenith angle θ_o determined at the various hours of each measurement day for the longitude and latitude coordinates of the ARM-SGP-CRF site; (v) the multimodal size-distribution curves defined at step (3) for the various aerosol types monitored during the experiment and listed in Table 8.7a; (vi) the optical characteristics of the multimodal size-distribution curves determined at step (2); and (vii) the non-Lambertian surface albedo models VS2 and VS3 chosen at step (5).
The diurnally averaged DARF terms ΔDF_{ToA}, ΔDF_{BoA}, and ΔDF_{Atm} were then calculated using the DARF-PROC procedure for the 19 measurement days of the campaign and the VS2 and VS3 surface albedo models, obtaining the values given in Table 8.7b. In Fig. 8.50, they are plotted versus the daily mean values of $\tau_a(0.55\ \mu m)$ listed in Table 8.6a, to provide the corresponding DARF efficiency parameters. It can be seen that the evaluations of ΔDF_{ToA},

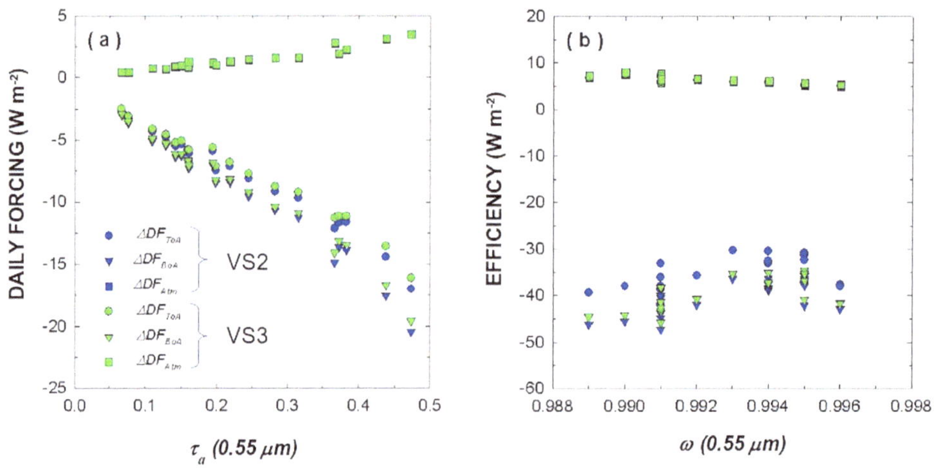

Fig. 8.50. Left part (a): Scatter plots of the daily mean values of DARF terms ΔDF_{ToA} (circles), ΔDF_{BoA} (down triangles), and ΔDF_{Atm} (squares) plotted versus the daily mean aerosol optical thickness $\tau_a(0.55\ \mu\text{m})$, determined on the 19 measurements days of the DOE/ARM/AIOP field campaign conducted in May 2003 and reported in Table 8.7a, for the VS2 (blue symbols) and VS3 (green symbols) surface albedo models of Tomasi et al. (2013), used to represent the surface reflectance characteristics of north-central Oklahoma in May. Right part (b): Scatter plots of the daily mean values of DARF efficiency parameters E_{ToA} (circles), E_{BoA} (down triangles), and E_{Atm} (squares) versus the columnar aerosol single-scattering albedo $\omega(0.55\ \mu\text{m})$, determined for the 19 measurements days of the DOE/ARM/AIOP field campaign reported in Table 8.7a, and the VS2 (blue symbols) and VS3 (green symbols) surface albedo models of Tomasi et al. (2013).

ΔDF_{BoA}, and ΔDF_{Atm} determined for the VS2 model differ only slightly from those obtained using model VS3. The daily values of ΔDF_{ToA} and ΔDF_{BoA} decrease almost linearly as $\tau_a(0.55\ \mu\text{m})$ increases from 0.06 to nearly 0.48, with values of around -5 W/m^2 for the lowest $\tau_a(0.55\ \mu\text{m})$ and around -20 W/m^2 for the highest $\tau_a(0.55\ \mu\text{m})$. The marked homogeneity of the results presumably arises from the very similar characteristics of aerosol optical parameters and, in particular, from those of the columnar aerosol SSA. Correspondingly, the estimates of diurnally averaged ΔDF_{Atm} made on the various measurement days turn out to increase slowly from $+0.4$ to $+3.5$ W/m^2 over the 0.06–0.48 range of $\tau_a(0.55\ \mu\text{m})$.

(7) The daily mean values of DARF efficiency parameters E_{ToA}, E_{BoA} and E_{Atm} were then calculated for all the 19 measurement days by dividing the values of ΔDF_{ToA}, ΔDF_{BoA}, and ΔDF_{Atm} by the corresponding daily mean values of $\tau_a(0.55\ \mu\text{m})$. They are plotted versus the columnar $\omega(0.55\ \mu\text{m})$ in the right part of Fig. 8.50, showing that E_{ToA} and E_{BoA} assume very similar values, ranging mainly between about -30 and -47 W/m^2. Considering that the range of $\omega(0.55\ \mu\text{m})$ is very narrow, the data provided no clear information on the DARF efficiency dependence on SSA. The daily evaluations of E_{Atm} obtained in terms of Eq. (8.4) also turn out to be very stable throughout the measurement period, varying between $+5.0$ and $+8.0$ W/m^2.

8.4 Conclusions

As clearly shown in Figs 8.42 and 8.43 concerning the polar aerosol radiative impact on climate, the daily mean values of the DARF efficiency terms at the ToA- and BoA-levels and within the atmosphere were found to vary considerably as a function of surface albedo characteristics. In fact, the changes in E_{ToA}, E_{BoA}, and E_{Atm} are only partially explained in terms of variations in the columnar aerosol optical characteristics, being mainly attributable to the variations in surface reflectance conditions. Because of the complex exchange processes involving the short-wave radiation, the DARF efficiency terms E_{ToA}, E_{BoA}, and E_{Atm} were found to vary most greatly when passing from the low oceanic ice-free surface albedo conditions to the very high ones of the Arctic and Antarctic areas covered by snow fields and glaciers.

In the other mid-latitude scenarios considered in the CLEARCOLUMN, PRIN-2004, AEROCLOUDS, Aerosols99, and DOE/ARM/AIOP field experiments, DARF efficiency evaluations were found to vary less drastically as a function of the columnar $\omega(0.55\ \mu m)$ for all the aerosol types suspended over oceanic and land surfaces. The efficiency estimates were mainly obtained in the said experiments for two general aerosol classes of (i) maritime particles, including a few cases of pure marine aerosol, and numerous different cases of maritime aerosol mixed with continental clean aerosol, continental polluted aerosol, Saharan dust, and forest fire smoke; and (ii) continental aerosol, including a few cases of continental clean aerosol, and numerous other mixed cases with predominant contents of continental polluted and dust aerosol originating from anthropogenic sources, continental aerosol with forest fire smoke, or mineral dust transported from the Saharan region. The evaluations of E_{ToA}, E_{BoA}, and E_{Atm} for the two general classes of aerosol particles over sea and land surfaces are plotted versus the columnar $\omega(0.55\ \mu m)$ in Figs 8.51 and 8.52.

Figure 8.51 shows that the daily mean values of E_{ToA}, E_{BoA}, and E_{Atm} determined for both pure and mixed marine aerosol loads are highly scattered because of the differences in the particle size-distribution shape-parameters, optical characteristics of particulate matter, duration of the sunlit period, and surface albedo properties assumed in the DARF calculations. The surface albedo features were represented using models OS1, OS2, OS3, and OS4 over the sea surface, giving values of broadband albedo $A(\theta_o = 60°)$ decreasing from 0.193 to 0.105, and the vegetation-covered surface albedo models VS1, VS2, VS3, and VS4 over land, giving values of broadband albedo $A(\theta_o = 60°)$ increasing from 0.155 to 0.306. It can be noted that parameters E_{ToA}, E_{BoA}, and E_{Atm} are greatly dispersed and exhibit average trends over sea that considerably differ from those over land:

(1) Values of efficiency E_{ToA} are highly dispersed over the 0.8–1.0 range of $\omega(0.55\ \mu m)$, giving a regression line over sea determined with standard error of estimate SEE $= \pm 25.2$ W/m^2, presenting an increase $\Delta E_{ToA} = +6.9$ W/m^2 over the 0.8–1.0 range of $\omega(0.55\ \mu m)$, and a regression line over land determined with SEE $= \pm 13.5$ W/m^2, presenting an average decrease $\Delta E_{ToA} = -42.4$ W/m^2 over the 0.8–1.0 range of $\omega(0.55\ \mu m)$.
(2) Daily mean estimates of E_{BoA} are also largely dispersed over the 0.8–1.0 range of $\omega(0.55\ \mu m)$, giving increasing regression lines, with $\Delta E_{ToA} = +78.1$ W/m^2

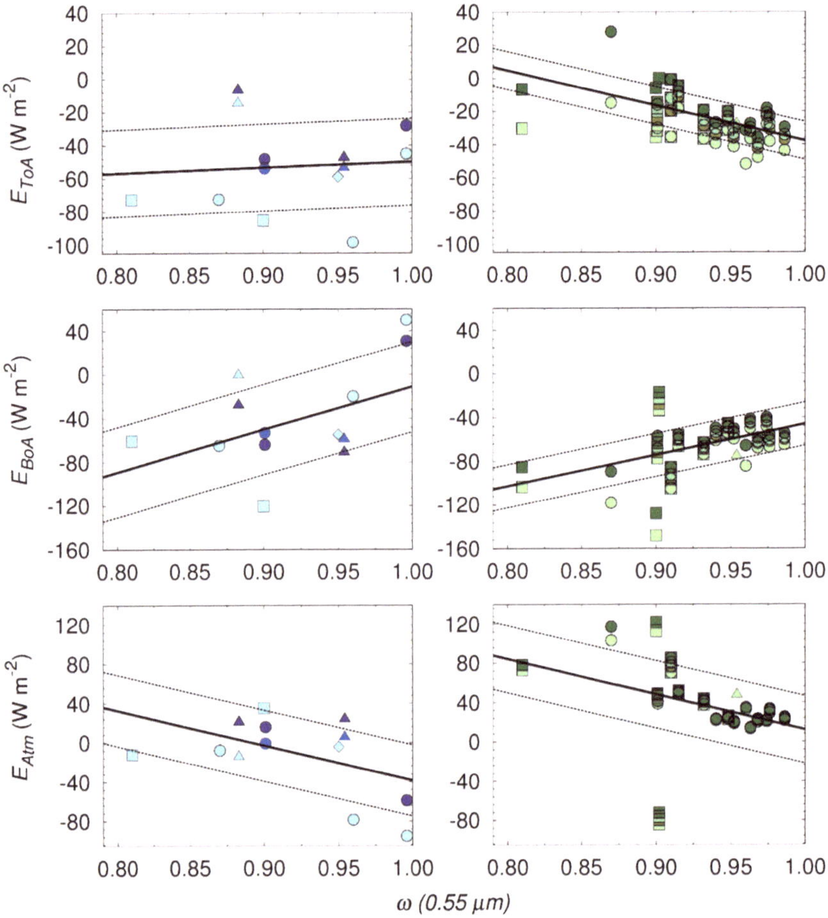

Fig. 8.51. Scatter plots of the daily mean values of DARF efficiency parameters E_{ToA} at the ToA-level, E_{BoA} at the BoA-level, and E_{Atm} within the atmosphere versus the columnar aerosol single-scattering albedo $\omega(0.55\ \mu m)$ over oceanic surfaces (left) and vegetation-covered land surfaces (right), as determined for (i) pure marine and mixed marine/continental clean aerosol (circles), (ii) mixed maritime/continental polluted aerosol (squares), (iii) mixed maritime/Saharan dust (triangles), and (iv) mixed maritime/forest fire smoke (diamonds). The surface albedo models used in the calculations are indicated using differently colored symbols for OS1 (cyan), OS2 (sky-blue), OS3 (dark blue), OS4 (indigo), VS1 (light green), VS2 (olive-green), VS3 (emerald-green), and VS4 (dark green).

and SEE = ±43.0 W/m^2 over sea, and ΔE_{ToA} = +56.8 W/m^2 and SEE = ±22.0 W/m^2 over land.

(3) Daily mean estimates of E_{Atm} are, once again, highly dispersed over the 0.8–1.0 range of $\omega(0.55\ \mu m)$, and present decreasing regression lines with ΔE_{ToA} = −71.2 W/m^2 and SEE = ±38.1 W/m^2 over sea, and ΔE_{ToA} = −72.2 W/m^2 and SEE = ±36.1 W/m^2 over land.

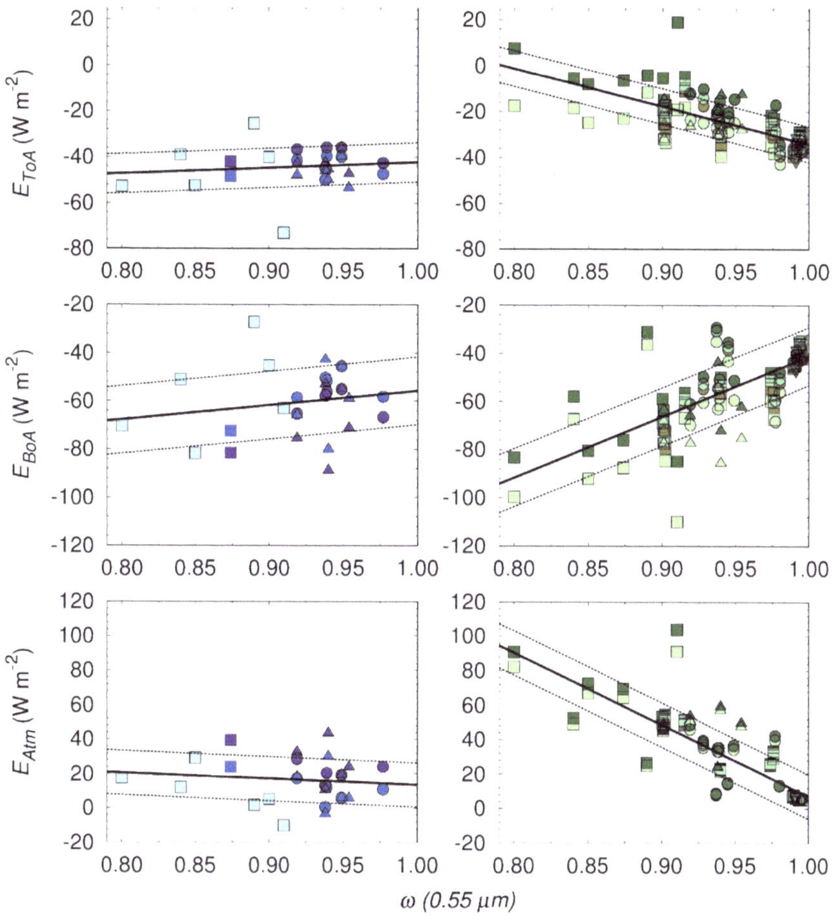

Fig. 8.52. Scatter plots of the daily mean values of DARF efficiency parameters E_{ToA} at the ToA-level, E_{BoA} at the BoA-level, and E_{Atm} in the atmosphere versus the columnar aerosol single-scattering albedo $\omega(0.55\ \mu m)$ over oceanic surfaces (left) and vegetation-covered land surfaces (right), as determined for (i) continental clean aerosol (circles), (ii) continental polluted/anthropogenic aerosol (squares), (iii) continental aerosol mixed with forest fire smoke (diamonds), (iv) mixed continental/Saharan dust aerosol (upward triangles), and (v) continental clean aerosol mixed with mineral dust (downward triangles). The surface albedo models used in the calculations are indicated using differently colored symbols for OS1 (cyan), OS2 (sky-blue), OS3 (dark blue), OS4 (indigo), VS1 (light green), VS2 (olive-green), VS3 (emerald-green), and VS4 (dark green).

In view of the great dispersion of the present estimates of E_{ToA}, E_{BoA}, and E_{Atm} and pronounced variations in results, the above regression lines are judged to yield only an average measure of the general trend of these DARF efficiency parameters varying over sea and land, due to the fact that the relative content of absorbing particles decreases gradually as SSA characteristics of columnar aerosol increase until approaching the ideal case of $\omega(0.55\ \mu m) = 1.0$ for a total content of pure maritime aerosol particles.

Figure 8.52 shows the daily mean values of E_{ToA}, E_{BoA}, and E_{Atm} found for continental clean aerosol mixed with particles originating from natural (mineral dust, forest fire smoke) and anthropogenic sources. The values are strongly scattered because of the different shapes of the particle size-distribution curves, the widely varying optical parameters of the particulate matter components, and the large variations in the surface albedo characteristics, which were assumed to calculate the DARF effects over sea and land surfaces. The surface reflectance features were represented by using the four oceanic surface albedo models OS1, OS2, OS3, and OS4, and the four vegetation-covered surface albedo models VS proposed by Tomasi et al. (2013). Parameters E_{ToA}, E_{BoA}, and E_{Atm} turn out to be highly dispersed also for such a complex aerosol particle class, presenting very different average trends over sea and over land, resulting in regression lines over sea with rather low slopes and regression lines over land with considerably higher slope coefficients. In particular, it was found that:

(1) The daily mean values of E_{ToA} give regression lines with $\Delta E_{ToA} = +4.9$ W/m^2 over the 0.8–1.0 range of $\omega(0.55\ \mu m)$ and SEE $= \pm 8.3$ W/m^2 over sea, and with $\Delta E_{ToA} = -33.0$ W/m^2 and SEE $= \pm 10.7$ W/m^2 over land.
(2) The daily mean values of E_{BoA} are greatly dispersed over the 0.8–1.0 range of $\omega(0.55\ \mu m)$, giving increasing regression lines in both cases, with $\Delta E_{ToA} = +12.0$ W/m^2 and SEE $= \pm 14.0$ W/m^2 over sea, and with $\Delta E_{ToA} = +50.5$ W/m^2 and SEE $= \pm 16.6$ W/m^2 over land.
(3) The daily mean values of E_{Atm} are also greatly dispersed over the 0.8–1.0 range of $\omega(0.55\ \mu m)$ for both sea and land surfaces, presenting in both cases decreasing regression lines with $\Delta E_{ToA} = -7.0$ W/m^2 and SEE $= \pm 12.5$ W/m^2 over sea, and $\Delta E_{ToA} = -83.5$ W/m^2 and SEE $= \pm 22.9$ W/m^2 over land.

Because these results were obtained for atmospheric loads consisting of continental particles of different origins, it is evident that the regression lines in Fig. 8.52 give only average estimates of the general trends of the three DARF efficiency parameters to vary over sea and land as $\omega(0.55\ \mu m)$ increases while the relative content of absorbing particles due to anthropogenic pollution and biomass-burning activities gradually diminishes until approaching the ideal case of $\omega(0.55\ \mu m) = 1.0$, for a null content of absorbing aerosol particles in the vertical atmospheric column. However, the widely dispersed features of the present results lead us to obtain rather great SEE values. This clearly indicates that accurate and realistic evaluations of the daily mean DARF effects and the corresponding efficiency parameters can only be achieved in field experiments, by using calculation procedures similar to the present DARF-PROC procedure for complete sets of field measurements performed over the entire sunlit period. Such measurements have to provide accurate time-patterns of the spectral values of $\tau_a(\lambda)$ and the microphysical and radiative parameters of columnar aerosol and need to be used with realistic non-Lambertian surface albedo models derived from local satellite observations and/or through radiative transfer applicative studies.

Acknowledgements

The authors gratefully acknowledge the NOAA/OAR/ESRL PSD (Boulder, Colorado, US) for the use of the NCEP Reanalysis data provided by their Web site at *www.esrl.noaa.gov/psd/*.

References

Anderson, T.L., Charlson, R.J., Bellouin, N., Boucher, O., Chin, M., Christopher, S.A., Haywood, J., Kaufman, Y.J., Kinne, S., Ogren, J.A., Remer, L.A., Takemura, T., Tanré, D., Torres, O., Trepte, C.R., Wielicki, B.A., Winker, D.M. and Yu, H. (2005) 'An "A-Train" strategy for quantifying direct climate forcing by anthropogenic aerosols', *Bull. Amer. Meteor. Soc.*, **86**, 1795–809, doi:10.1175/BAMS-86-12-1795.

Andreae, M.O. and Rosenfeld, D. (2008) 'Aerosol–cloud–precipitation interactions. Part 1. The nature and sources of cloud-active aerosols', *Earth Sci. Rev.*, **89**, 13–41, doi:10.1016/j.earscirev.2008.03.001.

Ångström, A. (1964) 'The parameters of atmospheric turbidity', *Tellus*, **16**, 64–75, doi:10.1111/j.2153-3490.1964.tb00144.x.

Arimoto, R., Hogan, A., Grube, P., Davis, D., Webb, J., Schloesslin, C., Sage, S. and Raccah, F. (2004) 'Major ions and radionuclides in aerosol particles from the South Pole during ISCAT-2000', *Atmos. Environ.*, **38**, 5473–84, doi:10.1016/j.atmosenv.2004.01.049.

Bates, T., Kapustin, V., Quinn, P., Covert, D., Coffman, D., Mari, C., Durkee, P., De Bruyn, W. and Saltzman, E. (1998a) 'Processes controlling the distribution of aerosol particles in the lower marine boundary layer during the First Aerosol Characterization Experiment (ACE 1)', *J. Geophys. Res.*, **103**, 16369–83, doi:10.1029/97JD03720.

Bates, T.S., Huebert, B.J., Gras, J.L., Griffiths, F.B. and Durkee, P.A. (1998b) 'International Global Atmospheric Chemistry (IGAC) Project's first Aerosol Characterization Experiment (ACE 1) overview', *J. Geophys. Res.*, **103**, 16297–318, doi:10.1029/97JD03741.

Bates, T.S., Quinn, P.K., Coffman, D.J., Johnson, J.E., Miller, T.L., Covert, D.S., Wiedensohler, A., Leinert, S., Nowak, A. and Neusüss, C. (2001) 'Regional physical and chemical properties of the marine boundary layer aerosol across the Atlantic during Aerosols99: An overview', *J. Geophys. Res.*, **106**, 20767–82, doi:10.1029/2000JD900578.

Bergstrom, R.W. and Russell, P.B. (1999) 'Estimation of aerosol direct radiative effects over the mid-latitude North Atlantic from satellite and *in situ* measurements', *Geophys. Res. Lett.*, **26**, 1731–4, doi:10.1029/1999GL900330.

Bodhaine, B.A. (1995) 'Aerosol absorption measurements at Barrow, Mauna Loa and the South Pole', *J. Geophys. Res.*, **100**, 8967–75, doi:10.1029/95JD00513.

Bush, B.C. and Valero, F.P.J. (2002) 'Spectral aerosol radiative forcing at the surface during the Indian Ocean Experiment (INDOEX)', *J. Geophys. Res.*, **107**, 8003, doi:10.1029/2000JD000020.

Bush, B.C. and Valero, F.P.J. (2003) 'Surface aerosol radiative forcing at Gosan during the ACE-Asia campaign', *J. Geophys. Res.*, **108**, 8660, doi:10.1029/2002JD003233.

Campanelli, M., Estellés, V., Tomasi, C., Nakajima, T., Malvestuto, V. and Martínez-Lozano, J.A. (2007) 'Application of the SKYRAD Improved Langley plot method for the *in situ* calibration of CIMEL Sun–sky photometers', *Appl. Opt.*, **46**, 2688–702, doi:10.1364/AO.46.002688.

Charlson, R.J., Langner, J., Rodhe, H., Leovy, C.B. and Warren, S.G. (1991) 'Perturbation of the northern hemisphere radiative balance by backscattering from anthropogenic sulfate aerosols', *Tellus*, **43 B**, 152–63, doi:10.1034/j.1600-0870.1991.00013.x.

Charlson, R.J., Schwartz, S.E., Hales, J.M., Cess, R.D., Coakley, J.A., Jr., Hansen, J.E. and Hofmann, D.J. (1992) 'Climate forcing by anthropogenic aerosols', *Science*, **255**, 423–30, doi:10.1126/science.255.5043.423.

Christopher, S.A. and Zhang, J. (2002) 'Daytime variation of shortwave direct radiative forcing of biomass burning aerosols from GOES-8 Imager', *J. Atmos. Sci.*, **59**, 681–91, doi:10.1175/1520-0469(2002)059<0681:DVOSDR>2.0CO:2.

Christopher, S.A., Li, X., Welch, R.M., Reid, J.S., Hobbs, P.V., Eck, T.F. and Holben, B. (2000) 'Estimation of surface and top-of-atmosphere shortwave irradiance in biomass burning regions during SCAR-B', *J. Appl. Meteorol.*, **39**, 1742–53, doi:10.1175/1520-0450-39.10.1742.

Christopher, S.A., Wang, J., Ji, Q. and Tsay, S.-C. (2003) 'Estimation of diurnal shortwave dust aerosol radiative forcing during PRIDE', *J. Geophys. Res.*, **108**, 8956, doi:10.1029/2002JD002787.

Christopher, S.A., Zhang, J., Kaufman, Y.J. and Remer, L.A. (2006) 'Satellite-based assessment of top of atmosphere anthropogenic aerosol radiative forcing over cloud-free oceans', *Geophys. Res. Lett.*, **33**, L15816, doi:10.1029/2005GL025535.

Chung, S.H. and Seinfeld, J.H. (2005) 'Climate response of direct radiative forcing of anthropogenic black carbon', *J. Geophys. Res.*, **110**, D11102, doi:10.1029/2004JD005441.

Chylek, P. and Coakley, J.A., Jr. (1974) 'Aerosols and climate', *Science*, **183**, 75–7, doi:10.1126/science.183.4120.75.

Chylek, P. and Wong, J. (1995) 'Effect of absorbing aerosols on global radiation budget', *Geophys. Res. Lett.*, **22**, 929–31, doi:10.1029/95GL00800.

Coakley, J.A., Jr. and Chylek, P. (1975) 'The two-stream approximation in radiative transfer: Including the angle of the incident radiation', *J. Atmos. Sci.*, **32**, 409–18, doi:10.1175/1520-0469(1975)032<0409:TTSAIR>2.0.CO;2.

Conant, W.C. (2000) 'An observational approach for determining aerosol surface radiative forcing: Results from the first field phase of INDOEX', *J. Geophys. Res.*, **105**, 15347–60, doi:10.1029/1999JD901166.

Draxler, R.R. and Hess, G.D. (1998) 'An overview of the HYSPLIT_4 modelling system for trajectories, dispersion, and deposition', *Australian Meteorol. Magazine*, **47**, 295–308.

Dubovik, O., Lapyonok, T., Kaufman, Y.J., Chin, M., Ginoux, P., Kahn, R.A. and Sinyuk, A. (2008) 'Retrieving global aerosol sources from satellites using inverse modelling', *Atmos. Chem. Phys.*, **8**, 209–50, doi:10.5194/acp-8-209-2008.

Dubovik, O., Lapyonok, T., Kaufman, Y.J., Chin, M., Ginoux, P., Remer, L.A. and Holben, B.N. (2004) 'Inversion of global distribution of aerosol sources using MODIS and AERONET data', *Optica Pura y Aplicada*, **37**, 3349–58.

Eck, T.F., Holben, B.N., Reid, J.S., Dubovik, O., Smirnov, A., O'Neill, N.T., Slutsker, I. and Kinne, S. (1999) 'Wavelength dependence of the optical depth of biomass burning, urban, and desert dust aerosols', *J. Geophys. Res.*, **104**, 31333–49, doi:10.1029/1999JD900923.

Fattori, I., Becagli, S., Bellandi, S., Castellano, E., Innocenti, M., Mannini, A., Severi, M., Vitale, V. and Udisti, R. (2005) 'Chemical composition and physical features of summer aerosol at Terra Nova Bay and Dome C, Antarctica', *J. Environ. Monit.*, **7**, 1265–74.

Ferrare, R., Feingold, G., Ghan, S., Ogren, J., Schmid, B., Schwartz, S.E. and Sheridan, P. (2006) 'Preface to special section: Atmospheric Radiation Measurement Program

May 2003 Intensive Operations Period examining aerosol properties and radiative influences', *J. Geophys. Res.*, **111**, D05S01, doi:10.1029/2005JD006908.

Gobbi, G.P., Angelini, F., Bonasoni, P., Verza, G.P., Marinoni, A. and Barnaba, F. (2010) 'Sunphotometry of the 2006–2007 aerosol optical/radiative properties at the Himalayan Nepal Climate Observatory-Pyramid (5079 m a.s.l.)', *Atmos. Chem. Phys.*, **10**, 11209–221, doi:10.5194/acp-10-11209-2010.

Hale, G.M. and Querry, M.R. (1973) 'Optical constants of water in the 200-nm to 200-μm wavelength region', *Appl. Opt.*, **12**, 555–63, doi:10.1364/AO.12.000555.

Hallar, A.G., Strawa, A.W., Schmid, B., Andrews, E., Ogren, J., Sheridan, P., Ferrare, R., Covert, D., Elleman, R., Jonsson, H., Bokarius, K. and Luu, A. (2006) 'Atmospheric radiation measurements aerosol intensive operating period: Comparison of aerosol scattering during coordinated flights', *J. Geophys. Res.*, **111**, D05S09, doi:10.1029/2005JD006250.

Hansen, J., Sato, M. and Ruedy, R. (1997) 'Radiative forcing and climate response', *J. Geophys. Res.*, **102**, 6831–64, doi:10.1029/96JD03436.

Haywood, J.M. and Shine, K.P. (1995) 'The effect of anthropogenic sulfate and soot aerosol on the clear sky planetary radiation budget', *Geophys. Res. Lett.*, **22**, 603–6, doi:10.1029/95GL00075.

Herber, A., Thomason, L.W., Gernandt, H., Leiterer, U., Nagel, D., Schulz, K., Kaptur, J., Albrecht, T. and Notholt, J. (2002) 'Continuous day and night aerosol optical depth observations in the Arctic between 1991 and 1999', *J. Geophys. Res.*, **107**, 4097, doi:10.1029/2001JD000536.

Herber, A., Thomason, L.W., Radionov, V.F. and Leiterer, U. (1993) 'Comparison of trends in the tropospheric and stratospheric aerosol optical depths in the Antarctic', *J. Geophys. Res.*, **98**, 18441–7, doi:10.1029/93JD01666.

Hess, M., Koepke, P. and Schult, I. (1998) 'Optical properties of aerosols and clouds: The software package OPAC', *Bull. Amer. Meteor. Soc.*, **79**, 831–44, doi:10.1175/1520-0477(1998)079<0831:OPOAAC>2.0.CO;2.

Hignett, P., Taylor, J.P., Francis, P.N. and Glew, M.D. (1999) 'Comparison of observed and modeled direct aerosol forcing during TARFOX', *J. Geophys. Res.*, **104**, 2279–87, doi:10.1029/98JD02021.

Hillamo, R., Allegrini, I., Sparapani, R. and Kerminen, V.-M. (1998) 'Mass size distributions and precursors gas concentrations of major inorganic ions in Antarctic aerosol', *Int. J. Env. Anal. Chem.*, **71**, 353–72.

Hirdman, D., Sodemann, H., Eckhardt, S., Burkhart, J.F., Jefferson, A., Mefford, T., Quinn, P.K., Sharma, S., Ström, J. and Stohl, A. (2010) 'Source identification of short-lived air pollutants in the Arctic using statistical analysis of measurement data and particle dispersion model output', *Atmos. Chem. Phys.*, **10**, 669–93, doi:10.5194/acp-10-669-2010.

Holben, B.N., Eck, T.F., Slutsker, I., Tanré, D., Buis, J.P., Setzer, A., Vermote, E., Reagan, J.A., Kaufman, Y.J., Nakajima, T., Lavenu, F., Jankowiak, I. and Smirnov, A. (1998) 'AERONET: A federated instrument network and data archive for aerosol characterization', *Remote Sens. Environ.*, **66**, 1–16, doi:S0034-4257(98)00031-5.

Hummel, J.R., Shettle, E.P. and Longtin, D.R. (1988) 'A new background stratospheric aerosol model for use in atmospheric radiation models', *Scientific Report No. 8*, AFGL-TR-88-0166, Air Force Geophysics Laboratory, Hanscom Air Force Base, Massachusetts, 01731-5000, 30 July.

Kahn, R.A. (2012) 'Reducing the uncertainties in direct aerosol radiative forcing', *Surveys Geophys.*, **33**, 701–21, doi:10.1007/s10712-011-9153-z.

Kalnay, E., Kanamitsu, M., Kistler, R., Collins, W., Deaven, D., Gandin, L., Iredell, M., Saha, S., White, G., Woollen, J., Zhu, Y., Leetmaa, A., Reynolds, R., Chelliah,

M., Ebisuzaki, W., Higgins, W., Janowiak, J., Mo, K.C., Ropelewski, C., Wang, J., Jenne, R. and Joseph, D. (1996) 'The NCEP/NCAR 40-Year Reanalysis Project', *Bull. Amer. Meteor. Soc.*, **77**, 437–71, doi:10.1175/1520-0477(1996)077<0437:TNYRP> 2.0.CO;2.

Kaufman, Y.J., Martins, J.V., Remer, L.A., Schoeberl, M.R. and Yamasoe, M.A. (2002) 'Satellite retrieval of aerosol absorption over the oceans using sunglint', *Geophys. Res. Lett.*, **29**, 1928, doi:10.1029/2002GL015403.

Kim, D.-H., Sohn, B.-J., Nakajima, T. and Takamura, T. (2005a) 'Aerosol radiative forcing over east Asia determined from ground-based solar radiation measurements', *J. Geophys. Res.*, **110**, D10S22, doi:10.1029/2004JD004678.

Kim, S.-W., Jefferson, A., Yoon, S.-C., Dutton, E.G., Ogren, J.A., Valero, F.P.J., Kim, J. and Holben, B.N. (2005b) 'Comparisons of aerosol optical depth and surface shortwave irradiance and their effect on the aerosol surface radiative forcing estimation', *J. Geophys. Res.*, **110**, D07204, doi:10.1029/2004JD004989.

King, M.D. (1982) 'Sensitivity of constrained linear inversions to the selection of the Lagrange multiplier', *J. Atmos. Sci.*, **39**, 1356–69, doi:10.1175/1520-0469(1982)039 <1356:SOCLIT>2.0.CO;2.

Kinne, S. and Pueschel, R. (2001) 'Aerosol radiative forcing for Asian continental outflow', *Atmos. Environ.*, **35**, 5019–28, doi:10.1016/S1352-2310(01)00329-6.

Kopp, G. and Lean, J.L. (2011) 'A new, lower value of total solar irradiance: Evidence and climate significance', *Geophys. Res. Lett.*, **38**, L01706, doi:10.10129/2010GL045777.

Lanconelli, C. (2007) 'Evaluation of the effects produced by surface reflectance anisotropy in the direct radiative forcing induced by atmospheric aerosols', Ph.D. thesis, University of Ferrara, 19th cycle of the Research Doctorate in Physics, February 5.

Li, F., Vogelmann, A.M. and Ramanathan, V. (2004) 'Dust aerosol radiative forcing measured from space over the Western Africa', *J. Climate*, **17**, 2558–71, doi:10.1175/1520-0442(2004)017<2558:Sdirect aerosol radiative forcing (DARF)M>2.0.CO;2.

Liao, H., Adams, P.J., Seinfeld, J.H., Mickley, L.J. and Jacob, D.J. (2003) 'Interactions between tropospheric chemistry and aerosols in a unified GCM simulation', *J. Geophys. Res.*, **108**, 4001, doi:10.1029/2001JD001260.

Liao, H., Seinfeld, J.H., Adams, P.J. and Mickley, L.J. (2004) 'Global radiative forcing of coupled tropospheric ozone and aerosols in a unified general circulation model', *J. Geophys. Res.*, **109**, D16207, doi:10.1029/2001JD002004.

Liu, H., Pinker, R.T. and Holben, B.N. (2005) 'A global view of aerosols from merged transport models, satellite, and ground observations', *J. Geophys. Res.*, **110**, D10S15, doi:10.1029/2004JD004695, 2005.

Loeb, N.G. and Manalo-Smith, N. (2005) 'Top-of-atmosphere direct radiative effect of aerosols over global oceans from merged CERES and MODIS observations', *J. Climate*, **18**, 3506–26, doi:10.1175/JCLI3504.1.

Loeb, N.G. and Su, W. (2010) 'Direct aerosol radiative forcing uncertainty based on a radiative perturbation analysis', *J. Climate*, **23**, 5288–93, doi:10.1175/2010JCLI3543.1.

Lupi, A., Tomasi, C., Orsini, A., Cacciari, A., Vitale, V., Georgiadis, T., Casacchia, R., Salvatori, R. and Salvi, S. (2001) 'Spectral curves of surface reflectance in some Antarctic regions', *Il Nuovo Cimento*, **24 C**, 313–27.

Marcq, S., Laj, P., Roger, J.C., Villani, P., Sellegri, K., Bonasoni, P., Marinoni, A., Cristofanelli, P., Verza, G.P. and Bergin, M. (2010) 'Aerosol optical properties and radiative forcing in the high Himalaya based on measurements at the Nepal Climate Observatory-Pyramid site (5079 m a.s.l.)', *Atmos. Chem. Phys.*, **10**, 5859–72, doi:10.5194/acp-10-5859-2010.

Marinoni, A., Cristofanelli, P., Laj, P., Duchi, R., Calzolari, F., Decesari, S., Sellegri, K., Vuillermoz, E., Verza, G.P., Villani, P. and Bonasoni, P. (2010) 'Aerosol mass and black carbon concentrations, a two year record at NCO-P (5079 m, Southern Himalayas)', *Atmos. Chem. Phys.*, **10**, 8551–62, doi:10.5194/acp-10-8551-2010.

Mazzola, M., Lanconelli, C., Lupi, A., Busetto, M., Vitale, V. and Tomasi, C. (2010) 'Columnar aerosol optical properties in the Po Valley, Italy, from MFRSR data', *J. Geophys. Res.*, **115**, D17206, doi:10.10129/2009JD013310.

Mazzola, M., Stone, R.S., Herber, A., Tomasi, C., Lupi, A., Vitale, V., Lanconelli, C., Toledano, C., Cachorro, V.E., O'Neill, N.T., Shiobara, M., Aaltonen, V., Stebel, K., Zielinski, T., Petelski, T., Ortiz de Galisteo, J.P., Torres, B., Berjon, A., Goloub, P., Li, Z., Blarel, L., Abboud, I., Cuevas, E., Stock, M., Schulz, K.-H. and Virkkula, A. (2012) 'Evaluation of sun photometer capabilities for the retrievals of aerosol optical depth at high latitudes: The POLAR-AOD intercomparison campaigns', *Atmos. Environ.*, **52**, 4–17, doi:10.1016/j.atmosenv.2011.07.042.

Mie, G. (1908) 'Beiträge zur Optik trüber Medien, speziell kolloidaler Metallösungen', *Annalen der Physik, Vierte Folge*, **25**, 377–445.

Nakajima, T., Yoon, S.-C., Ramanathan, V., Shi, G.-Y., Takemura, T., Higurashi, A., Takemura, T., Aoki, K., Sohn, B.-J., Kim, S.-W., Tsuruta, H., Sugimoto, N., Shimizu, A., Tanimoto, H., Sawa, Y., Lin, N.-H., Lee, C.-T., Goto, D. and Schutgens, N. (2007) 'Overview of the Atmospheric Brown Cloud East Asian Regional Experiment 2005 and a study of the aerosol direct radiative forcing in east Asia', *J. Geophys. Res.*, **112**, D24S91, doi:10.1029/2007JD009009.

Nemesure, S., Wagener, R. and Schwartz, S.E. (1995) 'Direct shortwave forcing of climate by anthropogenic sulfate aerosol: Sensitivity to particle size, composition, and relative humidity', *J. Geophys. Res.*, **100**, 26105–116, doi:10.1029/95JD02897.

Palmer, K.F. and Williams, D. (1975) 'Optical constants of sulphuric acid; application to the clouds of Venus?', *Appl. Opt.*, **14**, 208–19, doi:10.1364/AO.14.000208.

Patadia, F., Gupta, P. and Christopher, S.A. (2008) 'First observational estimates of global clear sky shortwave aerosol direct radiative effect over land', *J. Geophys. Res.*, **35**, L04810, doi:10.0129/2007GL032314.

Perrone, M.R., Santese, M., Tafuro, A.M., Holben, B. and Smirnov, A. (2005) 'Aerosol load characterization over south-east Italy by one year of AERONET Sun-photometer measurements', *Atmos. Res.*, **75**, 111–33, doi:10.1016/j.atmosres.2004.12.003.

Pósfai, M. and Buseck, P.R. (2010) 'Nature and climate effects of individual tropospheric aerosol particles', *Ann. Rev. Earth Planet. Sci.*, **38**, 17–43, doi:10.1146/annurev.earth.031208.100032.

Posse, P. and von Hoyningen-Huene, W. (1996) 'New method for retrieval of particle size distributions from combined data of spectral extinction and aureole scattering', *J. Aerosol Sci.*, **27, Suppl. 1**, S567–8, doi:10.1016/0021-8502(96)00356-4.

Pueschel, R.F., Kinne, S.A., Russell, P.B., Snetsinger, K.G. and Livingston, J.M. (1993) 'Effects of the 1991 Pinatubo volcanic eruption on the physical and radiative properties of stratospheric aerosols', in Keevallik, S. and Kärner, O. (eds), *IRS '92: Current Problems in Atmospheric Radiation, Proceedings of the International Radiation Symposium*, Tallinn (Estonia), August 2–8, 1992, Hampton, Virginia (USA), A. Deepak Publishing, pp. 183–6.

Pueschel, R.F., Snetsinger, K.G., Goodman, J.K., Toon, O.B., Ferry, G.V., Oberbeck, V.R., Livingston, J.M., Verma, S., Fong, W., Starr, W.L. and Chan, K.R. (1989) 'Condensed nitrate, sulfate, and chloride in Antarctic stratospheric aerosols', *J. Geophys. Res.*, **94**, 11271–84, doi:10.1029/JD094iD09p11271.

Quinn, P.K., Coffman, D.J., Bates, T.S., Miller, T.L., Johnson, J.E., Voss, K., Welton, E.J. and Neusüss, C. (2001) 'Dominant aerosol chemical components and their contribution to extinction during the Aerosols99 cruise across the Atlantic', *J. Geophys. Res.*, **106**, 20783–809, doi:10.1029/2000JD900577.

Radionov, V.F., Lamakin, M.V. and Herber, A. (2002) 'Changes in the aerosol optical depth of the Antarctic atmosphere', *Izvestia, Atmos. Oc. Phys.*, **38**, 179–83.

Ramana, M.V., Ramanathan, V., Podgorny, I.A., Pradhan, B.B. and Shrestha, B. (2004) 'The direct observations of large aerosol radiative forcing in Himalayan region', *Geophys. Res. Lett.*, **31**, L05111, doi:10.1029/2003GL018824.

Ramanathan, V., Crutzen, P.J., Kiehl, J.T. and Rosenfeld, D. (2001a) 'Aerosols, climate and the hydrological cycle', *Science*, **294**, 2119–24, doi:10.1126/science.1064034.

Ramanathan, V., Crutzen, P.J., Lelieveld, J., Mitra, A.P., Althausen, D., Anderson, J., Andreae, M.O., Cantrell, W., Cass, G.R., Chung, C.E., Clarke, A.D., Coakley, J.A., Collins, W.D., Conant, W.C., Dulac, F., Heintzenberg, J., Heymsfield, A.J., Holben, B., Howell, S., Hudson, J., Jayaraman, A., Kiehl, J.T., Krishnamurti, T.N., Lubin, D., McFarquhar, G., Novakov, T., Ogren, J.A., Podgorny, I.A., Prather, K., Priestley, K., Prospero, J.M., Quinn, P.K., Rajeev, K., Rasch, P., Rupert, S., Sadourny, R., Satheesh, S.K., Shaw, G.E., Sheridan, P. and Valero, F.P.J. (2001b) 'Indian Ocean Experiment: An integrated analysis of the climate forcing and effects of the great Indo-Asian haze', *J. Geophys. Res.*, **106**, 28371–98, doi:10.1029/2001JD900133.

Russell, P.B. and Heintzenberg, J. (2000) 'An overview of the ACE 2 clear sky column closure experiment (CLEARCOLUMN)', *Tellus*, **52B**, 463–83, doi:10.1034/j.1600-0889.2000.00013.x.

Russell, P.B., Hobbs, P.V. and Stowe, L.L. (1999a) 'Aerosol properties and radiative effects in the United States East Coast haze plume: An overview of the Tropospheric Aerosol Radiative Forcing Observational Experiment (TARFOX)', *J. Geophys. Res.*, **104**, 2213–22, doi:10.1029/1998JD200028.

Russell, P.B., Livingston, J.M., Hignett, P., Kinne, S., Wong, J., Chien, A., Bergstrom, R., Durkee, P. and Hobbs, P.V. (1999b) 'Aerosol-induced radiative flux changes off the United States mid-Atlantic coast: Comparison of values calculated from sunphotometer and *in situ* data with those measured by airborne pyranometer', *J. Geophys. Res.*, **104**, 2289–307, doi:10.1029/1998JD200025.

Russell, P.B., Redemann, J., Schmid, B., Bergstrom, R.W., Livingston, J.M., Mcintosh, D.M., Ramirez, S.A., Hartley, S., Hobbs, P.V., Quinn, P.K., Carrico, C.M., Rood, M.J., Ström, E.Ö., Noone, K.J., von Hoyningen-Huene, W. and Remer, L. (2002) 'Comparison of aerosol single scattering albedos derived by diverse techniques in two north Atlantic experiments', *J. Atmos. Sci.*, **59**, 609–19, doi:10.1175/1520-0469(2002)059 <0609:COASSA>2.0.CO;2

Satheesh, S.K. (2002) 'Radiative forcing by aerosols over Bay of Bengal region', *Geophys. Res. Lett.*, **29**, 2083, doi:10.1029/2002GL015334.

Satheesh, S.K. and Ramanathan, V. (2000) 'Large differences in tropical aerosol forcing at the top of the atmosphere and Earth's surface', *Nature*, **405**, 60–3, doi:10.1038/35011039.

Satheesh, S.K. and Srinivasan, J. (2002) 'Enhanced aerosol loading over Arabian Sea during the pre-monsoon season: Natural or anthropogenic?', *Geophys. Res. Lett.*, **29**, 1874, doi:10.1029/2002GL015687.

Schwartz, S.E. (1996) 'The whitehouse effect: Shortwave radiative forcing of climate by anthropogenic aerosols', *J. Aerosol Sci.*, **27**, 359–82, doi:10.1016/0021-8502(95)00533-1.

Seinfeld, J.H. and Pandis, S.N. (1998) *Atmospheric Chemistry and Physics, from Air Pollution to Climate Change*, 1st edn, New York (USA), John Wiley and Sons.

Seinfeld, J.H. and Pandis, S.N. (2006) *Atmospheric Chemistry and Physics, from Air Pollution to Climate Change*, 2nd edn, New York (USA), John Wiley and Sons.

Seinfeld, J.H., Carmichael, G.R., Arimoto, R., Conant, W.C., Brechtel, F.J., Bates, T.S., Cahill, T.A., Clarke, A.D., Doherty, S.J., Flatau, P.J., Huebert, B.J., Kim, J., Markowicz, K.M., Quinn, P.K., Russell, L.M., Russell, P.B., Shimizu, A., Shinozuka, Y., Song, C.H., Tang, Y., Uno, I., Vogelmann, A.M., Weber, R.J., Woo, J.-H. and Zhang, X.Y. (2004a) 'ACE-Asia: Regional climatic and atmospheric chemical effects of Asian dust and pollution', *Bull. Amer. Meteor. Soc.*, **85**, 367–80, doi:10.1175/BAMS-85-3-367.

Seinfeld, J.H., Kahn, R.A., Anderson, T.L., Charlson, R.J., Davies, R., Diner, D.J., Schwartz, S.E. and Wielicki, B. (2004b) 'Scientific objectives, measurement needs, and challenges motivating the PARAGON aerosol initiative', *Bull. Amer. Meteor. Soc.*, **85**, 1503–9, doi:10.1175/BAMS-85-10-1503.

Shaw, G.E. (1995) 'The arctic haze phenomenon', *Bull. Amer. Meteor. Soc.*, **76**, 2403–13, doi:10.1175/1520-0477(1995)076<2403:TAHP>2.0.CO;2.

Shettle, E.P. and Fenn, R.W. (1979) 'Models for the aerosols of the lower atmosphere and the effects of humidity variations on their optical properties', *Environ. Res. Papers*, No. 676, Air Force Geophys. Lab., AFGL-Techn. Rep. 79-0214, Hanscom AFB, MA.

Stone, R.S. (2002) 'Monitoring aerosol optical depth at Barrow, Alaska, and South Pole: Historical overview, recent results and future goals', *SIF Conf. Proc.*, **80**, 123–44.

Stone, R.S., Anderson, G.P., Andrews, E., Dutton, E.G., Shettle, E.P. and Berk, A. (2007) 'Incursions and radiative impact of Asian dust in northern Alaska', *Geophys. Res. Lett.*, **34**, L14815, doi:10.1029/2007GL029878.

Strawa, A.W., Elleman, R., Hallar, A.G., Covert, D., Ricci, K., Provencal, R., Owano, T.W., Jonsson, H.H., Schmid, B., Luu, A.P., Bokarius, K. and Andrews, E. (2006) 'Comparison of *in situ* aerosol extinction and scattering coefficient measurements made during the Aerosol Intensive Operating Period', *J. Geophys. Res.*, **111**, D05S03, doi:10.1029/2005JD006056.

Sumanth, E., Mallikarjuna, K., Stephen, J., Moole, M., Vinoj, V., Satheesh, S.K. and Moorthy, K.K. (2004) 'Measurements of aerosol optical depths and black carbon over Bay of Bengal during post-monsoon season', *Geophys. Res. Lett.*, **31**, L16115, doi:10.1029/2004GL020681.

Tafuro, A.M., Kinne, S., De Tomasi, F. and Perrone, M.R. (2007) 'Annual cycle of aerosol direct radiative effect over southeast Italy and sensitivity studies', *J. Geophys. Res.*, **112**, D20202, doi:10.1029/2006JD008265.

Tahnk, W.R. and Coakley, J.A., Jr. (2002) 'Aerosol optical depth and direct radiative forcing for INDOEX derived from AVHRR: Observations, January–March 1996–2000', *J. Geophys. Res.*, **107**, 8010, doi:10.1029/2000JD000183.

Takemura, T., Nakajima, T., Dubovik, O., Holben, B.N. and Kinne, S. (2002) 'Single-scattering albedo and radiative forcing of various aerosol species with a global three-dimensional model', *J. Climate*, **15**, 333–52, doi:10.1175/1520-0442(2002)015 <0333:SSAARF>2.0.CO;2.

Takemura, T., Nozawa, T., Emori, S., Nakajima, T.Y. and Nakajima, T. (2005) 'Simulation of climate response to aerosol direct and indirect effects with aerosol transport-radiation model', *J. Geophys. Res.*, **110**, D02202, doi:10.1029/2004JD005029.

Tomasi, C., Lanconelli, C., Lupi, A. and Mazzola, M. (2013) 'Dependence of direct aerosol radiative forcing on the optical properties of atmospheric aerosol and underlying surface', in Kokhanovsky, A.A. (ed.), *Light Scattering Reviews, Vol. 8, Radiative Transfer and Light Scattering*, Springer Praxis Books/Environmental Sciences, Chapter 11, pp. 505–626.

Tomasi, C., Lupi, A., Mazzola, M., Stone, R.S., Dutton, E.G., Herber, A., Radionov, V.F., Holben, B.N., Sorokin, M.G., Sakerin, S.M., Terpugova, S.A., Sobolesky, P.S., Lanconelli, C., Petkov, B.H., Busetto, M. and Vitale, V. (2012) 'An update on polar aerosol optical properties using POLAR-AOD and other measurements performed during the International Polar Year', *Atmos. Environ.*, **52 C**, 29–47, doi:10.1016/j.atmosenv.2012.02.055.

Tomasi, C., Vitale, V. and De Santis, L.V. (1998a) 'Relative optical mass functions for air, water vapour, ozone and nitrogen dioxide in atmospheric models presenting different latitudinal and seasonal conditions', *Meteorol. Atmos. Phys.*, **65**, 11–30, doi:10.1007/BF01030266.

Tomasi, C., Vitale, V. and Tarozzi, L. (1997) 'Sun-photometric measurements of atmospheric turbidity variations caused by the Pinatubo aerosol cloud in the Himalayan region during the summer periods of 1991 and 1992', *Il Nuovo Cimento*, **20 C**, 61–88.

Tomasi, C., Vitale, V., Cacciari, A. and Carboni, E. (1998b) 'Multiwavelength sunphotometer measurements of background aerosol optical depth at the Schneefernerhaus Observatory (2665 m) during the RAD-I-CAL 96 (ACE-2) intercomparison campaign', in *Booklet of Proceedings of the 25th International Conference of Alpine Meteorology (ICAM)*, Turin (Italy), September 14–19, pp. 101–5.

Tomasi, C., Vitale, V., Lupi, A., Cacciari, A., Marani, S. and Bonafè, U. (2003) 'Marine and continental aerosol effects on the upwelling solar radiation flux in Southern Portugal during the ACE-2 experiment', *Ann. Geophys.*, **46**, 467–79.

Tomasi, C., Vitale, V., Lupi, A., Di Carmine, C., Campanelli, M., Herber, A., Treffeisen, R., Stone, R.S., Andrews, E., Sharma, S., Radionov, V., von Hoyningen-Huene, W., Stebel, K., Hansen, G.H., Myhre, C.L., Wehrli, C., Aaltonen, V., Lihavainen, H., Virkkula, A., Hillamo, R., Ström, J., Toledano, C., Cachorro, V., Ortiz, P., de Frutos, A., Blindheim, S., Frioud, M., Gausa, M., Zielinski, T., Petelski, T. and Yamanouchi, T. (2007) 'Aerosols in polar regions: A historical overview based on optical depth and *in situ* observations', *J. Geophys. Res.*, **112**, D16205, doi:10.1029/2007JD008432.

Torres, O., Bhartia, P., Sinyuk, A., Welton, E. and Holben, B. (2005) 'Total Ozone Mapping Spectrometer measurements of aerosol absorption from space: Comparison to SAFARI 2000 ground-based observations', *J. Geophys. Res.*, **110**, D10S18, doi:10.1029/2004JD004611.

Tsigaridis, K., Krol, M., Dentener, F.J., Balkanski, Y., Lathière, J., Metzger, S., Hauglustaine, D.A. and Kanakidou, M. (2006) 'Change in global aerosol composition since preindustrial times', *Atmos. Chem. Phys.*, **6**, 5143–62, doi:10.5194/acp-6-5143-2006.

Tuncel, G., Aras, N.K. and Zoller, W.H. (1989) 'Temporal variations and sources of elements in the South Pole atmosphere. 1. Nonenriched and moderately enriched elements', *J. Geophys. Res.*, **94**, 13025–38, doi:10.1029/JD094iD10p13025.

Vermote, E.F., Tanré, D., Deuzé, J.L., Herman, M. and Morcrette, J.-J. (1997) 'Second simulation of the satellite signal in the solar spectrum (6S): An overview', *IEEE Trans. Geosci. Remote Sens.*, **35**, 675–86, doi:10.1109/36.581987.

Verver, G., Raes, F., Vogelezang, D. and Johnson, D. (2000) 'The second Aerosol Characterization Experiment (ACE-2): Meteorological and chemical context', *Tellus*, **52B**, 126–40.

Vitale, V., Tomasi, C., Lupi, A., Cacciari, A. and Marani, S. (2000a) 'Retrieval of columnar aerosol size distributions and radiative-forcing evaluations from the sun-photometric measurements taken during the CLEARCOLUMN (ACE 2) experiment', *Atmos. Environ.*, **34**, 5095–105, doi:10.1016/S1352-2310(00)00269-7.

Vitale, V., Tomasi, C., von Hoyningen-Huene, W., Bonafé, U., Marani, S., Lupi, A., Cacciari, A. and Ruggeri, P. (2000b) 'Spectral measurements of aerosol particle extinction in the 0.4–3.7 μm wavelength range, performed at Sagres with the IR-RAD sun-radiometer', *Tellus*, **52 B**, 716–33, doi:10.1034/j.1600-0889.2000.00028.x

Volz, F.E. (1972a) 'Infrared absorption by atmospheric aerosol substances', *J. Geophys. Res.*, **77**, 1017–31, doi:10.1029/JC077i006p01017.

Volz, F.E. (1972b) 'Infrared refractive index of atmospheric aerosol substances', *Appl. Opt.*, **11**, 755–9, doi:10.1364/AO.11.000755.

Volz, F.E. (1973) 'Infrared optical constants of ammonium sulfate, Sahara dust, volcanic pumice, and flyash', *Appl. Opt.*, **12**, 564–8, doi:10.1364/AO.12.000564.

Volz, F.E. (1974) 'Economical multispectral sun photometer for measurements of aerosol extinction from 0.44 μm to 1.6 μm and precipitable water', *Appl. Opt.*, **13**, 1732–3, doi:10.1364/AO.13.001732.

von Hoyningen-Huene, W., Schmidt, T., Herber, A. and Silva, A.M. (1998) 'Climate-relevant columnar optical aerosol parameters obtained during ACE-2 CLEARCOLUMN and their influence on the shortwave radiative balance', *J. Aerosol Sci.*, **29**, S269–70.

Voss, K.J., Welton, E.J., Quinn, P.K., Frouin, R., Miller, M. and Reynolds, R.M. (2001a) 'Aerosol optical depth measurements during the Aerosols99 experiment', *J. Geophys. Res.*, **106**, 20811–19, doi:10.1029/2000JD900783.

Voss, K.J., Welton, E.J., Quinn, P.K., Johnson, J., Thompson, A.M. and Gordon, H.R. (2001b) 'Lidar measurements during Aerosols99', *J. Geophys. Res.*, **106**, 20821–31, doi:10.1029/2001JD900217.

Weller, R. and Lampert, A. (2008) 'Optical properties and sulfate scattering efficiency of boundary layer aerosol at coastal Neumayer Station, Antarctica', *J. Geophys. Res.*, **113**, 03304, doi:10.1029/2008JD009962.

Weller, R. and Wagenbach, D. (2007) 'Year-round chemical aerosol records in continental Antarctica obtained by automatic samplings', *Tellus*, **59B**, 755–65, doi:10.1111/j.1600-0889.2007.00293.x.

Weller, R., Wöltjen, J., Piel, C., Resenberg, R., Wagenbach, D., König-Langlo, G. and Kriews, M. (2008) 'Seasonal variability of crustal and marine trace elements in the aerosol at Neumayer Station, Antarctica', *Tellus*, **60B**, 742–52, doi:10.1111/j.1600-0889.2008.00372.x.

WMO (1986) *Atmospheric Ozone 1985*, WMO Global Ozone Research and Monitoring Project, Report No. 16, Geneva (Switzerland).

Wolff, E.W. and Cachier, H. (1998) 'Concentrations and seasonal cycle of black carbon in aerosol at a coastal Antarctic station', *J. Geophys. Res.*, **103**, 11033–41, doi:10.1029/97JD01363.

Xu, J., Bergin, M.H., Greenwald, R. and Russell, P.B. (2003) 'Direct aerosol radiative forcing in the Yangtze delta region of China: Observation and model estimation', *J. Geophys. Res.*, **108**, 4060, doi:10.1029/2002JD002550.

Yu, H., Kaufman, Y.J., Chin, M., Feingold, G., Remer, L.A., Anderson, T.L., Balkanski, Y., Bellouin, N., Boucher, O., Christopher, S., DeCola, P., Kahn, R., Koch, D., Loeb, N., Reddy, M.S., Schulz, M., Takemura, T. and Zhou, M. (2006) 'A review of measurement-based assessments of the aerosol direct radiative effect and forcing', *Atmos. Chem. Phys.*, **6**, 613–66, doi:10.5194/acp-6-613-2006.

Zhao, T.X.-P., Yu, H., Laszlo, I., Chin, M. and Conant, W.C. (2008) 'Derivation of component aerosol direct radiative forcing at the top of atmosphere for clear-sky oceans', *J. Quant. Spectrosc. Rad. Transfer*, **109**, 1162–86, doi:10.1016/j.jqsrt.2007.10.006.

Index

MIX
Papier aus verantwortungsvollen Quellen
Paper from responsible sources
FSC® C105338

If you have any concerns about our products,
you can contact us on
ProductSafety@springernature.com

In case Publisher is established outside the EU,
the EU authorized representative is:
Springer Nature Customer Service Center GmbH
Europaplatz 3, 69115 Heidelberg, Germany

Printed by Libri Plureos GmbH
in Hamburg, Germany